Approximate Chemical Shifts of Hydrogens in ¹H-NMR Spectra.

Type of Hydrogen	Chemical Shift (δ)	Type of Hydrogen	Chemical Shift (δ)
—C—CH₃	0.9	Br—CH₃	2.7
C≡C—CH₃	1.6	Cl—CH₃	3.0
C=C—H	1.8	O—CH₃	3.3
N—H	1–3	O=C—O—CH₃	3.7
O—H	2–5	O₂N—CH₃	4.1
O=C—O—CH₃	2.0	F—CH₃	4.2
O=C—O—CH₃	2.2	C=C—H	5.5–6.5
N—CH₃	2.2	Ph—H	7–8
I—CH₃	2.2	O=C—H	10
N≡C—CH₃	2.2	O=C—O—H	12
Ph—CH₃	2.3		

Note that these positions are only approximate. Furthermore, most of these positions are given for CH₃ groups. CH₂ groups appear farther downfield by about 0.3 ppm and CH groups by about 0.7 ppm.

D0184747

Important Absorption Bands in the Infrared Spectral Region.

Position (cm⁻¹)	Group	Comments
3550–3200	—O—H	strong, very broad band
3400–3250	—N—H	weaker and less broad than O—H; NH₂ shows two bands, NH shows one
3300	≡C—H	sharp, C is *sp* hybridized
3100–3000	=C—H	C is *sp²* hybridized
3000–2850	—C—H	C is *sp³* hybridized; 3000 cm⁻¹ is a convenient dividing line between this type of C—H bond and the type above
2830–2700	O=C—H	two bands
2260–2200	—C≡N	medium
2150–2100	—C≡C—	weak
1820–1650	O=C	strong, exact position depends on substituents; see Table 12.1
1660–1640	C=C	often weak
1600–1450	(benzene ring)	four bands of variable intensity
1550 and 1380	—NO₂	two strong bands
1300–1000	—C—O—	strong
900–675	(benzene ring)	strong

Barcode overleaf →

Organic Chemistry

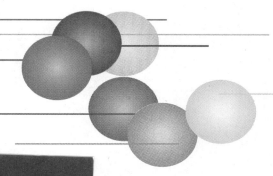

Organic Chemistry

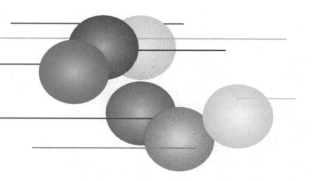

Joseph M. Hornback

University of Denver

Brooks/Cole Publishing Company

I(T)P®   An International Thomson Publishing Company

*Pacific Grove • Albany • Belmont • Bonn • Boston • Cincinnati • Detroit •
Johannesburg • London • Madrid • Melbourne • Mexico City • New York •
Paris • Singapore • Tokyo • Toronto • Washington*

Senior Developmental Editor: *Keith Dodson*
Marketing Team: *Caroline Croley/Jean Thompson*
Editorial Assistant: *Georgia Jurickovich*
Production Editor: *Tessa A. McGlasson*
Production: *GTS Graphics*
Manuscript Editor: *Barbara Willette*
Interior Design: *Geri Davis*

Interior Illustration: *GTS Graphics*
Cover Design: *Roy R. Neuhaus*
Cover Photo or Illustration: *Kenneth Eward/BioGrafx*
Indexer: *George Olshevsky*
Typesetting and Prepress: *GTS Graphics*
Printing and Binding: *World Color/Taunton*

For more information, contact:

BROOKS/COLE PUBLISHING COMPANY
511 Forest Lodge Road
Pacific Grove, CA 93950
USA

International Thomson Publishing Europe
Berkshire House 168-173
High Holborn
London WC1V 7AA
England

Thomas Nelson Australia
102 Dodds Street
South Melbourne, 3205
Victoria, Australia

Nelson Canada
1120 Birchmount Road
Scarborough, Ontario
Canada M1K 5G4

International Thomson Editores
Seneca 53
Col. Polanco
11560 México, D. F., México

International Thomson Publishing GmbH
Königswinterer Strasse 418
53227 Bonn
Germany

International Thomson Publishing Asia
221 Henderson Road
#05-10 Henderson Building
Singapore 0315

International Thomson Publishing Japan
Hirakawacho Kyowa Building, 3F
2-2-1 Hirakawacho
Chiyoda-ku, Tokyo 102
Japan

Printed in the United States of America

10 9 8 7 6 5 4 3 2 1

03825476

Library of Congress Cataloging-in-Publication Data

Hornback, Joseph M.
    Organic chemistry/Joseph M. Hornback.
        p.   cm.
    Includes index.
    ISBN 0-534-35254-5
    1. Chemistry, Organic   I. Title
QD253.H66   1998
547--dc21

97-46609
CIP

To Melani, Joe, Pat and Jordan,

who bring meaning and joy to my life.

About the Author

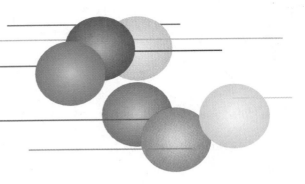

Joseph M. Hornback was born and raised in southwestern Ohio. He received a B.S. in Chemistry, magna cum laude, from the University of Notre Dame in 1965. He then attended the Ohio State University on an NSF traineeship and received his Ph.D. in 1968, working with Paul G. Gassman. He next moved to the University of Wisconsin at Madison, where he worked with Howard E. Zimmerman on an NIH postdoctoral fellowship.

In 1970, he joined the faculty of the Department of Chemistry at the University of Denver, where he has remained. His research interests are in the areas of synthetic organic chemistry and organic photochemistry. He has served in a number of administrative positions, including Associate Dean for Undergraduate Studies; Associate Dean for Natural Sciences, Mathematics, and Engineering; and Director of the Honors Program. But his first love has always been teaching, and he has taught organic chemistry nearly every term, even when he was in administration. He has received the Natural Sciences Award for Excellence in Teaching and the Outstanding Academic Advising Award.

Joe is married and has three children, two sons and a daughter, of whom he is very proud. He enjoys sports and outdoor activities, especially fishing and golf.

Brief Contents

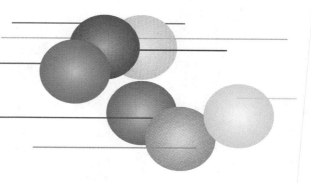

Contents

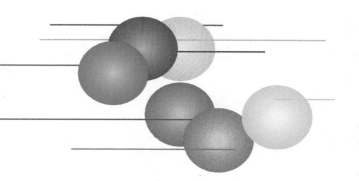

7 Nucleophilic Substitution Reactions   *257*

8 Elimination Reactions   *309*

9 Synthetic Uses of Substitution and Elimination Reactions   *345*

11 Functional Groups and Nomenclature II   *461*

12 Structure Determination by Spectroscopy I: Infrared and Nuclear Magnetic Resonance Spectroscopy   *497*

13 Structure Determination by Spectroscopy II: Ultraviolet-Visible Spectroscopy and Mass Spectrometry   *601*

We will use resonance often, so it is important to understand exactly what it means. In the case of nitromethane, resonance does not mean that the structure of the molecule is flipping back and forth between the two structures. It does not mean that the bond to one of those oxygens is a double bond part of the time and a single bond the rest of the time. Instead, the structure is static, an average of the individual structures. The double-headed **resonance arrow** is used only for resonance and should not be confused with the **equilibrium arrows,** which are used to represent an equilibrium reaction in which the molecules are indeed changing structures.

$$\longleftrightarrow \qquad\qquad \rightleftharpoons$$

resonance arrow   equilibrium arrows

Resonance must be used whenever more than one reasonable Lewis structure can be written for a molecule, provided that the Lewis structures have *identical positions of all atoms*. Only the positions of unshared electrons and multiple bonds are changed in writing different resonance structures. When a better picture of bonding is developed in Chapter 3, we will get a better understanding of what resonance means and when it must be used.

Problem

Problem 1.11   Experimental evidence indicates that the two oxygens of acetate ion are identical. Use resonance to explain this observation.

acetate ion

1.10   Polar Bonds

In calculations of formal charge, it is assumed that a shared pair of electrons spends equal time around both of the atoms that form the bond. This is true only if the atoms are the same. More commonly, the bond is between two different atoms, and the electrons are not necessarily shared equally. The atom that has a greater share of the electrons can be determined from the **electronegativities** (electron-attracting abilities) of the atoms involved in the bonds. The electron density represented by a pair of shared electrons is greater around the atom with the greater electronegativity.

Table 1.2 lists the electronegativities of the atoms that are of most interest to organic chemists. Remember that electronegativities increase from left to right (excluding the noble gases) in a period of the periodic table and also from bottom to top in a column. For our purposes we can consider carbon and hydrogen to have roughly the same electronegativity. The order of electronegativities of the atoms of most interest to us is as follows:

$$H \approx C < Br < Cl = N < O < F$$

increasing electronegativity $\longrightarrow$

Table 1.2   Pauling Scale of Electronegativity Values for Some Elements

H 2.1						
Li 1.0	Be 1.5	B 2.0	C 2.5	N 3.0	O 3.5	F 4.0
Na 0.9	Mg 1.2		Si 1.8	P 2.1	S 2.5	Cl 3.0
						Br 2.8
						I 2.5

Since the electrons in a covalent bond spend more time around the more electronegative atom, this atom has more negative charge around it than the formal charge indicates. Consider the hydrogen chloride molecule. The formal charges on both atoms in the Lewis structure are zero. However, chlorine is more electronegative than hydrogen. Therefore the shared electrons spend more time around the chlorine than around the hydrogen. The chlorine has more electron density around it than the Lewis structure indicates and thus has some negative charge associated with it. Similarly, the hydrogen has less electron density and some positive charge. Since the electrons of the covalent bond are still shared, although unequally, the amount of charge on each atom is less than 1. There is a partial negative charge on the chlorine and a partial positive charge on the hydrogen. The lowercase Greek letter delta (δ) followed by the sign of the charge is used to represent such partial charges:

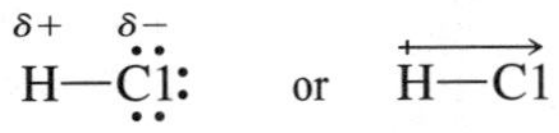

$$\overset{\delta+}{H}\!\!-\!\!\overset{\delta-}{\underset{\cdot\cdot}{\overset{\cdot\cdot}{Cl}}}\!: \quad \text{or} \quad \overset{\longleftarrow}{H\!-\!Cl}$$

When unequal electronegativities of two atoms involved in a bond result in charge separation as described above, we say that the bond is **polar.** Hydrogen chloride has a polar bond. The charge separation results in a **dipole,** that is, a positive and a negative "pole" in the molecule. The product of the amount of charge separation (e) times the distance of the charge separation (d) is called the **dipole moment** (μ).

$$\mu = (e)(d)$$

The dipole moment is a vector, that is, it has direction and magnitude. The dipole moment is usually represented as an arrow pointing from the positive end to the negative end of the dipole, as shown above for hydrogen chloride.

The amount of the charge, e, is on the order of 10^{-10} electrostatic units (esu), and the distance, d, is on the order of 10^{-10} m, so μ is on the order of 10^{-20} esu m. The debye unit (D) is defined as 1×10^{-20} esu m. The dipole moment of HCl is 1.1 D.

The direction and relative magnitude of the dipole of any bond can be predicted from the electronegativities of the atoms involved in the bond. For example, the electrons of a bond between oxygen and carbon will spend more time around the oxygen because it is more electronegative, and therefore the oxygen will be the negative end of the bond dipole. In the case of a bond between carbon and hydrogen, the two elements are of similar electronegativities. Therefore the electrons are shared nearly equally, resulting in a **nonpolar bond.**

Problem

Problem 1.12   Show the direction of the dipoles of these bonds:

   a)   C—N   **b)**   O—N   **c)**   O—Cl   **d)**   C—Cl

   e)   B—O   **f)**   C—Mg

The concept of bond polarities is very important, since much of chemistry, both the physical properties of compounds and their chemical reactions, depends on the interaction of charges. For example, a reagent that is seeking positive charge will likely be attracted to the carbon of a carbon–oxygen bond.

So far, we are able to predict the dipoles of individual bonds. The overall dipole moment of a molecule is the vector sum of these individual bond dipoles. Before the bond dipoles can be used to predict the overall dipole moment of a molecule, the three-dimensional orientation of the bonds must be known. We need to know the shapes of molecules.

1.11   Shapes of Molecules

The shapes of molecules are determined by actual experiments, not by theoretical considerations. But we do not want to have to memorize the shape of each molecule. Instead, we would like to be able to look at a Lewis structure and be able to predict the shape of the molecule. There are several models that enable us to do this. One of the easiest to use is **valence shell electron pair repulsion theory,** which is often referred to by its acronym VSEPR (pronounced "vesper") theory. As the name implies, the theory states that pairs of electrons in the valence shell repel each other and therefore they try to stay as far apart as possible. You probably remember this theory from your general chemistry class. The parts of VSEPR theory that are most important to organic chemistry will be reviewed here. The rules that are needed are as follows:

> **RULE 1**
>
> **Pairs of electrons in the valence shell repel each other and therefore stay as far apart as possible.** For example, if four pairs of electrons are arranged around a central atom so that they are as far apart as possible, they will be located at the corners of a tetrahedron. The geometries resulting from other numbers of electron pairs, arranged as far apart as possible, are given in Figure 1.11.

When we attempt to show the shapes of molecules, we are faced with the problem of how to represent such three-dimensional objects on a piece of two-dimensional paper. Some bonds extend in front of the page, while others extend behind the page. Commonly, a heavy wedged line is used to represent a bond that extends in front of

Figure 1.11

Shapes of Molecules from VSEPR Theory.

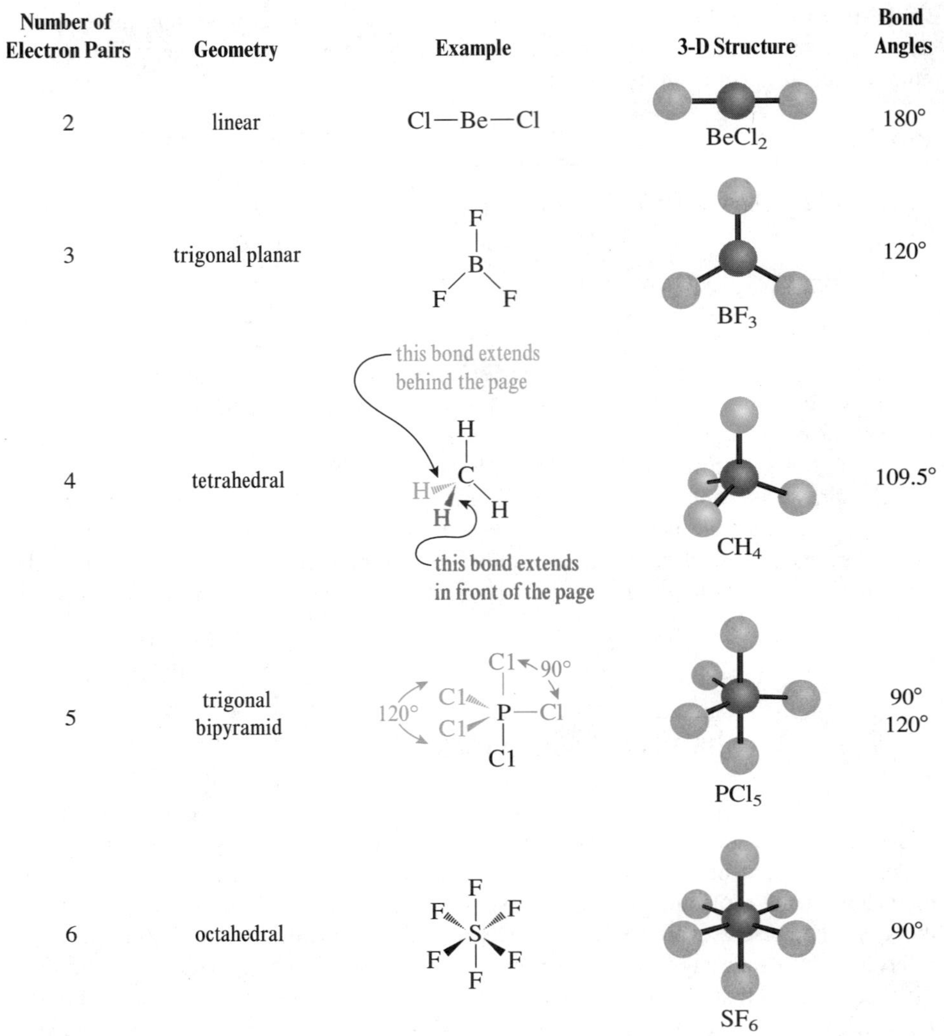

Number of Electron Pairs	Geometry	Example	3-D Structure	Bond Angles
2	linear	Cl—Be—Cl	$BeCl_2$	180°
3	trigonal planar	(BF₃ structure)	BF_3	120°
4	tetrahedral	(CH₄ structure)	CH_4	109.5°
5	trigonal bipyramid	(PCl₅ structure)	PCl_5	90° 120°
6	octahedral	(SF₆ structure)	SF_6	90°

this bond extends behind the page

this bond extends in front of the page

Cl 90°
120°

the plane of the page and a dashed wedge line is used to represent a bond that extends behind the plane of the page. These conventions are used to show the shapes of the molecules in Figure 1.11. Figure 1.12 shows some other pictures for the tetrahedral molecule CH_4. It is quite common for students to have difficulty picturing the actual three-dimensional shapes of the molecules from these two-dimensional drawings. It is a good idea to work with a set of models until the process becomes easier.

RULE 2
An unshared pair of electrons repels other pairs more than a shared pair of electrons does. This seems reasonable because a shared pair, which also spends time around another atom, should not offer as much charge concentration for repulsion as a pair of electrons that is not shared and therefore spends all of its time around the atom.

Figure 1.12
The Shape of CH₄.

This is a ball-and-stick model of methane. The carbon is gray, and the hydrogens are blue. The atoms are not shown to size.

This is a space-filling model for methane. The correct relative sizes of the atoms are shown.

Consider the shape of ammonia, NH₃, shown in Figure 1.13. Ammonia has four pairs of electrons in its valence shell: three shared pairs and one unshared pair. These four pairs of electrons have a basic tetrahedral arrangement. However, since the unshared pair repels more than the shared pairs do, the shared pairs are pushed closer together, and the bond angles are 107° rather than the exact tetrahedral bond angle of 109.5°. Usually, the location of the atoms in a molecule, not the position of electrons, is shown, so ammonia is termed a *pyramidal molecule,* that is, the nitrogen and the three hydrogens form a pyramid. Figure 1.13 also shows the shape of water. Again there are four pairs of electrons that have a basic tetrahedral arrangement. The two unshared pairs push the shared pairs even closer together. The result is that water is a bent molecule with a bond angle of 105°.

RULE 3
Double and triple bonds are treated as one shared pair of electrons.

ball-and-stick model   space-filling model

The three shared pairs of electrons and the unshared pair of ammonia are arranged in a tetrahedral manner around the nitrogen. The resulting geometry of the nitrogen and the three hydrogens is a pyramid. The HNH bond angle is less than 109.5° (the tetrahedral bond angle) because the unshared electron pair behaves as though it is larger than the electron pairs in the bonds.

ball-and-stick model   space-filling model

The two shared pairs of electrons and the two unshared pairs of water are also arranged in a tetrahedron. The resulting geometry of the three atoms is bent, with the HOH angle even smaller because of the repulsion of two unshared electron pairs.

Figure 1.13

The Shapes of the Ammonia and Water Molecules.

Several examples of the use of this rule are shown in Figure 1.14.

By using additional rules from VSEPR theory, it is possible to make more exact predictions about the shapes of molecules. For example, the HCO bond angle of CH_2O can be predicted to be slightly greater than 120°, as shown in Figure 1.14. However, we will not need to know geometries that accurately. It is enough to know that the geometry of CH_2O is approximately trigonal planar.

Problems

Problem 1.13   Predict the geometry at the carbon of these compounds:

a) H—C≡N:

b)

H—C—O—H

c)

Cl—C—H
   |
   Cl
   H

Problem 1.14   Predict the geometry of the following compounds at the indicated atoms:

a)

H—C—O—H
   |
   H

at the carbon
at the oxygen

b)

C=N—H

at the carbon
at the nitrogen

c)

C=O—H

at the carbon
at the oxygen

Exercises

Exercise 1.1   Build models of these compounds:

a) CH_4 b) BF_3 c) NH_3 d) H_2O e) PCl_5

f) C_2H_2 g) CH_2O

Exercise 1.2   Build models of the compounds shown in problem 1.14.

1.12   Dipole Moments

The overall dipole moment of a molecule is the vector sum of the individual bond dipoles. If the shape of a molecule is known, then vector addition can be used to predict the direction of the dipole moment of that molecule. Several examples are shown in Figure 1.15. The predictions agree with experimental results. For example, CO_2 is found by experiment to have a dipole moment of zero. Since the bonds are polar, this is possible only if the bond dipoles cancel out. Therefore CO_2 must be a linear molecule. On the other hand, since water has a dipole moment of 1.8 D, it cannot be a linear molecule.

Figure 1.14

The Shapes of C_2H_2, CO_2, and CH_2O.

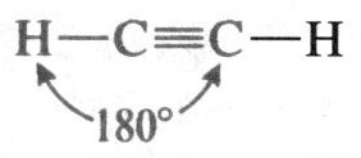

The triple bond counts as one pair of electrons for the purposes of VSEPR theory. Therefore there are two pairs around each carbon. The geometry is linear around each carbon and linear overall.

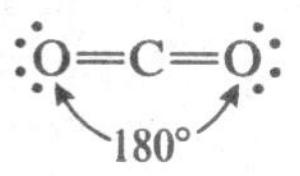

Each double bond counts as one pair of electrons. Therefore there are two pairs around the carbon, and the geometry is linear.

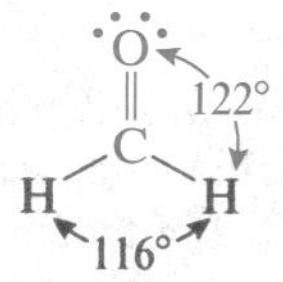

The double bond counts as one pair, so the geometry about the carbon is trigonal planar. The predicted bond angles are 120°. Actually, the HCO angle is 122°. Such minor deviations should be expected because the bonds are not the same and therefore do not necessarily have the same "size." In fact, a double bond to an oxygen might be expected to be "larger" than a single bond to a hydrogen.

Figure 1.15

Obtaining Dipole Moments from Bond Dipoles.

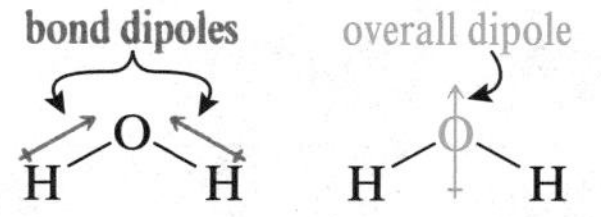

The **individual bond dipoles** of water are shown in the left structure. The vector sum shown in the right structure indicates the direction of the overall dipole moment of the molecule. Its magnitude is 1.8 D.

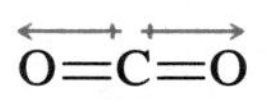

The individual bond dipoles of CO_2 point in opposite directions. These cancel on vector addition, so this compound has a dipole moment of zero.

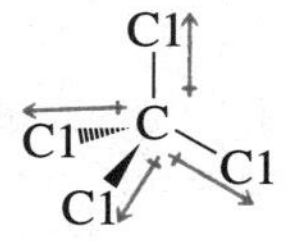

Although it is more difficult to see, the individual bond dipoles of CCl_4 also cancel on vector addition, and the overall dipole moment is zero.

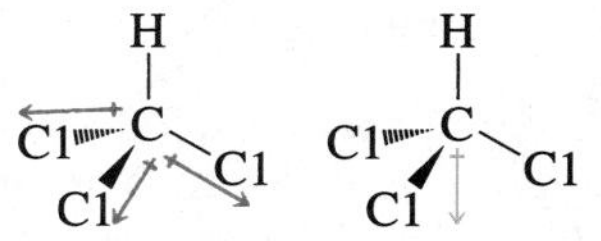

For $HCCl_3$ the **individual bond dipoles**, shown in the first structure, add to give an overall dipole moment of 1.9 D with the direction shown in the second structure.

Problems

Problem 1.15 Predict the direction of the dipole moment in this compound:

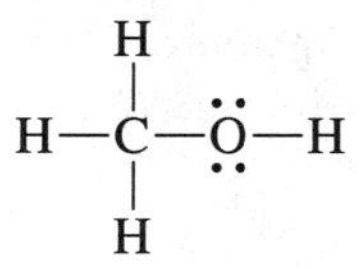

Solution

First we must determine the geometry of the molecule. In problem 1.14 we found that the geometry at this carbon, with four bonding pairs of electrons, is tetrahedral. The geometry of the electron pairs at the oxygen (two unshared pairs and two shared pairs) is also tetrahedral, so the geometry of the C—O—H atoms is bent. The C—H bonds are nonpolar, but the O—H and the O—C bonds are both polar with the negative end of the bond dipole at the oxygen because oxygen is more electronegative than either carbon or hydrogen. Vector addition gives the overall dipole moment shown below:

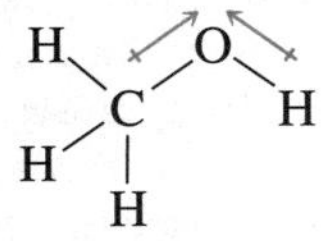

bond dipoles 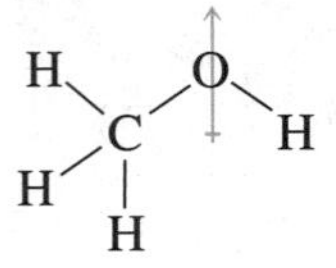   overall dipole

Problem 1.16   Predict the direction of the dipole moments of these compounds:

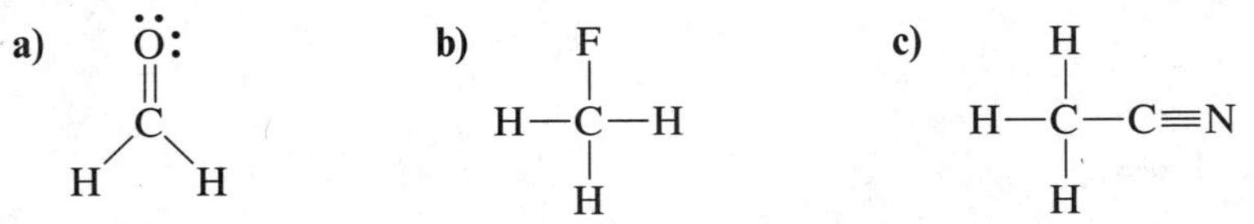

a)

b)

c)

1.13   Summary

After completing this chapter, you should be able to:

1. Write the best Lewis structure for any molecule or ion. This includes determining how many electrons are available, whether multiple bonds are necessary, and satisfying the octet rule if possible. For complex molecules, however, the connectivity must be known.

2. Calculate the formal charge on any atom in a Lewis structure. (In fact, you should be starting to recognize the formal charges on some atoms in some situations without doing a calculation.)

3. Estimate the stability of a Lewis structure by whether it satisfies the octet rule and by the number and the distribution of the formal charges in the structure.

4. Recognize some simple cases in which resonance is necessary to describe the actual structure of a molecule. However, a better understanding of resonance will have to wait until Chapter 3.

5. Arrange the atoms that are of most interest to organic chemistry in order of their electronegativities and assign the direction of the dipole of any bond involving these atoms.

6. Determine the shape of a molecule from its Lewis structure by using VSEPR theory.

7. Determine whether a compound is polar or not and assign the direction of its dipole moment.

In the next chapter the simple bonding picture that has been developed here will be used to help understand and predict which molecules are stable. We will also begin to learn how the structure of a molecule affects its physical and chemical properties.

End-of-Chapter Problems

1.17 Explain whether the bonds in these compounds would be ionic or covalent and show Lewis structures for them:

a) KCl   b) NCl_3

1.18 What is the formula for the simplest neutral compound formed from P and H? Show a Lewis structure for this compound and predict its shape.

1.19 Show Lewis structures for these compounds:

a) CH_5N b) C_2H_5Cl c) N_2 d) CH_3N
e) C_2H_3F f) CH_4S

1.20 Calculate the formal charges on all of the atoms, except hydrogens, in these compounds:

a) H—N̈—N≡N: b) H—N̈=N=N̈:

c) H—C̈—N≡N: d)   H   Ö:
          | | ‖
          H   H—C—C—Ö:
                                        |
                                        H

e)   H   f)   H
            | |
      H—Ö=C—H   H—B—H
                                        |
                                        H

1.21 Explain which of the two following structures would be more stable. Explain whether they represent isomers or are resonance structures.

1.22 Draw a Lewis structure for carbon monoxide (CO). Calculate the formal charges on the atoms and comment on the stability of this compound.

1.23 Use heavy and dashed wedged lines to show the shapes of the following molecules. Show the bond dipole of each polar bond and show the overall dipole of each molecule.

a)

$$H-\overset{\overset{\displaystyle H}{|}}{\underset{\underset{\displaystyle H}{|}}{C}}-\overset{..}{\underset{\underset{\displaystyle H}{|}}{N}}-H$$

b)

$$F-\overset{\overset{\displaystyle F}{|}}{\underset{\underset{\displaystyle H}{|}}{C}}-F$$

c)

$$H-\overset{\overset{\displaystyle H}{|}}{\underset{\underset{\displaystyle H}{|}}{C}}-\overset{\overset{\displaystyle \overset{..}{O}:}{\|}}{C}-H$$

1.24 Predict the geometry at each atom, except hydrogens, in these compounds:

a)

$$H-C=\overset{..}{\underset{\oplus}{O}}-\overset{\overset{\displaystyle H}{|}}{\underset{\underset{\displaystyle H}{|}}{C}}-H$$

b) $H-\overset{\overset{\displaystyle }{|}}{\underset{\underset{\displaystyle H}{|}}{C}}=C=\overset{\overset{\displaystyle }{|}}{\underset{\underset{\displaystyle H}{|}}{C}}-H$

1.25 **a)** Show the unshared electron pairs on the following anion. The S has a formal charge of -1, and the formal charges of the other atoms are zero.

$$H-\overset{\overset{\displaystyle H}{|}}{\underset{\underset{\displaystyle H}{|}}{C}}-C\overset{\overset{\displaystyle O}{\diagup}}{\underset{\underset{\displaystyle S}{\diagdown}}{}}$$

b) Draw a resonance structure for this ion.

1.26 Show a Lewis structure for C_2H_6O in which both carbons are bonded to the oxygen. What is the geometry of this molecule at the oxygen? Show the direction of the dipole for the molecule.

1.27 Show a Lewis structure for $NO_2^{\ominus}$. (Both oxygens are bonded to the nitrogen.) Show a resonance structure also.

1.28 A covalent ion can also have polar bonds. Consider the ammonium cation. How are its bonds polarized? Do you think that the N of the ammonium cation is more or less "electronegative" than the N of ammonia (NH_3)? Would the hydrogens of $NH_4^{\oplus}$ or NH_3 have a larger partial positive charge?

$$H-\overset{\overset{\displaystyle H}{|}}{\underset{\underset{\displaystyle H}{|}}{\overset{\oplus}{N}}}-H$$

ammonium cation

1.29 You need to know the melting point for $CaCl_2$ for a lab report you are writing. Your lab partner says that the *Handbook of Chemistry and Physics* lists this as 68°C. Do you think you should trust that your lab partner has looked up the value correctly or should you look it up for yourself?

1.30 Show a Lewis structure for $AlCl_4^{\ominus}$. What are the formal charges on the atoms of this anion? What is its shape?

1.31 Ammonium cyanate, which Wohler used in his preparation of urea, is composed of an ammonium cation ($NH_4^{\oplus}$) and a cyanate anion ($OCN^{\ominus}$). Show a Lewis structure for the cyanate anion. (Both O and N are bonded to C.) Which atom has the negative formal charge in your structure? What is the shape of the ion? Show a resonance structure for this ion.

1.32 **a)** Show a Lewis structure for urea, CH_4N_2O. Both N's and the O are bonded to the C. The H's are bonded to the N's. None of the atoms has a formal charge.

b) Show a Lewis structure of an isomer of urea that still has both N's and the O bonded to the C and has formal charges of zero at all atoms.

1.33 Phosphorus forms two compounds with chlorine, PCl_3 and PCl_5. The former follows the octet rule, but the latter does not. Show Lewis structures for each of these compounds. For the corresponding nitrogen compounds, explain why NCl_3 exists but NCl_5 does not.

1.34 On the basis of the rule that anything unusual about a structure must be shown explicitly, the nitrogen in the structure NH_3 is seen to have an unshared pair of electrons whether these electrons are shown or not. Suppose you wanted to discuss the unstable species NH_3 that was missing the unshared pair of electrons. How would you draw the structure so that it would be obvious to another person that the unshared pair was absent?

1.35 Ozone, O_3, is a form of oxygen found in the upper atmosphere. It has the connectivity O—O—O.

a) Show a Lewis structure for ozone.

b) Calculate the formal charge on each oxygen of ozone.

c) What is the shape of ozone?

d) Experimental observations show that both bonds of ozone are identical. Explain how this is possible.

1.36 **a)** Draw a Lewis structure for the carbonate anion CO_3^{2-}. Each oxygen is bonded only to the carbon.

b) Calculate the formal charge on each atom.

c) What is the shape of this species?

d) Experimental evidence shows that all of the oxygens are identical. Explain.

1.37 Consider the species CH_3 that has three normal carbon–hydrogen bonds and no other electrons on the carbon.

a) What is the charge of this species?

b) What is its geometry?

c) Discuss the stability of this species. Do you think it is more or less stable than the species shown in problem 1.5? Explain.

d) Show the Lewis structure of the product of the reaction of this species with hydroxide ion ($OH^{\ominus}$).

1.38 Explain how the dipole moments of FCl (0.9 D) and ICl (0.7 D) can be so similar.

1.39 Chlorine is more electronegative than phosphorus. Predict the dipole moment of PCl_5.

1.40 Although carbon–carbon double bonds are shorter than carbon–carbon single bonds, all of the carbon–carbon bonds of benzene are the same length. Explain.

benzene

c) $CH_3CH_2CH_2CO_2H$

Careful: how are the
O's bonded to the C?

d)

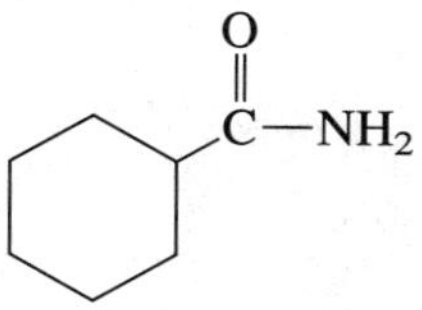

e) $CH_3\overset{O}{\overset{\|}{C}}OCH_3$

f)

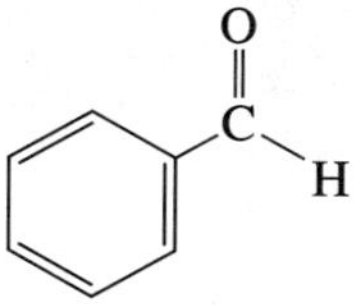

2.9 Summary

After completing this chapter, you should be able to:

1. Quickly recognize the common ways in which atoms are bonded in organic compounds. You should also recognize unusual bonding situations and be able to estimate the stability of molecules with such bonds.

2. Know the trends in bond strengths and bond lengths for the common bonds.

3. Recognize when compounds are structural isomers and be able to draw structural isomers for any formula.

4. Calculate the degree of unsaturation for a formula and use it to help draw structures for that formula.

5. Draw structures using any of the methods we have seen. You should also be able to examine a shorthand representation for a molecule and recognize all of its features.

6. Examine the structure of a compound and determine the various types of intermolecular forces that are operating. On this basis you should be able to crudely estimate the physical properties of the compound.

7. Recognize and name all of the important functional groups.

In the next chapter we are going to reexamine bonding, using orbitals. This will provide a better picture of why some bonds are stable and others are not.

End-of-Chapter Problems

2.16 Convert the following structures to shorthand representations that use lines for bonds:

a) $CH_3CH_2CH_2\overset{OH}{\overset{|}{C}H}CH_2CH_3$

b)

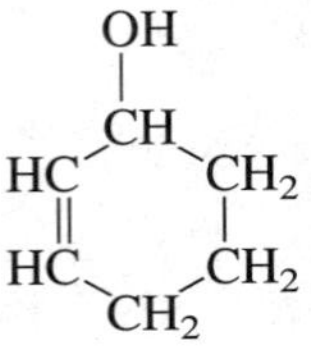

c)  $CH_3CCH_2CH_2CH_2Cl$

d)

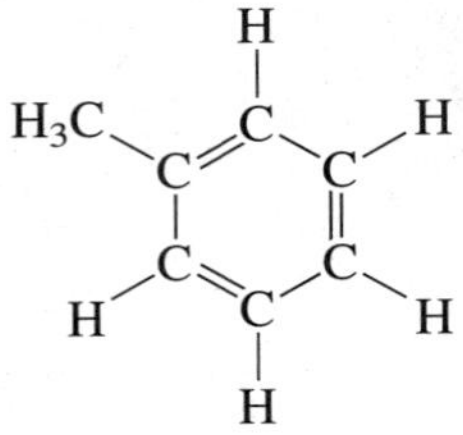

e) 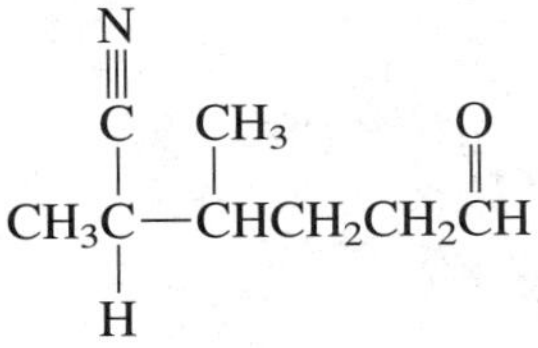 $CH_3C-CHCH_2CH_2CH$

f) 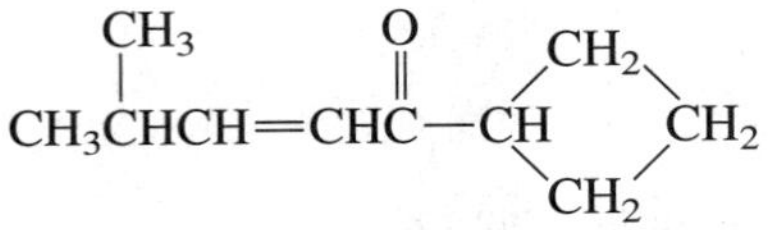 $CH_3CHCH=CHC-CH$

2.17 Convert the following shorthand representations to structures showing all of the atoms, bonds, and unshared electron pairs:

a)

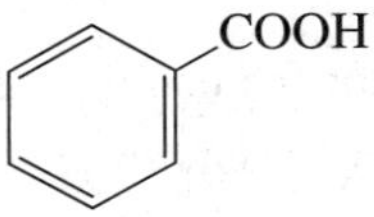

b)

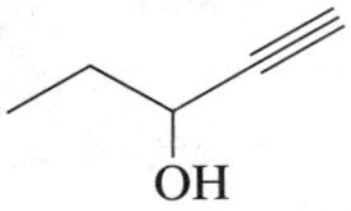

c)

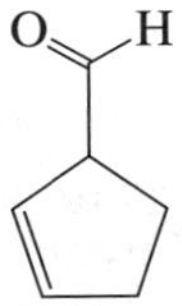

d)

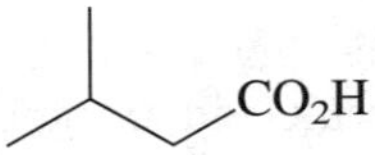

e)

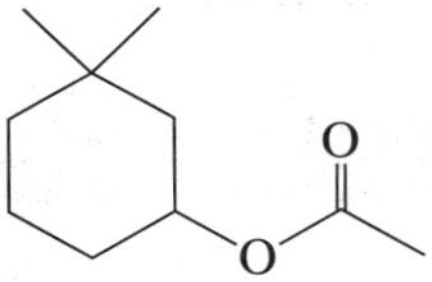

f)

2.18 Name the functional group(s) present in each of the compounds in problem 2.16.

2.19 Name the functional group(s) present in each of the compounds in problem 2.17.

2.20 Determine whether these structures represent the same compound or isomers:

a)

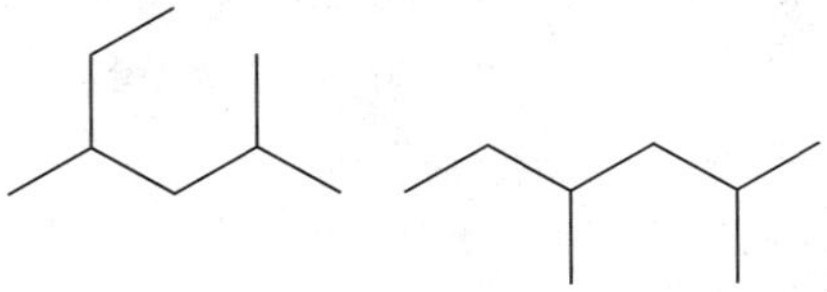

b)

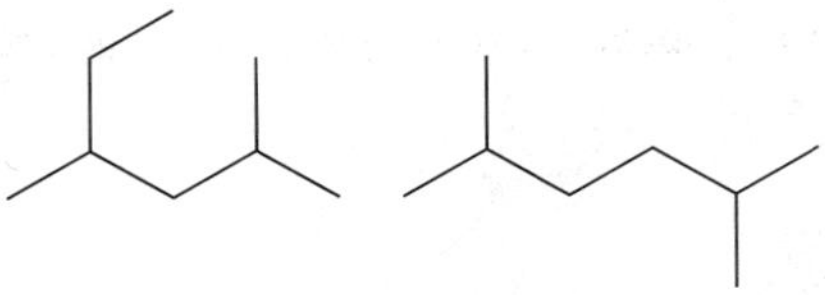

c)

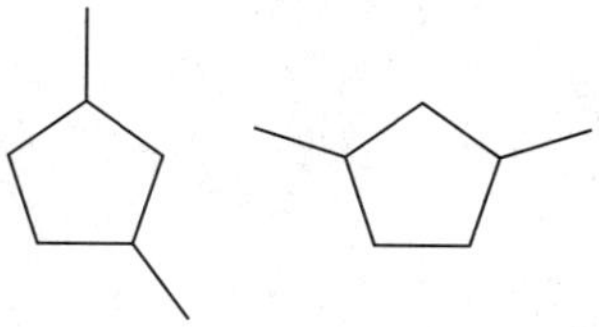

d) 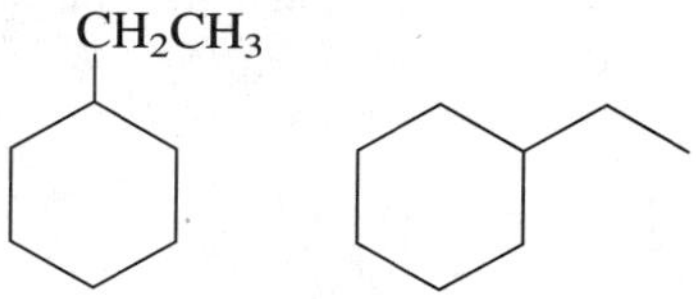 CH$_2$CH$_3$

e)

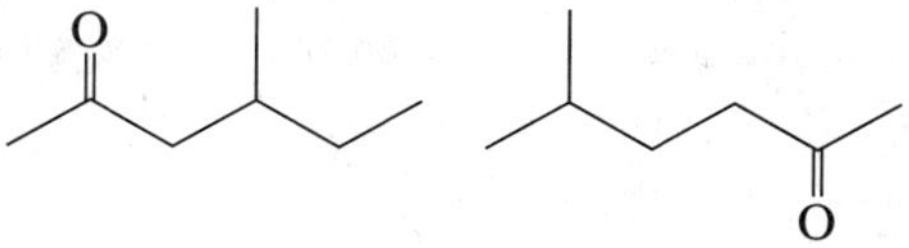

f) 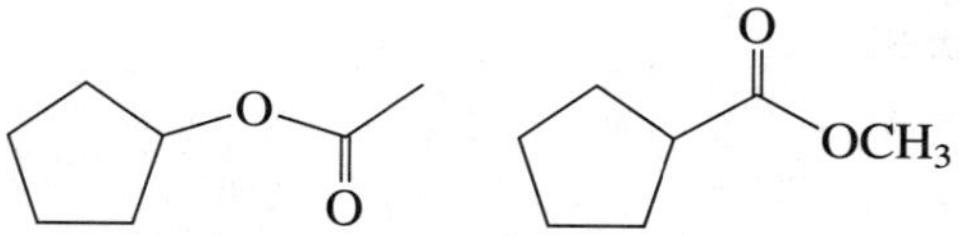 OCH$_3$

2.21 Draw all the isomers with each of these formulas. The total number for each is given in parentheses.

 a) C$_3$H$_8$O (3)   b) C$_4$H$_9$Cl (4)   c) C$_4$H$_8$ (5)

 d) C$_7$H$_{16}$ (9)

2.22 There are lots of isomers with the formula C$_4$H$_8$O.

 a) Draw three that have a carbon–oxygen double bond. What functional group is present in each of them?

 b) Draw three alcohols with this formula.

 c) Draw an ether with this formula that does not have a carbon–carbon double bond.

2.23 Four of the ten isomers of C$_5$H$_{10}$ are shown in Figure 2.5. Draw four other isomers with this formula.

2.24 Calculate the DU for each of these formulas and draw a structure that meets the listed restriction:

 a) C$_5$H$_9$NO (not an amine)

 b) C$_7$H$_{12}$O (a ketone)

 c) C$_6$H$_{14}$O (does not hydrogen bond)

 d) C$_7$H$_8$ (an aromatic compound)

2.25 Is it possible for C$_8$H$_{17}$N to have a nitrile as its functional group? Explain.

2.26 Show all the different hydrogen bonds that would occur in a mixture of

 a) CH$_3$CH$_2$OH and H$_2$O

b) CH_3OH and

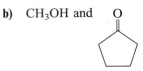

2.27 One of these isomeric alcohols has mp = 26°C and bp = 82°C; the other has mp = −90°C and bp = 117°C. Explain which isomer has the higher melting point and which has the higher boiling point.

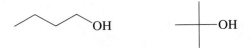

2.28 Explain the differences in the boiling points between the members of each of these pairs of compounds:

a) $CH_3(CH_2)_6CH_3$,   bp: 126°C
     $CH_3(CH_2)_8CH_3$,   bp: 174°C

b) $CH_3CH_2CH_2OH$,   bp: 97°C
     $CH_3CH_2OCH_3$,   bp: 11°C

c) $CH_3CH_2CH_3$,   bp: −42°C
     CH_3OCH_3,   bp: −23°C

2.29 Explain the difference in the melting points of these isomers:

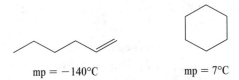

mp = −140°C   mp = 7°C

2.30 Which of these two salts would you expect to be more soluble in hexane (C_6H_{14})?

$\overset{\oplus}{NH_4}\ \overset{\ominus}{Cl}$ or $\overset{\oplus}{N}(CH_2CH_2CH_2CH_3)_4\ \overset{\ominus}{Cl}$

2.31 Benzene and hexane are both liquids at room temperature. Do you expect benzene and hexane to be miscible? Do you expect benzene and water to be miscible? Explain.

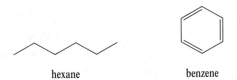

hexane   benzene

2.32 One of these isomers is miscible with water, and the other is nearly insoluble. Explain.

$\overset{\displaystyle O}{\overset{\displaystyle \|}{CH_3CH_2CH_2COH}}$ $\overset{\displaystyle O}{\overset{\displaystyle \|}{CH_3COCH_2CH_3}}$

2.33 Because of two hydrogen bonds, carboxylic acids show a very strong attractive force between two molecules that persists even in the gas phase. Show this hydrogen bonding between two carboxylic acid molecules.

2.34 Bond strengths can be used to estimate the relative stability of isomers that have different bonds. The isomer that has the larger total bond energy is more stable. One of the isomers below is more stable than the other. The less stable one is rapidly converted to the more stable one, so it cannot be isolated. On the basis of bond dissociation energies, which of these two isomers is more stable?

2.35 Bond strengths can also be used to estimate whether a reaction is energetically favorable or not—that is, whether the reactants or the products are more stable. Use bond dissociation energies to determine whether this reaction is energetically favorable. The bond dissociation energy for H_2 is 104 kcal/mol (435 kJ/mol).

$$CH_2=CH_2 + H-H \longrightarrow CH_3-CH_3$$

2.36 One of the isomers of C_5H_{12} reacts with Cl_2 in the presence of light to produce three isomers of $C_5H_{11}Cl$:

$$\underset{\text{one isomer}}{C_5H_{12}} + Cl_2 \xrightarrow{\text{light}} \underset{\text{three isomers}}{C_5H_{11}Cl} + HCl$$

This reaction replaces *any one* of the hydrogens of C_5H_{12} with a Cl. What are the structures of the C_5H_{12} isomer and the three $C_5H_{11}Cl$ isomers produced from it?

2.37 On reaction with Cl_2 in the presence of light, an unknown compound with the formula C_6H_{12} gives only one isomer of $C_6H_{11}Cl$ (see problem 2.36). What is the DU of the unknown compound? Show the structure of the unknown compound and the product of its reaction with Cl_2.

$$C_6H_{12} + Cl_2 \xrightarrow{\text{light}} \underset{\text{one isomer}}{C_6H_{11}Cl} + H-Cl$$

2.38 Explain how the dipole moment for CH_3Cl ($\mu = 1.9$ D) can be larger than the dipole moment for CH_3F ($\mu = 1.8$ D).

Orbitals and Bonding

3.1   Introduction

Chapter 1 described a covalent bond as a pair of electrons that is shared between two atoms. This chapter reexamines bonding in more depth. A model that uses electron orbitals to explain bonding and that provides a better understanding of bonds and reactivity is presented. The chapter begins with a review of atomic orbitals. Then a model in which bonding results from atomic orbitals interacting to form molecular orbitals is discussed. Because resonance is so important in organic chemistry, considerable attention is devoted to this topic. The idea of orbitals can help us to understand resonance better. Finally, a number of examples of how to use resonance and when it is important are presented.

Most of the subsequent chapters in this book use simple Lewis structures to represent molecules, as was done in Chapters 1 and 2. However, you should always have in mind the model that is presented in this chapter so that you can call on it whenever a better picture for bonds is needed to explain some observation.

Much of what is discussed in this chapter should be review of material that you learned in general chemistry. However, the presentation here concentrates more on the aspects of bonding that are important in organic chemistry.

3.2   Atomic Orbitals

You should recall from your general chemistry course that electrons have some of the properties of waves. Chemists use the equations of wave mechanics to describe these electron waves. Solving these wave equations for an electron moving around the nucleus of an atom gives solutions that lead to a series of **atomic orbitals.** These orbitals describe the location of the electron charge density when an electron occupies that orbital. Another way of stating this is that the shape of the orbital defines a

Quantum Numbers

You should remember from general chemistry that there are four quantum numbers. The first, called the **principal quantum number,** n, designates the shell for the orbital. It can take values of 1, 2, 3, and so on. The **azimuthal quantum number,** l, designates the subshell for the orbital. It defines the shape of the orbital and can have values of 0 (an s orbital), 1 (a p orbital), 2 (a d orbital), and so forth up to a value of $n - 1$. The **magnetic quantum number,** m_l, describes the orientation of the orbital in space. It can have integer values from $-l$ to $+l$, including zero. Thus for orbitals with $l = 1$, m_l can have values of -1, 0, and $+1$, so there are three p orbitals. The **spin quantum number,** m_s, designates the spin of the electron. It can have only two values, $+\frac{1}{2}$ and $-\frac{1}{2}$.

region about the nucleus where, if the orbital contains an electron, the probability of finding that electron is very high. It is simply the space where the electron spends most of its time.

Figure 3.1 shows the shapes of the orbitals that will be of most concern here. The first orbital pictured is the lowest-energy orbital, the **1s orbital.** It has a spherical shape, as do all s orbitals. Also shown in the figure is the mathematical sign (plus) of the orbital. *This sign does not relate to charge.* It simply indicates that the wave function (the solution to the wave equation) has a positive mathematical value throughout the entire orbital. These math signs are needed when atomic orbitals are combined to make molecular orbitals.

In order of increasing energy the next orbital is the **2s orbital.** Like the 1s orbital, it has a spherical shape. It is larger than the 1s orbital—that is, it extends farther out from the nucleus. Interestingly, the 2s orbital has a region where the mathematical sign of the wave function is positive and another region where the sign is negative. It also has a region where the value of the wave function equals zero. Such a region (in this case a spherical surface) is called a **node.** The probability of finding an electron at a particular point is proportional to the square of the value for the wave function at that point. Whether the wave function has a positive or a negative value at that point does not matter, since the square will always be positive. However, the value of the square of the wave function at a node is zero. This means that the electron density at a node is zero. In general, *the more nodes an orbital contains, the higher energy the orbital is*. Nodes are also very important in molecular orbitals.

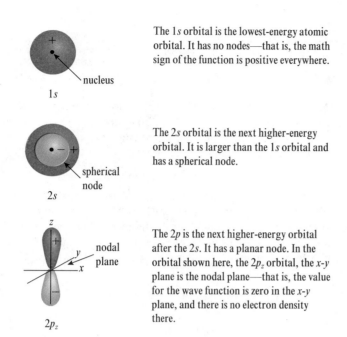

The 1s orbital is the lowest-energy atomic orbital. It has no nodes—that is, the math sign of the function is positive everywhere.

The 2s orbital is the next higher-energy orbital. It is larger than the 1s orbital and has a spherical node.

The 2p is the next higher-energy orbital after the 2s. It has a planar node. In the orbital shown here, the $2p_z$ orbital, the x-y plane is the nodal plane—that is, the value for the wave function is zero in the x-y plane, and there is no electron density there.

Figure 3.1

Shapes of Atomic Orbitals.

The next higher-energy orbital after the 2s orbital is the **2p orbital,** also shown in Figure 3.1. This orbital is not spherically symmetrical like the s orbitals. Its overall shape is something like a dumbbell, with regions of high probability of finding the electron (regions of high electron density) on opposite sides of the nucleus. One lobe of the orbital has a positive math sign and the other has a negative math sign, with a planar node in between. If, as shown in Figure 3.1, the lobes of the orbital are directed along the z-axis, then the orbital is called the $2p_z$ orbital and the plane formed by the x-axis and the y-axis is the plane of the node. There are three 2p orbitals, all of the same energy. Orbitals with the same energy are termed *degenerate.* The three 2p orbitals are mutually perpendicular. If one is directed along the z-axis ($2p_z$), as shown in Figure 3.1, the other two are directed along the x-axis and the y-axis ($2p_x$ and $2p_y$).

The orbitals pictured in Figure 3.1 are the ones of most interest to organic chemistry, since the atoms that are most commonly encountered in organic compounds are H, C, and other second-period atoms. However, atoms belonging to the third or higher periods of the periodic table are sometimes encountered. Orbitals in the third shell have two nodes. The **3s** and **3p orbitals** look similar to the 2s and 2p orbitals except that they are larger and have an additional spherical node. The spherical nodes do not affect the picture for bonding, so 3s and 3p orbitals (or any other s and p orbitals) can be treated similarly to 2s and 2p orbitals when they form molecular orbitals. In addition, the atoms of the third period have **3d orbitals** available. We will deal with bonding involving d orbitals later, as necessary.

Problem

Problem 3.1   Draw a 3s atomic orbital and compare it to a 2s orbital.

The electron configuration of an atom shows how the electrons are distributed among its orbitals. The electrons are arranged in the orbitals so that the overall energy of the system is minimized. The following set of rules can be used to quickly derive the electron configuration for an atom:

1.   Each electron is placed in the lowest-energy orbital available.

2.   Each orbital can contain a maximum of two electrons, and these must have opposite spins. This is a result of the **Pauli exclusion principle,** which states that no two electrons can have all four quantum numbers the same. Since two electrons in the same orbital must have three of the quantum numbers the same, the fourth quantum number (the spin quantum number) must be different.

3.   When degenerate orbitals are available, the electrons first occupy them singly,

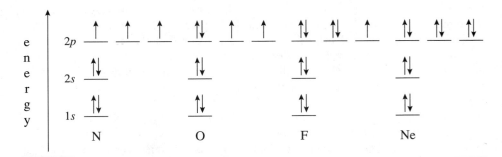

Figure 3.2

Atomic Orbital Energy Level Diagrams.

To simplify these diagrams, the orbitals have been shown at the same energies for different atoms. Actually, the energy of an orbital decreases as the number of protons in the atom increases. Thus the 2p orbitals of fluorine are lower in energy than the 2p orbitals of oxygen.

with the same (parallel) spins. After all of the degenerate orbitals contain one electron, additional electrons with opposite spins are added to each. This is known as **Hund's rule.**

Figure 3.2 shows the electron configurations for the atoms in the first two rows of the periodic table. This energy level diagram shows the atomic orbitals in order of increasing energy (the lowest-energy orbital is at the bottom) and shows the electrons occupying these orbitals as arrows (the direction of the arrow indicates the spin). This figure can easily be constructed from the above rules. For example, carbon has six electrons. The first two go into the lowest available orbital, the 1s orbital, (rule 1), with opposite spins (rule 2). Likewise, the next two electrons go into the 2s orbital. The last two electrons must be placed in the 2p orbitals. Since there are three degenerate 2p orbitals, these two electrons are placed in different orbitals, with the same spin (rule 3).

What is the reason behind Hund's rule? Let's consider carbon again. Recall that electrons, having the same charge, repel each other. This electron–electron repulsion is minimized if the electrons are in different regions of space, resulting in a lower-energy situation. If the two electrons in the 2p orbitals of carbon have the same spin, they must be in different 2p orbitals. (If the electrons were together in the same 2p orbital, the Pauli exclusion principle would be violated.) Being in different 2p orbitals keeps the electrons in different regions of space.

Figure 3.2 shows the electron configurations for the **ground state** or the **lowest-energy state** for the atoms. Any other electron arrangement is higher in energy and is termed an **excited state.** For example, a carbon atom with two electrons in the 1s orbital, one electron in the 2s orbital, and three electrons in the 2p orbitals is in an excited state. Such an excited carbon atom will exist for only a very short period of time—only until it can find a way to get rid of that extra energy and return the third 2p electron to the 2s orbital.

Problems

Problem 3.2　Show an atomic orbital energy level diagram for these atoms.

　a)　Si　　　**b)**　Al　　　**c)**　Cl

Problem 3.3　Explain whether the electron arrangement for these atoms is the ground state or an excited state.

a)　　　　　　　　　　nitrogen　　　　　　**b)**　　　　　　　　　carbon

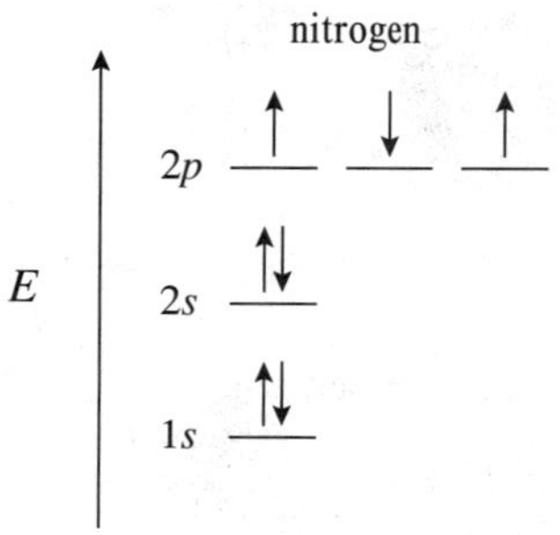

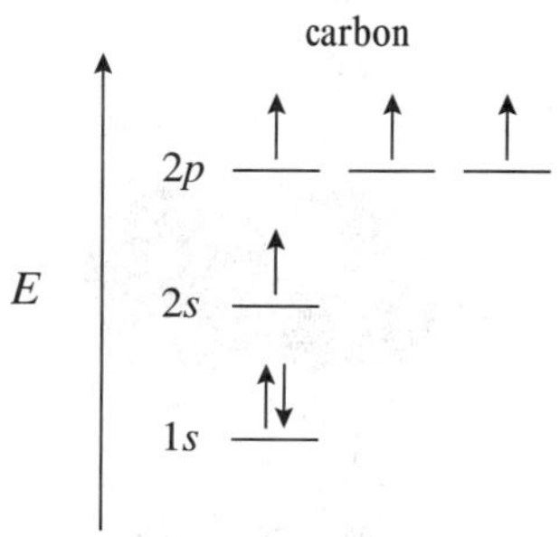

3.3   Molecular Orbitals

In isolated atoms the electrons are in the atomic orbitals (AOs) of that atom. What happens to the electrons when atoms come together to form bonds? In the simple Lewis model, some of the electrons are pictured as being shared between atoms. In the orbital model, these shared electrons are pictured as being in orbitals that extend around more than one atom. Such orbitals are called **molecular orbitals** (MOs).

The shapes of the MOs indicate the areas of high electron density in the molecule. MOs are useful in understanding reactions because orbitals must overlap to form any new bond. Thus MOs help us see how molecules must approach each other for their orbitals to overlap in forming new bonds. Molecular orbitals are especially useful in the discussion of spectroscopy (Chapters 12 and 13), aromatic compounds (Chapter 17), and pericyclic reactions (Chapter 20).

Most of what was presented previously about AOs also applies to MOs. The shape of the MO, which describes a region around the nuclei of the bonding atoms where the probability of finding an electron is high, is important, as is the energy of the MO. Let's examine H_2, a very simple molecule, to see what happens.

Let's consider the shape of the MO first. The simplest picture considers molecular orbitals as resulting from the overlap of atomic orbitals. When atoms are separated by their usual bonding distance, their AOs overlap. Where this overlap occurs, either the electron waves reinforce and the electron density increases, or the electron waves cancel and the electron density decreases. The left-hand side of Figure 3.3 shows the overlap of the $1s$ atomic orbitals on two different hydrogens (H_a and H_b) when these hydrogens are separated by their normal bonding distance. The *two* atomic orbitals interact to produce *two* molecular orbitals. The MOs result from a linear combination of the AOs (called the **LCAO approximation**). Simply, this means that the AOs are either added ($1s_a + 1s_b$) or subtracted ($1s_a - 1s_b$) to get the MOs.

In Figure 3.3 the AOs have the same math sign in the region of overlap. In the $1s_a + 1s_b$ combination the magnitude of the wave function increases in the overlap region, and the electron density also increases here. This increase of electron density between the nuclei results in a more stable orbital—the MO is lower in energy than the AOs. Such MOs are called **bonding MOs.** In the $1s_a - 1s_b$ combination the mag-

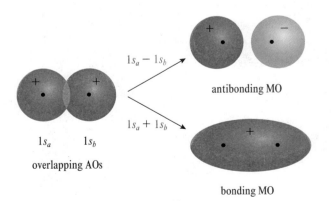

$1s_a - 1s_b$

antibonding MO

$1s_a$ $1s_b$

overlapping AOs

$1s_a + 1s_b$

bonding MO

Figure 3.3

Combination of Hydrogen 1s Atomic Orbitals to Form Molecular Orbitals.

nitude of the wave function decreases in the region between the nuclei and actually cancels along a plane perpendicular to a line connecting the two nuclei. This results in a node and a decrease in electron density between the nuclei. The resulting MO is less stable than the AOs and is called an **antibonding MO.**

Figure 3.4 shows the energies of these MOs. The energies of the $1s$ AOs of the separated hydrogen atoms are shown on the left and right sides of the diagram. The energies of the MOs are shown in the center of the diagram. The bonding MO is lower in energy than the $1s$ AOs by an amount of energy shown as ΔE in the diagram. In the simple picture we are using, the antibonding MO is higher in energy than the AOs by ΔE also. (A more sophisticated treatment shows that the antibonding MO is actually higher in energy than the $1s$ orbital by more than ΔE.)

The molecular orbitals are filled with electrons according to the same rules that were used to put electrons in the atomic orbitals of atoms. In this case there are two electrons. These fill the bonding MO. The stability of the molecule is determined by the total energy of the electrons. In the case of H_2 the molecule is more stable than the separated atoms by $2(\Delta E)$. In other words, it would be necessary to add $2(\Delta E)$ of energy to the H_2 molecule to break the covalent bond. As is the case here, the antibonding MOs usually do not have any electrons in them, and they do not affect the energy of the molecule. But they are real and can be occupied by electrons in some situations, such as certain types of spectroscopy and some chemical reactions.

The bond formed between two hydrogen atoms is one example of a **sigma (σ) bond.** In general, the MO of a sigma bond is shaped so that it is symmetrical about the internuclear axis (a line connecting the two nuclei). In other words, if a plane cuts through the MO, perpendicular to a line connecting the nuclei, the intersection of the plane and the MO is a circle. This is shown in Figure 3.5. In a sigma bond, rotation of the AO of one atom of the bond around the internuclear axis does not affect the overlap of the AOs, so it does not affect the energy of the bond.

The MO picture for H_2 agrees completely with the Lewis structure. It shows a pair of electrons shared between the two atoms (in an MO that extends over both atoms), resulting in a molecule that is more stable than the separated atoms. It also explains why the valence for hydrogen is 1. There is no room for additional electrons unless they are placed in the antibonding MO, a destabilizing situation. MO theory can also help us to understand why a molecule such as He_2 does not exist. Like hydrogen, both heliums would use $1s$ AOs to form a bonding and an antibonding MO. However, there would be four electrons to place in these MOs, two from each helium atom.

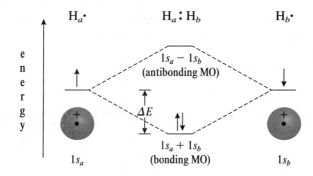

Figure 3.4

Energy Level Diagram for H_2 Molecular Orbitals.

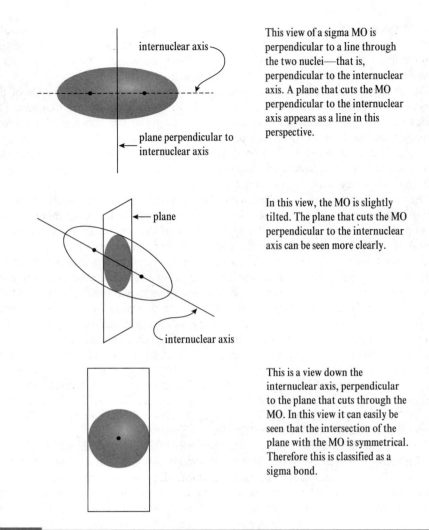

This view of a sigma MO is perpendicular to a line through the two nuclei—that is, perpendicular to the internuclear axis. A plane that cuts the MO perpendicular to the internuclear axis appears as a line in this perspective.

In this view, the MO is slightly tilted. The plane that cuts the MO perpendicular to the internuclear axis can be seen more clearly.

This is a view down the internuclear axis, perpendicular to the plane that cuts through the MO. In this view it can easily be seen that the intersection of the plane with the MO is symmetrical. Therefore this is classified as a sigma bond.

Figure 3.5

Several Views of a Sigma Bond.

Therefore both the bonding MO and the antibonding MO would be filled with electrons. The stabilization of the electrons in the bonding MO would be more than offset by the destabilization of the electrons in the antibonding MO. The electrons are actually more stable in the separated atoms.

Problem

Problem 3.4   Show an energy level diagram for the MOs for He_2 and show how the electrons would be arranged in these MOs.

What happens with more complicated molecules, which have more than two atoms? To be most accurate, MOs that extend over all the atoms in the molecule should be used. However, such MOs are more complex than are needed for most situations. Instead, we use the approximation that each MO is localized around only two atoms. These **localized MOs** are much easier to visualize and correspond well to the picture of a bond as a shared pair of electrons. Later, some special situations in which it is necessary to use **delocalized MOs** that extend around more than two atoms will be examined.

If the approximation that MOs are localized between two atoms is used, then molecular orbital theory is not as complicated as it might seem. The same thing happens every time. The following generalizations can be made:

1. Two AOs on the bonding atoms overlap to produce two MOs. In more complicated situations the number of MOs produced equals the number of AOs initially involved.

2. One combination of the two AOs results in a lower-energy, bonding MO; the other combination results in a higher-energy, antibonding MO with a node between the nuclei. The math signs are a convenient way to keep track of the nodes.

3. The same rules are used to assign electrons to MOs as are used to assign electrons to AOs. Usually, there are just enough electrons to fill the bonding MOs, and the antibonding MOs remain empty.

4. The bond energy is approximately equal to the total amount of energy by which the electrons are lowered in energy in comparison to the electrons in the AOs—that is, the number of electrons times ΔE.

5. The magnitude of ΔE increases with increasing overlap of the AOs. However, if the atoms get too close together, repulsion between the nuclei starts to dominate, and the overall energy increases very rapidly.

By using these rules, more complex situations can also be analyzed.

3.4   Single Bonds and sp^3 Hybridization

Let's consider methane (CH_4), a simple organic molecule. First, the Lewis structure, which shows four CH bonds, should be examined.

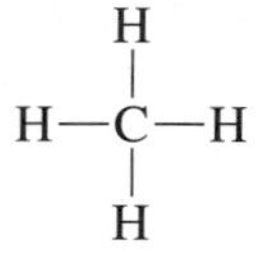

$$
\begin{array}{c}
\quad H \\
\quad | \\
H - C - H \\
\quad | \\
\quad H
\end{array}
$$

methane

In the MO picture, there will be a bonding MO (and an antibonding MO) for each bond in the Lewis structure. Furthermore, the MO model must be in accord with experimental observations. Experiments have shown that the bonds in methane are all identical, with tetrahedral geometry. Therefore methane must have four equivalent bonding MOs, with a tetrahedral arrangement.

Each MO results from an AO on the carbon overlapping with an AO on one of the hydrogens. Only AOs from the valence shell of an atom are used. Each hydrogen uses a $1s$ AO. The four orbitals that are available from the carbon are its $2s$ and three $2p$ AOs. However, there are problems if these carbon orbitals are used directly. First, while four equivalent MOs are needed, an MO resulting from a carbon $2s$ AO overlapping with a hydrogen $1s$ AO would obviously be different from an MO resulting from a carbon $2p$ AO overlapping with a hydrogen $1s$ AO. Second, the carbon $2p$ orbitals do not have the correct geometry to point directly at hydrogens that are tetrahedrally arranged around the carbon. This would result in poor overlap between the AOs, a small value for ΔE, and weak bonds.

You probably remember the solution to these problems from general chemistry. The $2s$ and the three $2p$ AOs are mathematically combined in a kind of averaging process to produce four equivalent **hybridized AOs.** Since they result from combining one s orbital and three p orbitals, the new AOs are said to be sp^3 **hybridized.** The hybrid AOs are ideal for bonding. They are all equivalent, and they have tetrahedral geometry. In addition, each has a large lobe of the orbital pointed in the direction where the other atom of the bond will be and only a small lobe on the other side of the carbon. This directionality of the hybrid AO allows for maximum overlap when it interacts with another AO to form an MO. The hybridization process and the sp^3 hybrid AOs are shown in Figure 3.6.

We are now ready to see how the MOs for methane are formed. The process is just what we have seen before. As shown in Figure 3.7, an sp^3 hybrid AO on carbon overlaps with a $1s$ AO on hydrogen to produce a lower-energy, bonding MO and a higher-energy, antibonding MO. These are both sigma MOs. (An asterisk is usually used to designate the antibonding MO, as in σ^*.) The two electrons of the bond are found in the bonding MO. An identical process occurs for the other three bonds. The four bonding MOs are pictured in Figure 3.7.

Let's consider ethane, C_2H_6, a slightly more complex example, shown in Figure 3.8. The geometry of the molecule tells the hybridization of the atoms involved. In

The $2s$ and one $2p$ AO.   One sp^3 hybrid AO.

Part of the $2s$ and part of the $2p$ orbitals are added. Where both AOs are plus, the resulting orbital has a larger lobe. Where the p orbital is minus and the s orbital is plus, the size of the lobe decreases.

The four sp^3 hybrid AOs point to the corners of a tetrahedron. The small back lobes of the AOs have been omitted for clarity.

Figure 3.6

sp^3 **Hybrid Atomic Orbitals.**

Proton Transfer: A Simple Reaction

4.1 Introduction

This chapter discusses a simple reaction: the acid–base reaction or transfer of a proton. Although acid–base reactions are simple, they are very important in organic chemistry because more complicated reactions often involve one or more proton transfer steps. In addition, much of what we learn about this reaction can be applied to other reactions in later chapters.

Acid–base reactions are equilibria, so the concept of equilibrium will be reviewed. The rates of these reactions will also be discussed. The largest part of the chapter is a discussion of how the structure of the acid or base affects its acid or base strength. We will learn to predict what happens to the strength of an acid when its structure is changed. Not only will this help us to remember how strong an acid (or base) is and how to estimate the strength of an acid or base we have not seen before, but it is of additional importance because exactly the same reasoning will be used in later chapters to predict the effects of structural changes on the rates and equilibria of other reactions.

4.2 Definitions

There are several definitions of acids and bases. According to the **Bronsted-Lowry definition,** an **acid** is a proton donor and a **base** is a proton acceptor. Any compound that has a hydrogen can potentially act as a Bronsted-Lowry acid (though the strength of the acid can vary enormously). Therefore H—A is used as a general representation for an acid. To accept a proton, most bases have an unshared pair of electrons that can be used to form a bond to the proton. Thus B: is used to represent a general base. The general equation for an acid–base reaction is

$$B: + H—A \rightleftharpoons \overset{\oplus}{B}—H + \overset{\ominus}{:A}$$

base   acid   conjugate   conjugate
                                    acid   base

equilibrium
arrows

In this reaction, a proton is transferred from the acid to the base. The unshared pair of electrons on the base is used to form the new bond to the proton while the electrons of the H—A bond remain with A as an unshared pair. Previously, arrows have been used to show electron reorganization in resonance structures. Organic chemists also use these arrows to show electron movement in reactions. The arrows are a kind of bookkeeping device that helps us keep track of electrons as the Lewis structures of the reactants are converted to the Lewis structures of the products.

Problem

Problem 4.1   Indicate whether each of these species can act as an acid, a base, or both:

a) H—N—H
         |
         H

b) H—O—H

c) H—C—H with H above and H below (H—C—H with H top and bottom)

d) :Cl:⁻

e) H—C—O—H with H above and H below

f) H—C—C—C—H with O double bonded to middle C, and H's on terminal carbons

g) H—O—C—O:⁻ with O double bonded to C

Solution

a)   Because NH_3 contains hydrogens, it can act as an acid, and because it has an unshared pair of electrons, it can act as a base. We tend to think of it as a base because that is how it reacts with water. However, in the presence of a strong base, NH_3 can react as an acid. Since many organic reactions involve strongly basic reagents, we need to be aware of the potential of any hydrogen-containing species to donate a proton.

Acid–base reactions are reversible or equilibrium processes. In the reverse reaction, $BH^{\oplus}$ acts as the acid and $A^{\ominus}$ is the base. Therefore $BH^{\oplus}$ is called the **conju-**

gate acid of the base B, and A$^\ominus$ is the **conjugate base** of the acid HA. The charges in a specific acid–base reaction may be different from those shown in the above general equation. The proton is positive, so one unit of positive charge is transferred from the acid to the base in the reaction. The initial charge of HA and B can vary, but B is always one unit more negative than BH and HA is always one unit more positive than A.

Problems

Problem 4.2   Show the conjugate acids of these species:

a) $CH_3-\overset{..}{\underset{..}{O}}-H$ b) $H-\overset{..}{\underset{..}{O}}\colon^{\ominus}$ c) $CH_3-\overset{..}{N}H_2$

Problem 4.3   Show the conjugate bases of these species:

a) $H-\overset{..}{\underset{..}{O}}-H$ b) $H-\overset{..}{\underset{|}{O}}\!\overset{\oplus}{}-H$ c) $H-\overset{..}{\underset{|}{N}}-H$ d) $H-\overset{H}{\underset{H}{C}}-\overset{H}{\underset{H}{C}}-H$

Note that a compound that has both a hydrogen and an unshared pair of electrons can potentially react as either an acid or a base, depending on the reaction conditions. Water, ammonia, and alcohols are examples of compounds that react as acids in the presence of strong bases and as bases in the presence of stronger acids. Some specific examples of acid–base reactions are shown in the following equations. Water is the base in the first equation and the acid in the second equation.

base	acid		conjugate acid		conjugate base

$H_2\overset{..}{O}\colon$ + $H-\overset{..}{\underset{..}{Cl}}\colon$ $\rightleftharpoons$ $H_3\overset{..}{O}\oplus$ + $\colon\!\overset{..}{\underset{..}{Cl}}\colon^{\ominus}$

$CH_3CH_2-\overset{..}{\underset{..}{O}}\colon^{\ominus}$ + $H-\overset{..}{\underset{|}{O}}\colon$ $\rightleftharpoons$ $CH_3CH_2-\overset{..}{\underset{..}{O}}-H$ + $^{\ominus}\colon\!\overset{..}{\underset{..}{O}}-H$

$\overset{..}{N}H_3$ + $H-\overset{\overset{\textstyle\cdot\overset{..}{O}\cdot}{\|}}{\underset{..}{O}}-\overset{}{C}-CH_3$ $\rightleftharpoons$ $\overset{\oplus}{N}H_4$ + $^{\ominus}\colon\!\overset{\overset{\textstyle\cdot\overset{..}{O}\cdot}{\|}}{\underset{..}{O}}-C-CH_3$

Problem

Problem 4.4   Complete these acid–base equations. Use the curved arrow method to show the electron movement in the reactions.

base	acid	conjugate acid	conjugate base

a) $:\overset{\ominus}{\underset{\cdot\cdot}{N}}H_2$ + $H_2\overset{\cdot\cdot}{\underset{\cdot\cdot}{O}}:$ ⇌

b) $CH_3\overset{\cdot\cdot}{\underset{\cdot\cdot}{O}}:{}^{\ominus}$ + $H_3O^{\oplus}$ ⇌

According to the **Lewis definition,** an acid is an electron pair acceptor and a base is an electron pair donor. All Bronsted-Lowry bases are also Lewis bases. However, Lewis acids include many species that are not proton acids; instead of $H^{\oplus}$, they have some other electron-deficient species that acts as the electron pair acceptor. An example of a Lewis acid–base reaction is provided by the following equation. In this reaction the boron of BF_3 is electron-deficient (it has only six electrons in its valence shell). The oxygen of the ether is a Lewis base and uses a pair of electrons to form a bond to the boron, thus completing boron's octet.

Lewis acid   Lewis base

Problem

Problem 4.5   Indicate whether each of these species is a Lewis acid, a Lewis base, or both:

a) $H-\overset{\overset{\displaystyle H}{|}}{\underset{\underset{\displaystyle H}{|}}{C}}{}^{\oplus}$

b) $H-\overset{\cdot\cdot}{\underset{\cdot\cdot}{O}}-H$

c) $H-\overset{\overset{\displaystyle H}{|}}{\underset{\underset{\displaystyle H}{|}}{B}}$

d) $CH_3-\overset{\cdot\cdot}{\underset{\underset{\displaystyle H}{|}}{N}}-H$

e) $:\overset{\cdot\cdot}{\underset{\cdot\cdot}{C}l}-\overset{\overset{\displaystyle :\overset{\cdot\cdot}{\underset{\cdot\cdot}{C}l}:}{|}}{\underset{\underset{\displaystyle :\overset{\cdot\cdot}{\underset{\cdot\cdot}{C}l}:}{|}}{Al}}$

To avoid confusion, when the term *acid* or *base* is used in this text, it refers to a proton acid or base—that is, a Bronsted-Lowry acid or base. The term *Lewis acid* or *Lewis base* will be used when the discussion specifically concerns this type of acid or base.

4.3   The Acid–Base Equilibrium

The reaction of an acid with a base is in equilibrium with the conjugate base and conjugate acid products. The equilibrium constants, termed *acid dissociation* or *acidity constants,* for the reactions of many acids with water as the base (and solvent) have been determined. They can be found in various reference books. Some selected acidity constants are listed in Table 4.2 in Section 4.10. Let's see how acidity constants are defined.

Consider the following acid–base equilibrium, the ionization (dissociation) of acetic acid in water:

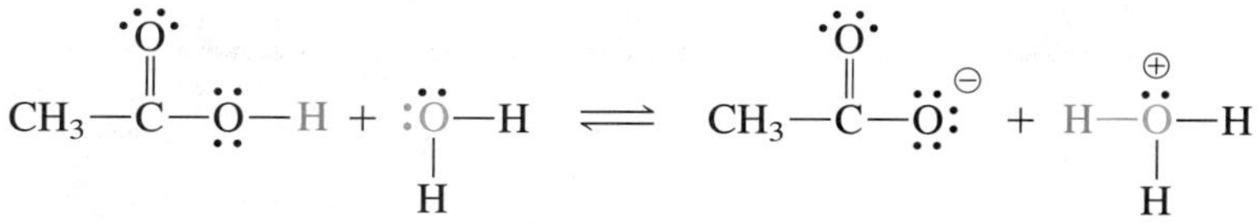

$$CH_3-\overset{\overset{\displaystyle \cdot\cdot\overset{\cdot\cdot}{O}\cdot}{\|}}{C}-\ddot{\underset{\cdot\cdot}{O}}-H \;+\; \underset{\underset{H}{|}}{:\ddot{O}}-H \;\rightleftharpoons\; CH_3-\overset{\overset{\displaystyle \cdot\cdot\overset{\cdot\cdot}{O}\cdot}{\|}}{C}-\overset{\ominus}{\ddot{\underset{\cdot\cdot}{O}}:} \;+\; \underset{\underset{H}{|}}{H-\overset{\oplus}{\ddot{O}}}-H$$

The equilibrium constant for this reaction is

$$K = \frac{[CH_3CO_2^{\ominus}][H_3O^{\oplus}]}{[CH_3CO_2H][H_2O]}$$

Since water is also the solvent, it is present in large excess, and its concentration is approximately constant during the reaction. Therefore a new equilibrium constant, the **acidity constant** (K_a), is used. For the above reaction the equation for K_a is

$$K_a = K[H_2O] = \frac{[CH_3CO_2^{\ominus}][H_3O^{\oplus}]}{[CH_3CO_2H]} = 1.8 \times 10^{-5}$$

The acidity constant is a measure of the strength of an acid. If the acidity constant for a particular acid is near 1, about equal amounts of the acid and its conjugate base are present at equilibrium. A strong acid, which dissociates nearly completely in water, has an acidity constant significantly greater than 1. A weak acid, which is only slightly dissociated in water, has an equilibrium constant significantly less than 1. The acidity constant for acetic acid is 1.8×10^{-5}—only a small amount of acetic acid actually ionizes in water. It is a weak acid.

The acidity constants that are encountered in organic chemistry vary widely, from greater than 10^{10} to less than 10^{-50}. Because of this wide range, it is convenient to use a logarithmic scale to express these values, as is done with pH. Therefore pK_a is defined as

$$pK_a = -\log K_a$$

Problem

Problem 4.6   If the pK_a of a compound is 10, what is its K_a?

Solution

The calculation is easy when pK_a is an integer. If pK_a = integer, then K_a =

$1 \times 10^{-\text{integer}}$. So if $pK_a = 10$, $K_a = 1 \times 10^{-10}$. It is also quite easy to go in the reverse direction when $K_a = 1 \times 10^{x}$ because then $pK_a = -x$.

It is important to be able to quickly recognize the strength of an acid from its K_a or pK_a value. As the strength of the acid increases, the K_a increases and the pK_a decreases (becomes more negative). Base strengths can be determined from the K_a (or pK_a) of the conjugate acid. A strong acid has a weak conjugate base, and a weak acid has a strong conjugate base. These relationships can be summarized as follows:

Strong acid	$K_a > 1$	Negative pK_a	Weak conjugate base
Weak acid	$K_a < 1$	Positive pK_a	Strong conjugate base

Problems

Problem 4.7 Indicate whether these compounds are weaker or stronger acids than water (the K_a for water is 1.8×10^{-16}; the pK_a is 15.74):

a) $HClO_4$ ($K_a = 10^{10}$)

b) $HC{\equiv}CH$ ($pK_a = 25$)

c) $HOC(=O)OH$ ($pK_a = 6.35$)

d) CH_3CH_3 ($K_a = 10^{-50}$)

Problem 4.8 Indicate whether these species are weaker or stronger bases than hydroxide ion. The K_a or pK_a values are for the conjugate acids.

a) $:\ddot{N}H_2^{-}$ ($K_a = 10^{-38}$)

b) $CH_3CH_2\ddot{C}H_2^{-}$ ($pK_a = 50$)

c) $\ddot{N}H_3$ ($pK_a = 9.24$)

d) $:\ddot{Cl}:^{-}$ ($pK_a = -7$)

Consider the general acid–base reaction (charges omitted)

$$H{-}A + B: \rightleftharpoons A: + H{-}B$$

The equilibrium constant (K) for this reaction is

$$K = \frac{K_a \text{ (for HA)}}{K_a \text{ (for HB)}}$$

If HA is a stronger acid than HB, then K is greater than 1 and the right-hand side of the equation is favored at equilibrium; that is, the concentrations of the products (A and HB) are greater than the concentrations of the reactants (HA and B). If HB is a stronger acid than HA, then K is less than 1 and the equilibrium lies to the left. In general, *the equilibrium favors the formation of the weaker acid and the weaker base.*

(Because the stronger acid has the weaker conjugate base, the weaker base is on the same side of the equation as the weaker acid.)

As an example, consider the following reaction:

$$H—Br + H_2O \; \rightleftharpoons \; \overset{\ominus}{Br} + \overset{\oplus}{H_3O}$$

HBr is a stronger acid ($K_a = 10^9$ or $pK_a = -9$) than $H_3O^{\oplus}$ ($K_a = 55$ or $pK_a = -1.74$), so this equilibrium lies to the right. From the K_as the equilibrium constant for the reaction can be calculated to be 1.8×10^7. This equilibrium constant is so large that the amount of HBr remaining cannot be measured by most experimental techniques. For all practical purposes the equilibrium lies completely to the right. Strong acids, such as HBr, are said to be completely dissociated in water. Differences in the strengths of very strong acids cannot be detected in water. Equilibrium constants as large as the one above cannot be measured directly and must be determined by some indirect means. Therefore the pK_a values of the very strong acids (and also the very weak acids) listed in Table 4.2 are only approximate. However, they are usually accurate enough for predictions to be made about the position of the equilibrium in an acid–base reaction.

Base Dissociation Constants

The acidity constant, K_a, which is a measure of acid strength, also provides a measure of the strength of the conjugate base because a strong acid has a weak conjugate base and a weak acid has a strong conjugate base. The strength of a base can also be described by its **base dissociation constant, K_b**. The base dissociation constant is the equilibrium constant for the reaction of the base with water as both the acid and the solvent. It is defined in a manner analogous to K_a. This is best seen with an example.

Consider acetate anion, the conjugate base of acetic acid, reacting as a base in aqueous solution:

$$\underset{\underset{CH_3CO^{\ominus}}{}}{\overset{\overset{O}{\|}}{}} + H_2O \; \rightleftharpoons \; \underset{\underset{CH_3COH}{}}{\overset{\overset{O}{\|}}{}} + {}^{\ominus}OH$$

The equilibrium constant for this reaction is

$$K = \frac{[CH_3CO_2H][^{\ominus}OH]}{[CH_3CO_2{}^{\ominus}][H_2O]}$$

Just as is the case for acidity constants, water is the solvent and is present in large excess, so its concentration does not change significantly during the reaction and can be considered to be constant. The base dissociation constant is defined according to the following equation. The value of K_b for acetate anion is 5.6×10^{-10}.

$$K_b = K[H_2O] = \frac{[CH_3CO_2H][^{\ominus}OH]}{[CH_3CO_2{}^{\ominus}]} = 5.6 \times 10^{-10}$$

continued

ELABORATION

There is a simple relationship between K_b for a base and K_a for its conjugate acid. For acetic acid, the conjugate acid of the base acetate anion, the acidity constant is

$$K_a = \frac{[CH_3CO_2^{\ominus}][H_3O^{\oplus}]}{[CH_3CO_2H]}$$

The product of K_a times K_b is given by the following equation.

$$K_aK_b = \left(\frac{[\cancel{CH_3CO_2^{\ominus}}][H_3O^{\oplus}]}{[\cancel{CH_3CO_2H}]}\right)\left(\frac{[\cancel{CH_3CO_2H}][^{\ominus}OH]}{[\cancel{CH_3CO_2^{\ominus}}]}\right)$$

$$K_aK_b = [H_3O^{\oplus}][^{\ominus}OH] = K_w = 1 \times 10^{-14}$$

Note that the concentrations of acetic acid and of acetate anion cancel out, so only the concentrations of hydronium ion and hydroxide ion remain. In aqueous solution the product of these two concentrations is a constant, $K_w = 1 \times 10^{-14}$. Thus if either K_a or K_b is known, the value for the other can be calculated. The K_a for acetic acid is 1.8×10^{-5}, so K_b is calculated to be 5.6×10^{-10}.

$$K_b = \frac{1 \times 10^{-14}}{K_a} = \frac{1 \times 10^{-14}}{1.8 \times 10^{-5}} = 5.6 \times 10^{-10}$$

As was done with K_a, the base dissociation constant can also be expressed on a logarithmic scale:

$$pK_b = -\log K_b$$

The relationship between the constants is then

$$pK_a + pK_b = pK_w = 14$$

In using base dissociation constants, remember that K_b increases and pK_b decreases (becomes more negative) as the base strength increases.

Strong base	$K_b > 1$	Negative pK_b	Weak conjugate acid
Weak base	$K_b < 1$	Positive pK_b	Strong conjugate acid

Figure 4.1 shows a scale of the strengths of some of the acids that are commonly encountered in organic chemistry. At the top of the scale are *strong acids* such as perchloric ($HClO_4$), hydrobromic (HBr), and sulfuric (H_2SO_4) acids, all of which are significantly stronger than $H_3O^{\oplus}$ and are completely ionized in water. Acids that are stronger than H_2O but weaker than $H_3O^{\oplus}$ are partially dissociated in water and are termed *weak acids* in this text. Included in this group are carboxylic acids and the

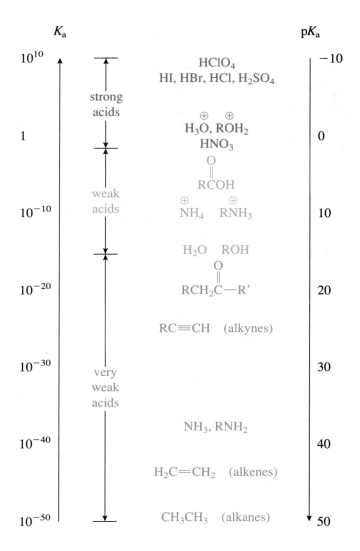

Figure 4.1

Scale of Selected Acid Strengths.

conjugate acids of ammonia and the amines. Acids that are comparable to or weaker than H_2O are essentially undissociated in water and are termed *very weak acids*. However, they can react as acids in the presence of very strong bases in solvents that are less acidic than water. The majority of organic compounds fall into this group, ranging from alcohols (pK_as about 16), which are similar in acid strength to water, to aldehydes and ketones (pK_as about 20), which are a little less acidic than water, to alkanes (pK_as about 50), which are extremely weak acids.

Problems

Problem 4.9 Using the information available in Figure 4.1, predict the position of the equilibrium in these reactions; that is, predict whether there is a higher concentration

of reactants or products present at equilibrium.

a) $CH_3\overset{..}{\overset{\ominus}{C}}H_2$ + H—C≡C—H ⇌ CH_3CH_3 + H—C≡C:$^{\ominus}$

b) $CH_3\overset{..}{N}H_2$ + $CH_3\overset{O}{\overset{\|}{C}}\overset{..}{\overset{\ominus}{O}}$: ⇌ $CH_3\overset{..}{N}H$ + $CH_3\overset{O}{\overset{\|}{C}}\overset{..}{O}H$

c) $\overset{\ominus}{\overset{..}{C}}H_2\overset{O}{\overset{\|}{C}}CH_3$ + H—$\overset{..}{C}l$: ⇌ $CH_3\overset{O}{\overset{\|}{C}}CH_3$ + :$\overset{..}{\overset{..}{C}l}$:$^{\ominus}$

Solution

a)   From Figure 4.1, HC≡CH is a stronger acid than CH_3CH_3. Since the equilibrium favors the weaker acid, more of the products are present when this reaction reaches equilibrium.

Problem 4.10   Use the information in Figure 4.1 to predict the positions of the equilibria in the reactions in problem 4.4.

_____

An equilibrium constant for a reaction is related to the free energy change in that reaction by the equation

$$\Delta G° = -RT \ln K$$

standard free ↗   gas ↗   ↖ absolute
energy change   constant   temperature

For a reaction to be spontaneous, K must be greater than 1—that is, $\Delta G°$ must be negative. Organic chemists find diagrams that show the free energies of the reactants and products on the same scale to be very useful. Figure 4.2 shows such diagrams for two general reactions. In the diagram on the left-hand side of this figure, the reac-

Figure 4.2

Free Energies of Reactants and Products.

The left-hand side illustrates the case in which the products are more stable than the reactants and are favored at equilibrium. The right-hand side illustrates the opposite case, in which the reactants are more stable and are favored.

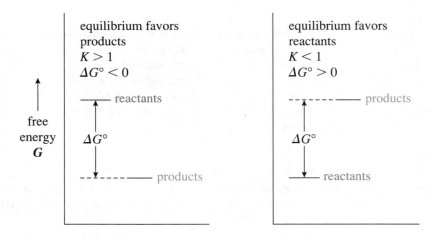

reactants ⇌ products

equilibrium favors products
$K > 1$
$\Delta G° < 0$

equilibrium favors reactants
$K < 1$
$\Delta G° > 0$

tants are higher in energy than the products. Therefore $\Delta G°$ ($G_{products} - G_{reactants}$) is negative and K is greater than 1. The reaction proceeds spontaneously to the products—that is, it proceeds spontaneously to the lower energy or more stable compound(s). The situation is reversed in the diagram on the right-hand side of the figure. The reactants are more stable, $\Delta G°$ is positive, K is less than 1, and the reactants are favored at equilibrium.

Problem

Problem 4.11   Draw diagrams like that in Figure 4.2 for the reactions in problem 4.9.

4.4 Rate of the Acid–Base Reaction

As you should recall from general chemistry, a favorable equilibrium constant is not sufficient to ensure that a reaction will occur. In addition, the rate of the reaction must be fast enough that the reaction occurs in a reasonable period of time. The reaction rate depends on a number of factors. First, the reactants, in this case the acid and the base, must collide. In this collision the molecules must be oriented properly so that the orbitals that will form the new bond can begin to overlap. The orientation required for the orbitals of the reactants is called the **stereoelectronic requirement** of the reaction. ("Stereo" means dealing with the three dimensions of space.) In the acid–base reaction the collision must occur so that the atomic orbital of the base that is occupied by the unshared pair electrons can begin to overlap with the $1s$ orbital of the acidic hydrogen. In the case of the reaction of ammonia with acetic acid the stereoelectronic requirement can be pictured as follows:

If, instead, the collision were to occur so that the orbital on the nitrogen bumped into some other part of the acetic acid molecule, that collision would not lead to an acid–base reaction.

You might also recall that there is an energy barrier that separates the reactants and the products in most reactions. Therefore a final requirement for a collision to lead to reaction is that the collision provides enough energy to surmount this energy barrier. This extra energy is called the **activation energy** for the reaction. Activation energy will be considered in more detail when more complicated reactions are discussed. For now, it is enough to note that the activation energy for acid–base reactions is usually small.

A diagram that shows how the free energy of the system changes as a reaction proceeds is often very informative. Such a diagram has the free energy, G, on the y-axis, just as in Figure 4.2, but has a measure of the progress of the reaction along the x-axis, so the reactants are shown on the left-hand side of the diagram and the products on the right-hand side. Figure 4.3 shows such a diagram for the ionization of HBr in water. Because HBr is a strong acid, $\Delta G°$ for this reaction is negative and $K_a > 1$. The products are favored at equilibrium because they are more stable than the reactants. The line connecting the energy of the reactants to that of the products shows how the energy changes during the reaction. As the reaction proceeds, there is initially a slight increase in energy, followed by a decrease as the reaction proceeds on toward the products. The size of the slight energy barrier that the reaction must pass over is called the activation energy, $\Delta G^{\ddagger}$. In this case, $\Delta G^{\ddagger}$ is very small, so the energy barrier is easily overcome—the reaction is very fast. These energy versus reaction progress diagrams are very useful and will be encountered often in subsequent chapters.

Because the activation energies are small and the stereoelectronic requirements are not difficult to meet, most acid–base reactions are very fast in comparison to other types of organic reactions. Therefore it is usually not necessary to be concerned with the rates of acid–base reactions. In organic reactions that involve several steps, including an acid–base step, one of the other steps in the reaction usually controls the rate.

Problem

Problem 4.12   Show a free energy versus reaction progress diagram for the following reactions.

a) CH$_3$$\overset{\displaystyle O}{\overset{\|}{C}}$OH + H$_2$O $\rightleftharpoons$ CH$_3$$\overset{\displaystyle O}{\overset{\|}{C}}O^{\ominus}$ + H$_3$O$^{\oplus}$

b) HCl + NH$_3$ $\rightleftharpoons$ Cl$^{\ominus}$ + $^{\oplus}$NH$_4$

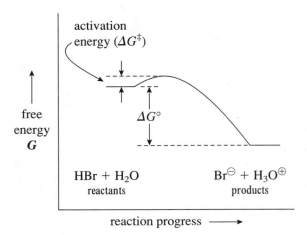

Figure 4.3

Free Energy versus Reaction Progress Diagram for the Reaction of HBr with H$_2$O.

Solution

a) First we must determine whether the reactants or the products are more stable. Acetic acid is a weak acid (see Figure 4.1), so the equilibrium favors the reactants. In other words, $K_a < 1$ and $\Delta G° > 0$. The diagram is just the reverse of that shown in Figure 4.3.

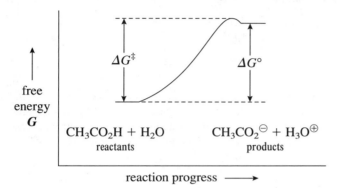

Note that the magnitude of the activation energy, $\Delta G^{\ddagger}$, is slightly larger than that of $\Delta G°$ in this case. This means that there is only a small activation barrier for the reverse reaction.

4.5   Effect of the Atom Bonded to the Hydrogen on Acidity

It is important to understand the various factors that determine the strength of an acid—that is, why one compound is a stronger acid than another. Such an understanding makes it much easier to remember whether a particular compound is a strong acid or a weak acid. In addition, it then becomes possible to make qualitative predictions about what a change in the structure of a compound will do to its K_a value. Therefore most of the remainder of this chapter is a discussion of how the structure of the acid affects its strength. The effects to be discussed are the atom to which the hydrogen is bonded, nearby charges or polar bonds, hydrogen bonding, the hybridization of the atom to which the hydrogen is bonded, and resonance. First, let's consider the effect of the atom bonded to the acidic hydrogen.

Consider the following acid–base reactions:

$$NH_3 + H_2O \;\rightleftharpoons\; NH_2^{\ominus} + H_3O^{\oplus} \qquad pK_a = 38$$

$$CH_4 + H_2O \;\rightleftharpoons\; CH_3^{\ominus} + H_3O^{\oplus} \qquad pK_a = 50$$

The hydrogen that is bonded to carbon is 10^{12} times less acidic than the hydrogen that is bonded to nitrogen. This can be explained by considering the electronegativities of the carbon and the nitrogen. Nitrogen is more electronegative than carbon. The unshared pair of electrons on the nitrogen of $NH_2^{\ominus}$ is in a lower-energy orbital, and are more stable, than the unshared pair of electrons on the carbon of $CH_3^{\ominus}$. Therefore the $NH_2^{\ominus}$ has less inclination to use its more stable pair of electrons to form a bond to an $H^{\oplus}$, so it is a weaker base than $CH_3^{\ominus}$, and NH_3 is a stronger acid than the related carbon compounds. Because of this increasing stabilization of the conjugate base as the electronegativity of the atom increases, acidity increases from left to right in a row of the periodic table.

$$CH_4 \quad < \quad NH_3 \quad < \quad H_2O \quad < \quad HF$$
$$pK_a \quad\ 50 \qquad\quad 38 \qquad\quad 15.7 \qquad\quad 3$$

$\longrightarrow$
increasing acidity

Many useful predictions can be made on the basis of this order. For example, suppose methanol, CH_3OH, reacts with a strong base. Which hydrogen is removed? Which hydrogen is more acidic? Since oxygen is more electronegative than carbon, the hydrogen on the oxygen is considerably more acidic than the hydrogens on the carbon. The reaction that occurs is

$$CH_3 - \overset{..}{\underset{..}{O}} - H + :B^{\ominus} \rightleftharpoons CH_3 - \overset{\ominus}{\underset{..}{\overset{..}{O}}} : + H - B$$

In general, oxygen acids (H bonded to O) will be stronger than carbon acids (H bonded to C), other things being equal.

When comparisons are made within a column of the periodic table, it is found that electronegativity is no longer the controlling factor. Acidity increases from top to bottom in a column of the periodic table, while electronegativity decreases.

$$HF \quad < \quad HCl \quad < \quad HBr \quad < \quad HI$$
$$pK_a \quad\ 3 \qquad\quad -7 \qquad\quad -9 \qquad\quad -10$$

$$H_2O \quad < \quad H_2S$$
$$pK_a \quad\ 15.7 \qquad\quad 7$$

$\longrightarrow$
increasing acidity

Part of the explanation for this trend is that atoms that are lower in a column form weaker bonds to hydrogen because their larger atomic orbitals do not overlap as well with the small hydrogen $1s$ orbital. These weaker bonds make removing the proton easier.

Problems

Problem 4.13   Which species is a stronger acid?

   a) H_2S or HCl **b)** $\overset{\oplus}{PH_4}$ or $\overset{\oplus}{NH_4}$ **c)** CH_3CH_3 or H_2S

Problem 4.14   Which anion is the stronger base?

   a) $HO^{\ominus}$ or $HS^{\ominus}$ **b)** $CH_3\overset{\ominus}{NH}$ or $CH_3\overset{\ominus}{O}$

Problem 4.15 Which is the most acidic hydrogen in each of these compounds?

a) $CH_3-\overset{\overset{\displaystyle H}{|}}{N}-CH_2CH_3$ b) CH_3CH_2OH c) CH_3SH

Solution

a) If none of the other effects described in subsequent sections are operating, a hydrogen on a nitrogen is more acidic than a hydrogen on a carbon because the nitrogen is more electronegative than the carbon.

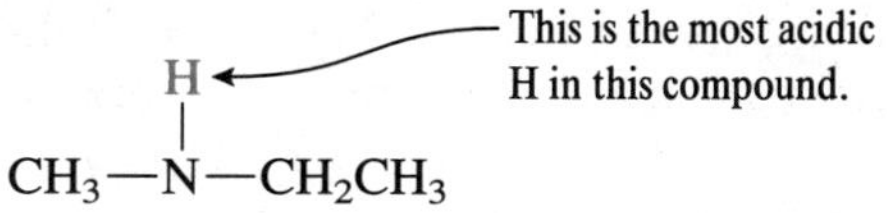

This is the most acidic
H in this compound.

$CH_3-\overset{\overset{\displaystyle H}{|}}{N}-CH_2CH_3$

4.6   Inductive Effects

The effect of a nearby dipole in a molecule on a reaction elsewhere in that molecule is termed an **inductive effect.** Consider two carboxylic acids, acetic acid and chloroacetic acid:

CH_3COOH acetic acid $pK_a = 4.76$

$ClCH_2COOH$ chloroacetic acid $pK_a = 2.86$

Replacing one of the hydrogens of acetic acid with a chlorine results in an increase in acid strength by almost a factor of 100. Remember that the acidic hydrogen in these molecules is the one bonded to the oxygen, so the chlorine is actually exerting its effect from several bonds away from the reacting bond. Figure 4.4 shows how the presence of the dipole of the C—Cl bond increases the energy of the acid while it

In chloroacetic acid the interaction of the dipole of the Cl—C bond with the dipole of the carboxylic acid group is destabilizing because the positive end of one dipole is closer to the positive end of the other. The interaction of these like charges is repulsive and increases the energy of the molecule.

In the conjugate base of chloroacetic acid the interaction of the positive end of the Cl—C dipole with the negative charge of the ionized carboxylate group is a stabilizing interaction. The energy of the anion is lowered by this effect.

Figure 4.4

Charge Interactions in Chloroacetic Acid and Its Conjugate Base.

lowers the energy of the conjugate base. The inductive effect of the chlorine destabilizes the acid and stabilizes the conjugate base. Figure 4.5 shows how this inductive effect changes $\Delta G°$ for the acid–base reaction of chloroacetic acid as compared to acetic acid. Chloroacetic acid is destabilized relative to acetic acid by the inductive effect of the chlorine. In contrast, the conjugate base of chloroacetic acid is stabilized relative to the conjugate base of acetic acid. Therefore $\Delta G°_2$ for the acid–base reaction of chloroacetic acid is less than $\Delta G°_1$ for that of acetic acid. K_a is larger for chloroacetic acid—it is a stronger acid than acetic acid.

Organic chemists often talk about the inductive effect of a group. Chlorine is an electron-withdrawing group relative to hydrogen because of its inductive effect, that is, it pulls more electron density away from its bond partner than does hydrogen. Because most functional groups contain atoms that are more electronegative than hydrogen, they are also inductive electron-withdrawing groups. The following is a partial list of inductive electron-withdrawing groups:

$$-\overset{\oplus}{N}R_3 \qquad -\overset{\oplus}{N}\underset{O_{\ominus}}{\overset{O}{\diagup}} \qquad -\overset{O}{\underset{O}{\overset{\|}{S}}}-R \qquad -C\equiv N$$

$$-\overset{O}{\overset{\|}{C}}-OH \qquad -\overset{O}{\overset{\|}{C}}-R \qquad -OH \qquad -OR$$

$$-F \qquad -Cl \qquad -Br \qquad -I$$

inductive electron-withdrawing groups

In contrast, only a few groups are electron-donating relative to hydrogen because of their inductive effects. Two of these are electron-rich because of their negatively charged oxygen atoms. In addition, alkyl groups, such as CH_3 and CH_2CH_3, behave

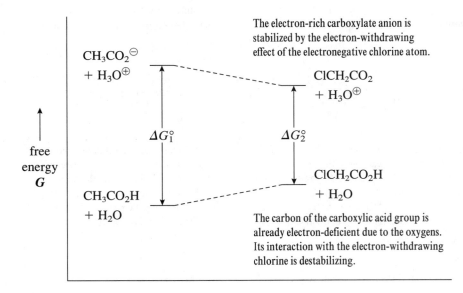

Figure 4.5

Inductive Effect on the Free-Energy Changes in an Acid–Base Reaction.

as weak electron-donating groups in many situations:

inductive electron-donating groups

Replacing a hydrogen with an electron-withdrawing group will destabilize an electron-poor site or stabilize an electron-rich site. Since the conjugate base is always more electron-rich than the acid, replacing a hydrogen with an electron-withdrawing group results in a stronger acid. Replacing a hydrogen with an electron-donating group has the opposite effect. Table 4.1 provides some examples of inductive effects on the acidities of carboxylic acids.

Table 4.1   Inductive Effects on the Acid Strength of Some Carboxylic Acids.

Compound	pK_a	Comments
CH_3COH (O)	4.76	
CH_3CH_2COH (O)	4.88	CH_3 is a weak electron-donating group; acid strength decreases slightly.
$ClCH_2COH$ (O)	2.86	Cl is an electron-withdrawing group; acid strength increases.
FCH_2COH (O)	2.66	F is a stronger electron-withdrawing group than Cl because it is more electronegative.
ICH_2COH (O)	3.12	I is less electronegative and a weaker electron-withdrawing group than Cl.
$ClCH_2CH_2COH$ (O)	3.98	The Cl is farther from the reaction site; the inductive effect decreases rapidly with increasing distance.
Cl_3CCOH (O)	0.65	Three Cl's have a stronger inductive effect than one.
$HOC—COH$ (O O)	2.83	The CO_2H group is electron-withdrawing.
$^{\ominus}OC—COH$ (O O)	5.69	The $CO_2^{\ominus}$ group is electron-donating; this is a weaker acid than acetic acid.

Problem

Problem 4.16   Explain which compound is the stronger acid.

a) 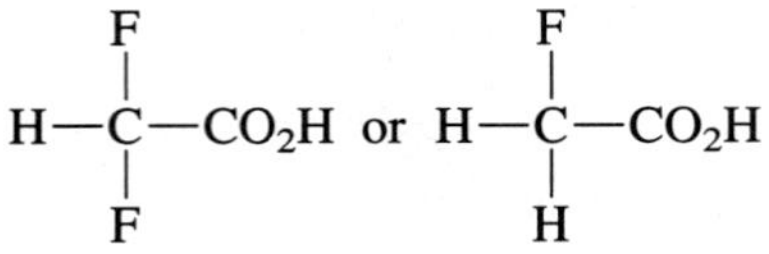
$$
\begin{array}{c}
\text{F} \\
| \\
\text{H}-\text{C}-\text{CO}_2\text{H} \\
| \\
\text{F}
\end{array}
\quad \text{or} \quad
\begin{array}{c}
\text{F} \\
| \\
\text{H}-\text{C}-\text{CO}_2\text{H} \\
| \\
\text{H}
\end{array}
$$

b) 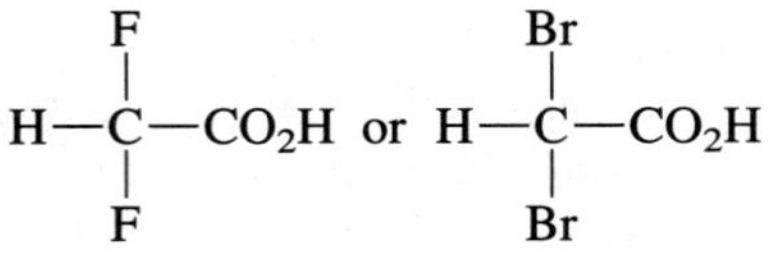
$$
\begin{array}{c}
\text{F} \\
| \\
\text{H}-\text{C}-\text{CO}_2\text{H} \\
| \\
\text{F}
\end{array}
\quad \text{or} \quad
\begin{array}{c}
\text{Br} \\
| \\
\text{H}-\text{C}-\text{CO}_2\text{H} \\
| \\
\text{Br}
\end{array}
$$

c) $CH_3OCH_2CO_2H$ or CH_3CO_2H

4.7   Hydrogen Bonding

If the acidic hydrogen is hydrogen bonded to another atom in the same molecule, the strength of the acid is decreased. It is more difficult for a base to remove the proton because the hydrogen bond must be broken in addition to the regular sigma bond to the hydrogen. However, this effect is complicated by the inductive effect of the group involved in the hydrogen bond. Consider the following examples.

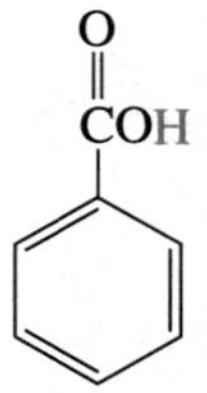

This compound is benzoic acid.

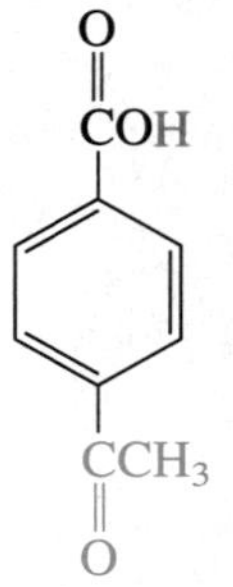

$pK_a = 4.19$

The $CH_3C{=}O$ group (the acetyl group) is electron-withdrawing. When it is substituted on the ring position opposite the carboxylic acid group, its inductive effect increases the strength of this acid as compared to benzoic acid.

$pK_a = 3.70$

intramolecular
hydrogen bond

$pK_a = 4.13$

If the acetyl group is substituted on the position adjacent to the carboxylic acid group, an even stronger acid should result because the inductive effect increases as the groups are brought closer together. However, this acid is weaker than the previous example; its pK_a is similar to that of benzoic acid. The acid-strengthening inductive effect is canceled by the acid-weakening effect of the hydrogen bond formed between the hydrogen of the carboxylic acid group and the oxygen of the acetyl group.

The inductive effect and the hydrogen-bonding effect in the last example are operating in opposite directions. The inductive effect increases the acid strength while the hydrogen bond decreases it. Although the two effects nearly cancel out in this particular case, this cannot usually be predicted in advance. In general, the direction of an effect—that is, whether it is acid-strengthening or acid-weakening—can readily be determined. However, it is much more difficult to estimate the magnitude of the effect. In a case such as the one above, in which the two effects are opposed, it is difficult to predict which one is larger, so it is not possible to predict whether the acid is stronger or weaker than the model compound, in which neither effect is present. Therefore most of our predictions will be qualitative rather than quantitative—we will be able to determine that one compound is a stronger acid (or reacts faster, etc.) than another, but we will not be able to predict exactly how much stronger (or faster).

4.8   Hybridization

As the following examples show, the hybridization of the atom bonded to the hydrogen has a large effect on the acidity of that hydrogen.

ethane
$pK_a = 50$

ethene
$pK_a = 44$

ethyne
$pK_a = 25$

In this series, as the hybridization changes from sp^3 in ethane to sp^2 in ethene and to sp in ethyne, the acidity increases and the pK_a decreases. This is because of the relative stability of the unshared electrons in the conjugate bases of each of these compounds.

Figure 4.6 shows the energies of the hybrid orbitals relative to the s and p orbitals from which they are formed. An sp hybrid orbital is composed of 50% p orbital and 50% s orbital. Therefore its energy is halfway between the energies of the s orbital and the p orbital. Similarly, the energy of an sp^2 orbital is higher than that of the s orbital by 67% of the difference between the energies of the p orbital and the s orbital,

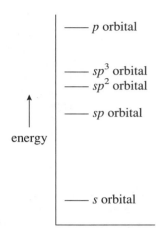

Figure 4.6

Energies of Hybrid Atomic Orbitals Relative to *s* and *p* Atomic Orbitals.

and the energy of an sp^3 orbital is higher than the energy of the *s* orbital by 75% of this difference.

The unshared pair of electrons of the conjugate base of ethane occupies an sp^3 atomic orbital on the carbon. The unshared pair of electrons of the conjugate base of ethene occupies an sp^2 atomic orbital. Because the sp^2 orbital is lower in energy, the unshared electrons in this orbital in the conjugate base of ethene are more stable (and less basic) than the electrons in the sp^3 orbital of the conjugate base of ethane. Thus ethene is a stronger acid than ethane. Since the electrons of the conjugate base of ethyne are even lower in energy in an sp orbital, ethyne is an even stronger acid.

This explanation using orbital energies can also be couched in terms of electronegativities. Electronegativity and orbital energy are directly related. For example, the reason why fluorine is more electronegative than oxygen is that the $2p$ orbital of fluorine is lower in energy than the $2p$ orbital of oxygen. Since an added electron would prefer to occupy the lower-energy orbital, fluorine is more electronegative. An

Calcium Carbide

Because of its sp hybridization, ethyne is a strong enough acid that its conjugate base can be readily generated by using one of the strong bases that are available in the laboratory. Calcium carbide, CaC_2, a relatively stable solid, can be viewed as a dianion of ethyne:

$$Ca^{2\oplus} \ {}^{\ominus}{:}C{\equiv}C{:}^{\ominus}$$

calcium carbide

Calcium carbide is prepared by the reaction of calcium oxide with carbon, in the form of coke, at very high temperatures:

$$CaO + 3\,C \ \underset{}{\overset{2000-3000°C}{\rightleftharpoons}} \ CaC_2 + CO$$

As expected, the carbide dianion is a very strong base and readily removes protons

sp orbital is lower in energy than an sp^2 orbital, so it might be said that an sp hybridized carbon is more electronegative than an sp^2 hybridized carbon. Therefore, using the same reasoning as in Section 4.5, a hydrogen that is bonded to an sp hybridized carbon should be more acidic than a hydrogen that is bonded to an sp^2 hybridized carbon.

Problem

Problem 4.17 Which is the most acidic hydrogen in $CH_3CH_2C{\equiv}CH$?

4.9   Resonance

Stabilization by resonance is used to explain many observations in organic chemistry. Resonance stabilization of a product can shift an equilibrium dramatically to the right, that is, to the product side of the reaction. Resonance can also lower the activation energy for a reaction, resulting in a considerable increase in reaction rate. What we learn here about how resonance affects the acid–base equilibrium will be very useful in discussions of both equilibria and rates of other reactions in subsequent chapters.

Consider the acid–base reactions of ethanol and acetic acid in water:

$$CH_3CH_2OH + H_2O \;\rightleftharpoons\; CH_3CH_2O^{\ominus} + H_3O^{\oplus} \qquad pK_a = 16$$

ethoxide ion

$$CH_3\overset{\overset{\textstyle O}{\|}}{C}OH \;+\; H_2O \;\rightleftharpoons\; CH_3\overset{\overset{\textstyle O}{\|}}{C}O^{\ominus} \;+\; H_3O^{\oplus} \qquad pK_a = 4.76$$

Acetic acid is a stronger acid than ethanol by a factor of about 10^{11}. In both compounds the acidic hydrogen is bonded to an oxygen. Replacing the CH_2 of ethanol

from water to produce ethyne (acetylene) gas:

$$Ca^{2\oplus} \;{}^{\ominus}{:}C{\equiv}C{:}^{\ominus} + 2\,H{-}OH \longrightarrow H{-}C{\equiv}C{-}H + Ca(OH)_2$$

Calcium carbide provides a fairly safe and easy-to-handle source of ethyne. Before the advent of battery-operated lights, portable lamps such as those used on bicycles, carriages, and miner's helmets were fueled by this material. Water was slowly dropped onto solid calcium carbide, and the ethyne that was generated was burned. This reaction is still used as a source of acetylene for welding torches.

Ethyne can be used as a starting material for the preparation of many industrially important organic compounds. Currently, it is more cost-effective to prepare these compounds from petroleum. However, as petroleum supplies dwindle in the future, ethyne prepared from coal via calcium carbide will become more economically attractive as a source of these compounds.

with a C=O results in an enormous increase in acidity. Part of this increase is due to the inductive effect of the oxygen of the carbonyl group, but the effect is much too large to be due only to this.

The other factor that is contributing to the dramatic increase in the acidity of acetic acid is resonance stabilization. Neither ethanol nor its conjugate base, which is called ethoxide ion, is stabilized by resonance. The following resonance structures can be written for acetic acid and its conjugate base, acetate anion:

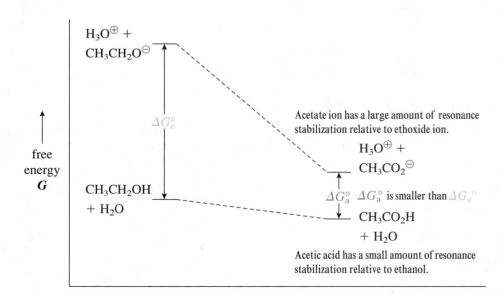

acetic acid   acetate anion

As was noted in Figure 3.16, acetic acid has only a small amount of resonance stabilization because the lower structure is only a minor contributor to the resonance hybrid. Acetate ion has a large amount of resonance stabilization because it has two equivalent contributors to the hybrid.

Figure 4.7 diagrams these energy changes. Because acetic acid has only a small amount of resonance stabilization, its energy is lowered only a little as compared to

Figure 4.7

Energy Diagram for the Reaction of Ethanol and Acetic Acid as Acids in Water.

refined by distillation. The various distillation cuts, based on boiling point, are shown in Table 5.2.

The most important use of alkanes is as fuels for combustion processes to produce heat and power:

$$C_nH_{2n+2} + (3n + 1)/2 \ O_2 \longrightarrow n \ CO_2 + (n + 1) \ H_2O + heat$$

Enormous amounts of natural gas and petroleum products are burned daily. Of course, in the combustion process, it is the heat produced that is more important, not the chemical products of the reaction, carbon dioxide and water. However, some scientists think that the production in this reaction of enormous quantities of carbon dioxide, a greenhouse gas, is initiating a global warming trend.

In the chemical laboratory the major use of alkanes is as solvents. Their property of being chemically unreactive makes them attractive solvents for many reactions because they will not interfere with the desired chemistry. However, their usefulness is limited because they will not dissolve highly polar or ionic compounds because of their completely nonpolar nature.

Table 5.1 Physical Properties of Some Alkanes.

Compound	Name	Melting Point (°C)	Boiling Point (°C)
CH_4	methane	−183	−162
CH_3CH_3	ethane	−183	−89
$CH_3CH_2CH_3$	propane	−187	−42
$CH_3CH_2CH_2CH_3$	butane	−138	−0.5
$CH_3CH_2CH_2CH_2CH_3$	pentane	−130	36
$CH_3CH_2CH_2CH_2CH_2CH_3$	hexane	−94	69
⬡	cyclohexane	6.5	80
$CH_3(CH_2)_5CH_3$	heptane	−91	98
$CH_3(CH_2)_6CH_3$	octane	−57	126
$CH_3(CH_2)_7CH_3$	nonane	−51	151
$CH_3(CH_2)_8CH_3$	decane	−30	174
$CH_3(CH_2)_{18}CH_3$	eicosane	36	345

Note the shorthand used to represent the longer chain alkanes: $CH_3 (CH_2)_5 CH_3$ means that there are five CH_2s between the CH_3s; the carbon chain has a total of seven carbons.

Table 5.2 Boiling Points for the Distillation Cuts of Petroleum.

Distillation Cut	Boiling Point	Comment
Natural gas	below 20°C	C_1-C_4 alkanes; used primarily in industry and as a fuel for heating homes
Light petroleum and ligroin	20−120°C	C_5-C_7 compounds; large amounts are heated to crack them to small alkenes, a major feedstock for the chemical industry
Gasoline	100−200°C	C_7-C_{10} compounds; contain a large fraction of straight-chain alkanes, which tend to detonate or "knock" when burned; heating over a catalyst breaks bonds, which reform to produce branched alkanes, which burn more smoothly
Kerosene	200–300°C	$C_{12}-C_{18}$ compounds; used as jet fuel and diesel fuel
Gas oil and lubricating oil	above 300°C	larger alkanes; used as heating and lubricating oils; large amounts are used to produce gasoline by heating over a catalyst in a process called cracking
Residue	nonvolatile	asphalt and bitumen

5.3   Common Nomenclature of Alkanes

The number of organic compounds is virtually limitless. Each needs a name that can be used in discussing the compound or writing about it. Furthermore, it is often necessary to look up the properties of a compound. (You should find out the physical properties of all the compounds that you use in the laboratory.) Very large tables of compounds and indexes use alphabetical listings of compound names.

In the early days of organic chemistry a newly discovered compound was often named according to the source from which it was isolated. The four-carbon carboxylic acid that was isolated from rancid butter was called butyric acid, after *butyrum,* the Latin word for butter. Even the early organic chemists realized the need to be systematic in naming compounds, so the four-carbon alkane that can be prepared from butyric acid was called butane. When an isomeric four-carbon alkane was discovered, it was called isobutane. Butane is a **straight-chain** or **unbranched** alkane. Such alkanes are termed **normal** and are sometimes written with a prefix *n-*. Another name, then, that you may encounter for butane is *n*-butane. (The *n-* is redundant because the absence of any prefix implies an unbranched alkane, but it is still encountered occasionally.) Isobutane is a **branched** alkane. There are three isomeric

The amount of heat produced by burning (heat of combustion) varies considerably with the type of fuel that is burned. Values for some fuels of interest, in kcal (or kJ) per gram, are as follows:

Energy Content of Fuels

Compound	Heat Evolved kcal/g (kJ/g)	Compound	Heat Evolved kcal/g (kJ/g)
H_2	34.2 (143)	coal	7.6 (32)
CH_4 methane	13.4 (56)	CH_3CH_2OH ethanol	7.2 (30)
$CH_3CH_2CH_2CH_3$ butane	12.0 (50)	CH_3OH methanol	5.5 (23)
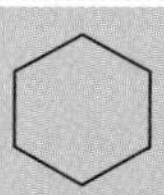 cyclohexane	11.2 (47)	wood	4.5 (19)
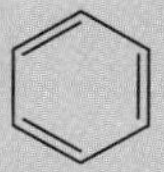 benzene	10.0 (42)	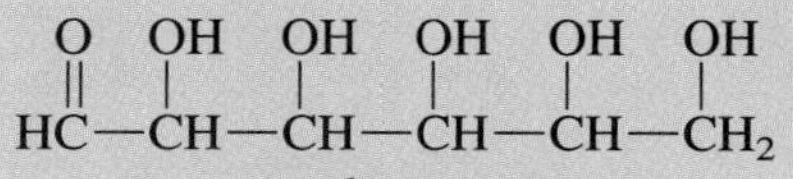 glucose	3.8 (16)

Some interesting observations can be made from these data. Hydrogen contains considerably more energy per gram than the other fuels. This is because the mass of a hydrogen atom is considerably less than that of a carbon or an oxygen atom, so 1 gram of hydrogen contains many more atoms than 1 gram of the other fuels. The energy contents of methane, butane, and cyclohexane, typical alkanes, are all similar. (There is a slight decrease in energy content as the percentage of hydrogen decreases from methane to butane to cyclohexane.) Part of the reason that benzene produces less energy than cyclohexane is that benzene has a large resonance stabilization energy. Since benzene is more stable, it produces less heat when burned (more on this in Chapter 17).

One definition of **oxidation** often used in organic chemistry is an increase in the oxygen content of a compound. Combustion is, then, an oxidation process. As can be seen from the heats of combustion of the alkanes as compared to those of ethanol, methanol, and glucose, increasing the initial oxygen content of a compound—that is, increasing its oxidation state—results in a lower energy content. Methanol, for example, can be viewed as resulting from partial oxidation of methane. In methanol there are only three carbon–hydrogen bonds to "burn," compared to four in methane, so the energy content per mole is less for methanol. (The energy content is even less on a per gram basis.)

Gasoline is composed primarily of alkanes, so its energy content is in the 11–12 kcal/g (46–50 kJ/g) region. This relatively high energy content makes gasoline an attractive fuel for automobiles. Of course, hydrogen is even better in terms of energy content and burns with less pollution. However, it is much more difficult to handle. Methanol has been proposed as an alternative to gasoline. One drawback is that its energy density is much less than gasoline, which means that the miles per gallon would drop substantially if methanol were used. The use of oxygenated gasoline, usually containing 10% or more of ethanol or methyl *tert*-butyl ether (MTBE), is mandated in many cities during the winter months because it purportedly reduces pollution due to carbon monoxide. If you live in such an area, you may notice a small decrease in your gas mileage during the winter.

Glucose is a common sugar. Its structure is similar to that of other sugars. Carbohydrates, such as starch, are composed of a large number of glucose units. Therefore the energy content of glucose is representative of the energy content of a major part of our food. The energy content of glucose is low because of its highly oxygenated structure. It supplies 3.8 kcal/g (16 kJ/g) of energy. Note that the "Calorie" (written with an uppercase C) that is used in nutrition is actually a kilocalorie, so sugars and carbohydrates have about 4 Calories per gram. You may have heard that alcoholic beverages are quite fattening. Ethanol, the alcohol contained in beverages, has nearly twice the energy density of sugar because it contains significantly less oxygen. It supplies 7 Calories per gram.

alkanes with five carbons. The unbranched one is called pentane. The one with a single branch is called isopentane. The remaining isomer requires a different prefix. It is called neopentane.

$$
\underset{\text{butyric acid}}{CH_3CH_2CH_2\overset{\overset{\displaystyle O}{\|}}{C}OH}
\qquad
\underset{\text{butane}}{CH_3CH_2CH_2CH_3}
\qquad
\underset{\text{isobutane}}{CH_3\overset{\overset{\displaystyle CH_3}{|}}{C}HCH_3}
$$

$$
\underset{\text{pentane}}{CH_3CH_2CH_2CH_2CH_3}
\qquad
\underset{\text{isopentane}}{CH_3\overset{\overset{\displaystyle CH_3}{|}}{C}HCH_2CH_3}
\qquad
\underset{\text{neopentane}}{\overset{\overset{\displaystyle CH_3}{|}}{CH_3\overset{}{C}CH_3}\atop CH_3}
$$

These names are called **common** or **trivial** names. As the size of the alkane increases, the number of isomers increases dramatically. It would be very cumbersome to continue this process of providing different prefixes for each isomer of a larger alkane. Furthermore, the task of learning all these prefixes would be daunting indeed. Decane has 75 isomers! Obviously, a systematic nomenclature is needed.

5.4   Systematic Nomenclature of Alkanes

A systematic method for naming alkanes (and other organic compounds) that is simple to use and minimizes memorization was developed by the International Union of Pure and Applied Chemistry and is called the IUPAC nomenclature. To make it easier for the chemists of that time to learn, it incorporated common nomenclature wherever possible.

Before we look at the steps for naming alkanes, let's see what an IUPAC name looks like. Basically, each alkane is considered a straight-chain carbon backbone on which various branches or groups are attached. The IUPAC name for isopentane is 2-methylbutane. This name consists of a root that designates the number of carbons in the backbone, a number and a prefix that designate the position of the branch on the backbone and the number of carbons in the branching group, and a suffix that designates the functional group.

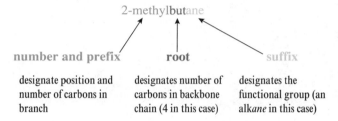

Now let's discuss how to name an alkane by the IUPAC method.

STEP 1

Find the longest continuous carbon chain in the compound. The number of carbons in this backbone determines the root (see Table 5.3). If there are two (or more) chains of equal length, choose the one with the greater number of branches. (This results in simpler branching groups that are easier to name.)

It is important to realize that the name of a compound must not depend on how that compound is drawn. When a drawing for a compound is encountered, the longest continuous chain will not necessarily be shown in a straight, horizontal line. Each structure must be carefully examined to identify the longest chain, no matter how it bends and curls around. Some examples, with the longest chain in red, are given below.

CH_3CH_2
|
$CH_3CHCH_2CH_3$

The longest chain has five carbons.
The root is pent-.

CH_2CH_3
|
$CH_3CH_2CHCHCH_3$
|
$CH_3CH_2CH_2$

The longest chain has seven carbons.
The root is hept-.

Table 5.3 Names for the Roots That Designate the Number of Carbons in the Backbone Chain.

Number of Carbons	Root	Number of Carbons	Root
1	meth-	11	undec-
2	eth-	12	dodec-
3	prop-	13	tridec-
4	but-	14	tetradec-
5	pent-	15	pentadec-
6	hex-	20	eicos-
7	hept-	21	heneicos-
8	oct-	22	docos-
9	non-	23	tricos-
10	dec-	30	triacont-

The roots for backbones containing one through four carbons come from the common names for these alkanes. You probably recognize most of these already. For backbones of five or more carbons, systematic roots (derived from Greek) are employed.

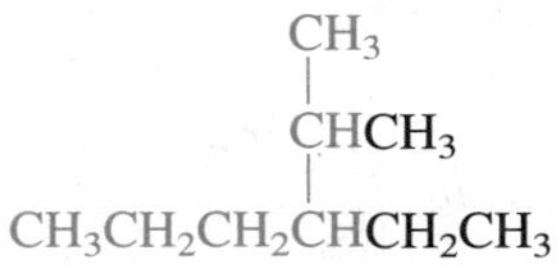

$$CH_3$$
$$|$$
$$CHCH_3$$
$$|$$
$$CH_3CH_2CH_2CHCH_2CH_3$$

correct

$$CH_3$$
$$|$$
$$CHCH_3$$
$$|$$
$$CH_3CH_2CH_2CHCH_2CH_3$$

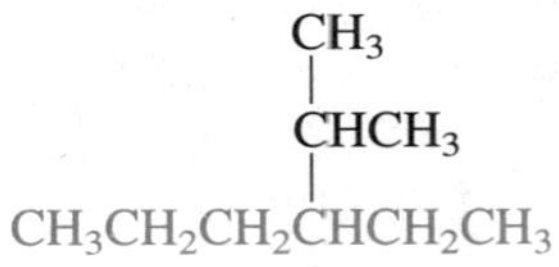

incorrect

There are two different six-carbon chains for this compound. Choose the chain as shown in the left structure because it has two branches rather than in the right one, which has only one branch. The root is hex-.

STEP 2
Attach the suffix. For alkanes, it is -ane.

Steps 1 and 2 are all that are needed to name unbranched alkanes. For example, the straight-chain alkane with seven carbons is heptane.

$$CH_3CH_2CH_2CH_2CH_2CH_2CH_3$$

heptane

To name branched alkanes, additional steps to name the branching group and locate it on the root chain are needed.

STEP 3

Number the carbons in the root chain. Start from the end that gives the lower number to the carbon where the first branch occurs. If both ends have the first branch at equal distances, choose the end that gives the lower number to the carbon where the next branch occurs.

Some examples are the following:

$$\underset{\substack{6 \quad\; 5 \quad\quad 4 \quad\; 3 \quad\;\; 2 \quad 1}}{CH_3CH_2CH_2CH}\overset{\underset{|}{CH_3}}{\underset{}{}}\!\!-\!\!\underset{}{CH}\overset{\underset{|}{CH_3}}{\underset{}{}}CH_3$$

For this structure, the correct numbering starts with the right carbon. Then the first branching group is attached to C-2. Incorrect numbering, starting with the left carbon, would result in the first branch being attached to C-4.

$$\underset{\substack{1 \quad 2 \quad\quad 3 \quad\; 4 \quad\quad\; 5 \quad\;\; 6 \quad\; 7 \quad 8}}{CH_3CHCH_2CHCH_2CH_2CHCH_3}$$

(with branches CH_3, CH_2CH_3, CH_3)

For this structure, the correct numbering starts with the left carbon. The first branch is attached to C-2, and the second branch is attached to C-4. If numbering were to begin with the right carbon, the first branch would be attached to C-2, but the second would be attached to C-5.

$$\underset{\substack{6 \quad\;\; 5 \quad\quad 4 \quad\;\; 3 \quad\; 2\; 1}}{CH_3CHCH_2CH_2CCH_3}$$

(with branches CH_3, CH_3; and CH_3 below C-2)

For this structure, the correct numbering starts with the right carbon. The first branch is at C-2, whether the starting point is the right carbon or the left carbon. However, the second branch is also at C-2 when the starting point is the right carbon. It occurs at C-5 when the starting point is the left carbon.

In other words, proceed inward, one carbon at a time, from each end of the root chain. At the first point of difference, where a group is attached to one carbon under consideration but not the other, choose the end nearer that carbon with the attached group to begin numbering.

STEP 4

Name the groups attached to the root. For straight-chain groups, use the same roots as before to designate the number of carbons and add the suffix -yl.

$$CH_3- \qquad\qquad CH_3CH_2- \qquad\qquad CH_3CH_2CH_2CH_2CH_2-$$

methyl   ethyl   pentyl

A group can be pictured as arising from an alkane by the removal of one hydrogen. It is not a complete compound by itself but requires something to be attached to the remaining valence.

> **STEP 5**
> Assemble the name as a single word in the following order: number, group, root, and suffix. Note that a hyphen is used to separate numbers from the name.

$$CH_3CH_2CH_2\overset{\overset{\displaystyle CH_3}{|}}{C}HCH_2CH_3 \qquad \text{3-methylhexane}$$
$$6 \quad 5 \quad 4 \quad 3 \quad 2 \quad 1$$

If several different groups are present, list them alphabetically. If there are several identical groups, use the prefixes di, tri, tetra, and so on to indicate how many are present. A number is required to indicate the position of each group. For example, 2,2,3-trimethyl indicates the presence of three methyl groups, two attached to C-2 of the backbone and one attached to C-3. Do not use the prefixes that denote the number of groups for alphabetizing purposes. For example, triethyl is listed under *e* and so comes before methyl in a name.

Let's use these steps to name the following compound:

$$CH_3\overset{\overset{\displaystyle CH_3}{|}}{C}HCH_2\overset{\overset{\displaystyle CH_2CH_3}{|}}{C}HCH_2CH_2\overset{\overset{\displaystyle CH_3}{|}}{C}HCH_3$$
$$1 \quad 2 \quad 3 \quad 4 \quad 5 \quad 6 \quad 7 \quad 8$$

4-ethyl 2,7-dimethyloctane

Previously, it was determined that the longest chain has eight carbons (so the compound is a substituted octane) and that numbering should begin with the leftmost carbon. There are three goups attached to the root chain: two methyl groups and an ethyl group. Therefore the systematic or IUPAC name is 4-ethyl-2,7-dimethyloctane. Note the use of hyphens to separate numbers from letters and the use of commas to separate a series of numbers. Also note that dimethyl is alphabetized under *m* (not *d*) and so is listed after ethyl in the name.

Problems

Problem 5.1   Provide IUPAC names for these alkanes.

a) $CH_3\overset{\overset{\displaystyle CH_3}{|}}{C}HCH_2CH_2CH_3$

b) $CH_3CH_2CH_2\overset{\overset{\displaystyle CH_3}{|}}{C}HCH_3$

c) $CH_3\overset{\overset{\displaystyle CH_3}{|}}{C}HCH_2\underset{\underset{\displaystyle CH_2CH_3}{|}}{C}HCH_3$

d) $CH_3CH_2\overset{\overset{\displaystyle CH_2CH_3}{|}}{}CHCH_2\underset{\underset{\displaystyle CH_3 \quad CH_2CH_2CH_3}{| \qquad |}}{C}CH_2CH_2CH_2CH_3$

e)

f)

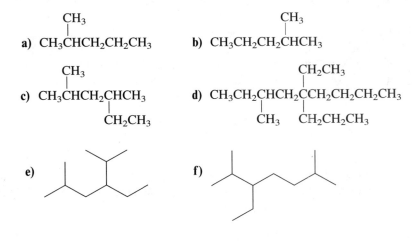

Problem 5.2 Draw the structures of these compounds.

a) 2,2,5-trimethylheptane

b) 4-methyloctane

c) 2,4-dimethyl-5-propyldecane

Solution

a) When the name of a compound is given, drawing the structure is usually straightforward. First draw the chain of carbons indicated by the root. In this case it is a seven-carbon chain. Number the chain starting at either end and add the appropriate groups at the appropriate positions. In this case we add three methyl groups: two on C-2 and one on C-5.

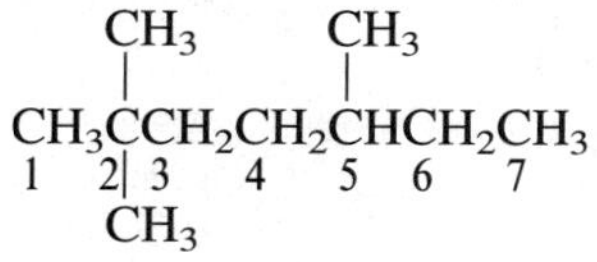

$$\begin{array}{ccccccc} & CH_3 & & CH_3 & & & \\ & | & & | & & & \\ CH_3 & C & CH_2CH_2 & CH & CH_2CH_3 \\ 1 & 2| & 3 \quad 4 & 5 & 6 \quad 7 \\ & CH_3 & & & & & \end{array}$$

Problem 5.3 What is wrong with these names? Provide the correct name for each.

a) 2-ethylpentane

b) 5,5-dimethyl-3-ethylhexane

c) 2-dimethylpentane

Solution

a) First, let's draw the structure suggested by the name:

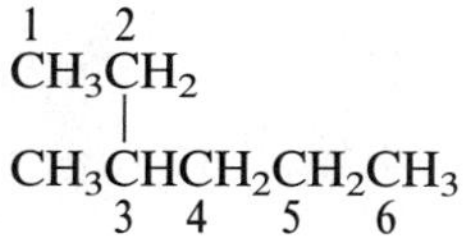

$$\begin{array}{cc} 1 & 2 \\ CH_3CH_2 & \\ | & \\ CH_3CHCH_2CH_2CH_3 \\ 3 \quad 4 \quad 5 \quad 6 \end{array}$$

Examination of this structure shows that the longest chain has six carbons rather than five. Therefore the correct name is 3-methylhexane.

Often groups attached to the root chain are branched rather than the straight-chain ones encountered so far. Such complex groups are named in a fashion similar to that used to name compounds, but with a -yl ending to indicate that they are not complete compounds and need to be attached to something to complete their bonding. Naming complex groups is described in Figure 5.1.

Problems

Problem 5.4 Provide names for these complex groups.

$$\begin{array}{cc} CH_3 & CH_2CH_2CH_3 \\ | & | \\ a) \ -CH_2CH_2CH_2CHCH_3 \qquad b) \ -CH_2CHCH_3 \end{array}$$

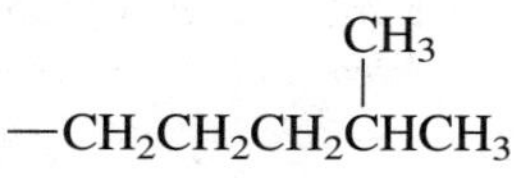

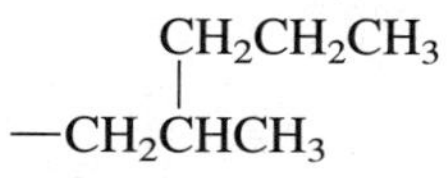

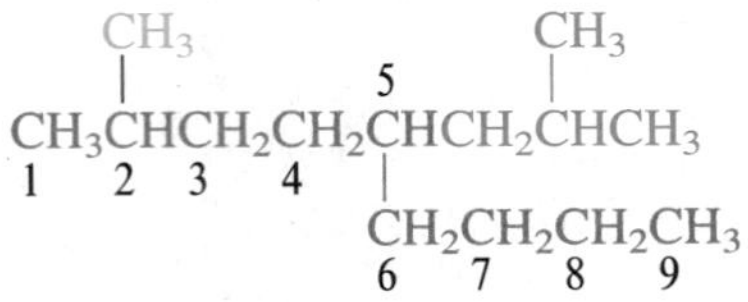

$$\underset{1}{CH_3}\underset{2}{CH}\underset{3}{CH_2}\underset{4}{CH_2}\overset{5}{\underset{}{CH}}CH_2CHCH_3$$

$$\begin{matrix} CH_3 & & CH_3 \\ | & \quad 5 \quad & | \\ CH_3CHCH_2CH_2CHCH_2CHCH_3 \\ 1 \quad 2 \quad 3 \quad 4 \quad | \\ \quad\quad\quad\quad CH_2CH_2CH_2CH_3 \\ \quad\quad\quad\quad 6 \quad 7 \quad 8 \quad 9 \end{matrix}$$

$$\begin{matrix} CH_3 \\ | \\ -CH_2CHCH_3 \\ 1 \quad 2 \quad 3 \end{matrix}$$

The longest chain in this compound (in red) has nine carbons. Numbering begins at the left carbon because the first group is then encountered at position 2. The group (in blue) bonded to position 5 has a branch. The compound is named as a substituted nonane, but how is the complex group at position 5 named?

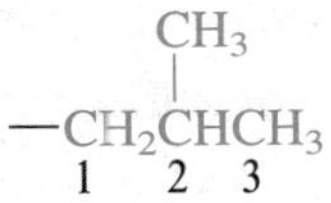

Here is the group. It is named in a manner similar to that used to name alkanes. The longest chain that begins with the carbon attached to the main chain is chosen. As before, the root is determined by the number of carbons in this longest chain (three in this case). The suffix -yl is used to indicate that this is a group rather than a complete compound. So this group is named as a substituted propyl group. Numbering the root chain of a complex group is easy. The carbon attached to the main chain is number 1. The name of the complex group is 2-methylpropyl-.

The name for the complex group is placed in parentheses to avoid confusion when the entire compound is named. The number designating the position of the complex group on the main chain is placed outside the parentheses. Therefore the name of the compound is

2-methyl-5-(2-methylpropyl)nonane.

Figure 5.1

Naming a Compound with a Complex Group.

$$\begin{matrix} & CH_3 & & & CH_3 \\ & | & & & | \\ \text{c)} & -CHCH_2CH_3 & \quad \text{d)} & -CH_2CCH_3 \\ & & & & | \\ & & & & CH_3 \end{matrix}$$

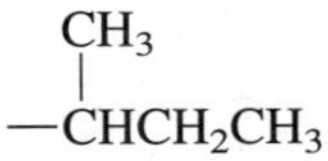

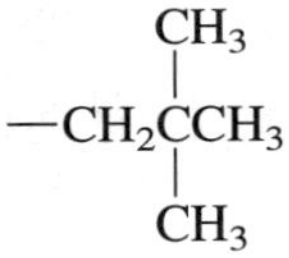

Solution

a) In naming complex groups, it is important to remember that the chain of the group must begin at the carbon attached to the main chain and that this carbon receives number 1. Therefore first find the longest chain that begins at the carbon attached to the main chain. Then number the carbons, also beginning at the carbon attached to the main chain.

$$\begin{matrix} & & & CH_3 \\ & & & | \\ -CH_2CH_2CH_2CHCH_3 \\ 1 \quad 2 \quad 3 \quad 4 \quad 5 \end{matrix}$$

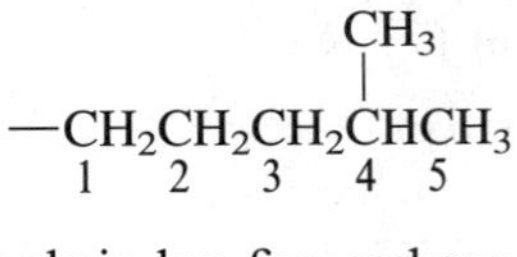

In this case the longest chain has five carbons, and there is a methyl group on C-4, so the name of the group is (4-methylpentyl). Do not forget to put the complex group name in parentheses when the full name of the compound is written.

Problem 5.5 Name these compounds.

a) CH$_3$CH$_2$CH$_2$CHCH$_2$CH$_2$CH$_3$
$\qquad\qquad\quad$ |
$\qquad\qquad\quad$ CHCH$_3$
$\qquad\qquad\quad$ |
$\qquad\qquad\quad$ CH$_3$

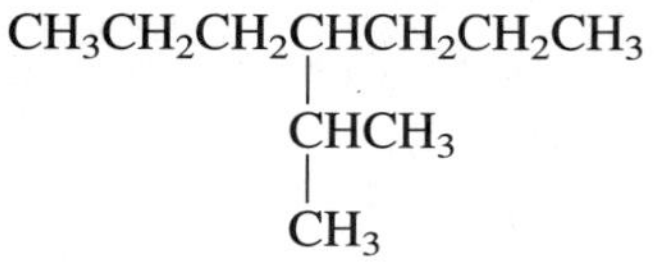

b)

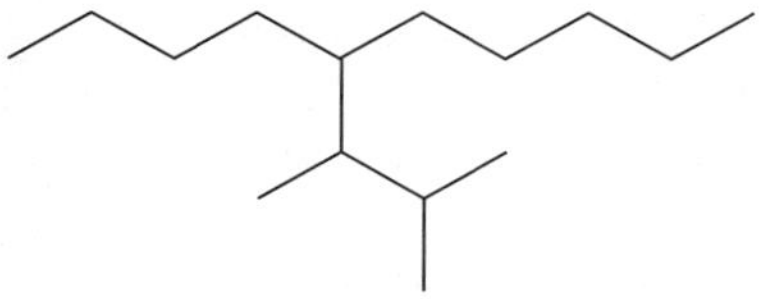

Problem 5.6 Draw structures for these compounds.

a) 4-(1,1-dimethylethyl)decane

b) 3-ethyl-7-methyl-5-(1-methylpropyl)undecane

Solution

a) The compound is a decane, so there are ten carbons in the main chain. On C-4 there is an ethyl group that has two methyl groups attached to the carbon that is attached to the decane chain:

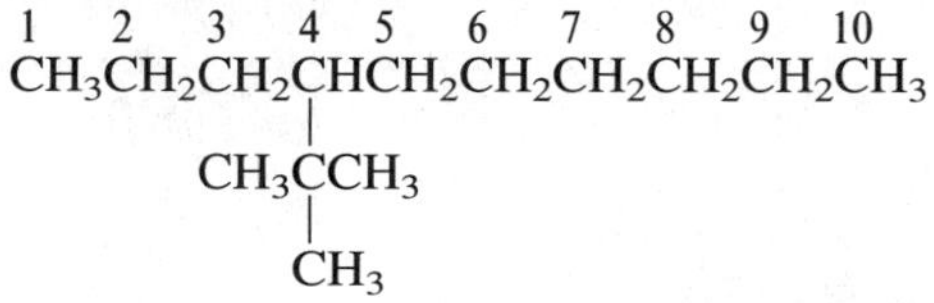

$\quad$1$\quad$2$\quad$3$\quad$4$\quad$5$\quad$6$\quad$7$\quad$8$\quad$9$\quad$10
CH$_3$CH$_2$CH$_2$CHCH$_2$CH$_2$CH$_2$CH$_2$CH$_2$CH$_3$
$\qquad\qquad\quad$ |
$\qquad\qquad\quad$ CH$_3$CCH$_3$
$\qquad\qquad\qquad\quad$ |
$\qquad\qquad\qquad\quad$ CH$_3$

There is some additional terminology that permeates organic chemistry and must be part of every organic chemist's vocabulary. The chemical reactivity of a carbon or of a functional group attached to that carbon often varies according to the number of other carbons bonded to it. In discussions of chemical reactivity, therefore, it is often useful to distinguish carbons according to how many other carbons are bonded to them. A **primary carbon** is bonded to one other carbon, a **secondary carbon** is bonded to two carbons, a **tertiary carbon** is bonded to three carbons, and a **quaternary carbon** is bonded to four carbons. The following compound contains each of these types of carbons.

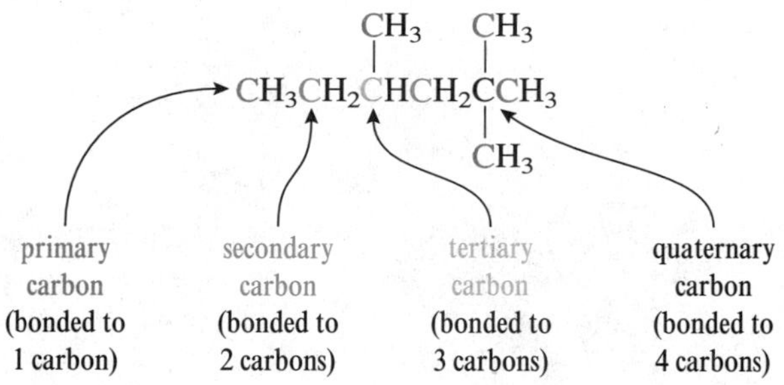

CH$_3$$\qquad$CH$_3$
$\quad$|$\qquad\quad$|
CH$_3$CH$_2$CHCH$_2$CCH$_3$
$\qquad\qquad\qquad$|
$\qquad\qquad\qquad$CH$_3$

| primary carbon (bonded to 1 carbon) | secondary carbon (bonded to 2 carbons) | tertiary carbon (bonded to 3 carbons) | quaternary carbon (bonded to 4 carbons) |

Problem

Problem 5.7   Designate each carbon of these compounds as primary, secondary, tertiary, or quaternary.

a) 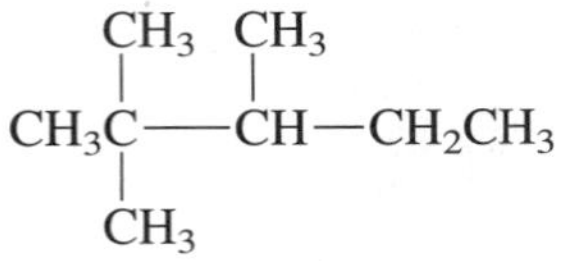 $\underset{\displaystyle CH_3}{\underset{|}{CH_3C}}{\overset{\displaystyle CH_3\ CH_3}{\overset{|\quad\ \ |}{}}}\!\!-\!\!CH\!-\!CH_2CH_3$

b)

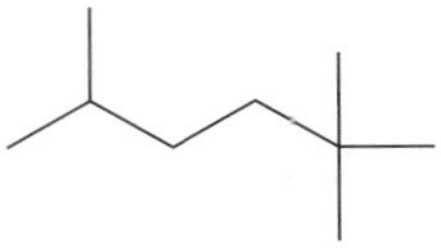

c) 2,3-dimethylpentane

Finally, in addition to the systematic method for naming groups that we have just seen, we will still encounter some common group names. IUPAC nomenclature allows the use of these common group names as part of the systematic name. (IUPAC nomen-

Table 5.4   Systematic Names, Common Names, and Abbreviations for Some Groups. The common names may be used in IUPAC nomenclature.

Group	Systematic Name	Common Name	Abbreviation
CH_3-	methyl		Me
CH_3CH_2-	ethyl		Et
$CH_3CH_2CH_2-$	propyl		Pr
$\underset{CH_3CH-}{\overset{CH_3}{|}}$	1-methylethyl	isopropyl	i-Pr
$CH_3CH_2CH_2CH_2-$	butyl		Bu
$\underset{CH_3CHCH_2-}{\overset{CH_3}{|}}$	2-methylpropyl	isobutyl	i-Bu
$\underset{CH_3CH_2CH-}{\overset{CH_3}{|}}$	1-methylpropyl	sec-butyl	sec-Bu (or s-Bu)
$CH_3-\underset{CH_3}{\overset{CH_3}{\underset{|}{\overset{|}{C}}}}-$	1,1-dimethylethyl	tert-butyl	tert-Bu (or t-Bu)
$-CH_2-$		methylene	
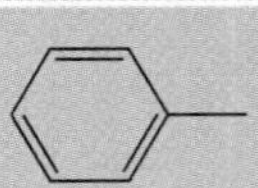 ⬡—	phenyl		Ph

clature does allow different names for the same compound; however, no two compounds may have the same name.) Table 5.4 shows a number of groups, along with their systematic names, their common names and abbreviations that are sometimes used to represent the groups when writing structures. (The abbreviations are not used in nomenclature.) There are four butyl groups in the table. To help remember them, note that the *sec*-butyl group is attached via a secondary carbon and the *tert*-butyl group is attached via a tertiary carbon. The italicized prefixes *sec*- and *tert*- are not used in alphabetizing; *sec*-butyl is alphabetized under *b*. However, isobutyl is alphabetized under *i*.

Problems

Problem 5.8   Name this compound using the common name for the group.

$$CH_3$$
$$|$$
$$CHCH_3$$
$$|$$
$$CH_3CH_2CH_2CHCH_2CH_2CH_3$$

Problem 5.9   Draw the structure of 4-*tert*-butyl-2,3-dimethyloctane.

5.5   Systematic Nomenclature of Cycloalkanes

The procedure used to name a cycloalkane by the IUPAC method is very similar to that used for alkanes. Here, the root designates the number of carbons in the ring and the prefix cyclo- is attached to indicate that the compound contains a ring. The rules for numbering the ring carbons are as follows:

No number is needed if only one group is attached to the ring.

For rings with multiple substituents, begin numbering at one substituent and proceed in the direction that gives the lowest numbers to the remaining substituents.

Some examples are the following:

cycloheptane
(There are 7 carbons in the ring.)

isopropylcyclopentane or
(1-methylethyl) cyclopentane
(No number is needed to locate the isopropyl group because all positions of the ring are identical.)

1,3-dimethylcyclohexane
(To keep the numbers as low as possible, begin at one methyl group—either one in this case—and proceed by the shortest possible path to the other methyl group.)

In cases in which the alkyl chain has more carbons than the ring, then the compound is named as an alkane with the ring as a substituent group with a -yl suffix.

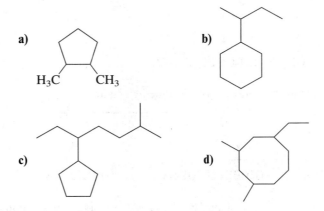

1-cyclobutyl-2-methylpentane

(There are five carbons in the longest alkyl chain, while the ring has only four carbons. Therefore the ring is named as a substituent group on the alkane chain.)

Problems

Problem 5.10   Name the following compounds.

a)

H₃C   CH₃

b)

c)

d)

Problem 5.11   Draw structures for these compounds.

a)   1,1-dimethylcyclohexane   b)   ethylcyclopropane

5.6   Alkenes

Alkenes have one or more carbon–carbon double bonds. In the discussion in Chapter 2 about calculating the degree of unsaturation of a compound it was shown that each double bond present in an alkene results in a decrease of two hydrogens in the formula when compared to an alkane with the same number of carbons. The term **unsaturated** is used to describe alkenes and actually has a chemical derivation. It is possible to cause compounds containing double or triple bonds to react with hydrogen gas to form alkanes. When the compound will no longer react with hydrogen, it is said to be **saturated.** A compound that will react with hydrogen, such as an alkene or an alkyne, is, then, unsaturated. You have probably heard these terms used a lot in association with cooking oils or margarines. A polyunsaturated oil is composed of compounds that contain several carbon–carbon double bonds. A saturated fat, by contrast, is composed of similar compounds that have no double bonds.

Alkenes are named similarly to alkanes, with the following modifications:

1. The longest continuous chain that includes both carbons of the double bond provides the root.

2. The suffix used for an alkene is -ene. Names for compounds with more than one double bond use the suffixes -diene, -triene, and so on.

3. The root is numbered from the end that gives the lower number to the first carbon of the double bond. The number of this first carbon is used in the name to designate the position of the double bond.

Some examples are the following:

$$CH_3CH \!=\! CHCHCH_3$$
$$\overset{\displaystyle CH_3}{|}$$
$$\begin{matrix}1 & 2 & 3 & 4 & 5\end{matrix}$$

4-methyl-2-pentene
(The double bond, not the methyl group, determines the numbering.)

3-ethyl-4-methylcyclohexene
(Numbering begins with the carbons of the double bond having numbers 1 and 2 and proceeds in the direction to give the lowest numbers to the remaining substituents.)

$$CH_3CH_2CH_2C \!=\! CHCH \!=\! CH_2$$
$$\begin{matrix}7 & 6 & 5 & 4 & 3 & 2 & 1\end{matrix}$$
(with CHCH_3 and CH_3 substituent)

4-isopropyl-1,3-heptadiene
(Note the *a* in hept*a*diene. This is added to the root whenever the first letter of the suffix is a consonant to make the name easier to pronounce.)

$$CH_3CH_2CCH_2CH_2CH_3$$
$$\overset{\displaystyle 1\, CH_2}{\|}$$
$$\begin{matrix}2 & 3 & 4 & 5\end{matrix}$$

2-ethyl-1-pentene
(Both carbons of the double bond must be part of the root chain.)

Problems

Problem 5.12 Name these compounds.

a) $CH_2 \!=\! CHCHCH_2CHCH_3$
with CH_3 above and $CH_2CH_2CH_3$ below

b)

c)

d)

Solution

a)   Choose the longest chain that contains both of the carbons of the double bond and has the most branches:

$$\underset{\underset{\overset{|}{CH_2CH_2CH_3}}{\overset{}{}}}{\overset{\overset{CH_3}{|}}{\underset{1\qquad2\ \ \ 3\ \ \ 4\ \ \ 5|\ \ \ 6}{CH_2{=}CHCHCH_2CHCH_3}}}$$

Number it beginning at the left end so that the carbons of the double bond have lower numbers. It has a propyl group on C-3 and a methyl group on C-5, so the name is 5-methyl-3-propyl-1-hexene.

Problem 5.13   Draw structures for these compounds.

a)   3-ethyl-3-hexene   b)   cyclobutene

c)   3-propyl-1,4-cyclohexadiene

There are a few common names that are often encountered for alkenes and groups containing double bonds. These are listed in Table 5.5.

The polarity of an alkene is not much different from the polarity of an alkane. Therefore the physical properties of an alkene are similar to those of the corresponding alkane. For example, 1-pentene melts at $-138°C$ and boils at $30°C$, values that are comparable to those of pentane, which melts at $-130°C$ and boils at $36°C$.

However, the chemical properties of an alkene are dramatically affected by the presence of the double bond. Recall that a carbon–carbon pi bond is considerably

Table 5.5   Common Names for Some Compounds and Groups Containing Double Bonds

Compound or Group	Common Name
$CH_2{=}CH_2$	ethylene
$CH_2{=}CHCH_3$	propylene
$CH_2{=}CHCH_2CH_3$	butylene
$\underset{CH_2{=}CCH_3}{\overset{\overset{CH_3}{|}}{}}$	isobutylene
$CH_2{=}C{=}CH_2$	allene
$CH_2{=}CH{-}$	vinyl group
$CH_2{=}CHCH_2{-}$	allyl group

single bonds are in the 3 to 5 kcal/mol (12–21 kJ/mol) range, so rotations about most single bonds are fast at room temperature. Therefore, different conformations cannot be separated or isolated and are not usually considered to be isomers. We say that there is *free rotation* about single bonds.

Next, let's consider the case of propane. Conformational analysis can be done about either of the two identical carbon–carbon bonds. Again there are two limiting conformations: staggered and eclipsed. The only difference between this example and the analysis of ethane done previously is that here there are two hydrogen–hydrogen eclipsing interactions and one hydrogen–methyl eclipsing interaction.

Propane that is eclipsed about one CC bond has one CH$_3$/H eclipsing interaction and

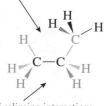

two H/H eclipsing interactions.

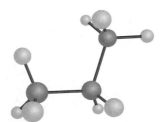
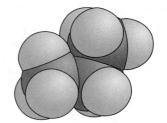

It is to be expected that the repulsion between the eclipsed methyl and hydrogen is slightly larger than that between two hydrogens. In fact, the eclipsed conformation of propane is 3.3 kcal/mol (13.8 kJ/mol) higher in energy than the staggered conformation. If each of the H with H interactions contributes about 1 kcal/mol (4 kJ/mol) to this value, as was the case for ethane, then the eclipsing interaction between the methyl and the hydrogen must contribute about 1.3 kcal/mol (5.4 kJ/mol) in repulsion energy. This is due primarily to torsional strain between the C—H bond and the C—C bond, along with a smaller contribution due to steric strain between the hydrogen on the carbon and the hydrogens on the methyl group. The energy versus dihedral angle plot for propane looks just like the one for ethane except that the energy difference between the staggered and eclipsed conformations is 3.3 kcal/mol.

Problem

Problem 6.5   Draw a plot of energy versus dihedral angle for the conformations of propane about one of the C—C bonds.

Butane provides a more complex example. Here there are two different types of carbon–carbon bonds. Analysis of the conformations available by rotation about the bond between carbon 1 and carbon 2 (or carbon 3 and carbon 4) is very similar to the analysis of propane, with the difference that there is an ethyl group on one carbon rather than a methyl group. However, conformational analysis about the bond between carbon 2 and carbon 3 provides a more interesting situation. In this case, each carbon has one methyl group and two hydrogens. The various conformations lead to a number of different interactions.

Conformational analysis about the C-2—C-3 bond of butane is more complex than previous examples because each carbon has a methyl and two hydrogens bonded to it. There are several different interactions that occur in the conformations.

Examples of the various conformations of butane are shown below:

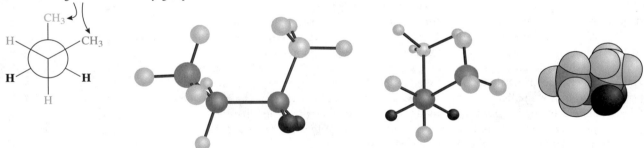

Above is the anti conformation of butane. (Two groups, the methyl groups in this case, are said to be anti if the dihedral angle between them is 180°.) It is the most stable conformation because it is staggered and has the two large methyl groups as far apart as possible.

steric crowding between these methyl groups

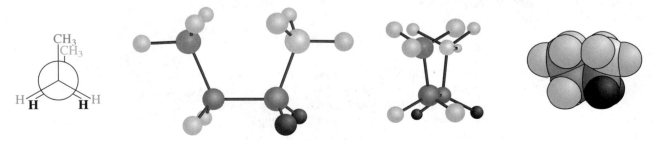

Above is another staggered conformation of butane, called the gauche conformation. (Two groups are said to be gauche when the dihedral angle between them is 60°.) It is 0.8 kcal/mol (3.3 kJ/mol) higher in energy than the anti conformation. Since it is staggered, it has no torsional strain. Its higher energy is due to a small amount of steric strain caused by interaction between the bulky methyl groups, which are a little too close together.

The eclipsed conformation above has two H with H eclipsing interactions and one CH_3 with CH_3 eclipsing interaction. As expected, the eclipsing interaction between the methyl groups is more destabilizing than that between the hydrogens. This conformation is about 4.5 kcal/mol (19 kJ/mol) less stable than the anti conformation. If the two H with H interactions contribute 1 kcal/mol each to this number, as was the case for ethane, then the CH_3 with CH_3 interaction must contribute the remaining 2.5 kcal/mol (10.5 kJ/mol). This value is due to both torsional and steric strain.

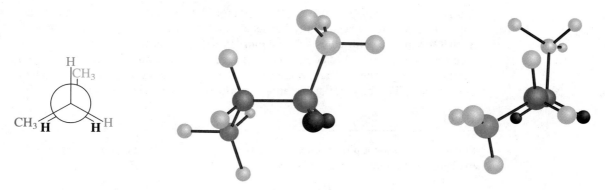

This eclipsed conformation has two CH₃ with H interactions and one H with H interaction. If a CH₃ with H interaction here costs the same amount of energy as it did in the case of propane (1.3 kcal/mol), then this conformation would be expected to be destabilized by $2(1.3) + 1 = 3.6$ kcal/mol (15 kJ/mol). This value is in reasonable agreement with the experimental value of 3.7 kcal/mol (15.5 kJ/mol).

A plot of the energies of these conformations, with the methyl groups used as the markers for the dihedral angles, is shown in Figure 6.6. The lowest-energy minimum on this plot is located at the anti conformation. There are also minima at the

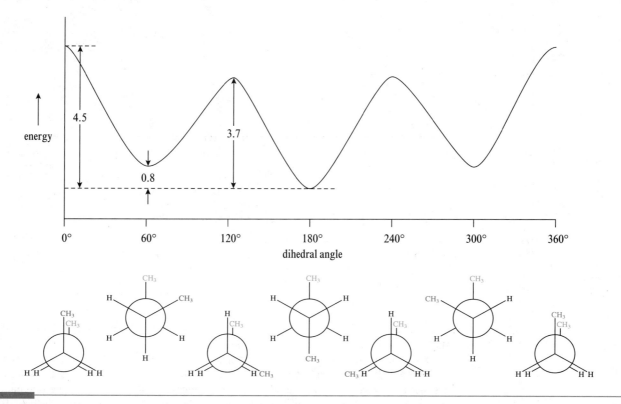

Figure 6.6

Plot of Energy versus Dihedral Angle for Conformations of Butane.

Energies are in units of kcal/mol.

two gauche conformations that are 0.8 kcal/mol (3.3 kJ/mol) less stable than the anti conformation. The highest-energy barrier on the plot, at the conformation where the methyl groups are eclipsed, is 4.5 kcal/mol (19 kJ/mol) higher in energy than the anti conformation. Therefore rotation about this carbon–carbon bond is fast at room temperature. The gauche and anti conformations are in equilibrium, about 70% of the molecules having the more stable anti conformation at any given instant.

Problem

Problem 6.6   Draw a plot of energy versus dihedral angle for the conformations of 2-methylbutane about the C2—C3 bond.

$$CH_3$$
$$|$$
$$CH_3-CH-CH_2-CH_3$$
$$1 \quad\ 2 \quad\ 3 \quad\ 4$$

Solution

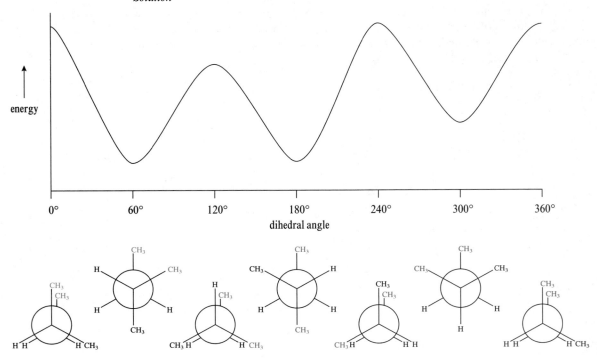

First, let's examine the three eclipsed conformations. The 0° and 240° conformations both have one CH_3 with CH_3, one CH_3 with H and one H with H eclipsing interaction, so they have the same amount of strain energy. The 120° conformation has three CH_3 with H eclipsing interactions, so it has less strain energy than the others—it is at lower energy. The 60° and 180° staggered conformations have one gauche interaction and therefore are lower in energy than the 300° staggered conformation, which has two gauche interactions.

Analysis of the conformations of other alkanes can be done in a very similar manner. For any linear alkane the most stable conformation is the so-called zigzag conformation, which is anti about all of the carbon–carbon bonds. The following is the zigzag conformation for hexane:

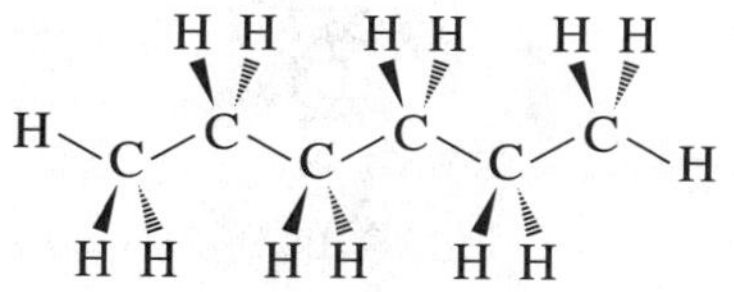

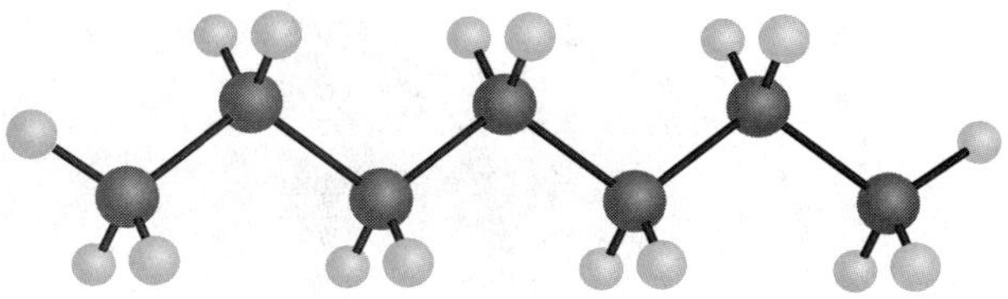

Free rotation about the carbon–carbon bonds generates a large number of other conformations that are gauche about one or more of these bonds. Although the zigzag conformation is the most stable one, many of the others are only slightly higher in energy and are readily attainable at room temperature. The shape of an individual molecule changes rapidly, twisting and turning among these various possibilities. Finally, it should be noted that the presence of polar substituents can dramatically affect and complicate conformational preferences because of interactions among their dipoles and hydrogen bonding.

6.5   Conformations of Cyclic Molecules

Because their carbon chains are confined in rings, cycloalkanes are much less flexible than noncyclic (or acyclic) alkanes. The number of conformations available is dramatically reduced. Furthermore, cycloalkanes are often held in shapes that cause them to have considerable strain energy. One way to measure this strain energy in the laboratory is to burn the alkane in a calorimeter and measure the amount of heat that is produced (the heat of combustion). The heat of combustion must first be corrected for the number of carbons and hydrogens in the ring. For example, cyclohexane (C_6H_{12}) has twice as many atoms as cyclopropane (C_3H_6) and would be expected to produce twice as much heat on a per mole basis, other factors being equal. The easiest way to correct for this is to divide the heat of combustion per mole by the number of CH_2 groups in the ring. This gives a heat of combustion "per mole of CH_2 group." This value can then be compared to that of a standard compound that has no strain, such as a long-chain normal alkane.

Such heat of combustion values for the cycloalkanes ranging in size from 3 to 9 carbons are provided in Table 6.1. The lowest heat of combustion per CH_2 group is found for cyclohexane, which is usually considered to be free of strain. The extra heat produced by the other cycloalkanes can then be attributed to their strain energies. For example, the heat of combustion per CH_2 for cyclopropane is 166.3 kcal/mol (696 kJ/mol), and that for cyclohexane is 157.4 kcal/mol (659 kJ/mol). The difference, 8.9 kcal/mol (37 kJ/mol), is the amount of strain energy per CH_2 group for cyclopropane. The total strain energy for cyclopropane is three times this value,

Table 6.1   Heats of Combustion and Strain Energies of Some Cycloalkanes.

Ring Size	Heat of Combustion per CH_2	Strain Energy per CH_2	Total Ring Strain
3	166.3 (695.8)	8.9 (37.2)	26.7 (111.6)
4	163.9 (685.8)	6.5 (27.2)	26.0 (108.8)
5	158.7 (664.0)	1.3 (5.4)	6.5 (27.0)
6	157.4 (658.6)	0	0
7	158.3 (662.3)	0.9 (3.7)	6.3 (25.9)
8	158.6 (663.6)	1.2 (5.0)	9.6 (40.0)
9	158.8 (664.4)	1.4 (5.8)	12.6 (52.2)

Units are kcal/mol. Units for values in parentheses are kJ/mol. To obtain the values listed in the second column, the total heats of combustion were divided by the number of carbons in the ring. Values in the third column were obtained by subtracting 157.4 (the heat of combustion per CH_2 for cyclohexane) from the values in the second column. Values in the fourth column were calculated by multiplying values in the third column by the number of carbons in the ring.

26.7 kcal/mol (111 kJ/mol), because cyclopropane has three CH_2 groups. Values for the strain energy per CH_2 group and total ring strain for the other cycloalkanes are also provided in Table 6.1. Note that cyclopropane and cyclobutane have large amounts of strain compared to cyclohexane, while the other cycloalkanes have much smaller amounts.

By examining the conformations of the cycloalkanes, we will be able to determine the origin of these strain energies. Let's begin with the smallest one, cyclopropane, and see what is the cause of the large amount of strain energy that it has.

The three carbons of cyclopropane form an equilateral triangle. The C—C—C angle is 60°. However, if the hybridization at each carbon is sp^3, as would be expected from the structure, then the angles between the sp^3 hybrid AOs are 109.5°.

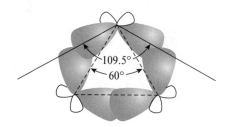

The orbitals forming
the C—C bonds of cyclopropane

Because the angle between the sp^3 hybrid AOs on one carbon is wider than the angle between the carbons to which these orbitals are bonding, the orbitals cannot point directly toward the carbons. Instead, as shown here, they point slightly outside of a line connecting the nuclei. Since the orbitals of a bond do not point directly at each other, the amount of overlap is decreased. This causes the C—C bonds of cyclopropane to be weaker than normal C—C bonds. This type of destabilization is called **angle strain.**

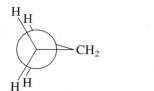

In addition to angle strain, cyclopropane has a significant amount of torsional strain. This can best be seen by looking at a Newman projection down any of the C—C bonds. As can be seen in this diagram, each C—C bond is held in an eclipsed conformation by the rigidity of the molecule.

From heat of combustion data, cyclopropane has 26.7 kcal/mol (111.6 kJ/mol) of strain energy. Most of this strain is due to angle strain, but the contribution due to torsional strain is also significant. As we will see later, this strain energy causes cyclopropane to be more reactive than a normal alkane or cycloalkane. However, even though cyclopropane rings are reactive, they are fairly common in organic chemistry.

Exercise

Exercise 6.3   Build a model of cyclopropane and examine its strain.

If the four carbons of cyclobutane lie in a plane, then its carbons form a square. The angles of a square are 90° so it is expected that cyclobutane also has some angle strain, although not as much as cyclopropane. Planar cyclobutane would also be eclipsed about each C—C bond and would have considerable torsional strain. As the carbons of cyclobutane are distorted from planarity, torsional strain decreases while angle strain increases. For small distortions the increase in angle strain is less than the decrease in torsional strain. Therefore the lowest-energy conformation of cyclobutane is slightly nonplanar (the angle between the planes of the carbons is about 35°), as shown in the diagram on the following page. Cyclobutane has some angle strain and some torsional strain contributing to its total strain energy of 26.0 kcal/mol (108.8 kJ/mol). A Newman projection shows how twisting the carbons out of planarity results in less torsional strain. The hydrogens are no longer exactly eclipsed. Although cyclobutane has less strain than cyclopropane, cyclobutane rings are less common than cyclopropane rings because, as we shall see later, they are more difficult to prepare.

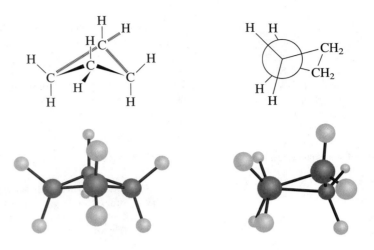

Exercise

Exercise 6.4 Build a model of cyclobutane. Examine the various types of strain present in the planar and nonplanar geometries.

The angles of a regular pentagon are 108°. Therefore planar cyclopentane would have little or no angle strain. However, like planar cyclobutane, it would have considerable torsional strain because each C—C bond would be held in an eclipsed conformation. It is to be expected, then, that cyclopentane will distort from planarity to relieve this torsional strain. In one low-energy conformation, one carbon folds out of the plane so that the overall shape is somewhat like an envelope. This relieves most of the torsional strain without increasing the angle strain significantly. Overall, cyclopentane has very little strain, 6.5 kcal/mol (27 kJ/mol). It is a very common ring system and is widely distributed among naturally occurring compounds.

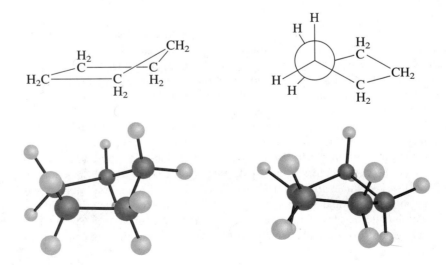

Exercise
_____

Exercise 6.5   Build a model of cyclopentane. Examine the various types of strain present in the planar and nonplanar geometries.

_____

6.6   Conformations of Cyclohexane

The cyclohexane ring is very important because it is virtually strain-free. This is one of the reasons why compounds containing six-membered rings are very common. If cyclohexane were planar, its C—C—C angles would be 120°—too large for the 109.5° angle of sp^3 hybrid AOs. However, the angles of the ring decrease as it becomes nonplanar. There are two nonplanar conformations, called the **chair conformation** and the **boat conformation,** that are completely free of angle strain. These conformations are shown in Figures 6.7 and 6.8, respectively. The chair conformer of cyclohexane is perfectly staggered about all of the C—C bonds and so has no torsional strain either—it is strain-free. The boat conformer, on the other hand, has both steric strain, due to interactions of the flagpole hydrogens, and torsional strain. It is about 6 kcal/mol (25 kJ/mol) less stable than the chair conformer. Some of the steric and torsional strain of the boat can be relieved by twisting. The twist boat conformation is about 5 kcal/mol (21 kJ/mol) less stable than the chair conformation.

Figure 6.7 also shows that there are two different types of hydrogens, called **axial** hydrogens and **equatorial** hydrogens, in the chair conformer of cyclohexane. The axial C—H bonds are parallel to the axis of the ring; the equatorial C—H bonds project outward from the ring around its "equator." Steps to help you learn to draw the chair conformation of cyclohexane, including the axial and equatorial hydrogens, are provided in Figure 6.9.

The chair conformation of cyclohexane is not rigid. It can convert to a boat conformation, as shown in the following diagrams (not all the hydrogens are shown for clarity):

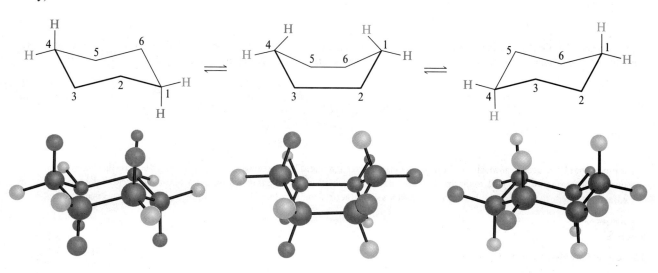

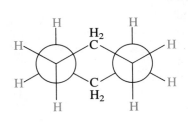

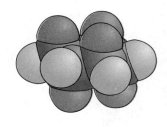

This is the chair conformation of cyclohexane. All of the C—C—C bond angles are 109.5°, so it has no angle strain. In addition, it has no torsional strain because all of the C—H bonds are perfectly staggered. This can best be seen by examining a Newman projection down C—C bonds on opposite sides of the ring:

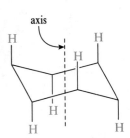
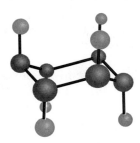
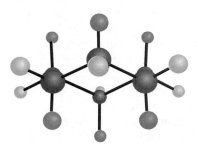

The staggered arrangement of all the bonds can be seen clearly in the Newman projection. This same picture is seen when the projection is viewed down any C—C bond. All the C—C bonds in the molecule are in conformations in which the hydrogens are perfectly staggered.

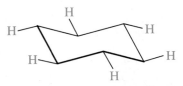

Careful examination of cyclohexane in the chair conformation shows that there are two different types of hydrogens. The bonds to one type, shown above, are parallel to the axis of the ring. These are called axial hydrogens. Note how the axial bonds alternate up and down around the ring.

The other type of hydrogens, shown above, are directed outward from the ring. They are called equatorial hydrogens because they lie around the "equator" of the ring. Now go back to the first structure in this figure, in which both types of hydrogens are shown, and identify the axial hydrogens (red) and the equatorial hydrogens (blue). Also examine the view of the axial and equatorial hydrogens provided by the Newman projection.

Figure 6.7

The Chair Conformation of Cyclohexane.

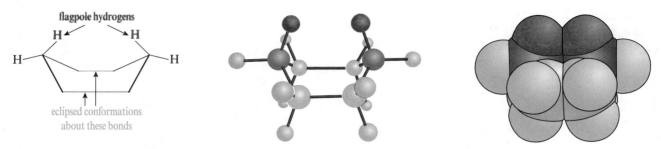

This is the boat conformation of cyclohexane. (In the first diagram, not all the hydrogens are shown for clarity.) Like the chair conformation, all of the C—C—C bond angles are 109.5°, so the boat has no angle strain. However, it does have other types of strain. The **two red hydrogens**, called flagpole hydrogens, approach each other too closely and cause some steric strain. In addition, the conformations about the gold bonds are eclipsed. This can be seen more easily in the Newman projection down these bonds:

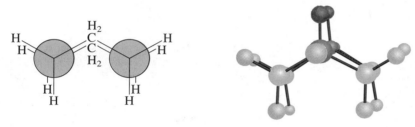

The Newman projection shows that two bonds of the boat conformation are eclipsed. The torsional strain due to these eclipsing interactions and the steric strain due to the interaction of the flagpole hydrogens make the boat conformation higher in energy than the chair conformation. The boat conformation is flexible enough to twist somewhat to slightly decrease its overall strain energy.

In the twist boat conformation the "bow" and the "stern" of the boat have been twisted slightly. While this decreases the flagpole interaction and relieves some of the torsional strain, angle strain is introduced. Overall, the twist boat conformation is a little more stable than the boat conformation but not nearly as stable as the chair conformation.

Figure 6.8

The Boat Conformation of Cyclohexane.

Figure 6.9

Steps for Drawing Chair
Cyclohexane.

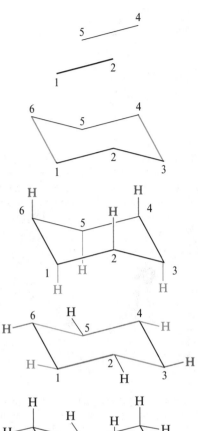

To draw a good cyclohexane chair conformation, first draw
two parallel lines, sloping upward and slightly offset. These
are the C—C bonds between carbons 1 and 2 and carbons 4
and 5.

Add carbon 6 to the left and above these lines and carbon 3
to the right and slightly below and draw the C—C bonds.
Note how bonds on opposite sides of the ring are drawn
parallel (the C-1 — C-6 bond is parallel to the C-3 — C-4
bond, etc.)

Add the axial C—H bonds. These are drawn **up from
carbons 2, 4, and 6** and down from carbons 1, 3, and 5.

The equatorial C—H bonds should point outward from
the ring. They are drawn parallel to the C—C bond once
removed, that is, the equatorial C—H bond on C-1 is drawn
parallel to the C-2— C-3 (or C-5— C-6) bond, and the
C—H bond on C-3 is drawn **parallel to the C-1 — C-2 (or
C-4— C-5) bond**, etc.

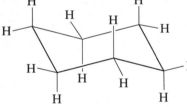

The completed structure should look like this.

In this illustration, C-1 flips up to give a boat. Then C-4 can flip down to produce
another chair conformation. When opposite carbons flip like this, all axial and equatorial bonds interconvert; that is, all hydrogens that were axial are converted to equatorial, and all hydrogens that were equatorial are converted to axial. This can be seen
in the above diagrams, in which the red hydrogens, which are axial in the left chair,
are converted to equatorial hydrogens in the right chair. The energy required for this
ring-flipping process is shown in Figure 6.10. The highest barrier, called the half-chair
conformation, is about 10.5 kcal/mol (44 kJ/mol) higher in energy than the chair conformation. Again, the 20 kcal/mol (83.7 kJ/mol) of energy that is available at room
temperature provides plenty of energy to surmount this barrier; therefore this ring flipping is fast. It occurs about 100,000 times per second at room temperature.

Exercises

Exercise 6.6   Build a model of cyclohexane.

    a)   Examine the strain present when the geometry is planar.

The actual three-dimensional arrangement of groups around a chiral center is called the **absolute configuration.** Until a special X-ray technique was developed in 1951, it was not possible to determine the absolute configuration of any compound. Although samples of one enantiomer (or both) of a multitude of compounds were available, there was no experimental method to determine whether that enantiomer had the *R* or the *S* absolute configuration. This was not a major problem for organic chemists, though, because they were able to convert one chiral molecule to another, using reactions whose stereochemical effects were well known. Thus it was possible to relate the configuration of one compound to that of another. The **relative configurations** of the compounds were known. For example, if one enantiomer of 2-butanol is converted to 2-chlorobutane using a reaction that is known to put the chlorine exactly where the hydroxy group was, then the two compounds have the same relative configuration. If, as shown below, the starting material is (*R*)-2-butanol, then the product is (*R*)-2-chlorobutane:

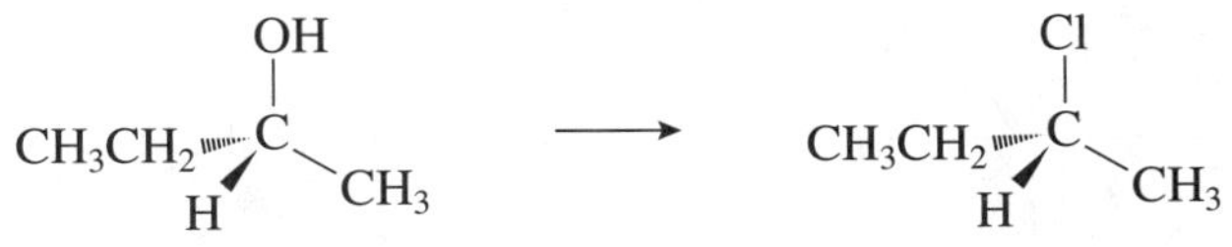

$$CH_3CH_2 - \underset{\underset{H}{|}}{\overset{\overset{OH}{|}}{C}} - CH_3 \quad \longrightarrow \quad CH_3CH_2 - \underset{\underset{H}{|}}{\overset{\overset{Cl}{|}}{C}} - CH_3$$

(*R*)-2-butanol   (*R*)-2-chlorobutane

If the absolute configuration of the starting 2-butanol enantiomer is not known, then the absolute configuration of the product 2-chlorobutane is not known either. However, because the reaction is known to put the chlorine exactly where the hydroxy group was, the two compounds must have the same relative configuration. Often, knowing the relative configurations of the compounds is enough to answer the chemical question under consideration. Of course, once the absolute configuration of one compound has been determined, the absolute configuration of any other compound whose configuration has been related to the first is also known.

6.13   Properties of Enantiomers

When do enantiomers have different properties? Again, it is helpful to draw analogies with everyday objects that are chiral. When do your hands have different properties? They are different when you put on a glove; they are different when you write; they are different when you shake hands. What do these objects or activities have in common? A glove, writing, and shaking hands are all chiral! Hands are different when they interact with one enantiomer of a chiral object or activity. Likewise, enantiomeric molecules are different when they are in a chiral environment. Most commonly, the chiral environment is the presence of one enantiomer of another chiral compound. Otherwise, their properties are identical. For example, the naturally occurring ketones (*R*)- and (*S*)-carvone have identical melting points, boiling points, solubilities in ethanol, and heats of combustion. However, they have different solubilities in one enantiomer of a chiral solvent, and they have different odors (the odor receptors in the nose are chiral). (*R*)-carvone smells like caraway and (*S*)-carvone smells like spearmint. An important difference is that *enantiomers have different rates of reaction with one enantiomer of a chiral reagent.*

(*S*)-carvone

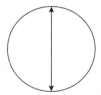

(*R*)-carvone

The most common method that is used to detect the presence of chiral molecules in a sample employs the interaction of plane-polarized light with the sample.

Regular light waves consist of electromagnetic fields that oscillate in all directions perpendicular to the direction of travel of the wave. If we could see these fields while viewing the light beam coming directly at us, the oscillations would occur along the arrows.

When a regular light beam is passed through a polarizer (polarized sunglasses will work), all of the light waves, except those whose electromagnetic fields oscillate in a single direction, are filtered out. The result is a beam of **plane-polarized light.** The oscillations all occur in a single plane as shown.

When plane-polarized light is passed through a sample containing one enantiomer of a chiral compound, the plane of polarization of the light is rotated. (Samples that rotate plane-polarized light are said to be **optically active.**) A schematic diagram of a simple instrument, called a polarimeter, that can detect this rotation is shown in Figure 6.14. In this instrument the organic compound to be analyzed is placed in the sample tube, either as a pure liquid or as solution in an achiral solvent. When the plane-polarized light passes through the sample, the plane of polarization is rotated. The magnitude of the observed rotation, α, in degrees, is measured by the analyzing polarizer. If the beam has been rotated in a clockwise direction, α is assigned a positive value; if the beam has been rotated in a counterclockwise direction, α is assigned a negative value.

For a particular compound the observed rotation depends on the concentration of the compound, the path length of the sample tube, and the wavelength of the light that is used. Often the yellow light produced by a sodium lamp, called the sodium D line (wavelength = 589 nm), is used. The specific rotation, a constant characteristic of each chiral compound, can be calculated from the observed rotation obtained in the laboratory by the following equation.

specific rotation ⎯ (using sodium D line)   ⎯ observed rotation (degrees)

$$[\alpha]_D = \frac{\alpha}{(c)\,(l)}$$

concentration ⎯ (g/mL)   ⎯ length of sample tube (dm)

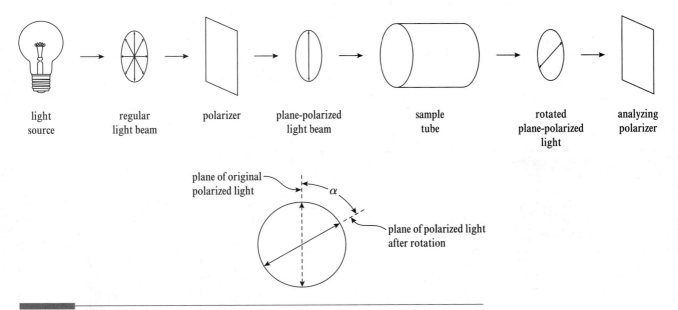

Figure 6.14

Schematic Diagram of a Polarimeter.

Enantiomers rotate plane-polarized light by identical magnitudes but in opposite directions. The enantiomer that rotates the light clockwise is called **dextrorotatory,** (*d*) or (+), while the one that rotates light counterclockwise is called **levorotatory,** (*l*) or (−). A common way to encounter a chiral compound is as an equal mixture of enantiomers, called a **racemic mixture** or a **racemate.** A racemic mixture, often designated as *d,l* or (±), does not rotate plane-polarized light because the rotation due to one enantiomer is canceled by that of the other.

There are methods that enable the absolute configuration of some compounds to be predicted from the direction of their rotations, but the process is quite complex. Therefore the observation that a compound rotates plane-polarized light will indicate to us only that it is chiral and that one enantiomer is present in excess of the other. If the sample contains only one enantiomer, then the specific rotation can be determined. The specific rotation is a constant that can be used to help identify the compound in the same manner as its melting point or boiling point. For example, the specific rotation of sucrose (table sugar) is $[\alpha]_D = +66.37°$, and that of (−)-2-butanol is $[\alpha]_D = -13.9°$. There is no way for us to tell that the (*l*)-enantiomer of 2-butanol actually has the *R* absolute configuration. In general, there is no relationship between the absolute configuration (*R* or *S*) and the direction of rotation of plane-polarized light (+ or −).

Problem

Problem 6.20   Consider the two enantiomers of 2-pentanol. Explain whether each of these statements is true, is false, or cannot be determined from this information.

   a)   (*R*)-2-Pentanol is a stronger acid than (*S*)-2-pentanol.

   b)   The two enantiomers have different boiling points.

c)   The two enantiomers have identical solubilities in water.

d)   (S)-2-Pentanol rotates plane-polarized light in the counterclockwise direction.

e)   (d)-2-Pentanol rotates plane-polarized light in the clockwise direction.

ELABORATION

The D, L Method for Designating Absolute Configuration

An older method for designating configuration of enantiomers uses the descriptors D and L. In this method the molecule is drawn as illustrated in the following examples:

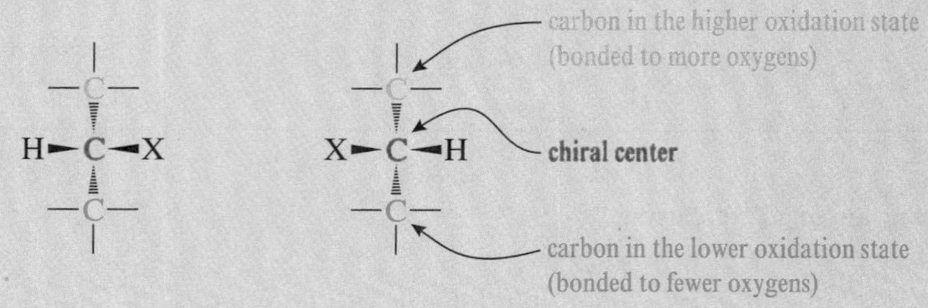

The main chain of carbons containing the chiral center is arranged vertically on the page with the carbon in the higher oxidation state (the one with more C—O bonds) on the top. The vertical bonds extend behind the plane of the page and the horizontal bonds project in front of the plane of the page. When the molecule is arranged in this manner, the enantiomer that has a hydrogen on the left and another group (X in the above drawings) on the right is designated as the D stereoisomer. If the hydrogen is on the right and X is on the left, it is the L stereoisomer.

Because of the difficulties that sometimes occur in assigning which carbon is more highly oxidized or that arise when there is no hydrogen on the chiral center, the D, L method can be ambiguous, and it has been replaced by the R, S method in most organic chemistry applications. However, the D and L descriptors are well-entrenched in biochemistry and biology and are still widely used to indicate the configuration of compounds such as sugars and amino acids. For example, the naturally occurring enantiomer of glyceraldehyde has the D configuration, and most naturally occurring amino acids have the L configuration.

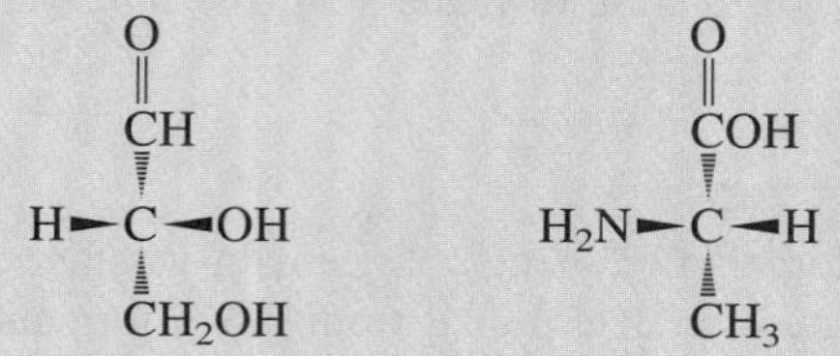

D-glyceraldehyde   L-alanine

Problem

Problem 6.21 Determine whether these compounds have the R or S absolute configuration.

   a) D-glyceraldehyde

   b) L-alanine

In the case of a compound that has several chiral centers, such as sugars, D or L refers only to the configuration of the chiral center closest to the carbon in the lower oxidation state, that is, the bottommost chiral center. For example, these compounds are simple sugars with two chiral centers:

      D-erythrose   D-threose   L-erythrose

If the hydroxy group on C-3, the bottommost chiral center, is on the right, then the compound has the D configuration. D-Erythrose has the hydroxy groups at both C-2 and C-3 on the right side. D-Threose is a diastereomer of D-erythrose. It has the same configuration as D-erythrose at C-3 but has the opposite configuration at C-2. Therefore while the configuration at C-3 in these examples is denoted by D or L, the configuration at C-2 relative to that at C-3 is provided by the name of the diastereomer (which would need to be memorized). In systematic nomenclature, both D-erythrose and D-threose have the same name: 2,3,4-trihydroxybutanal. They would be differentiated by designating the configurations at C-2 and C-3. One is (2R,3R)-2,3,4-trihydroxybutanal, and the other is (2S,3R)-2,3,4-trihydroxybutanal.

Problem

Problem 6.22 Which of these 2,3,4-trihydroxybutanals is D-erythrose and which is D-threose? What is the systematic name for L-erythrose?

It is worth emphasizing again that the terms D and L are used to designate absolute configurations, like the terms R and S. There is no general relationship between D and L and the direction in which a compound rotates plane-polarized light, which is designated by d (+) or l (−). For example, the amino acid L-alanine has a positive rotation, so it is the (d)-enantiomer. In contrast, another amino acid, L-phenylalanine, has a negative rotation and is the (l)-enantiomer.

6.14   Molecules with Multiple Chiral Centers

When a molecule has more than one chiral center, things become somewhat more complicated. Let's consider the simplest case: a molecule with two chiral centers. Each chiral center can have either the *R* or the *S* absolute configuration. There are four possible combinations: *RR*, *SS*, *RS*, and *SR*. The mirror image of a molecule, that is, its enantiomer, has the opposite configuration (inverted configuration) at all chiral centers. Therefore the *RR* and the *SS* stereoisomers are enantiomers, as are the *RS* and *SR* stereoisomers. The *RR* and the *RS* stereoisomers are not mirror images. Such non-mirror-image stereoisomers are called **diastereomers.** They have the opposite configuration at some, but not all, chiral centers.

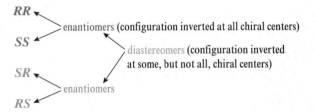

As we saw previously, enantiomers have identical properties unless they are placed in a chiral environment. Diastereomers, on the other hand, are not mirror images and have different properties in all environments. They have different physical properties and different chemical properties.

A similar analysis holds for molecules that have more than two chiral centers. Each chiral center may have either the *R* or the *S* configuration. The number of possible stereoisomers can be calculated by using simple probability theory. A molecule with a number of chiral centers equal to *n* has a maximum of 2^n stereoisomers:

$$\text{maximum number of stereoisomers} = 2^n \qquad (n = \text{number of chiral centers})$$

As an example, consider the case of 2-(methylamino)-1-phenyl-1-propanol:

$$
\begin{array}{ccc}
\text{OH} & \text{NHCH}_3 & \\
| & | & \\
\text{Ph}-\text{CH}-\text{CH}-\text{CH}_3 & & \\
1\ * & 2\ * & 3
\end{array}
$$

Carbon 1 and carbon 2 are both chiral centers, so there are $2^2 = 4$ stereoisomers. The (1*R*, 2*S*)-stereoisomer has a specific rotation of $-6.3°$ and is known as $(-)$-ephedrine. It occurs naturally in the Chinese *ma-huang* plant. It is a bronchodilator and is the decongestant used in many cold remedies. The enantiomer of $(-)$-ephedrine, the (1*S*,2*R*)-stereoisomer, does not occur naturally. It has the same melting point as $(-)$-ephedrine but different physiological properties and rotates plane-polarized light in the positive direction. The racemic mixture, (*d,l*)-ephedrine, packs better into a crystal lattice and has a higher melting point (76°C) than either enantiomer.

R —— Ph
HO—C—H
CH₃HN—C—H
              CH₃
S

S —— Ph
H—C—OH
H—C—NHCH₃
              CH₃
R

(1R,2S)-stereoisomer
(−)-ephedrine
mp: 40°C
$[\alpha]_D$: −6.3°

(1S,2R)-stereoisomer
(+)-ephedrine
mp: 40°C
$[\alpha]_D$: +6.3°

One way to produce a drawing of the enantiomer of (−)-ephedrine is to imagine a mirror placed just to the right of the preceding diagram and perpendicular to the page. When the mirror image of the original structure is drawn, the structure on the right, (+)-ephedrine, is produced. Another way to construct the enantiomer is to interchange two groups on each chiral center. (Recall that interchanging any two groups at a chiral center inverts the configuration at that chiral center.) Thus if the OH and H on carbon 1 of (−)-ephedrine are interchanged and the CH₃NH and H on carbon 2 are interchanged, then the diagram of the enantiomeric (+)-ephedrine on the right is again produced.

To draw a diastereomer of ephedrine, two groups on one chiral center are interchanged, while the configuration at the other chiral center is not changed. For example, if the H and OH of the preceding diagram of (−)-ephedrine are interchanged, the resulting stereoisomer is the (1S,2S)-stereoisomer. It is known as (+)-pseudoephedrine. Note that the two enantiomers of pseudoephedrine have very different physical constants from ephedrine. Their chemical and physiological properties are different also.

R —— Ph
HO—C—H
H—C—NHCH₃
              CH₃
R

S —— Ph
H—C—OH
CH₃HN—C—H
              CH₃
S

(1R,2R)-stereoisomer
(−)-pseudoephedrine
mp: 118°C
$[\alpha]_D$: −52°

(1S,2S)-stereoisomer
(+)-pseudoephedrine
mp: 118°C
$[\alpha]_D$: +52°

Problem

Problem 6.23 Draw all the stereoisomers of 2-bromo-3-chlorobutane and indicate whether they are enantiomers or diastereomers.

As another example, consider the compound cholesterol:

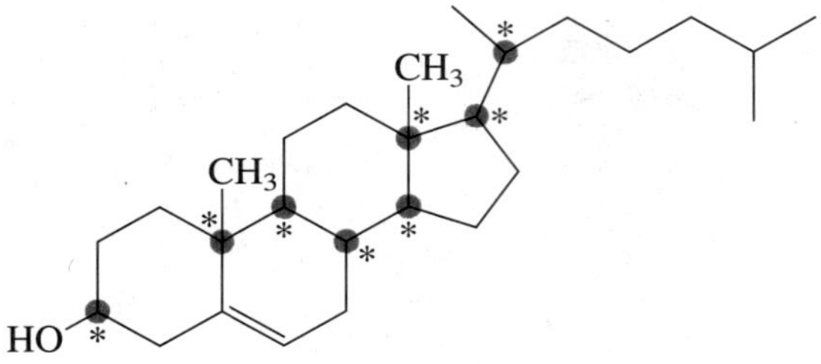

cholesterol

Cholesterol belongs to a class of natural products called steroids, which are characterized by the presence of one five-membered and three six-membered rings fused together in the same pattern as cholesterol. This structure has eight chiral centers, so there are $2^8 = 256$ stereoisomers. Only one of these is cholesterol. Another is the enantiomer of cholesterol, and the other 254 are diastereomers of cholesterol. Of the 256 possibilities, nature produces only one: cholesterol.

Problem

Problem 6.24   Label each chiral center in these compounds with an asterisk and calculate the maximum number of stereoisomers for each.

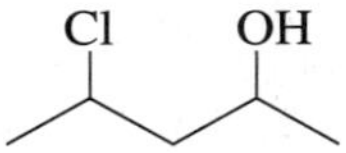

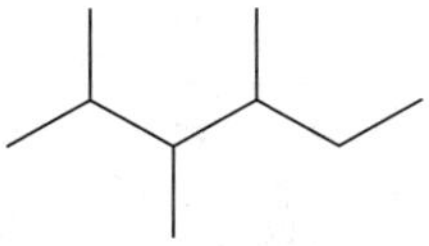

Sometimes there are fewer stereoisomers than predicted by the rule presented above. This occurs when identical chiral centers are symmetrically placed in a compound. As an example, consider the case of tartaric acid.

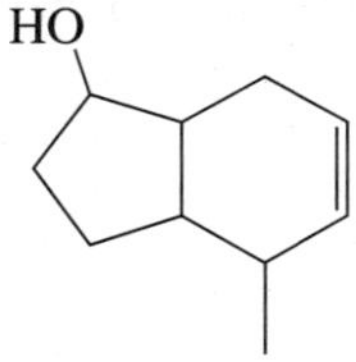

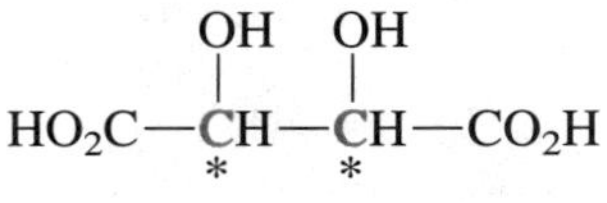

tartaric acid

There are two chiral centers, so the formula predicts a total of four stereoisomers. However, each of the chiral centers has identical groups attached to the carbon, so fewer than four stereoisomers actually exist. The analysis can be conducted in the same manner as was done previously for ephedrine. We start by drawing one of the stereoisomers, the (2R,3R)-isomer for example. Then the mirror image of this, the

(2*S*,3*S*)-stereoisomer is drawn. These two compounds are nonsuperimposable mirror images—enantiomers.

(2*R*,3*R*)-stereoisomer   (2*S*,3*S*)-stereoisomer
(+)-tartaric acid   (−)-tartaric acid
mp: 169–170°C   mp: 169–170°C
$[\alpha]_D$: +12.0° $[\alpha]_D$: −12.0°

(+)-Tartaric acid occurs naturally in fruits and plants. Its monopotassium salt is called cream of tartar and is a component of baking powder. (−)-Tartaric acid is much less common in nature and has been found only in the fruit of a single West African tree.

A diastereomer of these compounds is constructed by interchanging the H and OH on one of the chiral centers, as shown. The mirror image of this compound is also shown.

rotate 180°

plane of symmetry

(2*R*,3*S*)-stereoisomer   (2*S*,3*R*)-stereoisomer
meso-tartaric acid   [identical to the
mp: 159–160°C   (2*R*,3*S*)-stereoisomer]

Careful examination of these two molecules shows that they are, in fact, identical. This can be seen more easily if the (2*S*,3*R*)-stereoisomer is rotated 180° in the plane of the paper as shown. This produces the structure on the far right, which is identical to the (2*R*,3*S*)-stereoisomer on the left. Because the compound is superimposable on its mirror image, it is not chiral and does not rotate plane-polarized light. Another way to determine that this compound is not chiral is to note that it has a plane of symmetry that bisects the C-2—C-3 bond. Compounds such as this one, which contain chiral centers but are not chiral are called ***meso*-stereoisomers.** *meso*-Tartaric acid is human-made and does not occur in nature. Overall, then, tartaric acid has only three stereoisomers: the two enantiomers of the chiral diastereomer (often called the *d,l*-diastereomer) and the *meso*-diastereomer.

Because of its symmetry, numbering can begin at either end of tartaric acid. It is not surprising, then, that (2*R*,3*S*)-tartaric acid is the same as (2*S*,3*R*)-tartaric acid. One of the chiral centers is the mirror image of the other. That is why *meso*-tartaric acid has an internal plane of symmetry. As an everyday example, consider an idealized human figure. Although the figure has chiral parts such as hands and feet, they are present in pairs of left and right enantiomers arranged so that the figure has an internal plane of symmetry. Whenever we encounter a compound that has identical chiral centers, placed symmetrically, we should expect *meso*-stereoisomers to occur, resulting in the total number of stereoisomers being less than that predicted by the 2^n rule.

Problem

Problem 6.25   Draw all of the stereoisomers for 2,3-dichlorobutane. Indicate which rotate plane-polarized light and which are *meso*.

Exercise

Exercise 6.10   Build a model of *meso*-2,3-dichlorobutane. Build a model of its mirror image. Show that these models superimpose. Locate the plane of symmetry in one of the models.

6.15   Stereoisomers and Cyclic Compounds

The *cis*- and *trans*-stereoisomers of cyclic compounds that were presented previously are actually just special cases of the type of stereoisomers that we have just discussed. For example, consider the case of 1,2-dimethylcyclohexane:

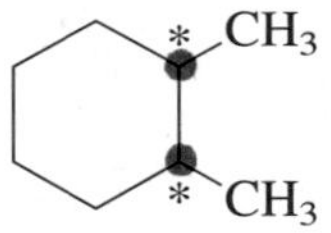

There are two chiral centers, so the formula predicts that there are four stereoisomers. However, since the chiral centers are identical, we expect that there are actually fewer than four stereoisomers. In fact, the analysis is identical to the one we just did for tartaric acid. There are three stereoisomers: a *d,l*-pair of enantiomers and a *meso*-diastereomer.

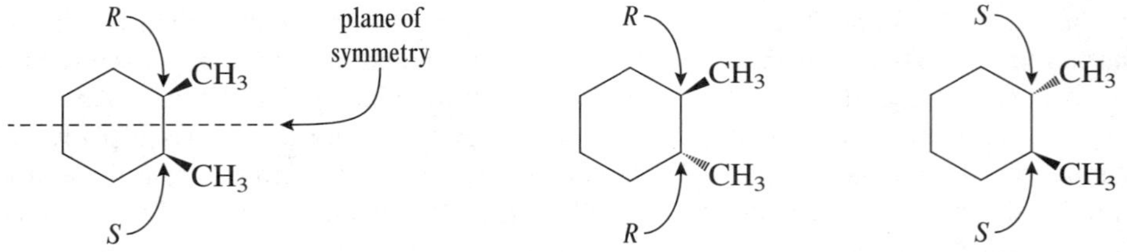

meso-diastereomer
(1*R*,2*S*) or (1*S*,2*R*)
cis-stereoisomer

(1*R*,2*R*)-stereoisomer

(1*S*,2*S*)-stereoisomer

enantiomers of the *trans*-diastereomer

The *cis*-diastereomer is *meso*. It has a plane of symmetry bisecting the ring bond between the two methyl groups. It is not chiral and does not rotate plane-polarized light. The *trans*-diastereomer exists as a pair of enantiomers. These stereochemical differences do not depend on the presence of the ring. In fact, suppose the ring is cleaved at the C—C bond opposite the one connecting the chiral centers. The resulting compound, 3,4-dimethylhexane, also has three stereoisomers: a *meso*-diastereomer and two enantiomers of a *d,l*-diastereomer. It is just easier to see that the *cis*- and *trans*-diastereomers of 1,2-dimethylcyclohexane are different than it is to see that the *meso*- and *d,l*-diastereomers of 3,4-dimethylhexane are different.

meso-3,4-dimethylhexane   (3R,4R)-dimethyl-   (3S,4S)-dimethyl-
                                      hexane   hexane

Although cyclohexane rings have chair shapes, rather than being flat, stereochemical analyses such as the preceding one can be done with drawings using planar rings. (This applies to most other rings also.) This is true because the chair conformations are interconverting rapidly and the "average" shape can be considered to be planar. Careful analysis of one chair conformation of *cis*-1,2-dimethylcyclohexane shows that it is chiral—it is not superimposable on its mirror image. However, the ring-flipped conformation is the enantiomer of the original conformation. Since the conformers interconvert rapidly at room temperature, the compound is not chiral. Again, this result is not unique to ring systems. Some of the conformations of *meso*-3,4-dimethylhexane are also chiral, but there is always an enantiomeric conformation and interconversion between them is rapid. In looking for internal planes of symmetry in such molecules, it is necessary to use the most symmetrical conformation.

Problems

Problem 6.26   Draw all of the stereoisomers of 1,2-dimethylcyclopropane. Explain which rotate plane-polarized light.

Problem 6.27   Explain whether or not these compounds rotate plane-polarized light.

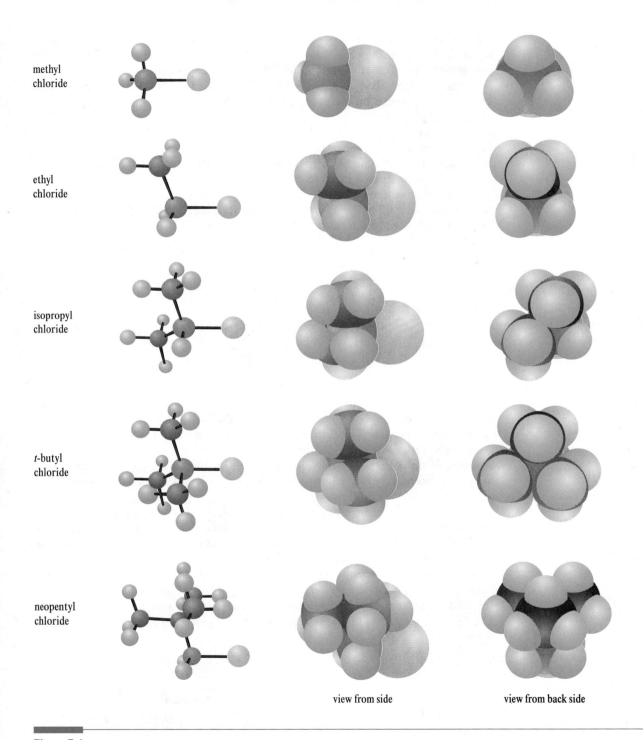

methyl
chloride

ethyl
chloride

isopropyl
chloride

t-butyl
chloride

neopentyl
chloride

view from side   view from back side

Figure 7.4

Some Alkyl Chlorides.

The chlorine is yellow, and the electrophilic carbon is red.

computer-generated space-filling pictures of these molecules, illustrating the increasing difficulty the nucleophile has in approaching the back side of the electrophilic carbon as the number of methyl groups bonded to that carbon increases.

Other primary alkyl groups have effects similar to that of the methyl group. Replacing a hydrogen on the electrophilic carbon of methyl chloride with an ethyl group rather than a methyl group causes only a slightly larger rate decrease (compare the relative rates of ethyl chloride and propyl chloride in Table 7.1). This indicates that, as far as this mechanism is concerned, an ethyl group is only slightly "larger" than a methyl group, a result that is consistent with the axial destabilization energies of these groups discussed in Chapter 6.

To summarize, the rate of the S_N2 reaction is controlled by steric factors at the electrophilic carbon. Steric hindrance slows the reaction. Based on the number of carbon groups attached to that carbon, the reactivity order is

tertiary ≪ secondary < primary < methyl

increasing S_N2 reaction rate

Rules in organic chemistry cannot be followed blindly, for there are often exceptions. Examination of Table 7.1 shows that neopentyl chloride, a primary chloride, reacts 2500 times more *slowly* than isopropyl chloride, a secondary chloride. Closer examination of neopentyl chloride reveals the reason for this discrepancy. While the electrophilic carbon is indeed primary, the group attached to it is an extremely bulky *tert*-butyl group. A single *tert*-butyl group hinders the back-side approach of the nucleophile even more than two methyl groups. Figure 7.4 shows a space-filling picture of this group.

Finally, Table 7.1 lists three primary alkyl chlorides—allyl chloride, benzyl chloride and chloroacetone—that react considerably faster than other primary alkyl chlorides. This increase in reaction rate is due to resonance stabilization of the transition state. Each of these compounds has a pi bond adjacent to the reactive site and forms a transition state that is conjugated. The *p* orbital that develops on the electrophilic carbon in the transition state overlaps with the *p* orbital of the adjacent pi bond. The stabilization due to the conjugated transition state results in a significantly faster reaction. The transition state for the reaction of allyl chloride with a nucleophile is shown as follows:

transition state

Problems

Problem 7.2 Explain which compound has a faster rate of S_N2 reaction.

a)   [structure: cyclohexane ring with H_3C and Cl on top carbon]   or   [structure: cyclohexane ring with H and Cl on top carbon]

b)   [structure: cyclopentane ring with Cl on top carbon and a carbon bearing two CH_3 groups]   or   [structure: cyclopentane ring with Cl on top carbon]

c) $Ph\overset{Cl}{\underset{|}{C}}HCH_3$ or $CH_3\overset{Cl}{\underset{|}{C}}HCH_3$

Problem 7.3 Arrange these compounds in order of decreasing S_N2 reaction rate.

$$CH_3CH_2CH_2CH_2\overset{Cl}{\underset{|}{C}}H_2 \qquad CH_3Cl \qquad CH_3CH_2\overset{CH_3}{\underset{|}{C}}HCH_2Cl \qquad CH_3CH_2CH_2\overset{Cl}{\underset{|}{C}}HCH_3$$

7.7 Unimolecular Nucleophilic Substitution

Now consider the reaction of acetate ion with *tert*-butyl chloride:

$$CH_3\overset{O}{\overset{||}{C}}-\overset{..}{\underset{..}{O}}\!:^{\ominus} \;+\; H_3C-\overset{CH_3}{\underset{\underset{CH_3}{|}}{C}}-Cl \;\xrightarrow{CH_3COH}\; H_3C-\overset{CH_3}{\underset{\underset{CH_3}{|}}{C}}-O-\overset{O}{\overset{||}{C}}CH_3 \;+\; :\overset{..}{\underset{..}{Cl}}\!:^{\ominus}$$

This reaction looks very similar to the reaction of hydroxide ion with methyl chloride presented earlier, but with the negative oxygen of the acetate anion acting as the nucleophile. However, investigation of this reaction in the laboratory has shown that the reaction rate depends only on the concentration of *tert*-butyl chloride (*t*-BuCl). It is totally independent of the concentration of acetate anion. The reaction follows the first-order rate law:

$$\text{rate} = k[t\text{-BuCl}].$$

Since the reaction follows a different rate law from the S_N2 mechanism, it must also proceed by a different mechanism.

The fact that the rate law depends only on the concentration of *tert*-butyl chloride means that only *tert*-butyl chloride is present in the transition state that determines the rate of the reaction. There must be more than one step in the mechanism because the acetate ion must not be involved until after the step with this transition state. Because only one molecule (*tert*-butyl chloride) is present in the step involving the transition state that determines the rate of the reaction, this step is said to be uni-molecular. The reaction is therefore described as a **unimolecular nucleophilic sub-stitution** reaction, or an S_N1 reaction.

The S_N1 mechanism proceeds by the following two steps:

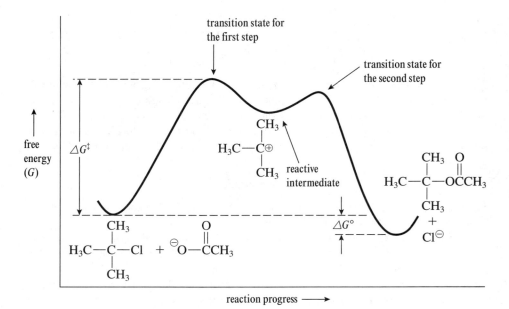

In this mechanism the bond to the chloride is broken in the first step, and the bond to the acetate is formed in the second step. A free energy versus reaction progress diagram for the S_N1 mechanism is shown in Figure 7.5. In this reaction, each of the two steps has an energy maximum or transition state separating its reactant and prod-uct, which are both at energy minima. The transition state for the first step is at higher energy in this case. Once a molecule makes it over the higher-energy barrier of the first step, it has enough energy to proceed rapidly over the lower-energy barrier of the second step. The first step is called **the rate-limiting** or **rate-determining step** because it determines the rate of the reaction. It acts as a kind of bottleneck for the

Figure 7.5

Free Energy versus Reaction Progress Diagram for the S_N1 Reaction of *tert*-Butyl Chloride (2-Chloro-2-methylpropane) and Acetate Anion.

reaction. The rate of this reaction should depend only on the concentration of *tert*-butyl chloride because it is the only molecule involved in the rate-determining step. Therefore this mechanism is consistent with the experimentally determined rate law.

In general, for a nonconcerted reaction, that is, a reaction that proceeds in several steps, the free energy versus reaction progress diagram has a separate transition state for each step. One or more intermediates are present along the reaction pathway, each of these located at an energy minimum. These intermediates may be located at relatively high energy and have only a transient existence, such as the carbocation formed in the S_N1 reaction, or they may be located at lower energy and have a longer lifetime. If one of the transition states is located at significantly higher energy than the others, then that step is the rate-determining step for the reaction. Molecules that become involved in the mechanism after the rate-determining step do not appear in the rate law for the reaction.

Problem

Problem 7.4   Draw a free energy versus reaction progress diagram for a reaction that occurs in two steps with a relatively stable intermediate and in which the transition state for the second step is the highest-energy transition state.

When the chloride ion leaves in the first step of the mechanism, a **reactive intermediate** is formed. This reactive intermediate is a high-energy, reactive species. Under most conditions it has a very short lifetime. However, it differs from a transition state in that it is located at a minimum on the energy curve. It has an activation barrier, although small, that must be surmounted for reaction in either the forward or reverse direction. Although its lifetime is short, it is significantly longer than that of a transition state. It may be possible, under certain circumstances, to obtain experimental observations of a reactive intermediate.

The particular reactive intermediate formed in this reaction is called a **carbocation**. It has a carbon with only three bonds and a positive charge. This carbon has only six electrons in its valence shell and is quite unstable because it does not satisfy the octet rule. It has trigonal planar geometry and sp^2 hybridization at the positively charged carbon:

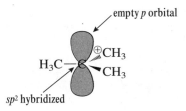

Carbocations are one of the most important types of reactive intermediates in organic chemistry. They are encountered in many reactions in addition to the S_N1 reaction.

Let's now turn our attention to the transition state for this reaction. What is the structure of the transition state? This is an important question because a better understanding of its structure will help in predicting how various factors affect its stability and therefore will aid in predicting how these same factors affect the rate of the reaction. The transition state has a structure that is intermediate between that of the reactant, *tert*-butyl chloride, and that of the product, the *tert*-butyl carbocation. It has the bond between the carbon and the chlorine partially broken and can be represented as shown in the following structures:

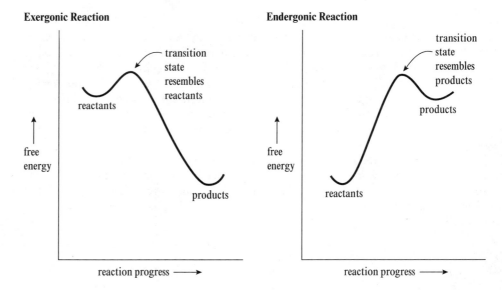

It has a partial positive charge on the carbon and a partial negative charge on the chlorine. The hybridization of the carbon is between that of the reactant, sp^3, and that of the carbocation, sp^2.

Exergonic Reaction

free energy

reaction progress ⟶

transition state resembles reactants

reactants

products

Endergonic Reaction

free energy

reaction progress ⟶

transition state resembles products

products

reactants

Figure 7.6

Using the Hammond Postulate to Predict the Structure of a Transition State.

For an exergonic reaction, in which the reactants are at higher energy than the products, the energy of the transition state is closer to that of the reactants than that of the products. Therefore the structure of the transition state resembles that of the reactants more than that of the products also. If a bond is forming in the reaction, that bond is less than half formed in the transition state, and if a bond is breaking, it is less than half broken.

For an endergonic reaction, in which the products are at higher energy than the reactants, the energy of the transition state is closer to that of the products. Therefore its structure is also closer to that of the products. Any bonds that are forming in the reaction are more than half formed, and any bonds that are breaking are more than half broken.

Is the bond in the transition state more or less than half broken? The Hammond postulate enables questions such as this to be answered. It states that the structure of the transition state for a reaction step is more similar to that of the species (reactant or product of that step) to which it is closer in energy. If the product of the step is higher in energy than the reactant, the structure of the transition state is more similar to that of the product than it is to that of the reactant. In contrast, if the reactant is higher in energy than the product, the structure of the transition state is more similar to that of the reactant (see Figure 7.6). Since the carbocation is much higher in energy than the starting alkyl halide (the slow step of the mechanism in Figure 7.5 corresponds to the case on the right in Figure 7.6), the structure of the transition state for the S_N1 reaction is closer to that of the carbocation; the bond is more than half broken.

Problem

Problem 7.5   Consider the free energy versus reaction progress diagram for the S_N2 reaction shown in Figure 7.1. Does the transition state for this reaction have the C—Cl bond less than half broken, approximately half broken, or more than half broken?

Solution

Although the reaction is slightly exergonic, there is not much difference in energy between the reactants and products. Therefore the C—Cl bond should be approximately half broken in the transition state.

7.8   Stereochemistry of the S_N1 Reaction

What happens in the S_N1 reaction if the leaving group is attached to a chiral carbon? A common result for the S_N1 reaction is racemization; that is, the product is formed with 50% inversion and 50% retention of configuration. An example, the reaction of (S)-1-chloro-1-phenylethane with water to give racemic 1-phenyl-1-ethanol, is illustrated in Figure 7.7. In this reaction the stereochemical integrity of the reactant is randomized on the pathway to the product. This usually means that there is some intermediate along the reaction pathway that is not chiral. In the case of the S_N1 reaction the carbocation intermediate is sp^2 hybridized and has trigonal planar geometry. Since planar carbons are not chiral, this explains why the reaction results in racemization.

Although many S_N1 reactions proceed with racemization, many others result in more inversion of configuration in the product than retention. This is a result of the extremely short lifetime of the carbocation. When the carbocation is first formed, the leaving group is still present on the side of the carbocation where it was originally attached, as shown in Figure 7.8. This species is called an **ion pair.** If the nucleophile attacks the ion pair, the leaving group is still blocking the front side of the carbocation and inversion is favored. After the leaving group has had time to diffuse away, generating a "free" carbocation, the nucleophile can attack equally well from either

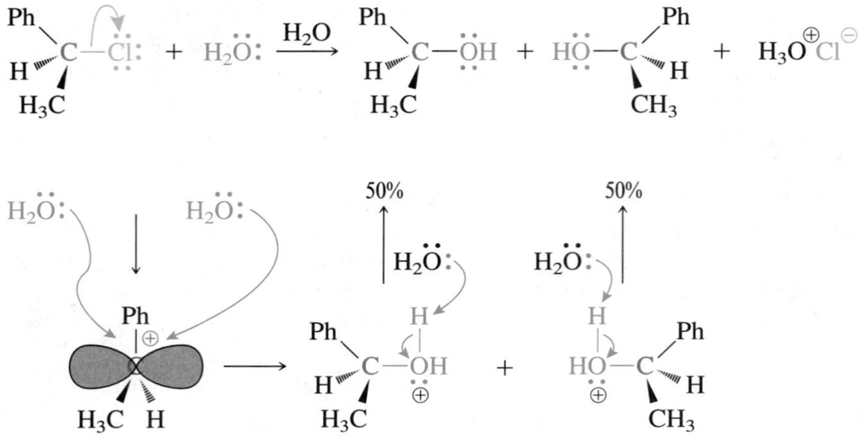

The carbocation is *sp²* hybridized and planar. It is not chiral, so the products formed from it must be racemic. The nucleophile—water in this case—can approach equally well from either side, resulting in the formation of equal amounts of the *R* and *S* enantiomers of the product. Note that it is water, not hydroxide ion, that acts as a nucleophile because the concentration of hydroxide ion in neutral water is extremely low.

When water acts as a nucleophile, the initial product is the conjugate acid of the alcohol. In the final step of the reaction a proton is removed from the protonated alcohol by a base in the solution. This base is often a molecule of the solvent—water in this case.

Figure 7.7

Stereochemistry of the S$_N$1 Reaction of (*S*)-1-Chloro-1-phenylethane in Aqueous Solution.

side, and equal amounts of inversion and retention result. As the lifetime of the carbocation increases, it will more likely reach the free stage, resulting in more complete racemization. The lifetime of the carbocation increases as its stability increases and also depends on the nucleophile and the solvent that are used in a particular reaction. The change of nucleophile and solvent is why the reaction of 1-chloro-1-phenylethane in water (Figure 7.7) gives a different stereochemical result than the reaction of the same compound in acetic acid containing potassium acetate (Figure 7.8). The carbocation has a longer lifetime under the reaction conditions of Figure 7.7 than under those of Figure 7.8 (see problem 7.14), allowing the chloride ion time to diffuse away before the nucleophile attacks, resulting in the formation of a racemic product. The shorter-lived carbocation of Figure 7.8 reacts partly at the ion-pair stage, resulting in more inversion than retention. In summary, S$_N$1 reactions occur with racemization, often accompanied by some excess inversion. We will not attempt to predict the exact amount of each enantiomer that is produced.

The reaction scheme and figures:

Ph–C(H)(CH$_3$)–Cl: → (CH$_3$CO$_2^\ominus$ / CH$_3$CO$_2$H) → CH$_3$CO–C(Ph)(H)(CH$_3$) 58% + H–C(Ph)(CH$_3$)–OCCH$_3$ 42%

16% / 84%

CH$_3$C–O$^\ominus$ C$^\oplus$(H$_3$C)(H) :Cl$^\ominus$
ion-pair

84% →

Ph–C$^\oplus$(H$_3$C)(H)   free carbocation

42% / 42%

The initial species formed in this S$_N$1 reaction, a carbocation with the chloride anion situated directly adjacent, is called an **ion pair.** There is a competition between reaction of the nucleophile with the ion pair and diffusion of the anion away from the carbocation. Under this particular set of reaction conditions, 16% of the ion pairs react with the nucleophile, acetate ion. Because the chloride is blocking one side of the ion pair, acetate can approach only from the opposite side. Therefore only the product with inverted configuration is formed when the nucleophile reacts at this stage.

For the other 84% of the ion pairs, the chloride ion diffuses away before the nucleophile attacks. The nucleophile can approach the resulting free carbocation equally well from either side, resulting in the formation of equal amounts of product with retained and inverted configuration (42% of each). The final result is 42% retention and 16% + 42% = 58% inversion. The reaction has occurred with racemization and some excess inversion.

Figure 7.8

Stereochemistry of the S$_N$1 Reaction of (S)-1-Chloro-1-phenylethane in Acetic Acid Containing Potassium Acetate.

Problem

Problem 7.6 Show the products, including stereochemistry, of these S$_N$1 reactions.

a) CH$_3$CH$_2$––C(CH$_3$)(Ph)–Cl + CH$_3$OH ⟶

b) H$_3$C–[cyclopentane]–CH$_3$, Cl + H$_2$O ⟶

c) H$_3$C–C(CH$_3$CH$_2$)(H)–CH$_2$–C(CH$_3$)$_2$–Cl + CH$_3$COH ⟶

Solution

b) Because a planar carbocation is formed at the reaction center, the nucleophile, H_2O, can bond to either side of the cyclopentane ring.

Because the two products, one with the methyl groups *cis* and the other with the methyl groups *trans*, are diastereomers, they have different energies and therefore are not necessarily formed in equal amounts even if all of the carbocation makes it to the "free" stage. However, their stabilities are not very different, so similar amounts of both are produced.

7.9 Effect of Substituents on the Rate of the S_N1 Reaction

How do the other groups bonded to the electrophilic carbon affect the rate of the S_N1 reaction? Table 7.2 lists the relative rates of the S_N1 reaction for a number of compounds, compared to the rate for isopropyl chloride taken as 1. Methyl chloride and ethyl chloride are not listed in the table because *methyl and simple primary alkyl chlorides do not react by the S_N1 mechanism.* Even under the most favorable S_N1 conditions, these unhindered compounds react by the S_N2 mechanism.

For the S_N1 reaction, formation of the carbocation is the rate-limiting step. We have already seen that the transition state for this step resembles the carbocation. Any

Table 7.2   Relative Rates of S_N1 Reactions for Selected Compounds.

Name	Structure	Relative Rate
isopropyl chloride	CH_3CH-Cl with CH_3 above	1
tert-butyl chloride	CH_3C-Cl with CH_3 above and CH_3 below	1×10^5
allyl chloride	$CH_2=CHCH_2-Cl$	3
benzyl chloride	$PhCH_2-Cl$	30
diphenylmethyl chloride	Ph_2CH-Cl	1×10^4
triphenylmethyl chloride	Ph_3C-Cl	1×10^9

change that makes the carbocation more stable will also make the transition state more stable, resulting in a faster reaction. Carbocation stability controls the rate of the S$_N$1 reaction. Many studies have provided the following order of carbocation stabilities:

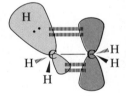

$$R-\overset{\displaystyle R}{\underset{\displaystyle R}{C}}\oplus \qquad R-\overset{\displaystyle R}{\underset{\displaystyle H}{C}}\oplus \qquad R-\overset{\displaystyle H}{\underset{\displaystyle H}{C}}\oplus \qquad H-\overset{\displaystyle H}{\underset{\displaystyle H}{C}}\oplus$$

tertiary   >   secondary   >   primary   >   methyl

←————————————————————

increasing carbocation stability

This stability order is important to remember because carbocations occur as intermediates in several other reactions.

This order shows that the substitution of a methyl group for a hydrogen on a carbocation results in considerable stabilization. (The substitution of other alkyl groups provides a similar stabilization.) This is due to the overlap of a sigma bonding MO from the adjacent carbon with the empty *p* orbital of the carbocation. This overlap forms a conjugated system and allows electron density to flow from the sigma bond to the electron-deficient carbon. This overlap can be illustrated for the ethyl cation as follows:

The sigma MO and the empty *p* AO are coplanar, so they overlap in a manner similar to a pi bond, even though they are not parallel. This overlap provides a path for the electrons of the sigma bond to be delocalized into the empty *p* orbital, thus helping to stabilize the carbocation. Other kinds of sigma bonds can interact with an empty *p* orbital in a similar fashion, as long as the sigma MO and the *p* AO are on adjacent carbons. Such a stabilizing interaction is termed **hyperconjugation.**

Further examination of Table 7.2 shows that allyl chloride and benzyl chloride have much faster rates for S$_N$1 reactions than would be expected for primary systems. Examination of the carbocations reveals that the reason for this enhanced reactivity is the significant resonance stabilization provided by the adjacent double bond or benzene ring. Resonance stabilization increases with the substitution of additional phenyl groups, as illustrated by the reaction rates of di- and triphenylmethyl chloride.

$$CH_2{=}CH{-}\overset{\oplus}{CH_2} \quad \longleftrightarrow \quad \overset{\oplus}{CH_2}{-}CH{=}CH_2$$

The Triphenylmethyl Carbocation

Most carbocations are quite unstable and have only a fleeting existence as intermediates in reactions such as the S_N1 substitution. However, some, such as the triphenylmethyl carbocation, are stable enough that they can exist in significant concentrations in solution or even can be isolated as salts.

When triphenylmethanol is dissolved in concentrated sulfuric acid, a solution with an intense yellow color is formed. The yellow species is the triphenylmethyl carbocation, formed by the following reaction:

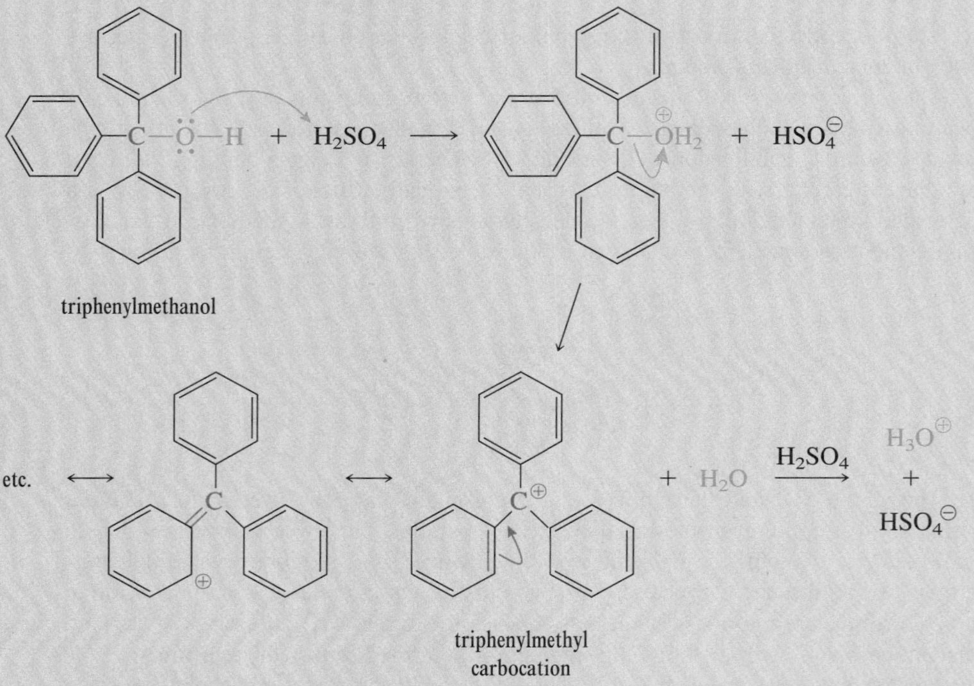

triphenylmethanol

triphenylmethyl
carbocation

The extensive resonance stabilization of this carbocation allows it to exist in solution as long as there is no good nucleophile around to react with it.

Other examples of the stability of this carbocation abound. Triphenylmethyl chloride forms conducting solutions in liquid sulfur dioxide because of cleavage of the carbon–chlorine bond (the first step of an S_N1 reaction):

If the anion is not very nucleophilic, solid salts containing the triphenylmethyl carbocation can actually be isolated. Thus the tetrafluoroborate salt, $Ph_3C^{\oplus}$ $BF_4^{\ominus}$, can be isolated and stored for years as a stable ionic solid and is even commercially available. The geometry of perchlorate salt, $Ph_3C^{\oplus}$ $ClO_4^{\ominus}$, has been determined by X-ray crystallography. The central carbon has planar geometry as expected for an sp^2 hybridized carbocation but the rings are twisted out of the plane because of the severe steric crowding that would occur if they were all planar, so the cation has a shape that resembles a propeller.

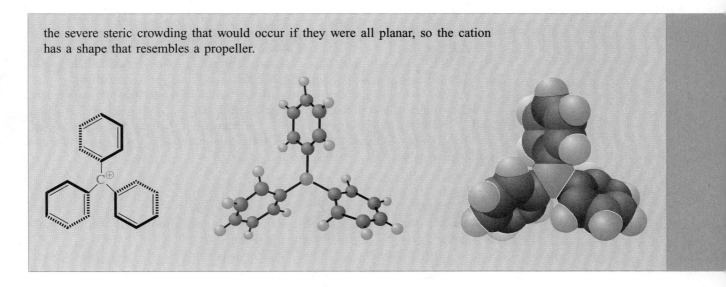

Problems

Problem 7.7 Explain which compound has a faster rate of S_N1 reaction.

a) [H₃C, Cl cyclopentane structure] or [Cl cyclopentane structure]

b) [Cl cyclopentene structure] or [Cl cyclopentene structure]

c) [CH₂=CH–CH(Cl)–CH₃ structure] or [(CH₃)₂CH–CH(Cl) structure]

d) [CH₂Cl benzyl structure] or [CH₂Cl para-OCH₃ benzyl structure]

Problem 7.8 Arrange these compounds in order of decreasing S_N1 reaction rate.

[structures: with Ph and Cl; with Cl; with Cl; with Ph, Cl, and Ph]

7.10   Leaving Groups

The bond to the leaving group is broken during the rate-determining step in both the S_N1 and S_N2 reactions. Therefore the structure of the leaving group affects the rates of both of these reactions. Although the only leaving group we have seen so far is chloride, there are others that can be used. In general, the more stable the leaving group is as a free species—that is, after it has left—the faster it will leave. This stability is also reflected in the basicity of the species: The more stable it is, the weaker base it is. In general, the leaving groups that are used in the S_N1 and the S_N2 reactions are weak bases. Table 7.3 lists the most important leaving groups and provides their relative reaction rates in an S_N1 reaction. Similar rate effects are found for S_N2 reactions.

As can be seen from Table 7.3, the leaving ability of the halides increases as one goes down the column of the periodic table; that is, $Cl^{\ominus}$ is the slowest, $Br^{\ominus}$ is faster, and $I^{\ominus}$ is the fastest. This order parallels the decrease in basicity that occurs as one proceeds down a column of the periodic table. Fluoride ion ($F^{\ominus}$) is so slow that it is not commonly used as a leaving group.

Problem

Problem 7.9 Explain whether these reactions would follow the S_N1 or the S_N2 mechanism and then explain which reaction is faster.

a) $\underset{\substack{Cl \\ |}}{CH_3CH_2} + {}^{\ominus}OH \xrightarrow[CH_3OH]{H_2O}$ or $\underset{\substack{O-SO_2CH_3 \\ |}}{CH_3CH_2} + {}^{\ominus}OH \xrightarrow[CH_3OH]{H_2O}$

b) $\underset{\substack{CH_3 \\ | \\ CH_3}}{CH_3C-Br} + CH_3\overset{O}{\overset{||}{C}}O^{\ominus} \xrightarrow{CH_3CO_2H}$ or $\underset{\substack{CH_3 \\ | \\ CH_3}}{CH_3C-I} + CH_3\overset{O}{\overset{||}{C}}O^{\ominus} \xrightarrow{CH_3CO_2H}$

Because alcohols are very common and easily prepared by a variety of methods, it would be very useful to be able to use OH as a leaving group in nucleophilic substitution reactions. However, $OH^{\ominus}$ is much too basic to act as a leaving group in S_N1 and S_N2 reactions. It is necessary to modify the OH, converting it to a better leaving group, to use alcohols as substrates for these reactions. Several methods have been developed to accomplish this goal. If the reaction of the alcohol is conducted in acidic solution, the oxygen becomes protonated, producing $ROH_2^{\oplus}$. The leaving group is now water, which is comparable to bromide in reactivity. (Of course, to use this leaving group, the nucleophile must be stable in acidic solution.) An example is provided in the following equation:

$$CH_3-\underset{\substack{CH_3 \\ | \\ CH_3}}{\overset{CH_3}{C}}-\ddot{O}H + H-\ddot{Cl}: \longrightarrow CH_3-\underset{\substack{| \\ CH_3}}{\overset{CH_3}{C}}-\overset{\oplus}{\ddot{O}}H_2 \longrightarrow CH_3-\underset{\substack{| \\ CH_3}}{\overset{CH_3}{\overset{\oplus}{C}}} \xrightarrow{:\ddot{Cl}:^{\ominus}} CH_3-\underset{\substack{| \\ CH_3}}{\overset{CH_3}{C}}-\ddot{Cl}: + H_2\ddot{O}:$$

Another method that can be used to transform the hydroxy group of an alcohol into a leaving group is to replace the hydrogen with some other group that significantly decreases the basicity of the oxygen. A group that is commonly used for this purpose is the SO_2R group. Replacing the hydrogen of the alcohol with this group produces a sulfonate ester, such as the mesylate or tosylate ester shown in Table 7.3. As can be seen by their resemblance to the bisulfate anion (the conjugate base of sul-

Table 7.3 Approximate Reactivities of Some Important Leaving Groups.

Structure	Leaving Group	Name	Relative Reactivity
R—Cl	Cl$^{\ominus}$	chloride	1
R—Br	Br$^{\ominus}$	bromide	10
R—O(H)(H)$^{\oplus}$	OH$_2$	water	10
R—I	I$^{\ominus}$	iodide	10^2
R—O—S(=O)(=O)—CH$_3$	$^{\ominus}$O—S(=O)(=O)—CH$_3$	mesylate (methanesulfonate)	10^4
R—O—S(=O)(=O)—C$_6$H$_4$—CH$_3$	$^{\ominus}$O—S(=O)(=O)—C$_6$H$_4$—CH$_3$	tosylate (*p*-toluenesulfonate)	10^4

furic acid), sulfonate anions are weak bases and excellent leaving groups.

$$^{\ominus}O-\overset{\overset{O}{\|}}{\underset{\underset{O}{\|}}{S}}-R \qquad ^{\ominus}O-\overset{\overset{O}{\|}}{\underset{\underset{O}{\|}}{S}}-OH$$

a sulfonate   bisulfate
anion   anion

Problem

Problem 7.10

a) Show all the steps in the mechanism for this reaction. Don't forget to use curved arrows to show the movement of electrons in each step of the mechanism.

$$\text{Ph}-\overset{\overset{CH_3}{|}}{\underset{\underset{CH_3}{|}}{C}}-\text{OH} \ + \ \text{HI} \ \longrightarrow \ \text{Ph}-\overset{\overset{CH_3}{|}}{\underset{\underset{CH_3}{|}}{C}}-\text{I} \ + \ \text{H}_2\text{O}$$

b) Show a free energy versus reaction progress diagram for this reaction.

b)

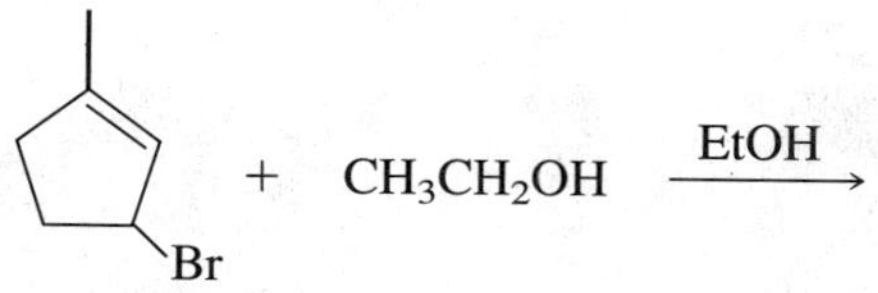

+ CH_3CH_2OH $\xrightarrow{EtOH}$

c)

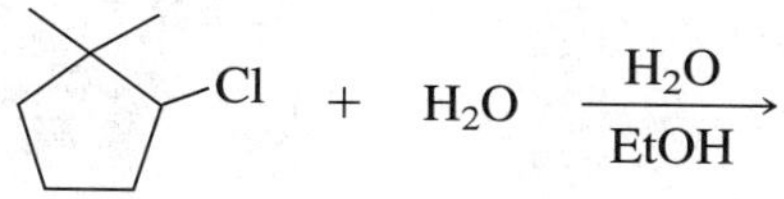

—Cl + H_2O $\xrightarrow[EtOH]{H_2O}$

Problem 7.21

a) Show all of the steps in the mechanism for this reaction. Don't forget to use curved arrows to show the movement of electrons in each step of the mechanism.

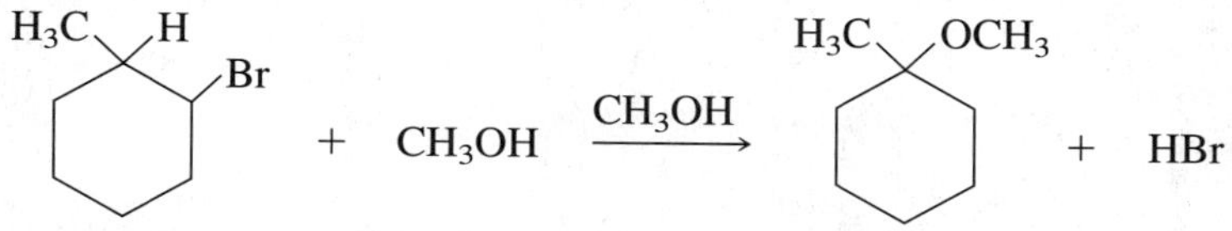

+ CH_3OH $\xrightarrow{CH_3OH}$

+ HBr

b) Show a free energy versus reaction progress diagram for this reaction.

7.16   Summary

Table 7.6 provides a summary of the most important features of the S_N1 and S_N2 reactions.

After completing this chapter you should be able to:

1. Write mechanisms for the S_N1 and S_N2 reactions.
2. Recognize the various nucleophiles and leaving groups and understand the factors that control their reactivities.
3. Understand the factors that control the rates of these reactions, such as steric effects, carbocation stabilities, the nucleophile, the leaving group, and solvent effects.
4. Be able to use these factors to predict whether a particular reaction will proceed by an S_N1 or an S_N2 mechanism and to predict what effect a change in reaction conditions will have on the reaction rate.
5. Show the products of any substitution reaction.
6. Show the stereochemistry of the products.
7. Recognize when a carbocation rearrangement is likely to occur and show the products expected from the rearrangement.
8. Show the structures of the products that result from the elimination reactions that compete with the substitution reactions.

Table 7.6   Summary of the S_N1 and S_N2 Reactions.

	S_N1	S_N2
Mechanism	two step $R—L \longrightarrow R^{\oplus} \longrightarrow R—Nu$	one step $R—L + Nu^{\ominus} \longrightarrow R—Nu + L^{\ominus}$
Kinetics	first-order rate $= k[R—L]$	second-order rate $= k[R—L][Nu^{\ominus}]$
Effects of Nucleophile	no effect on rate	stronger nucleophiles cause faster rate
Effect of Carbon Structure	tertiary > secondary resonance stabilization of $R^{\oplus}$ important	methyl > primary > secondary
Stereo-chemistry	racemization (possible excess inversion)	inversion
Effect of Solvent	favored by polar solvents	favored by aprotic solvents
Competing Reactions	elimination rearrangement	elimination

End-of-Chapter Problems

7.22   Which of these compounds would have a faster rate of S_N2 reaction?

a)   /\/\Br   or   /\/\I

b)   (allyl Cl)   or   (sec Cl)

c)   /\/\Br   or   /\/\Br (branched)

d)   /\/\Br (sec)   or   /\/\Br (branched)

e)   (ketone Cl)   or   /\/\Cl

7.23 Which of these compounds would have a faster rate of S$_N$1 reaction?

a)   (structure with Cl)   or   (structure with Cl)

b)   ⟶Cl   or   ⟶I

c)   Ph / Ph—Br with Ph   or   Ph / Ph—Br with CH$_3$

d)   (cyclohexene with Br)   or   (cyclohexane with Br)

e)   (ketone with Cl)   or   (structure with Cl)

7.24 Arrange these compounds in order of increasing S$_N$2 reaction rate.

(four structures: one with Cl, one with Br, one with Cl, one ketone with Br)

7.25 Arrange these compounds in order of increasing S$_N$1 reaction rate.

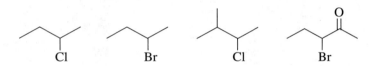

7.26 Show the products of these reactions and explain whether each would follow an S$_N$1 or an S$_N$2 mechanism.

a)   (structure with Cl)   + $^{\ominus}$OH $\xrightarrow{\text{DMF}}$

b)   (structure with Cl)   +   H$_2$O $\xrightarrow{\text{CH}_3\text{OH}}$

c)   ⟶Br   + $^{\ominus}$SH $\xrightarrow[\text{H}_2\text{O}]{\text{CH}_3\text{OH}}$

d)   (structure)—OTs   +   CH$_3$O$^{\ominus}$ $\xrightarrow{\text{CH}_3\text{OH}}$

e)   (structure)—Br   +   H$_2$O $\xrightarrow[\text{H}_2\text{O}]{\text{CH}_3\text{OH}}$

f)   (structure)—Br   +   CH$_3$CO$^{\ominus}$ (with O) $\xrightarrow{\text{DMSO}}$

7.27   Show the substitution products for these reactions. Don't forget to show the stereochemistry of the product, where appropriate.

a) $CH_3CH_2\text{·····}\underset{CH_3CH_2CH_2}{\overset{I}{\diagdown}}C\diagdown_H$ + $CH_3\overset{O}{\overset{\|}{C}}-\overset{..}{\underset{..}{O}}:^{\ominus}$ $\xrightarrow{\text{DMF}}$

b) $\underset{Ph}{\overset{CH_3CH_2\text{····}}{\diagdown}}C\diagup_{\displaystyle \overset{CH_3}{\underset{Br}{}}}$ + CH_3CH_2OH $\xrightarrow{\text{EtOH}}$

c) [structure: cyclohexane with Ph and Cl at top carbon, CH$_3$ at bottom]   + H_2O $\xrightarrow[\text{EtOH}]{H_2O}$

d) $\diagup\!\!\diagdown\!\!\diagup Cl$ + $:\overset{\ominus}{C}\equiv N:$ $\xrightarrow{\text{DMSO}}$

e) [structure: cyclopentane with OTs at top, CH$_3$ at bottom]   + $:\overset{\ominus..}{S}-CH_3$ $\xrightarrow{\text{acetone}}$

f)   Br$\diagdown\!\!\diagup\!\!\diagdown\!\!\diagup\!\!\diagdown$Cl + 1 $:\overset{\ominus..}{\underset{..}{O}}CH_3$ $\xrightarrow{CH_3OH}$

g) [structure: isopropyl group]$\diagdown$Br   + $CH_3\overset{O}{\overset{\|}{C}}-\overset{..}{\underset{..}{O}}:^{\ominus}$ CH_3CO_2H $\longrightarrow$

h) $CH_3-\underset{\underset{\displaystyle CH_3}{|}}{\overset{\overset{\displaystyle CH_3}{|}}{C}}-\underset{\overset{|}{Br}}{CH}-CH_2CH_3$ + CH_3OH $\xrightarrow{CH_3OH}$

i) [structure: cyclopentane with H$_3$C and OH at top carbon]   +   HCl $\xrightarrow{H_2O}$

j) $\diagup\!\!\diagdown\!\!\diagup$I   + NH_3 $\xrightarrow{\underset{CH_3OH}{H_2O}}$

k) $\underset{Cl}{\overset{Ph}{\diagdown}}C\overset{H}{\underset{CH_3}{\diagup}}$ + $:\overset{\ominus..}{S}H$ $\xrightarrow{\text{DMSO}}$

l)

$$\text{(CH}_3\text{CHCH}_2\text{CH}_3\text{)} \quad + \quad \text{Cl}-\overset{\overset{\displaystyle O}{\|}}{\underset{\underset{\displaystyle O}{\|}}{S}}-\text{CH}_3 \quad \xrightarrow{\text{pyridine}} \quad \xrightarrow[\text{DMF}]{\text{NaCl}}$$

(with OH on the carbon)

m) $CH_3CH_2CH_2CH_2Br \quad + \quad NaI \quad \xrightarrow{\text{acetone}}$

7.28 Explain whether each pair of reactions should follow an S_N1 or an S_N2 mechanism. Then explain which member of the pair should proceed at a faster rate.

a) $\longrightarrow\!\!\!\!\!\!\overset{|}{\underset{|}{C}}\!\!-Cl \; + \; CH_3OH \; \xrightarrow{CH_3OH} \;$ or $\; \longrightarrow\!\!\!\!\!\!\overset{|}{\underset{|}{C}}\!\!-Br \; + \; CH_3OH \; \xrightarrow{CH_3OH}$

b) $\diagdown\!\!\!\diagup\!\!\!\diagdown Br \; + \; {}^{\ominus}CN \; \xrightarrow{\text{acetone}} \;$ or $\; \diagdown\!\!\!\diagup\!\!\!\diagdown Br \; + \; {}^{\ominus}CN \; \xrightarrow{\text{methanol}}$

c)

$$\underset{\text{(phenyl)}}{\overset{\overset{\displaystyle OTs}{|}}{\underset{\underset{\displaystyle CHCH_3}{}}{}}} \; + \; CH_3CH_2OH \; \xrightarrow{EtOH} \;$$ or $$\underset{\text{(p-tolyl)}}{\overset{\overset{\displaystyle OTs}{|}}{\underset{\underset{\displaystyle CHCH_3}{}}{}}} \; + \; CH_3CH_2OH \; \xrightarrow{EtOH} \quad (CH_3)$$

d) $\overset{}{\diagdown\!\!\!\diagup}\!\!\diagdown I \; + \; CH_3S^{\ominus} \; \xrightarrow{CH_3OH} \;$ or $\; \diagdown\!\!\!\diagup\!\!\diagdown I \; + \; CH_3S^{\ominus} \; \xrightarrow{CH_3OH}$

e) $\diagup\!\!=\!\!\diagdown\!\!\!\diagup Br \; + \; CH_3OH \; \xrightarrow{CH_3OH} \;$ or $\; \diagdown\!\!\!\diagup\!\!\diagdown\!\!\!Br \; + \; CH_3OH \; \xrightarrow{CH_3OH}$

f) $\diagdown\!\!\!\diagup\!\!\!\diagdown OMs \; + \; {}^{\ominus}OH \; \xrightarrow{DMF} \;$ or $\; \diagdown\!\!\!\diagup\!\!\!\diagdown OMs \; + \; CH_3\overset{\overset{\displaystyle O}{\|}}{C}O^{\ominus} \; \xrightarrow{DMF}$

g) $\overset{\overset{\displaystyle Br}{|}}{\diagup\!\!\!\diagdown} \; + \; CH_3OH \; \xrightarrow{CH_3OH} \;$ or $\; \overset{\overset{\displaystyle Br}{|}}{\diagup\!\!\!\diagdown} \; + \; CH_3CH_2OH \; \xrightarrow{CH_3CH_2OH}$

h) $\overset{\ominus}{:\!\ddot{O}}\diagdown\!\!\!\diagup\!\!\!\diagdown Cl \; \xrightarrow{CH_3OH} \;$ or $\; \diagdown\!\!\!\diagup\!\!\!\diagdown Cl \; + \; CH_3O^{\ominus} \; \xrightarrow{CH_3OH}$

i) $\diagdown\!\!\!\diagup\!\!\!\diagdown Br \; + \; \underset{\ddot{N}}{\bigcirc} \; \xrightarrow{EtOH} \;$ or $\; \diagdown\!\!\!\diagup\!\!\!\diagdown Br \; + \; \underset{\ddot{N}}{\bigcirc}CH_3 \; \xrightarrow{EtOH}$

7.29   Show all of the steps in the mechanisms of these reactions.

a) $CH_3-\underset{\underset{CH_3}{|}}{\overset{\overset{CH_3}{|}}{C}}-Br$ + CH_3CH_2OH $\xrightarrow{\text{EtOH}}$ $CH_3-\underset{\underset{CH_3}{|}}{\overset{\overset{CH_3}{|}}{C}}-OCH_2CH_3$ + $CH_3CH_2\overset{\oplus}{O}H_2$ $\overset{\ominus}{Br}$

b) $CH_3CH_2CH_2\overset{\overset{Cl}{|}}{C}H_2$ + $CH_3O^{\ominus}$ $\xrightarrow{\text{DMSO}}$ $CH_3CH_2CH_2\overset{\overset{OCH_3}{|}}{C}H_2$ + $Cl^{\ominus}$

c) $CH_3CH_2\underset{\underset{CH_3}{|}}{\overset{\overset{OH}{|}}{C}}CH_3$ + $H_3O^{\oplus}$ $\overset{\ominus}{Br}$ $\xrightarrow{\text{H}_2\text{O}}$ $CH_3CH_2\underset{\underset{CH_3}{|}}{\overset{\overset{Br}{|}}{C}}CH_3$ + $2\ H_2O$

7.30   This reaction gives three substitution products (not counting *E/Z* isomers). Show the structures of these products and show the mechanism for their formation.

$+\ H_2O$ $\xrightarrow[\text{CH}_3\text{OH}]{\text{H}_2\text{O}}$

7.31   Explain why the trifluorosulfonate anion is a better leaving group than the mesylate anion.

$\overset{\ominus}{O}-\overset{\overset{O}{\|}}{\underset{\underset{O}{\|}}{S}}-CF_3$ trifluorosulfonate anion

7.32   When benzyl tosylate is heated in methanol, the product is benzyl methyl ether. When bromide ion is added to the reaction, the reaction proceeds at exactly the same rate, but the product is now benzyl bromide. Explain.

$PhCH_2-OTs$ $\xrightarrow{\text{CH}_3\text{OH}}$ $PhCH_2-OCH_3$

$\xrightarrow[\text{CH}_3\text{OH}]{Br^{\ominus}}$ $PhCH_2-Br$

7.33   The substitution reaction of bromomethane with hydroxide ion proceeds about 5000 times faster than the reaction of bromomethane with water. However, the substitution reaction of 2-bromo-2-methylpropane proceeds at the same rate with both of these nucleophiles. Explain.

7.34   Explain why this primary halide reacts very rapidly under conditions that are favorable for the S_N1 mechanism.

$$CH_3-O-CH_2-Cl$$

7.35 Ethers can be cleaved by treatment with strong acids. Show all of the steps in the mechanism for this reaction and explain why these products are formed rather than iodomethane and 2-methyl-2-butanol.

$$\underset{\underset{CH_3}{|}}{\overset{\overset{CH_3}{|}}{CH_3-C-O-CH_3}} \; + \; H-I \; \xrightarrow{\;H_2O\;} \; \underset{\underset{CH_3}{|}}{\overset{\overset{CH_3}{|}}{CH_3-C-I}} \; + \; H-O-CH_3$$

7.36 Which of these halides would give a precipitate more rapidly when reacted with $AgNO_3$ in ethanol?

a) ⊢Br   or   ⊢Cl   b) $\underset{Cl}{PhCHCH_3}$ or $\underset{Cl}{CH_3CHCH_3}$

c) (cyclopentane with Cl)   or   (cyclopentane with CH₃ and Cl)   d) (benzene with Br)   or   (benzene with CH_2Br)

7.37 Which of these halides would give a precipitate more rapidly when reacted with NaI in acetone?

a) ⊢Cl   or   ⪫Cl   b) (propyl-Cl)   or   (propyl-Br)

c) (cyclopentane with Cl)   or CH_3CH_2Cl d) (cyclopentane with Br)   or   (cyclopentane with Br and two CH₃)

7.38 Which of these alcohols reacts more rapidly with HCl and $ZnCl_2$ in H_2O (Lucas test)?

a) (butanol chain)OH   or   (chain with OH)

b) (cyclohexanol with OH)   or   (cyclopentane with OH and CH₃)   c) (chain)OH   or   (chain)OH

7.39 Explain why this secondary alcohol reacts with HCl and $ZnCl_2$ in H_2O (Lucas test) at about the same rate as a primary alcohol.

$$\underset{\overset{|}{OH}}{Cl-CH_2CHCH_3}$$

7.40   What reagent and solvent would you use to carry out the following transformations?

a) 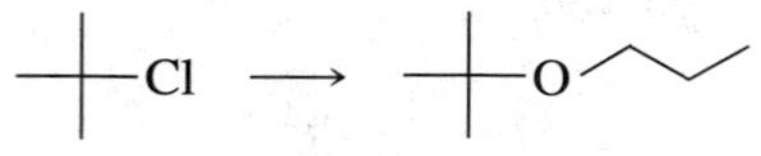 $\overset{\displaystyle|}{\underset{\displaystyle|}{C}}\!-\!Cl \longrightarrow \overset{\displaystyle|}{\underset{\displaystyle|}{C}}\!-\!O\diagdown\diagup$

b) $\underset{CH_3CH_2}{\overset{Ph}{\diagdown}}C\overset{Br}{\underset{H}{\diagup}} \longrightarrow \underset{CH_3CH_2}{\overset{Ph}{\diagdown}}C\overset{H}{\underset{OCCH_3}{\diagup}}$ with a C=O below (||O)

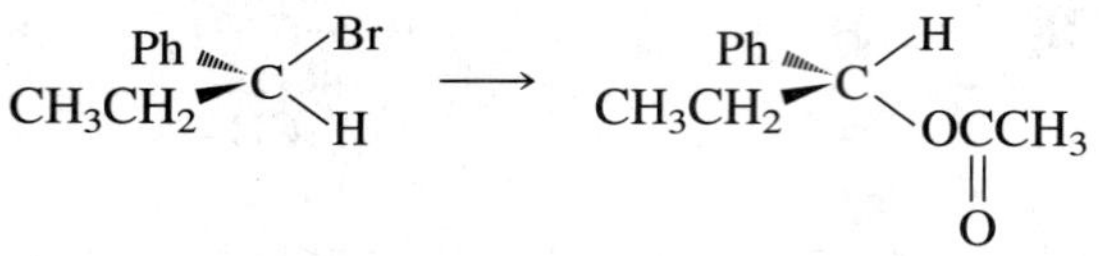

c) $\underset{CH_3CH_2}{\overset{Ph}{\diagdown}}C\overset{Br}{\underset{H}{\diagup}} \longrightarrow Ph-\overset{\overset{\displaystyle H}{|}}{\underset{\underset{\displaystyle CH_2CH_3}{|}}{C}}-O\overset{\overset{\displaystyle O}{||}}{C}CH_3$ (racemic)

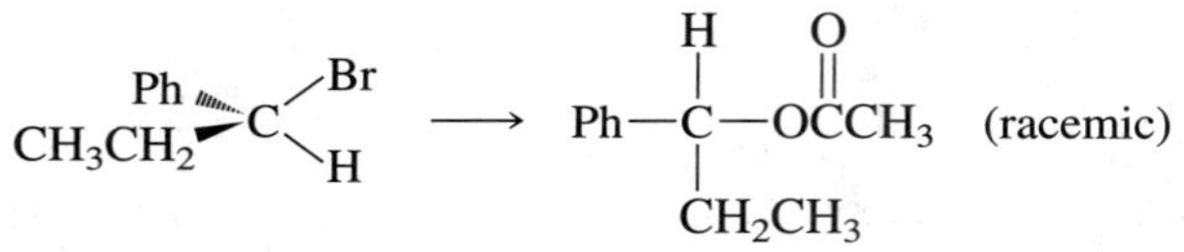

d) 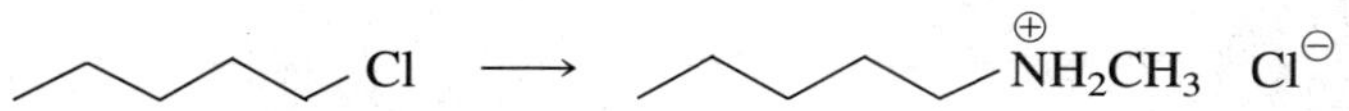 $\diagup\!\!\diagdown\!\!\diagup\!\!\diagdown\!Cl \longrightarrow \diagup\!\!\diagdown\!\!\diagup\!\!\diagdown\!\overset{\oplus}{N}H_2CH_3 \;\; Cl^{\ominus}$

e) 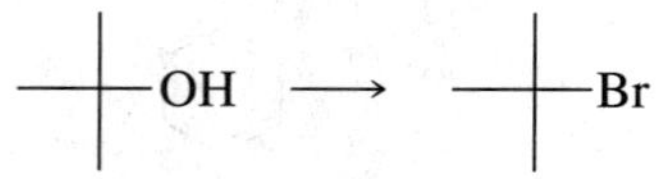 $\overset{\displaystyle|}{\underset{\displaystyle|}{C}}\!-\!OH \longrightarrow \overset{\displaystyle|}{\underset{\displaystyle|}{C}}\!-\!Br$

7.41   In most cases, $RO^{\ominus}$ cannot act as a leaving group in nucleophilic substitution reactions. Explain why the following reaction does occur:

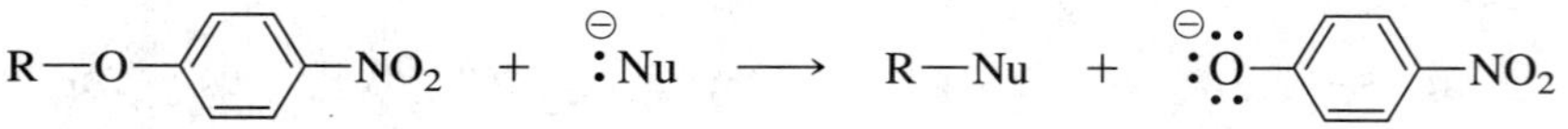

$$R-O-\!\!\!\!\bigcirc\!\!\!\!-NO_2 \;+\; {}^{\ominus}\!:\!Nu \longrightarrow R-Nu \;+\; {}^{\ominus}\!:\!\ddot{O}-\!\!\!\!\bigcirc\!\!\!\!-NO_2$$

7.42   Explain which of these reactions would have the faster rate.

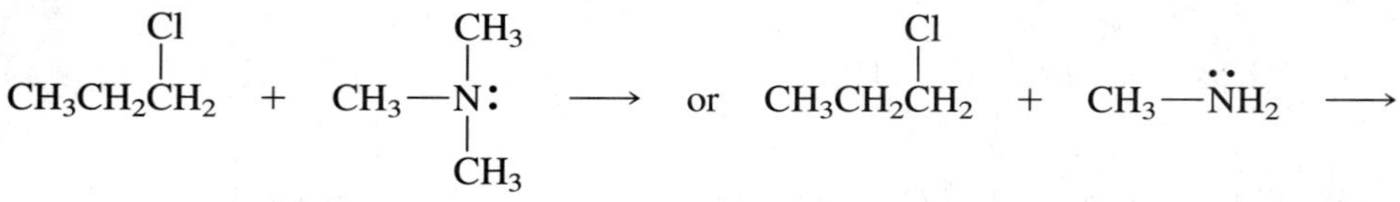

$$CH_3CH_2\overset{\overset{\displaystyle Cl}{|}}{C}H_2 \;+\; CH_3\!-\!\overset{\overset{\displaystyle CH_3}{|}}{\underset{\underset{\displaystyle CH_3}{|}}{N}}\!: \longrightarrow \quad or \quad CH_3CH_2\overset{\overset{\displaystyle Cl}{|}}{C}H_2 \;+\; CH_3\!-\!\ddot{N}H_2 \longrightarrow$$

7.43   Heating ethanol with sulfuric acid is one method used for the preparation of diethyl ether. Show all of the steps in the mechanism for this reaction.

$$2\,CH_3CH_2OH \xrightarrow{\;H_2SO_4\;} CH_3CH_2\!-\!O\!-\!CH_2CH_3 \;+\; H_2O$$

7.44   When an aqueous solution of (*R*)-2-butanol is treated with a catalytic amount of sulfuric acid, slow racemization of the alcohol occurs. Show all of the steps in the mechanism for this process.

7.45   Show all of the steps in the mechanism and explain the stereochemistry for this reaction.

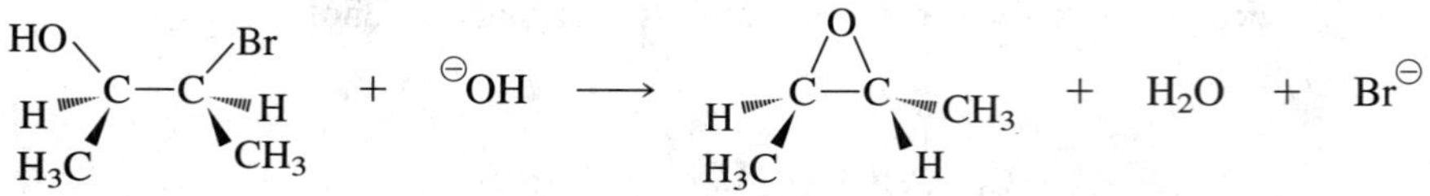

$$\underset{H_3C}{\overset{HO}{\diagdown}}\underset{H}{\overset{}{C}}\!-\!\underset{CH_3}{\overset{Br}{C}}\underset{}{\overset{}{H}} \;+\; {}^{\ominus}OH \longrightarrow \underset{H_3C}{\overset{}{H}}\!\overset{\overset{\displaystyle O}{\diagup\diagdown}}{C}\!-\!\underset{H}{\overset{CH_3}{C}} \;+\; H_2O \;+\; Br^{\ominus}$$

Solution

a) The bromine is bonded to a tertiary carbon, and there is not a strong base present, so the reaction will proceed by an S_N1/E1 mechanism. The substitution product should predominate. The E1 reaction follows Zaitsev's rule, so more 1-methylcyclohexene should be formed than methylenecyclohexane.

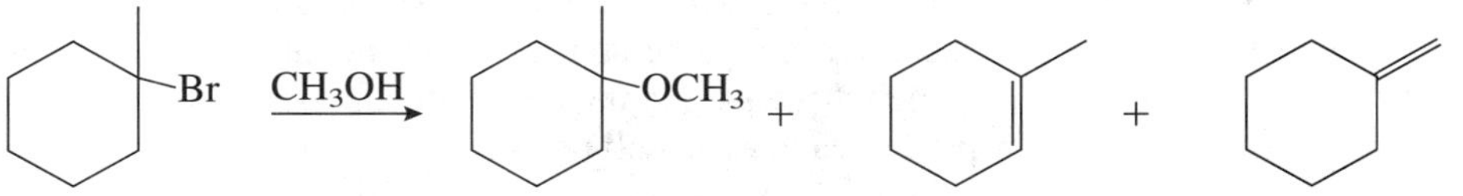

S_N1 major | 1-methylcyclohexene | methylenecyclohexane

E1

The presence of a good anion-stabilizing group, such as a carbonyl group, attached to the carbon from which the proton is lost makes the E1cb mechanism quite favorable. In such situations, even a poor leaving group, such as hydroxide ion, can be eliminated as shown in the following equation:

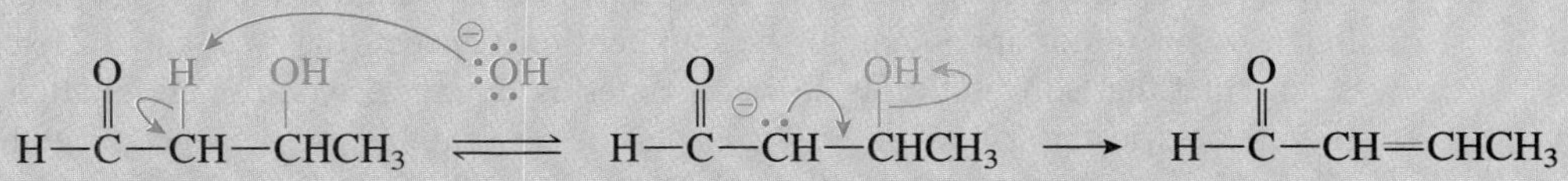

It is not always easy to distinguish an elimination reaction that is following the E1cb mechanism from one that follows the E2 pathway because the E1cb reaction usually exhibits second-order kinetics also. However, because the E1cb reaction is not concerted, there are no strict requirements concerning the stereochemistry of the reaction. In contrast to the preferred anti elimination that occurs in the E2 mechanism, E1cb reactions often produce a mixture of stereoisomers, as illustrated in the following equation:

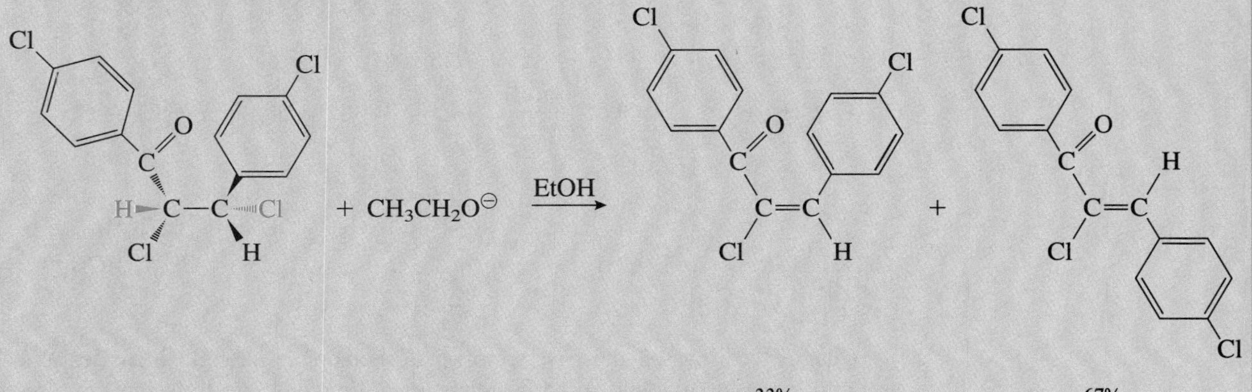

33% 67%

8.8   The Competition Between Elimination and Substitution

As the previous discussion has shown, there is a competition among four differ-ent mechanisms in these reactions: S_N1, S_N2, E1, and E2. The amount of each that occurs depends on multiple factors, such as the substrate, the identity of the base or nucleophile, and the solvent. When these factors favor a particular one of these mech-anisms, then it is possible to predict which one will predominate. However, when these factors favor different mechanisms, then predictions are very difficult. In fact, it is quite possible for several of the mechanisms to occur at the same time in a particu-lar reaction. However, when these reactions are used in the laboratory, we are usually attempting to prepare a particular substitution product or a particular elimination prod-uct. It is usually possible to choose conditions carefully so that one mechanism is faster than the others and the yield of the desired substitution or elimination product is maximized.

The following generalizations provide a useful summary of the factors that con-trol the competition among these mechanisms.

S_N2 and E2

These two pathways require the reaction of the carbon substrate with a nucle-ophile or a base in the rate-determining step of the reaction. So the presence of a good base or nucleophile favors these mechanisms over the S_N1 and E1 pair. Steric hindrance is an important factor in the competition between the S_N2 and E2 path-ways. The fact that substitution is slowed by steric hindrance, whereas elimination is not, is reflected in the following examples. In these examples the presence of ethox-ide ion, which is a strong base and a strong nucleophile, favors the S_N2/E2 mecha-nisms over the S_N1/E1 pathways. As the electrophilic carbon is changed from primary to secondary to tertiary, the mechanism changes from nearly complete S_N2 to com-plete E2 as the increasing steric hindrance at the electrophilic carbon slows the rate of the S_N2 reaction.

$$CH_3CH_2Br + {}^{\ominus}OCH_2CH_3 \xrightarrow{EtOH} CH_3CH_2OCH_2CH_3 + CH_2{=}CH_2$$
$$\qquad\qquad\qquad\qquad\qquad\qquad\qquad\quad 99\% \qquad\qquad\quad 1\%$$

$$\begin{array}{c} Br \\ | \\ CH_3CHCH_3 \end{array} + {}^{\ominus}OCH_2CH_3 \xrightarrow{EtOH} \begin{array}{c} CH_3 \\ | \\ CH_3CHOCH_2CH_3 \end{array} + CH_3CH{=}CH_2$$
$$\qquad\qquad\qquad\qquad\qquad\qquad\qquad\qquad\quad 20\% \qquad\qquad\quad 80\%$$

$$\begin{array}{c} Br \\ | \\ CH_3CCH_3 \\ | \\ CH_3 \end{array} + {}^{\ominus}OCH_2CH_3 \xrightarrow{EtOH} \begin{array}{c} CH_3 \\ | \\ CH_3C{=}CH_2 \end{array}$$
$$\qquad\qquad\qquad\qquad\qquad\qquad\qquad\qquad\qquad\qquad\quad 97\%$$

Chapter 7 discussed the observation that nucleophile strength increases as base strength increases (as long as the basic/nucleophilic atoms are from the same period of the periodic table). Therefore the rates of both S_N2 and E2 reactions increase as the strength of the base and nucleophile increases. However, many experimental observations have shown that a stronger base (which is also a stronger nucleophile) tends to give a higher ratio of elimination to substitution than does a weaker base (which is also a weaker nucleophile). For example, the reaction of 2-bromopropane with ethoxide ion, a strong base and a strong nucleophile, results in 20% substitution and 80% elimination. In contrast, the reaction of 2-bromobutane with acetate ion, a weaker base and a weaker nucleophile, results in a high yield of the substitution product.

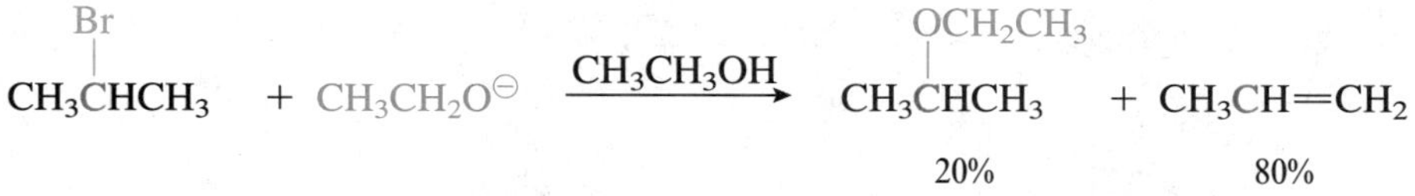

$$
\underset{\underset{\displaystyle CH_3\overset{\displaystyle |}{C}HCH_3}{}}{\overset{\displaystyle Br}{|}} + CH_3CH_2O^{\ominus} \xrightarrow{CH_3CH_3OH} \underset{\underset{20\%}{\displaystyle CH_3\overset{\displaystyle |}{C}HCH_3}}{\overset{\displaystyle OCH_2CH_3}{|}} + \underset{80\%}{CH_3CH=CH_2}
$$

$$
\underset{CH_3CH_2\overset{\displaystyle |}{C}HCH_3}{\overset{\displaystyle Br}{|}} + CH_3\overset{\displaystyle O}{\overset{\displaystyle \|}{C}}O^{\ominus} \xrightarrow{DMF} \underset{\underset{96\%}{CH_3CH_2\overset{\displaystyle |}{C}HCH_3}}{\overset{\displaystyle O}{\overset{\displaystyle \|}{\underset{\displaystyle |}{O\overset{}{C}CH_3}}}}
$$

The temperature of the reaction also affects the relative amounts of substitution and elimination. Usually, the percentage of elimination increases at higher temperatures. This results from the fact that elimination breaks the molecule into fragments and therefore is favored more by entropy than is substitution. As the temperature is increased, the entropy term in the equation $\Delta G = \Delta H - T\Delta S$ becomes more important and the amount of elimination increases. This effect is illustrated in the following equations:

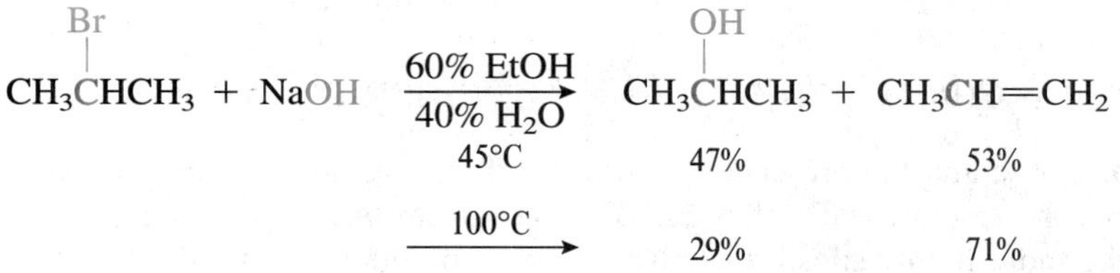

$$
\underset{CH_3\overset{\displaystyle |}{C}HCH_3}{\overset{\displaystyle Br}{|}} + NaOH \xrightarrow[\underset{45°C}{\frac{40\% \ H_2O}{}}]{60\% \ EtOH} \underset{\underset{47\%}{CH_3\overset{\displaystyle |}{C}HCH_3}}{\overset{\displaystyle OH}{|}} + \underset{53\%}{CH_3CH=CH_2}
$$

$$
\xrightarrow{100°C} \qquad 29\% \qquad\qquad 71\%
$$

In summary, in the competition between S_N2 and E2, nucleophiles that are weak bases, minimum steric hindrance, and lower temperatures are used to maximize substitution; strong bases, maximum steric hindrance, and higher temperatures are used to maximize elimination.

S_N1 and E1

Both of these mechanisms involve rate-determining formation of a carbocation, so they most commonly occur with tertiary (best) or secondary substrates in polar solvents and in the absence of strong bases and strong nucleophiles. Because the step that controls which product is formed occurs after the rate-determining step, it is much more difficult to influence the ratio of substitution to elimination here. In general, some elimination always accompanies an S_N1 reaction and must be tolerated. An example is provided in the equation in Figure 8.6.

Problem

Problem 8.14 Show the substitution and/or elimination products for these reactions. Explain which mechanisms are occurring and which product you expect to be the major one.

a) $CH_3CH_2CH_2CH_2Cl + CH_3O^{\ominus} \xrightarrow{CH_3OH}$

b) $CH_3\overset{\underset{|}{CH_3}}{\underset{\underset{|}{Br}}{C}}CH_2CH_3 + {}^{\ominus}OH \xrightarrow{CH_3CH_2OH}$

c) $CH_3CH_2\overset{\underset{|}{OTs}}{C}HCH_3 + CH_3CH_2\overset{\overset{O}{\|}}{C}O^{\ominus} \xrightarrow{CH_3CH_2CO_2H}$

d) $CH_3CH_2\overset{\underset{|}{OTs}}{C}HCH_3 + CH_3CH_2O^{\ominus} \xrightarrow{EtOH}$

e) $Ph\overset{\underset{|}{CH_3}}{\underset{\underset{|}{CH_3}}{C}}Cl + CH_3OH \xrightarrow{CH_3OH}$

Solution

a) The presence of methoxide ion ($CH_3O^{\ominus}$), which is both a strong base and a strong nucleophile, suggests that this reaction should follow the S_N2 or E2 mechanism. The absence of steric hindrance indicates that the reaction will follow the S_N2 mechanism. Not much elimination should occur.

$$CH_3CH_2CH_2CH_2Cl + CH_3O^{\ominus} \xrightarrow{CH_3OH} CH_3CH_2CH_2CH_2OCH_3$$

b) A strong base and the presence of steric hindrance at the reaction site indicate that the reaction will follow the E2 mechanism. According to Zaitsev's rule, the more highly substituted alkene should be the major product.

When substitution or elimination reactions are used in organic synthesis—that is, to prepare organic compounds—it is usually possible to control whether the substitution product or the elimination product is the major compound that is formed. From the viewpoint of the carbon substrate the use of these reactions in synthesis can be summarized in the following manner.

Primary Substrates: RCH_2L

Because of their low amount of steric hindrance, these compounds are excellent for S_N2 reactions with almost any nucleophile. However, they can be forced to follow a predominantly E2 pathway by the use of a strong base that is not a very good nucleophile because of its steric bulk. Most commonly, potassium *tert*-butoxide is used in this role, as shown in the following equation:

$$CH_3(CH_2)_5\overset{\overset{\displaystyle Br}{|}}{C}H_2CH_2 \;+\; CH_3\overset{\overset{\displaystyle CH_3}{|}}{\underset{\underset{\displaystyle CH_3}{|}}{C}}O^{\ominus}K^{\oplus} \;\xrightarrow{\;t\text{-BuOH}\;}\; CH_3(CH_2)_5CH{=}CH_2$$
$$ 85\%$$

Primary substrates do not follow the S_N1 or E1 mechanisms unless they are allylic or benzylic because primary carbocations are too high in energy.

Secondary Substrates: R_2CHL

Secondary substrates can potentially react by any of the four mechanisms. They give good yields of substitution products by the S_N2 mechanism when treated with nucleophiles that are not too basic. They give predominantly elimination products by the E2 mechanism when treated reacted with strong bases such as $OH^{\ominus}$ or $OR^{\ominus}$. (Many examples were provided in the preceding discussion.) Therefore if it is desired to prepare an ether from a secondary substrate, the strongly basic $OR^{\ominus}$ nucleophile must be avoided. The use of S_N1 conditions, with ROH as both nucleophile and solvent (called a **solvolysis reaction**), is usually successful, as illustrated in the following equation:

$$CH_3\overset{\overset{\displaystyle Br}{|}}{C}HCH_3 \;+\; CH_3CH_2OH \;\xrightarrow{\;EtOH\;}\; CH_3\overset{\overset{\displaystyle OCH_2CH_3}{|}}{C}HCH_3 \;+\; CH_3CH{=}CH_2$$
$$ 97\% 3\%$$

Tertiary Substrates: R_3CL

Tertiary substrates do not give S_N2 reactions because they are too hindered. Therefore if a substitution reaction is desired, it must be done under S_N1 conditions (polar solvent, absence of base). Acceptable yields are usually obtained, but some elimination by the E1 pathway usually occurs also. Excellent yields of elimination product can be obtained by the E2 mechanism if the substrate is treated with a strong base (usually $OH^{\ominus}$ or $OR^{\ominus}$). Examples are provided in the following equations:

$$CH_3{-}\overset{\overset{\displaystyle CH_3}{|}}{\underset{\underset{\displaystyle CH_3}{|}}{C}}{-}Br \;+\; CH_3CH_2OH \;\xrightarrow{\;EtOH\;}\; CH_3{-}\overset{\overset{\displaystyle CH_3}{|}}{\underset{\underset{\displaystyle CH_3}{|}}{C}}{-}OCH_2CH_3 \;+\; CH_3{-}\overset{\overset{\displaystyle CH_2}{\|}}{C}{-}CH_3$$
$$ 81\% 19\%$$

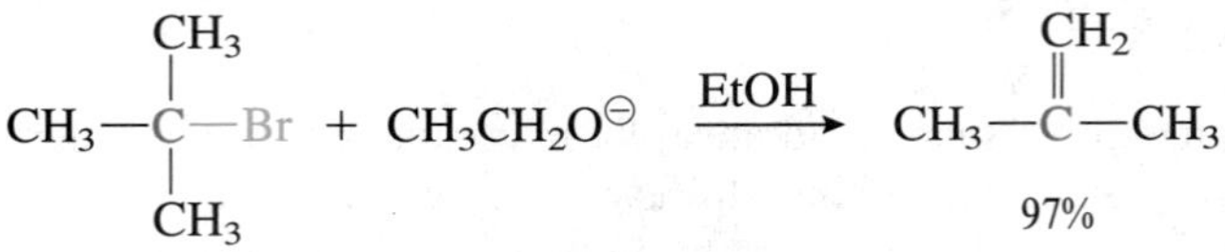

$$\underset{\overset{|}{CH_3}}{\overset{\overset{CH_3}{|}}{CH_3-C-Br}} + CH_3CH_2O^{\ominus} \xrightarrow{EtOH} \underset{97\%}{\overset{\overset{CH_2}{\|}}{CH_3-C-CH_3}}$$

Problem

Problem 8.15   Explain which mechanism is preferred in these reactions and show the major products.

a) 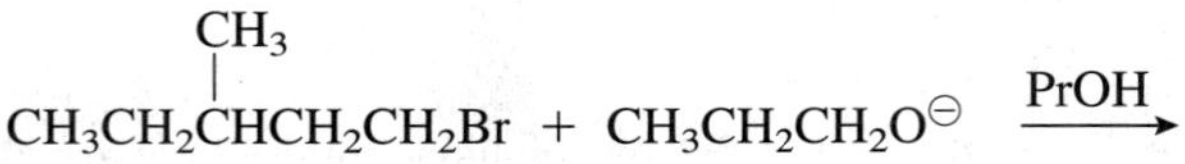 $\underset{\overset{|}{CH_3}}{CH_3CH_2CHCH_2CH_2Br} + CH_3CH_2CH_2O^{\ominus} \xrightarrow{PrOH}$

b) 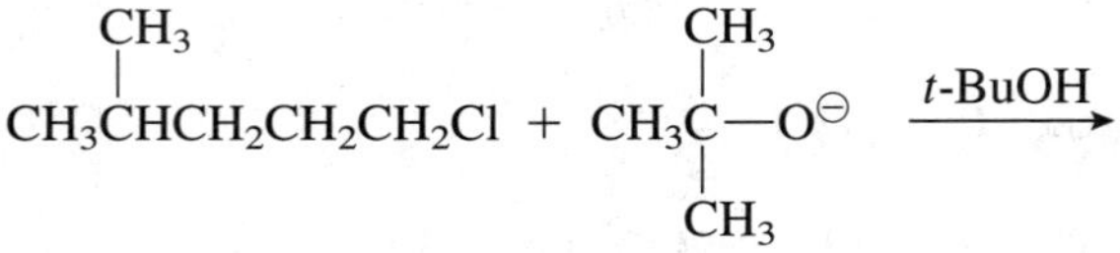 $\underset{\overset{|}{CH_3}}{CH_3CHCH_2CH_2CH_2Cl} + \underset{\overset{|}{CH_3}}{CH_3\overset{\overset{CH_3}{|}}{C}-O^{\ominus}} \xrightarrow{t\text{-BuOH}}$

c) + KOH $\xrightarrow[EtOH, \ reflux]{H_2O}$

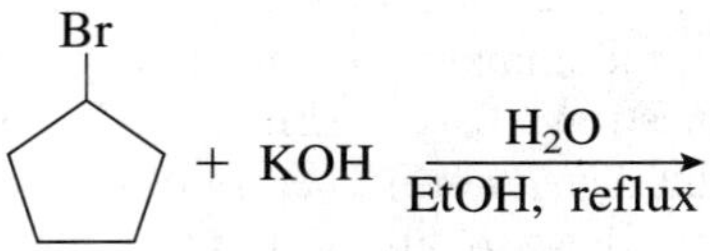

d) 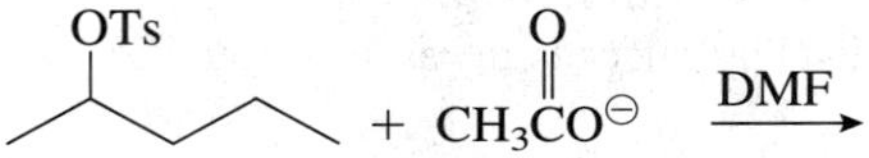 + $CH_3\overset{\overset{O}{\|}}{C}O^{\ominus} \xrightarrow{DMF}$

e) 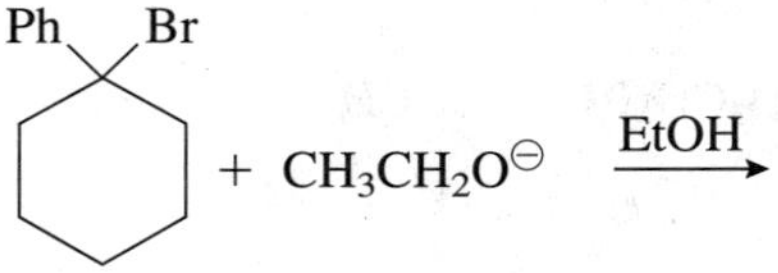 + $CH_3CH_2O^{\ominus} \xrightarrow{EtOH}$

f) 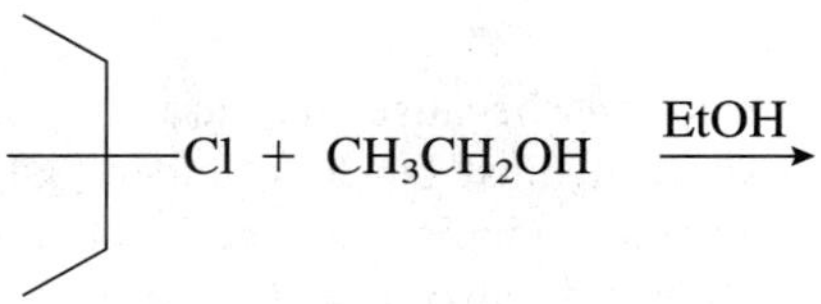 $-Cl + CH_3CH_2OH \xrightarrow{EtOH}$

g) 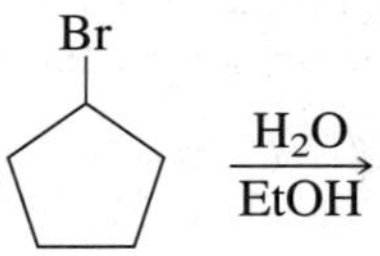 $\xrightarrow[EtOH]{H_2O}$

Solution

   b)   The leaving group is on a primary carbon, so the reaction must follow the

Biological Elimination Reactions

Elimination reactions occur in living organisms also. One important example is the conversion of 2-phosphoglycerate to phosphoenolpyruvate during the metabolism of glucose:

2-phosphoglycerate   phosphoenolpyruvate

This elimination is catalyzed by the enzyme enolase and follows an E1cb mechanism. The enzyme supplies a base to remove the acidic proton and generate a carbanion in the first step. In addition, a Mg^{2+} cation in the enzyme acts as a Lewis acid and bonds to the hydroxy group, making it a better leaving group.

Another example occurs in the citric acid cycle, where the enzyme aconitase catalyzes the elimination of water from citrate to produce aconitate:

citrate   aconitate

S_N2 or E2 mechanism. The sterically hindered base slows the S_N2 pathway so that the major product results from E2 elimination.

major (E2)   minor (S_N2)

f) The leaving group is on a tertiary carbon and there is no strong base present, so the reaction follows the S_N1 and E1 mechanisms. Substitution is usually the major product, but a significant amount of elimination product is also formed. According to Zaitsev's rule, the more highly substituted product should predominate among the alkenes.

$$\text{—}C\text{l} + CH_3CH_2OH \xrightarrow{\text{EtOH}} \text{—}OCH_2CH_3 +$$

major (S$_N$1)   Z + E   more minor (E1)
                        minor (E1)

Chapter 9 discusses the synthetic uses of all of these reactions in considerably more detail.

8.9   Summary

The two mechanisms for the elimination reactions that compete with substitutions are summarized in Table 8.1.

After completing this chapter you should be able to:

1. Provide a detailed description, including stereochemistry, for the E2 mechanism and summarize the conditions that favor its occurrence.

Table 8.1   Summary of Elimination Mechanisms.

	E1	E2
Mechanism	two step R—L $\longrightarrow$ R$^\oplus$ $\longrightarrow$ alkene	one step R—L + B$^\ominus$ $\longrightarrow$ alkene + BH + L$^\ominus$
Kinetics	first-order rate = k[RL]	second-order rate = k[RL][B$^\ominus$]
Competes with	S$_N$1	S$_N$2
Stereochemistry	no preferred conformation	*trans*-coplanar elimination (*trans*-diaxial in cyclohexanes)
Regiochemistry	Zaitsev's rule: favors more highly substituted alkene conjugated always preferred	Zaitsev's rule: favors more highly substituted alkene (Hofmann elimination follows Hofmann's rule: favors less highly substituted alkene) conjugated always preferred
Competition with Substitution	some E1 always accompanies S$_N$1	E2 favored by strong bases, steric hindrance, and higher temperatures

Therefore the proton on the nitrogen must be removed to use this nitrogen as a nucleophile. This hydrogen is relatively acidic (pK_a = 9.9) because of resonance stabilization of the conjugate base, similar to that shown for phthalimide. Hydroxide ion is a
strong enough base to remove this proton and generate the conjugate base of phthalimide. The reaction of this nucleophile with an alkyl halide or an alkyl sulfonate ester,
by an S_N2 mechanism, produces a substituted phthalimide with an alkyl group bonded
to the nitrogen. The electrons on the nitrogen of this alkylated phthalimide are not
nucleophilic, so there is no danger of multiple alkylation. The carbonyl groups are
then removed to give the desired primary amine. The process is outlined in Figure 9.5
and an additional example is provided by the following equation:

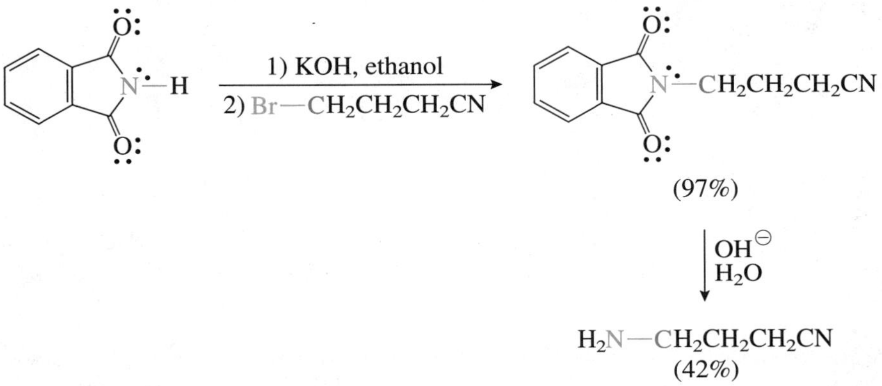

Ammonia acts as the nucleophile in an S_N2 reaction, replacing the bromine.

In an equilibrium process, the resulting ammonium salt can lose a proton to a base such as ammonia to produce a primary amine. The primary amine is a stronger base and therefore a better nucleophile than ammonia. Even when a large excess of ammonia is present, some of the primary amine reacts to produce a secondary amine.

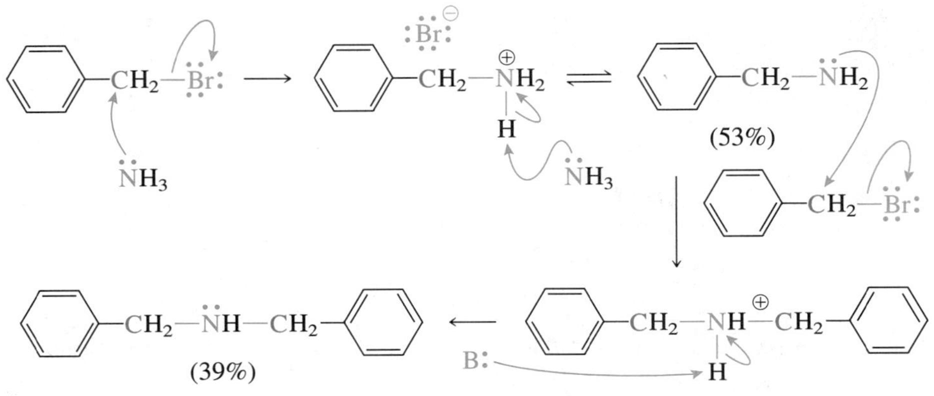

In this particular case, in which an eightfold excess of ammonia is used, the product mixture consists of 53% of the primary amine and 39% of the secondary amine.

Figure 9.4

Multiple Alkylation Using Ammonia as a Nucleophile.

Because of resonance, the electron pair on the nitrogen of phthalimide is not very basic or nucleophilic. The hydrogen on the nitrogen is much more acidic than a hydrogen of a normal amine because the conjugate base is stabilized by resonance. It is acidic enough to be removed completely by a base like hydroxide ion.

The resulting phthalimide anion is a good nucleophile in the S_N2 reaction.

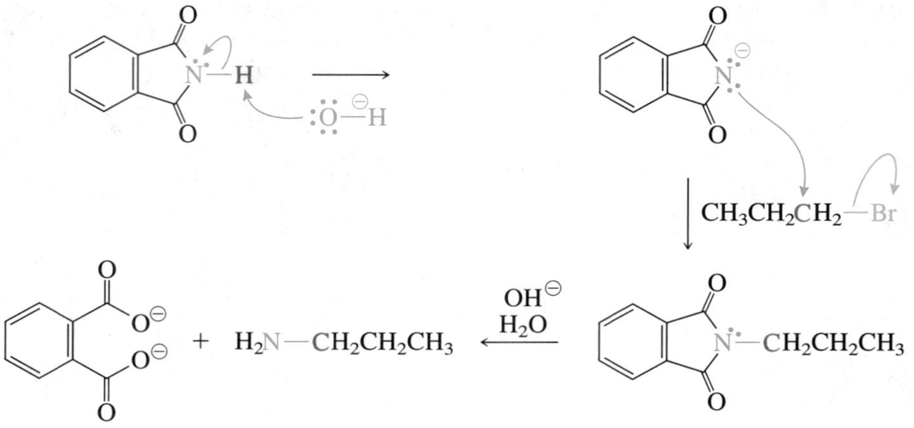

Figure 9.5

The Gabriel Synthesis of a Primary Amine.

The desired amine can be freed by reaction of the phthalimide with an aqueous base. (The reaction conditions are very similar to those of the ester cleavage shown in Figure 9.1.) Overall, phthalimide is a synthetic equivalent for ammonia.

Like phthalimide itself, the alkylated phthalimide is not nucleophilic, so there is no problem with multiple alkylation occurring.

Problems

Problem 9.19   Show the products of these reactions.

a)

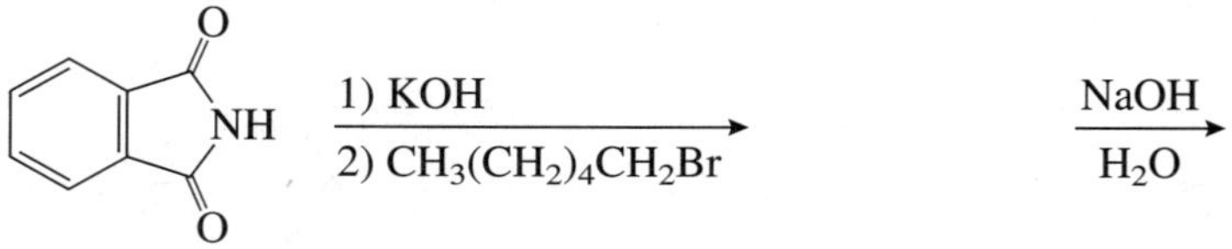

$$\xrightarrow[\text{2) } CH_3(CH_2)_4CH_2Br]{\text{1) KOH}} \qquad \xrightarrow[H_2O]{NaOH}$$

b)

$$\xrightarrow{\hspace{2cm}} \qquad \xrightarrow[H_2O]{NaOH}$$

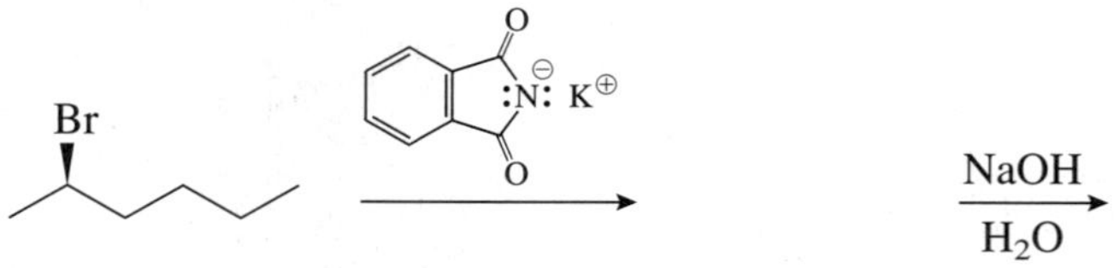

Problem 9.20   Suggest a method that could be used to prepare this amine from an alkyl halide.

$PhCH_2CH_2NH_2$

When we desire to attach a methyl group to a nucleophile in the laboratory, we often choose a simple reagent such as iodomethane or dimethyl sulfate. Living organisms cannot use reagents like these because they are too reactive and too indiscriminate. They will react with almost any nucleophile. Nature's iodomethane is a much more complex molecule called S-adenosylmethionine or SAM. The leaving group in SAM is a disubstituted sulfur atom and confers just the right reactivity on the compound. SAM is used to methylate the nitrogen of norepinephrine in the biosynthesis of epinephrine (adrenaline) and also serves as the methylating agent in the biosynthesis of the important lipid phosphatidylcholine (lecithin) from phosphatidylethanolamine.

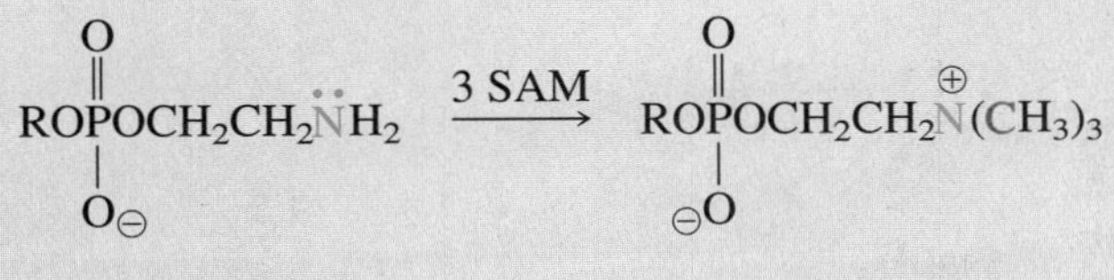

Biological Methylations

ELABORATION

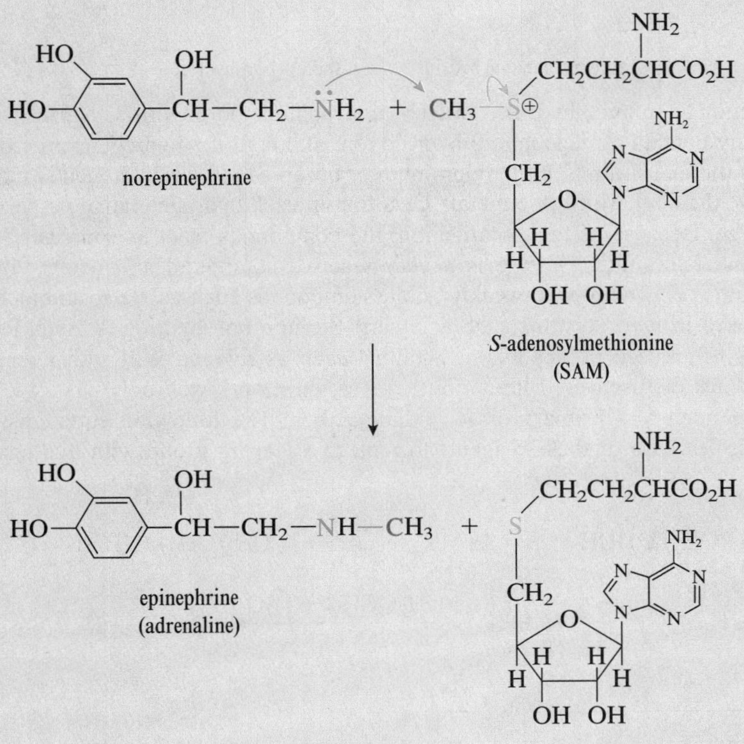

norepinephrine

S-adenosylmethionine
(SAM)

epinephrine
(adrenaline)

phosphatidylethanolamine   phosphatidylcholine

9.8   Preparation of Hydrocarbons

Hydrocarbons can be prepared by replacing a leaving group with a hydrogen, according to the following general equation. This requires a hydrogen with an unshared pair of electrons and a negative charge, that is, **hydride ion,** as the nucleophile.

$$H:^{\ominus} + R-L \longrightarrow R-H + :L^{\ominus}$$

hydride ion

$$Li^{\oplus} \quad H-\underset{\underset{H}{|}}{\overset{\overset{H}{|}}{Al}}{}^{\ominus}-H \qquad\qquad Na^{\oplus} \quad H-\underset{\underset{H}{|}}{\overset{\overset{H}{|}}{B}}{}^{\ominus}-H$$

lithium aluminum hydride   sodium borohydride

Both lithium aluminum hydride, $LiAlH_4$, and sodium borohydride, $NaBH_4$, react as though they contain a nucleophilic hydride ion, although the hydrogens are covalently bonded to the metal atoms, either aluminum or boron. However, hydrogen is more electronegative than either of these metals, resulting in each hydrogen having a partial negative charge. Because of this polarization, the compounds react as sources of hydride ion. Lithium aluminum hydride is a very reactive compound and reacts vigorously (often explosively) with even weakly acidic compounds such as water and alcohols. It must be used in inert solvents such as ethers. Sodium borohydride is much less reactive and is often used in alcohols or alkaline water as solvent. With either reagent the reactions have the usual S_N2 limitations; that is, they work well only when the leaving group is bonded to a primary or secondary carbon. The following equations provide examples of the use of these reagents to replace a leaving group with hydrogen:

$$CH_3(CH_2)_4\overset{\overset{Br}{|}}{C}HCH_3 \ + \ LiAlH_4 \ \xrightarrow{Et_2O} \ CH_3(CH_2)_4\overset{\overset{H}{|}}{C}HCH_3 \quad (92\%)$$

$$PhCH_2-Br \ + \ NaBH_4 \ \xrightarrow{(CH_3OCH_2CH_2)_2O} \ PhCH_2-H \quad (86\%)$$

Problem

Problem 9.21   Show the products of these reactions.

a) $\xrightarrow[\text{ether}]{LiAlH_4}$

b) $\xrightarrow[CH_3OH]{NaBH_4}$

9.9 Formation of Carbon–Carbon Bonds

Reactions that form carbon–carbon bonds are of great importance in organic synthesis because they enable smaller compounds to be converted to larger compounds. To form these bonds by nucleophilic substitution reactions requires a carbon nucleophile—a **carbanion** (carbon anion), as shown in the following general equation:

$$R': ^{\ominus} + R\!-\!L \longrightarrow R'\!-\!R + :L^{\ominus}$$

Two useful carbon nucleophiles are introduced in this section. Other important carbon nucleophiles are discussed in later chapters, especially Chapter 16.

The first of these carbon nucleophiles, **cyanide ion,** is a moderate base and a good nucleophile:

$$:\overset{\ominus}{C}\!\equiv\!N:$$

cyanide ion

Cyanide ion reacts by the S_N2 mechanism and aprotic solvents are often employed to increase its reactivity. Yields of substitution products are excellent when the leaving group is attached to a primary carbon. Because of competing elimination reactions, yields are lower, but still acceptable, for secondary substrates. As expected for an S_N2 process, the reaction does not work with tertiary substrates. Substitution with cyanide ion adds one carbon to the compound while also providing a new functional group for additional synthetic manipulation. Some examples are given in the following equations:

$$CH_3CH_2CH_2CH_2\!-\!Cl + Na^{\oplus}\ :\!\overset{\ominus}{C}\!\equiv\!N: \xrightarrow{\text{DMSO}} CH_3CH_2CH_2CH_2\!-\!C\!\equiv\!N: + NaCl \quad (92\%)$$

A second group of important carbon nucleophiles are the **acetylide anions**. These nucleophiles are generated by treating 1-alkynes with a very strong base, such as amide ion:

$$R\!-\!C\!\equiv\!C\!-\!H + :\overset{\ominus}{\underset{..}{N}}H_2 \rightleftharpoons R\!-\!C\!\equiv\!\overset{\ominus}{C}: + :\overset{..}{N}H_3$$

amide ion   an acetylide
                            anion

As was discussed in Chapter 4, a proton on a carbon–carbon triple bond is relatively acidic ($pK_a = 25$) because of the *sp* hybridization of the carbon to which it is bonded. The proton is acidic enough that it can be removed with some of the strong bases that are available to the organic chemist. Usually, sodium amide ($NaNH_2$), the conjugate base of ammonia ($pK_a = 38$), is used to remove the proton. Amide ion is a strong enough base so that the equilibrium in the above equation lies entirely to the right. (Note that carbanions generated by removing protons from sp^3 and sp^2 hybridized carbons are not generally available for use as nucleophiles in S_N2 reactions because the protons attached to them are not acidic enough to be removed in this manner.)

Since acetylide anions are strong nucleophiles, they react by the S_N2 mechanism. Good yields of substitution products are obtained only when the leaving group is attached to a primary carbon; secondary substrates give mainly elimination because the anion is also a strong base. Several examples are provided in the following equations. The last example shows how ethyne can be alkylated twice—both hydrogens can be replaced with alkyl groups in sequential steps!

$$H-C\equiv C-H \xrightarrow[NH_3 (1)]{NaNH_2} H-C\equiv C:^{\ominus} \xrightarrow{CH_3CH_2CH_2CH_2-Br} H-C\equiv C-CH_2CH_2CH_2CH_3 \quad (75\%)$$

$$\text{Ph}-C\equiv C-H \xrightarrow[\text{2) } CH_3CH_2-OTs]{\text{1) } NaNH_2, Bu_2O} \text{Ph}-C\equiv C-CH_2CH_3 \quad (77\%)$$

$$H-C\equiv C-H \xrightarrow[\text{2) } CH_3CH_2CH_2CH_2CH_2-Br]{\text{1) } NaNH_2, NH_3 (1)} H-C\equiv C-CH_2CH_2CH_2CH_2CH_3 \quad (83\%)$$

$$\left\downarrow \begin{array}{l} \text{1) } NaNH_2, Et_2O \\ \\ \text{2) } CH_3O-\overset{\overset{O}{\|}}{\underset{\underset{O}{\|}}{S}}-OCH_3 \end{array}\right.$$

$$H_3C-C\equiv C-CH_2CH_2CH_2CH_2CH_3 \quad (81\%)$$

Problems

Problem 9.22   Show the products of these reactions.

a) Ph—CH$_2$Cl $\xrightarrow[\text{DMSO}]{\text{NaCN}}$

b) $CH_3C\equiv C-H$ $\xrightarrow[\text{2) } CH_3CH_2CH_2Br]{\text{1) } NaNH_2, NH_3 (1)}$

c) (2-bromopentane) $\xrightarrow[\text{DMSO}]{\text{NaCN}}$

d) $HC\equiv CH$ $\xrightarrow[\text{2) } CH_3CH_2Br]{\text{1) } NaNH_2, NH_3 (1)}$ $\xrightarrow[\text{2) } CH_3I]{\text{1) } NaNH_2}$

e) Cl~~~Br $+$ 1 NaCN $\xrightarrow{\text{DMF}}$

f) (bromocyclopentane) $+$ $HC\equiv C:^{\ominus}$ $\xrightarrow{NH_3 (1)}$

Problem 9.23   Suggest methods for preparing these compounds from alkyl halides.

a)

b) $H—C≡C—CH_2CH_2\overset{\overset{\displaystyle CH_3}{|}}{C}HCH_3$

c) $CH_3—C≡C—CH_2Ph$

9.10   Phosphorus and Sulfur Nucleophiles

Sulfur occurs directly beneath oxygen in the periodic table. Therefore sulfur compounds are weaker bases but better nucleophiles than the corresponding oxygen compounds. Sulfur compounds are excellent nucleophiles in S_N2 reactions, and because they are relatively weak bases, elimination reactions are not usually a problem. Yields are good with primary and secondary substrates. For similar reasons, phosphorus compounds also give good yields when treated with primary and secondary substrates in S_N2 reactions. The following equations provide examples of the use of these nucleophiles:

$$CH_3—\overset{\cdot\cdot}{\underset{\cdot\cdot}{S}}\!\!:^{\ominus} \; + \; Cl—CH_2CH_2OH \; \xrightarrow{\text{ethanol}} \; CH_3—\overset{\cdot\cdot}{\underset{\cdot\cdot}{S}}—CH_2CH_2OH \; + \; Cl^{\ominus} \quad (80\%)$$

$$:\!\overset{\cdot\cdot}{\underset{\cdot\cdot}{S}}\!\!:^{2\ominus} \; + \; 2\,CH_3CH_2CH_2—Br \; \xrightarrow{\text{ethanol}} \; CH_3CH_2CH_2—S—CH_2CH_2CH_3 \; + \; 2\,Br^{\ominus} \quad (85\%)$$

$$Ph_3P: \; + \; CH_3—Br \; \xrightarrow{\text{benzene}} \; Ph_3\overset{\oplus}{P}—CH_3 \;\; \overset{\ominus}{Br} \quad (99\%)$$

triphenylphosphine   a phosphonium salt

$$Ph_3P: \; + \; PhCH{=}CH\overset{\overset{\displaystyle Cl}{|}}{C}H_2 \; \xrightarrow{\text{xylene}} \; PhCH{=}CH\overset{\overset{\displaystyle \overset{\oplus}{P}Ph_3}{|}}{C}H_2 \;\; Cl^{\ominus} \quad (92\%)$$

Triphenylphosphine is probably the most important phosphorus nucleophile for organic chemists because it produces phosphonium salts (see the last two equations above). These phosphonium salts are starting materials for an important preparation of alkenes that will be discussed in Chapter 14.

Problem

Problem 9.24   Show the products of these reactions.

a)

$Cl \; + \; PhS^{\ominus}Na^{\oplus} \; \xrightarrow{CH_3OH}$

b) Ph_3P + $CH_3CH_2CH_2Br$ $\xrightarrow{\text{benzene}}$

c) $CH_3CH_2CH_2CH_2S^{\ominus}$ + CH_3I $\xrightarrow{\text{EtOH}}$

d) $NaSCH_2CH_2SNa$ + $BrCH_2CH_2Br$ $\xrightarrow{CH_3OH}$

9.11   Ring Opening of Epoxides

Section 7.10 discussed the generation of a leaving group, water, from an alcohol by protonation of the oxygen of the hydroxy group. In a similar fashion, protonation of the oxygen of an ether also generates a leaving group—an alcohol in this case—as shown in the following equation:

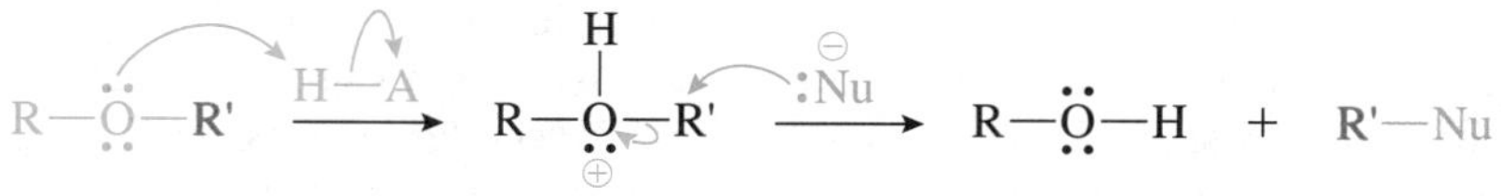

This reaction requires more vigorous conditions than the reaction of alcohols, resulting in low yields in many cases. For this reason the reaction is not commonly used in synthesis.

Sulfur Nucleophiles in Biochemistry

The fact that sulfur is an excellent nucleophile is used to advantage in biochemical reactions in living organisms. For example, the methylating agent S-adenosylmethionine or SAM (see the Elaboration on page 365) is biosynthesized by a nucleophilic substitution reaction of the sulfur of the amino acid methionine on adenosine triphosphate (ATP):

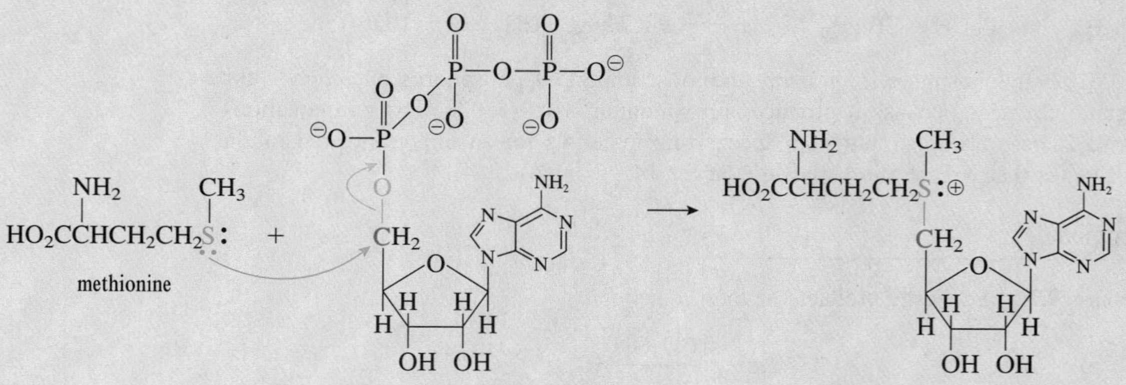

adenosine triphosphate
(ATP)

methionine

S-adenosylmethionine
(SAM)

An **epoxide** (also known as an oxirane) is a three-membered cyclic ether.

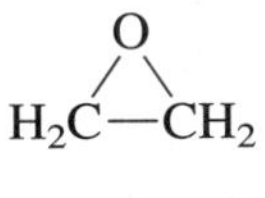

$$
\overset{\displaystyle O}{\underset{\displaystyle H_2C-CH_2}{\triangle}}
$$

an epoxide

Like cyclopropane, epoxides have a large amount of ring strain and are much more reactive than normal ethers. Because of this ring strain, one carbon–oxygen bond of an epoxide can be broken in a nucleophilic substitution reaction. The following equation shows an example:

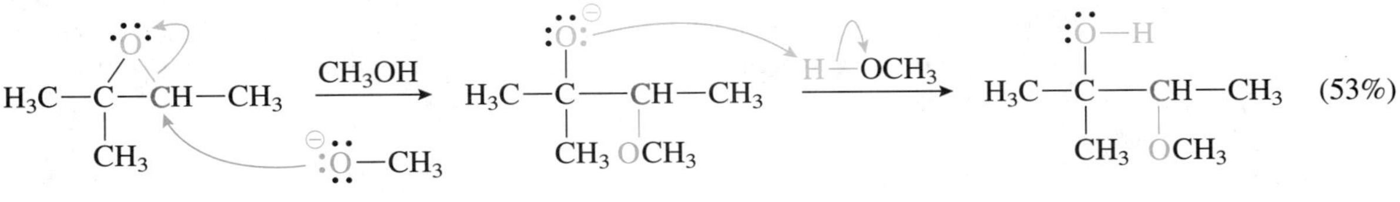

Both of the carbons of the epoxide ring are electrophilic, so at first glance, either might be expected to react with the nucleophile, methoxide ion. However, reactions of epoxides under basic or neutral conditions, as in this case, usually follow an S_N2 mechanism. Therefore the nucleophile reacts at the less hindered secondary carbon, with inversion of configuration.

In the preceding reaction the leaving group ($RO^{\ominus}$) is a very strong base. As was discussed in Chapter 7, $HO^{\ominus}$ and $RO^{\ominus}$ are much too basic to act as leaving groups

Sulfur nucleophiles also play an important role in dealkylation enzymes that help repair damaged DNA. The bases of DNA (see Chapter 26) contain oxygen and nitrogen atoms that are nucleophilic. Occasionally, one of these nucleophiles becomes alkylated by a random reaction of SAM or by some other alkylating agent from the environment. The alkylated base often does not hydrogen bond in the same manner as the unalkylated base, thus causing a different base-pairing interaction. For example, when the oxygen of a guanidine becomes alkylated, the product is highly mutagenic because, on replication, the alkylated guanidine causes the incorporation of thymine instead of cytosine. As is shown in the following equation, a dealkylation enzyme repairs the DNA by removing the alkyl group. The enzyme uses the thiol group of a cysteine amino acid residue as a nucleophile to displace the alkyl group from the base.

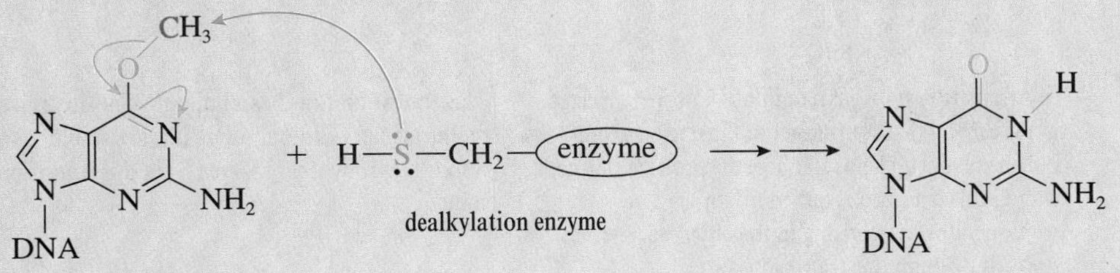

O-methylated guanidine
residue of DNA

dealkylation enzyme

guanidine residue of DNA

in normal nucleophilic substitution reactions. In the special case of epoxides, however, even $RO^{\ominus}$ can act as a leaving group because of the large amount of strain that is relieved when the carbon–oxygen bond is broken and the ring is opened.

Nucleophilic ring opening of epoxides can also be accomplished in acid solution. The oxygen is first protonated, making it a much better leaving group. Although these are typical S_N1 conditions, the actual mechanism is somewhere between S_N1 and S_N2—the reaction has characteristics of both mechanisms. The stereochemistry is that predicted for an S_N2 mechanism; the nucleophile approaches from the side opposite the leaving oxygen. The regiochemistry is that predicted for an S_N1 mechanism; the substitution occurs at the carbon that would be more stable as a carbocation. This often results in the carbon–oxygen bond that is broken under acidic conditions being different from the one that is broken under basic conditions, as can be seen by comparing the product in the following reaction with the one from the preceding reaction.

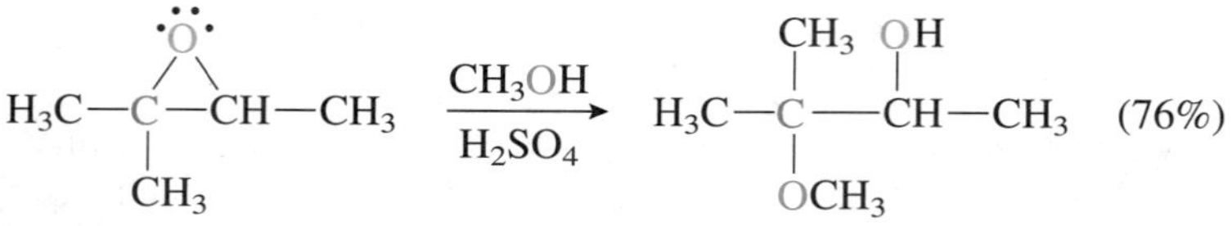

Another example, in which it can be seen that the reaction proceeds with inversion of configuration, is provided in the following equation:

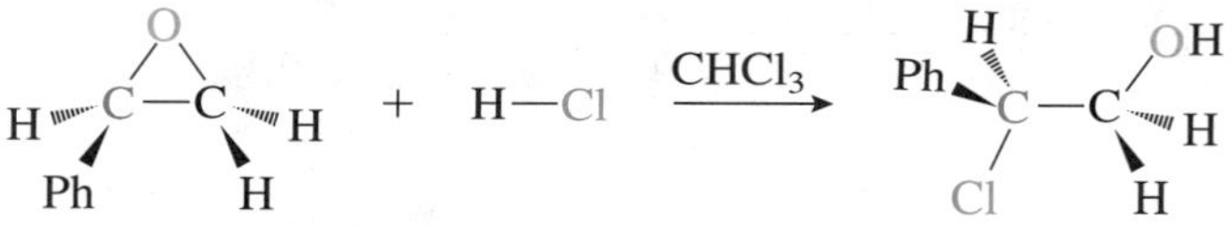

Because such reactions have features of both the S_N2 mechanism (stereochemistry) and the S_N1 mechanism (regiochemistry), they are said to follow a borderline S_N2 mechanism. The transition state geometry resembles that for an S_N2 reaction, but the bond to the leaving group is broken to a greater extent than the bond to the nucleophile is formed, resulting in considerable positive charge buildup on the carbon. Therefore the transition state that has this positive charge buildup on the carbon that would be the more stable carbocation is favored. The two possible transition states for the preceding reaction are as follows:

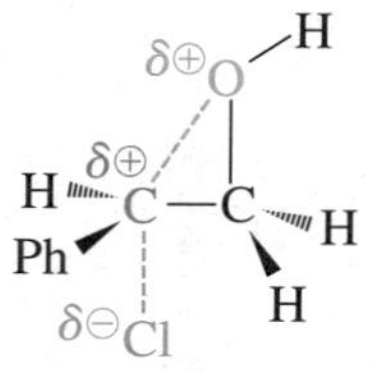

This transition state has a buildup of positive charge on the carbon that is attached to the phenyl group. The phenyl group helps stabilize the positive charge, making this transition state more stable. The reaction pathway resulting in the observed product proceeds through this transition state.

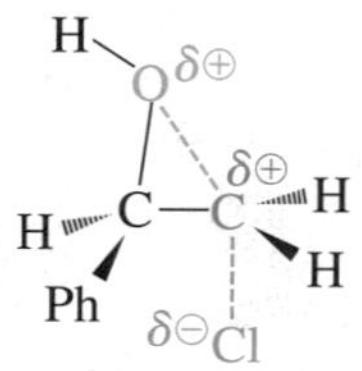

This transition state has a buildup of positive charge on the primary carbon, where it is less stable. As a result, no product is observed from this transition state.

Other examples of nucleophilic substitutions on epoxides are given in the following equations:

$\dfrac{H_2SO_4}{CH_3OH}$ (82%)

$\dfrac{KOH}{\substack{H_2O \\ DMSO}}$ (65%)

$H_3C-\overset{\overset{\displaystyle O}{|}}{CH}-CH_2$ $\xrightarrow[\text{ether}]{LiAlH_4}$ $H_3C-\overset{\overset{\displaystyle :\ddot{O}:^{\ominus}}{|}}{CH}-CH_3$ $\xrightarrow{H_2O}$ $H_3C-\overset{\overset{\displaystyle :\ddot{O}-H}{|}}{CH}-CH_3$ (88%)

Problem

Problem 9.25   Show the products of these reactions.

a)   + CH_3OH $\dfrac{H_2SO_4}{CH_3OH}$

b)   + $CH_3O^{\ominus}$ $\xrightarrow{CH_3OH}$

c) $\dfrac{1)\ LiAlH_4,\ ether}{2)\ H_3O^{\oplus}}$

9.12   Elimination Reactions

Elimination reactions are a useful method for the preparation of alkenes, provided that certain limitations are recognized. One problem is the competition between substitution and elimination. The majority of eliminations are done under conditions that favor the E2 mechanism. In these cases, steric hindrance can be used to slow the competing S_N2 pathway. Tertiary substrates and most secondary substrates give good yields of the elimination product when treated with strong bases. Sterically hindered bases can be employed with primary substrates to minimize substitution.

Another problem that occurs with eliminations is the regiochemistry of the reaction. As we saw in Chapter 8, most eliminations follow Zaitsev's rule and produce the

Uses of Epoxides in Industry

Epoxides are important intermediates in many industrial processes. For example, the reaction of the simplest epoxide, ethylene oxide, with water is employed to produce ethylene glycol, which is used in antifreeze and to prepare polymers such as Dacron. One method for the preparation of ethylene oxide employs an intramolecular nucleophilic substitution reaction of ethylene chlorohydrin:

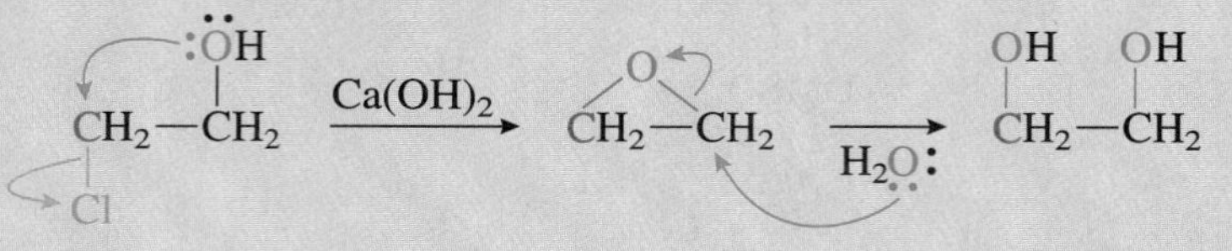

| ethylene chlorohydrin | ethylene oxide | ethylene glycol |

Nucleophilic cyanide ion can also be used to open the epoxide ring. This reaction was employed in a now obsolete pathway for the preparation of acrylonitrile, which is used to make Orlon:

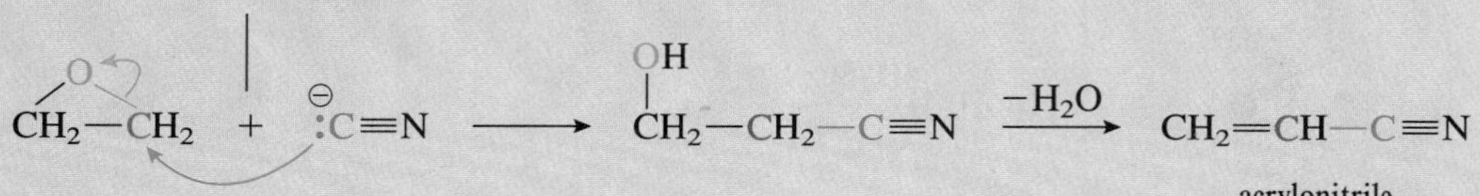

acrylonitrile

Propranolol, a drug that is used to lower blood pressure, is prepared from the epoxide epichlorohydrin. First, the oxygen of 1-naphthol displaces the chlorine of epichlorohydrin in an S_N2 reaction. Then the epoxide ring is opened by the nucleophilic nitrogen of isopropylamine in another S_N2 reaction to form propranolol.

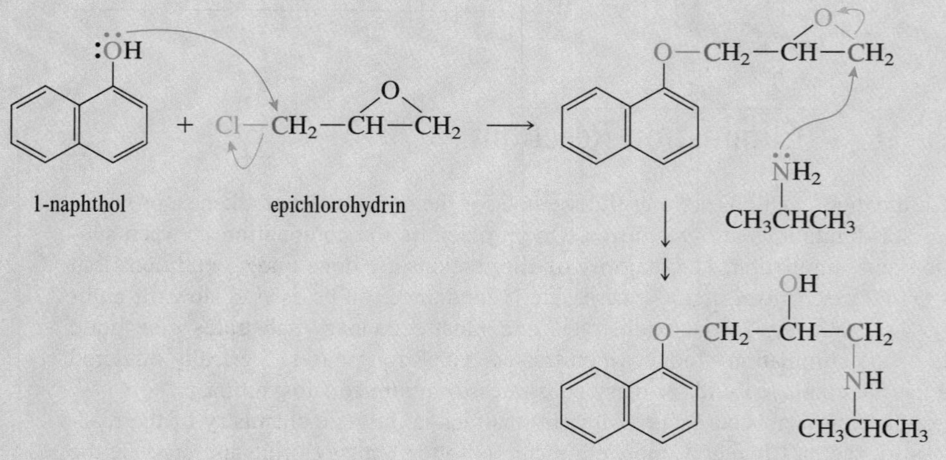

1-naphthol   epichlorohydrin

propranolol

more highly substituted alkene as the major product. However, a significant amount of the less highly substituted product is also formed. In addition, mixtures of *cis* and *trans* geometrical isomers are produced when possible, further complicating the product mixture. Because separating a mixture of such isomers is usually a difficult task, elimination reactions are often not the best way to prepare alkenes. (Other methods will be described in subsequent chapters.) However, if only one product can be formed, or if one is expected to greatly predominate in the reaction mixture, then these elimination reactions can be quite useful.

9.13   Elimination of Hydrogen Halides (Dehydrohalogenation)

Reaction of an alkyl halide with a strong base can be employed for the preparation of an alkene, provided that a mixture of isomers is not produced. The strong bases that are commonly used for these eliminations are sodium hydroxide, potassium hydroxide, sodium methoxide ($NaOCH_3$), and sodium ethoxide ($NaOCH_2CH_3$). Potassium *tert*-butoxide (*t*-BuOK) is especially useful with less hindered substrates to avoid competing substitution. Sulfonate esters can also be used as leaving groups. Several examples are shown in the following equations.

$$HC\equiv CCH_2\overset{\overset{\displaystyle OTs}{|}}{C}HCH_3 \;+\; KOH \;\;\xrightarrow{H_2O}\;\; HC\equiv C-CH=CHCH_3 \;\;(91\%)$$

$$H_3C-\overset{\overset{\displaystyle CH_3}{|}}{\underset{\underset{\displaystyle Br}{|}}{C}}-CH_3 \;+\; NaOCH_2CH_3 \;\;\xrightarrow{EtOH}\;\; \overset{\displaystyle CH_2}{\underset{\displaystyle H_3C \quad CH_3}{\overset{\|}{C}}} \;\;(97\%)$$

$$CH_3(CH_2)_5CH_2CH_2-Br \;+\; t\text{-BuOK} \;\;\xrightarrow{t\text{-BuOH}}\;\; CH_3(CH_2)_5CH=CH_2 \;\;(85\%)$$

Problems

Problem 9.26   Show the products of these reactions.

a) $CH_3CH_2CH_2\overset{\overset{\displaystyle OTs}{|}}{C}H_2 \;+\; t\text{-BuOK} \;\;\xrightarrow{t\text{-BuOH}}$

b) $+\; NaOCH_2CH_3 \;\;\xrightarrow{EtOH}$

c) $+\; CH_3ONa \;\;\xrightarrow{CH_3OH}$

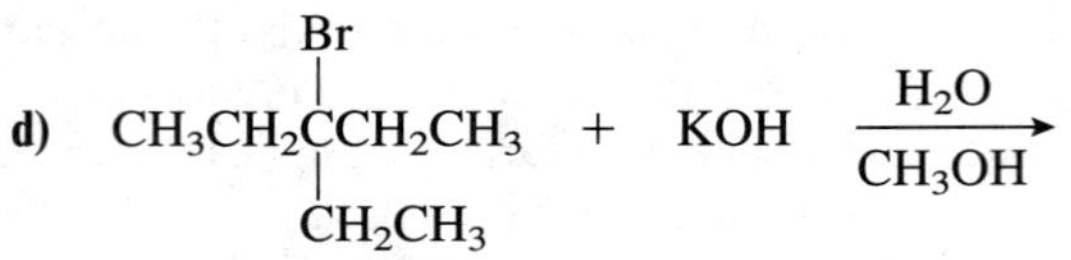

d) $\overset{\overset{\displaystyle Br}{|}}{CH_3CH_2CCH_2CH_3} \; + \; KOH \quad \xrightarrow[CH_3OH]{H_2O}$
$\underset{\displaystyle CH_2CH_3}{|}$

Problem 9.27 Explain whether these elimination reactions would be a good way to prepare these alkenes.

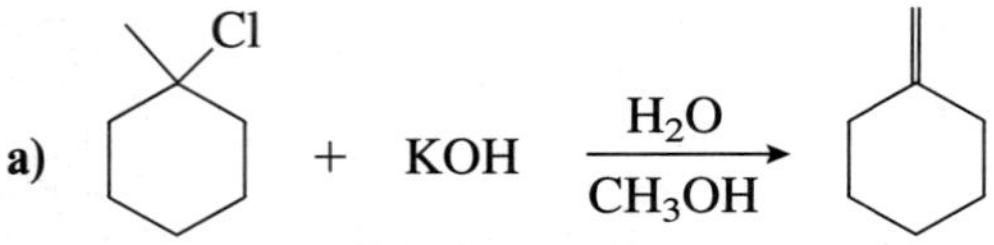

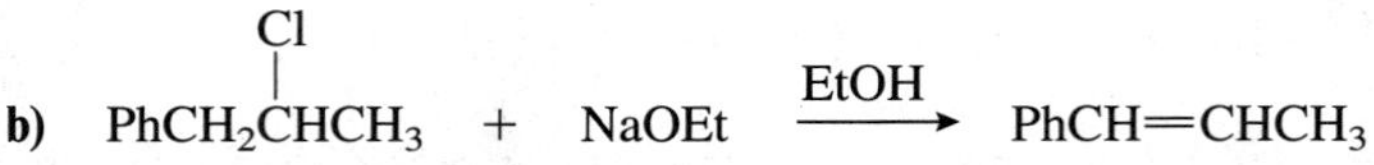

a)   + KOH $\xrightarrow[CH_3OH]{H_2O}$

b) $\underset{\underset{\displaystyle Cl}{|}}{PhCH_2CHCH_3} \; + \; NaOEt \quad \xrightarrow{EtOH} \quad PhCH{=}CHCH_3$

Problem 9.28 Explain which of these reactions would provide a better synthesis of 2-pentene.

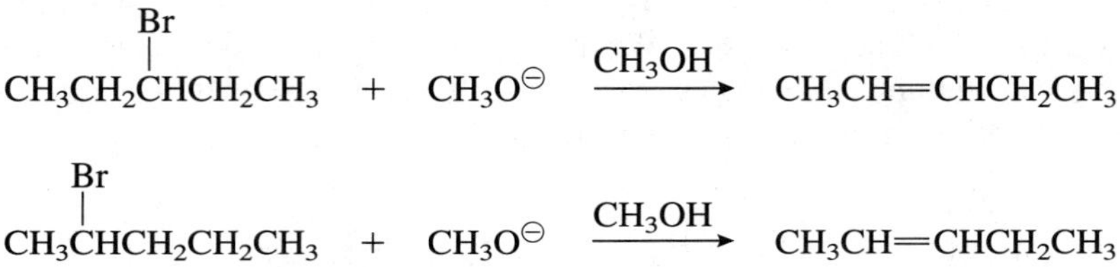

$\overset{\overset{\displaystyle Br}{|}}{CH_3CH_2CHCH_2CH_3} \; + \; CH_3O^{\ominus} \quad \xrightarrow{CH_3OH} \quad CH_3CH{=}CHCH_2CH_3$

$\overset{\overset{\displaystyle Br}{|}}{CH_3CHCH_2CH_2CH_3} \; + \; CH_3O^{\ominus} \quad \xrightarrow{CH_3OH} \quad CH_3CH{=}CHCH_2CH_3$

9.14 Preparation of Alkynes

Alkynes can be prepared from dihaloalkanes by elimination of two molecules of HX. This reaction requires very strongly basic conditions so potassium hydroxide at elevated temperatures or the stronger base sodium amide ($NaNH_2$) is commonly employed. Examples are provided by the following equations:

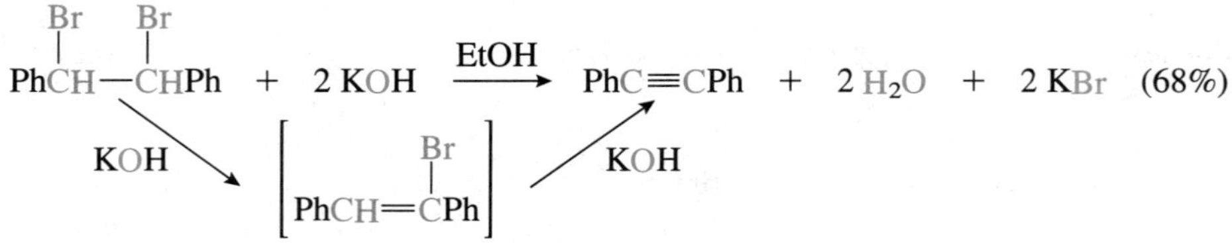

$\overset{Br\;\;\;Br}{\overset{|\;\;\;\;|}{PhCH{-}CHPh}} \; + \; 2\,KOH \quad \xrightarrow{EtOH} \quad PhC{\equiv}CPh \; + \; 2\,H_2O \; + \; 2\,KBr \quad (68\%)$

$\underset{\displaystyle KOH}{\searrow} \qquad \left[\overset{Br}{\overset{|}{PhCH{=}CPh}} \right] \qquad \underset{\displaystyle KOH}{\nearrow}$

a vinyl halide

$\overset{Br\;\;\;Br}{\overset{|\;\;\;\;|}{PhCH{-}CH_2}} \; + \; 3\,NaNH_2 \quad \xrightarrow{NH_3\,(1)} \quad PhC{\equiv}C\overset{\ominus}{:}\,\overset{\oplus}{Na} \; + \; 3\,NH_3 \; + \; 2\,NaBr$

$\downarrow H_2O$

$PhC{\equiv}CH \quad (50\%)$

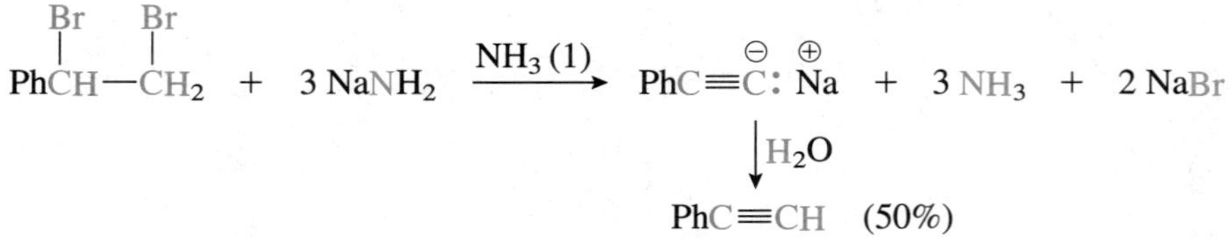

In these reactions, elimination of the first molecule of HX results in the formation of a vinyl halide—an alkene with a halogen bonded to one of the carbons of the double bond. A second, more difficult elimination (this is why the strong base is necessary) produces the triple bond. Therefore it is not surprising that vinyl halides can also be used to prepare alkynes, as shown in the following equations:

$$PhCH\!\!=\!\!\overset{\displaystyle Br}{\underset{\displaystyle |}{CH}} \quad + \quad KOH \quad \longrightarrow \quad PhC\!\!\equiv\!\!CH \quad + \quad H_2O \quad + \quad KBr \quad (67\%)$$

[cyclohexyl]—$CH_2\overset{Br}{\underset{|}{C}}\!\!=\!\!CH_2$ $\quad \dfrac{\text{1) NaNH}_2,\ \text{mineral oil}}{\text{2) HCl}} \quad$ [cyclohexyl]—$CH_2C\!\!\equiv\!\!CH$ $\quad (66\%)$

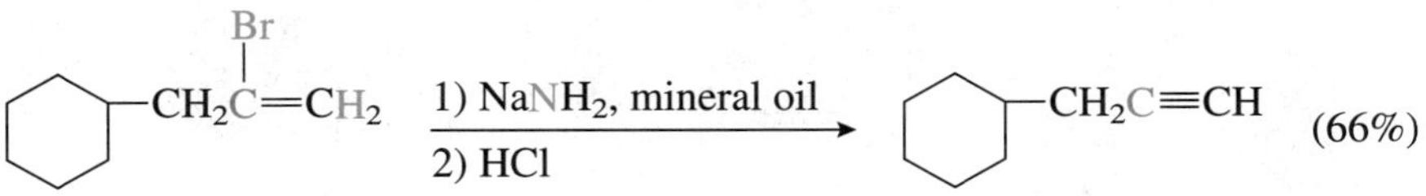

Problem

Problem 9.29 Show the products of these reactions.

a) 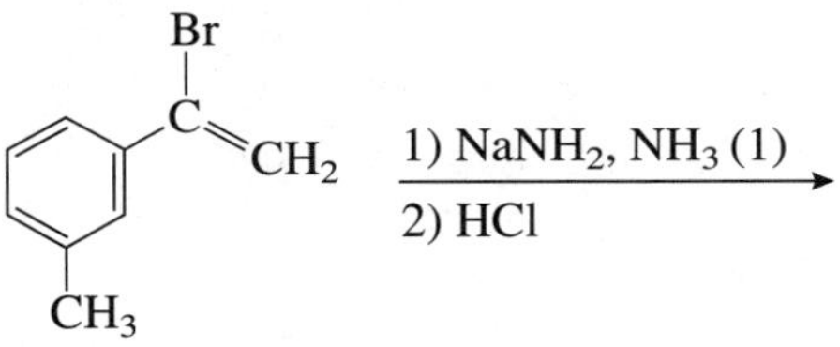 [structure: benzene ring with CH₃ meta substituent, bearing a C(Br)=CH₂ group] $\dfrac{\text{1) NaNH}_2,\ \text{NH}_3\ (1)}{\text{2) HCl}}$

b) 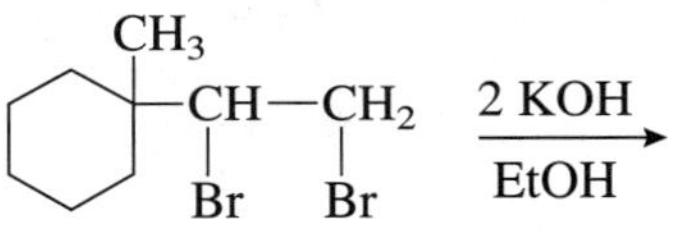 [structure: cyclohexyl—CH(CH₃)... with CH(Br)—CH₂(Br) chain] $\dfrac{\text{2 KOH}}{\text{EtOH}}$

9.15   Dehydration

Section 9.6 described the reaction of alcohols with the halogen acids, HX, to produce alkyl halides. If, instead of a halogen acid, a catalytic amount of sulfuric or phosphoric acid is used, the reaction takes a different pathway and an elimination product is formed. Since water is eliminated, the reaction is termed dehydration.

The mechanism for the dehydration of cyclohexanol to produce cyclohexene is shown in Figure 9.6. In general, these reactions follow the E1 mechanism, so tertiary alcohols are more reactive than secondary alcohols. (Note that this is one of the few cases in which the E1 mechanism is favored over the S_N1 mechanism.) At the carbocation stage, there is a competition between substitution and elimination. Under the conditions used for the dehydration reaction, elimination is favored, since there are no good nucleophiles present to cause substitution. The conjugate bases of sulfuric and phosphoric acids ($HSO_4^{\ominus}$ and $H_2PO_4^{\ominus}$) are not very nucleophilic. Only a small amount of acid is needed because the reaction is acid catalyzed; that is, the acid is regenerated in the final step of the mechanism.

9.39 Show how this synthesis might be accomplished:

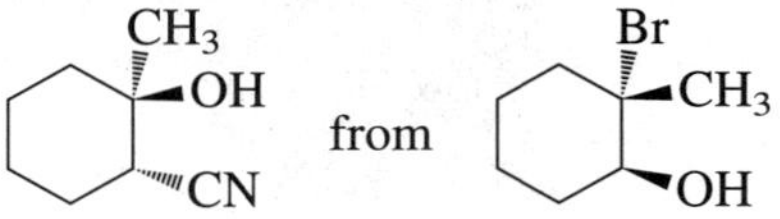

9.40 What is wrong with these reactions? Explain.

a)

b)

c)

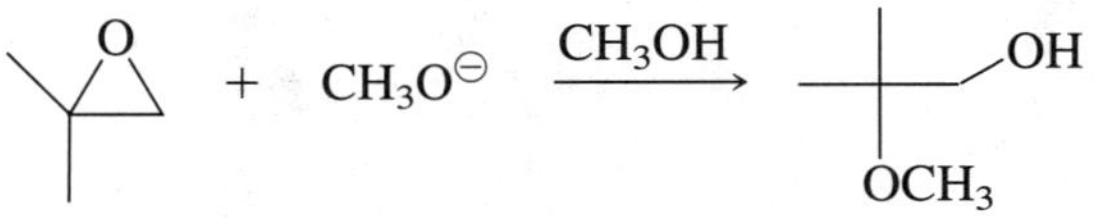

d)

9.41 What is wrong with these syntheses? Explain.

a)

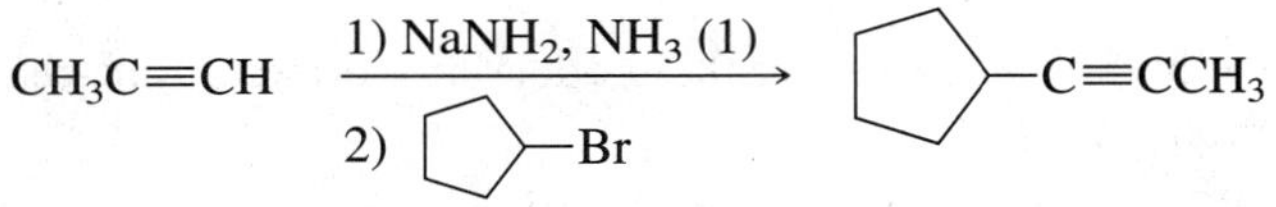

b) CH_3CH_2I + NH_3 $\xrightarrow{H_2O}$ $CH_3CH_2-\overset{\oplus}{N}H_3$ $I^{\ominus}$

c)

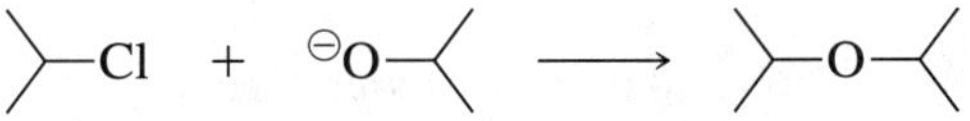

d)

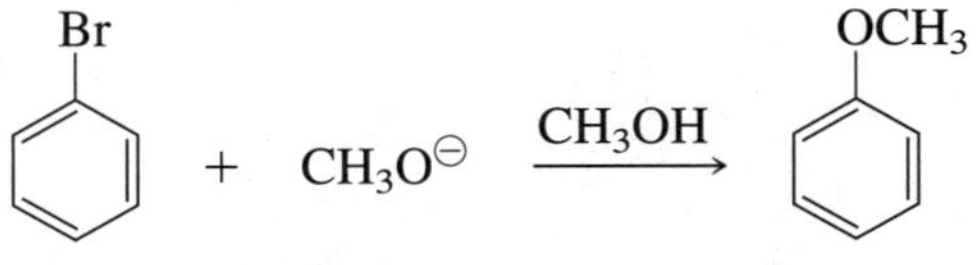

e)

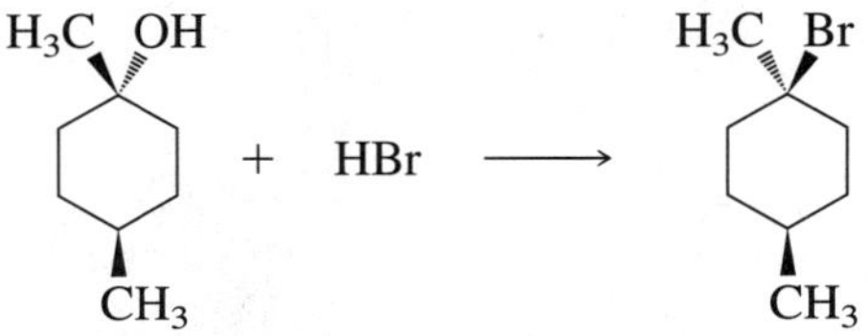

f)

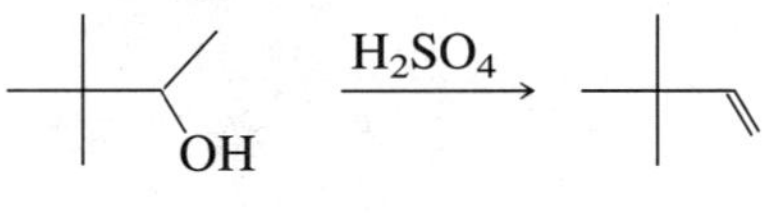

9.42 Show all the steps in the mechanisms for these reactions. Don't forget to use curved arrows to show the movement of electrons in each step.

a)

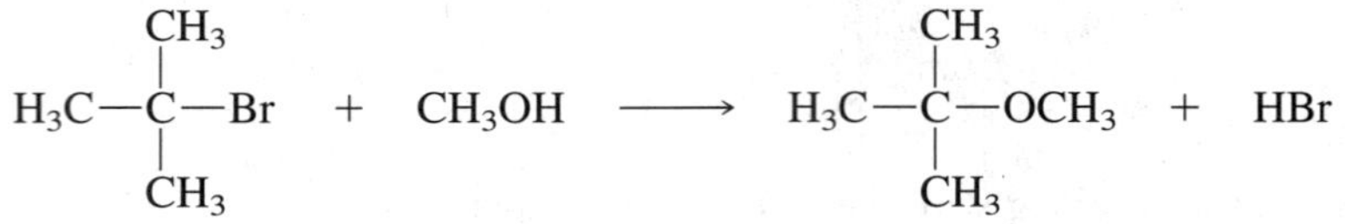

b) $2\ CH_3CH_2OH \xrightarrow{H_2SO_4} CH_3CH_2OCH_2CH_3\ +\ H_2O$

c)

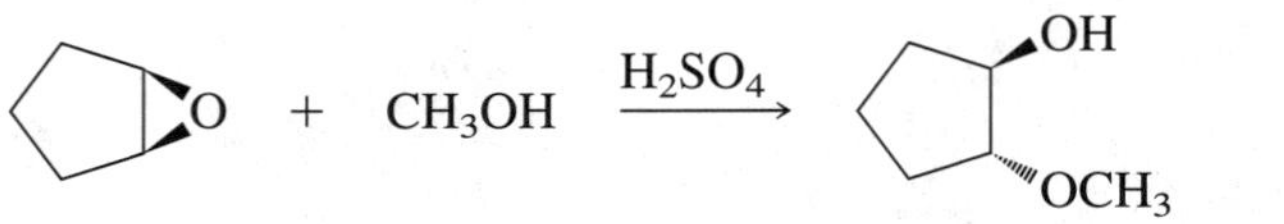

d)

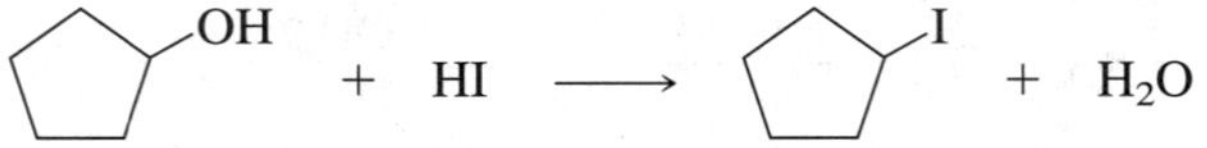

e)

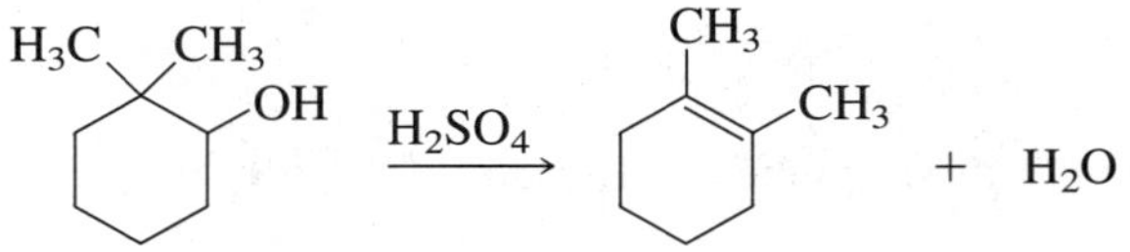

9.43 Explain how both enantiomers of the product are formed in the reaction shown in problem 9.42c.

9.44 Show all the steps in the mechanism for this reaction:

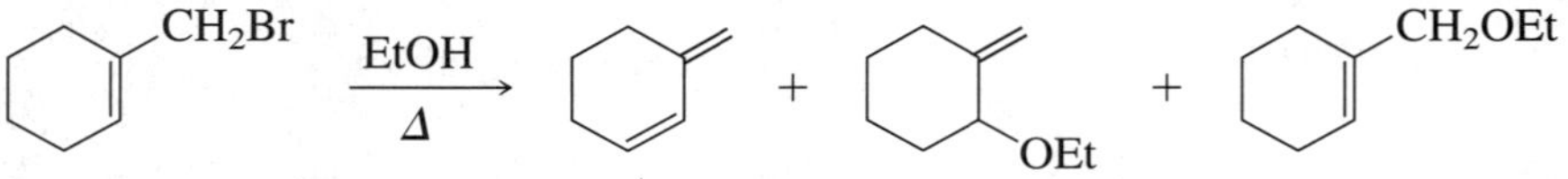

9.45 Classify these transformations as oxidations or reductions.

a) 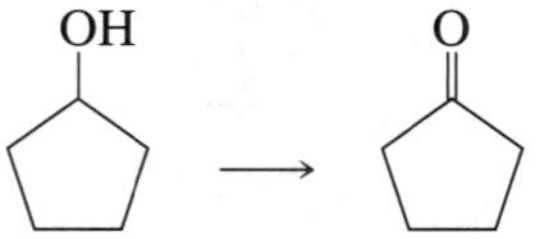   b)

c)    d) $CH_3CH_2\overset{O}{\overset{\|}{C}}OH \longrightarrow CH_3CH_2CH_2OH$

9.46 Show a mechanism for this reaction:

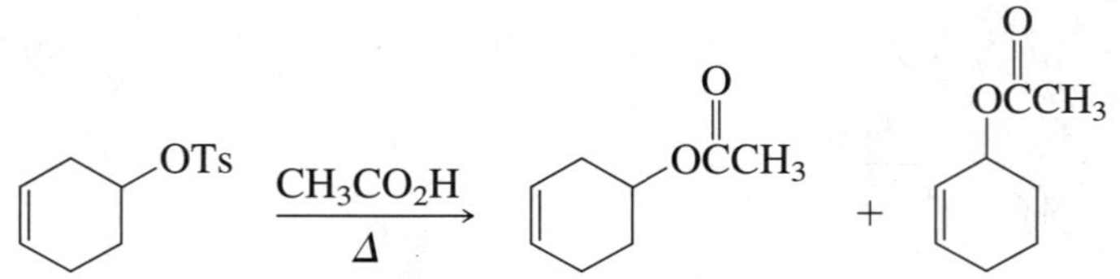

9.47 Show how mustard gas could be prepared from ethylene oxide and sodium sulfide (Na_2S).

ethylene oxide

9.48 Explain why one of the oxygens preferentially acts as the nucleophile in this reaction.

OH $\xrightarrow[\text{2) 1 } CH_3I]{\text{1) } K_2CO_3}$ OCH$_3$

CH$_2$OH   CH$_2$OH

9.49 Only one of the chlorines acts as a leaving group in this reaction. Explain.

Cl $\xrightarrow[\text{H}_2\text{O}]{\text{NaOH}}$

Cl

9.50 This reaction gives two substitution products. Show the structures for these products and provide a mechanism for their formation.

H$_3$C   Br $\xrightarrow[\Delta]{CH_3OH}$

9.51 Suggest a mechanism for this reaction.

CH$_3$ CH$_3$ $\xrightarrow{\Delta}$ CH$_3$

N   CH   N   CH$_3$

Cl   Cl

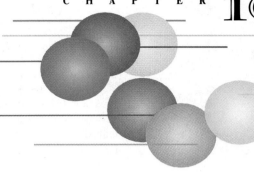

CHAPTER 10

Additions to Carbon–Carbon Double and Triple Bonds

10.1   Introduction

The reaction that is presented in this chapter results in the addition of two groups—an electrophile and a nucleophile—to the carbons of a carbon–carbon double or triple bond. In the following example of this **addition reaction,** a proton (the electrophile) and a chloride ion (the nucleophile) add to the carbons of ethene to produce chloroethane:

The features of this reaction that will be examined include the following:

the mechanism of the reaction;

the effect of substituents on the double bond on the rate of the reaction;

the regiochemistry of the reaction;

the stereochemistry of the reaction;

the various combinations of electrophiles and nucleophiles that undergo this reaction;

the variations in mechanism, regiochemistry, and stereochemistry that occur with these different reagents; and

similar reactions that occur with carbon–carbon triple bonds.

There are a large number of reactions in this chapter. You will find it much easier to remember them and the important details that pertain to each if you keep in mind the general mechanism and what changes are caused in the mechanism by the nature of the reagent.

401

10.2 The General Mechanism

The simplest version of the mechanism for this addition reaction occurs in two steps. First, the electrophile adds to the double bond, producing a carbocation intermediate. In the second step the nucleophile adds to the carbocation. This step is identical to the second step of the S_N1 reaction. Because the initial species that reacts with the double bond is an electrophile, the reaction is called an **electrophilic addition reaction.**

In the first step the nucleophile is the alkene, or, more specifically, the highest-energy pair of electrons of the alkene: the pi electrons of the double bond. These electrons, being in a bonding MO rather than in a higher energy nonbonding MO, are only weakly nucleophilic, so a relatively strong electrophile, such as $H^\oplus$ from a strong acid, is needed to react with them. Most of these addition reactions are run under acidic or neutral conditions to avoid destroying the electrophiles, which are all fairly strong Lewis acids.

The product of the first step is a high-energy carbocation intermediate. Forming this reactive intermediate is the slow step of the mechanism. As was the case in the S_N1 reaction, the transition state for this step resembles the carbocation (recall the Hammond postulate). Therefore structural features that stabilize the carbocation also stabilize the transition state leading to it, thus lowering the activation barrier and resulting in a faster rate of reaction. We know from Chapter 7 the various features that stabilize carbocations. The presence of electron-donating alkyl groups on the positive carbon stabilizes the carbocation; tertiary carbocations are more stable than secondary carbocations, and secondary carbocations are more stable than primary carbocations. Therefore alkyl substituents on the double bond accelerate the reaction. Resonance stabilization of the carbocation also speeds up the reaction. The presence of groups that withdraw electrons destabilizes the carbocation and results in a slower reaction rate. We also know that carbocations are prone to rearrange if a more stable cation can result. Rearrangements do occur in these electrophilic addition reactions, and so we must always examine the carbocation to see whether rearrangement is likely.

Since the second step is the same as the second step in the S_N1 mechanism, similar nucleophiles, such as H_2O and the halide ions, are found here. In addition, there are electrophiles and nucleophiles that we have not yet encountered that undergo this

Problem

Problem 10.1   Arrange these alkenes in order of increasing rate of reaction with HCl.

a) $CH_3CH_2CH=CH_2$ $CH_2=CH_2$ $CH_3CH_2\overset{\overset{\displaystyle CH_3}{|}}{C}=CH_2$

b) 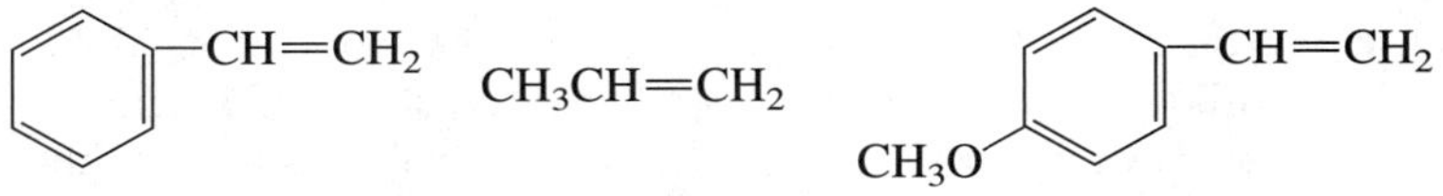 [benzene ring]$-CH=CH_2$ $CH_3CH=CH_2$ CH_3O-[benzene ring]$-CH=CH_2$

reaction. Some of these cause variations on the mechanism presented above. However, the general theme of the electrophile adding first and ultimate formation of a product with the electrophile bonded to one carbon of the initial double bond and the nucleophile bonded to the other remains unchanged. Let's look at the various combinations of electrophiles and nucleophiles that are commonly employed and see how the details of the reaction are affected in each case.

10.3   Addition of Hydrogen Halides

All of the halogen acids, HF, HCl, HBr, and HI, add to alkenes to give alkyl halides as shown in the following example, in which hydrogen chloride adds to 2-butene:

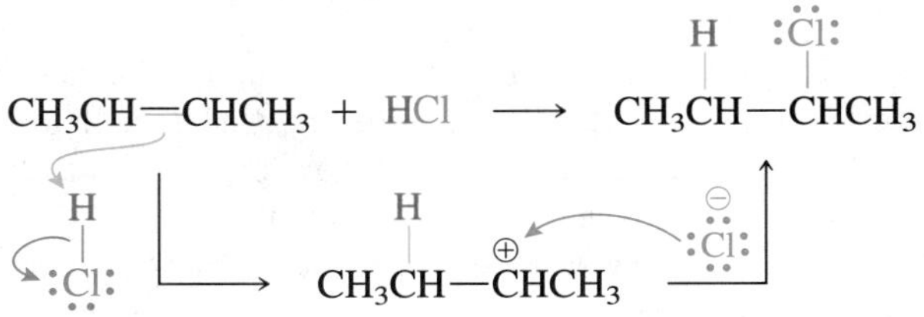

In this example the electrophile is a proton and the nucleophile is a chloride anion. The mechanism is just as described above; first the electrophilic proton adds to produce a carbocation intermediate, and then the chloride nucleophile bonds to the carbocation.

Because 2-butene is a symmetrical alkene, it does not matter which carbon initially bonds to the proton. Only one product, 2-chlorobutane, is possible. In the case of an unsymmetrical alkene—that is, one with different substituents on the two carbons of the double bond—two products could be formed, depending on which carbon bonds to the electrophile and which bonds to the nucleophile. For example, the reaction of hydrogen chloride with propene could produce 1-chloropropane or 2-chloropropane:

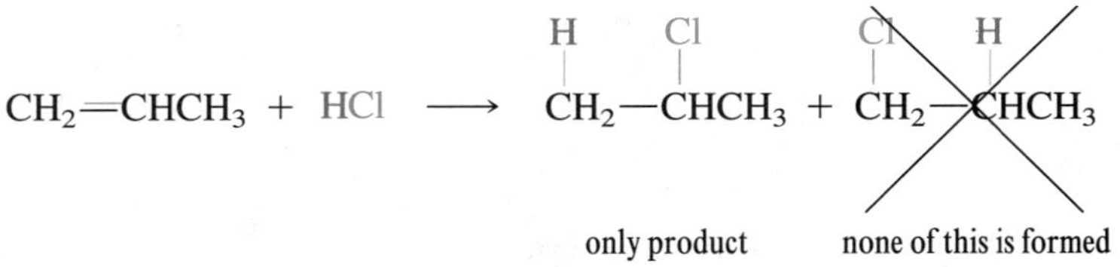

only product   none of this is formed

When this reaction is run in the laboratory, the only product formed is 2-chloro-propane. No 1-chloropropane is observed. A reaction such as this one that produces only one of two possible orientations of addition is termed a **regiospecific reaction.** (A reaction that produces predominantly one possible orientation but does form some of the product with the other orientation is termed a **regioselective reaction.**)

In 1869 a Russian chemist, Vladimir Markovnikov, studied the regiochemistry of a large number of these addition reactions. On the basis of his observations, he postulated an empirical rule that can be used to predict the orientation of additions to alkenes. **Markovnikov's rule** states that, in addition reactions of HX to alkenes, the H bonds to the carbon with more hydrogens (fewer alkyl substituents) and the X bonds to the carbon with fewer hydrogens (more alkyl substituents). As an example, Markovnikov's rule predicts that the addition of hydrogen bromide to 2-methylpropene should produce 2-bromo-2-methylpropane:

C-1 is bonded to
more hydrogens (two),
so the H should bond here

C-2 is bonded to
fewer hydrogens (none),
so the Br should bond here

When the reaction is run in the laboratory, only the product predicted by Markovnikov's rule is observed.

Markovnikov's rule was empirical, that is, it was based on observation only. At the time it was proposed, the concept of organic reaction mechanisms was far in the future, so it was impossible to provide a theoretical basis for why the H added to one carbon and the X to the other. Now that we know the mechanism for the reaction, it is easy to understand why these reactions are regiospecific. In fact, if we write the mechanism for the additions to give both possible products, we can predict the preferred product simply on the basis of what we already know about carbocations. It is much better to make predictions based on the mechanism of the reaction because cases that are exceptions to the empirical rule will be readily apparent.

The mechanisms for the two possible orientations of addition of HCl to propene are as follows:

a secondary carbocation

a primary carbocation

The first mechanism, which leads to the Markovnikov product 2-chloropropane, proceeds via a secondary carbocation. The second mechanism, which would lead to the unobserved product 1-chloropropane, proceeds via a primary carbocation. Because a secondary carbocation is lower in energy than a primary carbocation, the first mechanism should be preferred over the second mechanism. In fact, formation of the secondary carbocation is enough faster than formation of the primary carbocation that the first mechanism occurs to the exclusion of the second.

We can now see the reason behind Markovnikov's rule. If the proton adds to the carbon with more hydrogens, the positive carbon, where the nucleophile will ultimately bond, is bonded to more alkyl groups, resulting in a more stable carbocation. A more modern version of Markovnikov's rule, based on this mechanistic reasoning is: *The electrophile adds so as to form the more stable carbocation.* This means that the product has the electrophile attached to the carbon that would be less stable as a carbocation and the nucleophile attached to the carbon that would be more stable as a carbocation. Later, we will encounter exceptions to Markovnikov's rule. However, these exceptions are still in accord with this mechanistically based rule.

Some examples of additions of hydrogen halides are provided by the following equations. As you look at each example, try to predict the regiochemistry of the product before looking at the actual product that is formed.

$$CH_3CH{=}CH_2 + HF \xrightarrow{\text{no solvent}} CH_3\overset{\displaystyle F}{\underset{\displaystyle |}{C}}HCH_3 \quad (61\%)$$

$$Ph{-}\overset{\displaystyle CH_3}{\underset{\displaystyle |}{C}}{=}CH_2 + HCl \xrightarrow{\text{no solvent}} Ph{-}\overset{\displaystyle CH_3}{\underset{\displaystyle |}{\underset{\displaystyle Cl}{C}}}{-}CH_3 \quad (98\%)$$

$$CH_3CH_2CH_2CH_2CH{=}CH_2 + HBr \xrightarrow{H_2O} CH_3CH_2CH_2CH_2\overset{\displaystyle Br}{\underset{\displaystyle |}{C}}H{-}CH_3 \quad (88\%)$$

$$CH_3CH_2CH_2CH{=}CHCH_3 + HBr \xrightarrow{H_2O} CH_3CH_2CH_2\overset{\displaystyle Br}{\underset{\displaystyle |}{C}}H{-}CH_2CH_3 \quad (29\%)$$

$$+$$

$$CH_3CH_2CH_2CH_2{-}\overset{\displaystyle Br}{\underset{\displaystyle |}{C}}HCH_3 \quad (57\%)$$

Problems

Problem 10.2 Show the structures of the carbocations that are formed in the reaction of HBr with 2-hexene and explain why two products are formed.

Problem 10.3 Show the products of these reactions.

a)

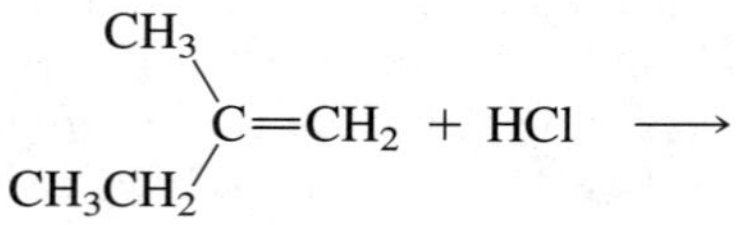

$$\begin{array}{c} CH_3 \\ \diagdown \\ C{=}CH_2 + HCl \longrightarrow \\ \diagup \\ CH_3CH_2 \end{array}$$

b) 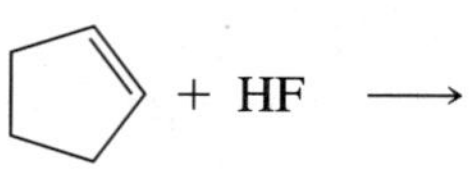 [cyclopentene] + HF $\longrightarrow$

c) $PhCH{=}CHCH_3 + HBr \longrightarrow$

d)

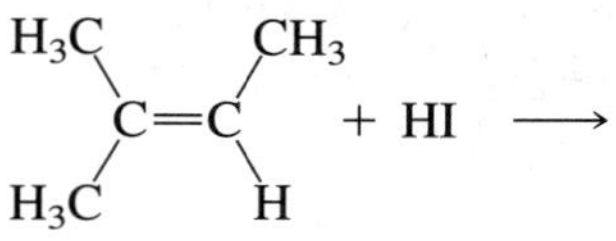

$$\begin{array}{c} H_3C \qquad CH_3 \\ \diagdown \quad \diagup \\ C{=}C \qquad + HI \longrightarrow \\ \diagup \quad \diagdown \\ H_3C \qquad H \end{array}$$

e) 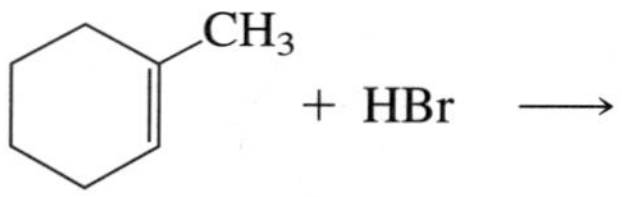 [cyclohexene with CH$_3$] + HBr $\longrightarrow$

f) 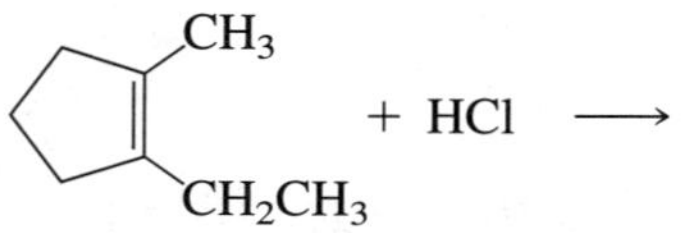 [cyclopentene with CH$_3$ and CH$_2$CH$_3$] + HCl $\longrightarrow$

Solution

c) Addition of $H^{\oplus}$ occurs so as to form the more stable carbocation. The two possible carbocations are shown in the following equations:

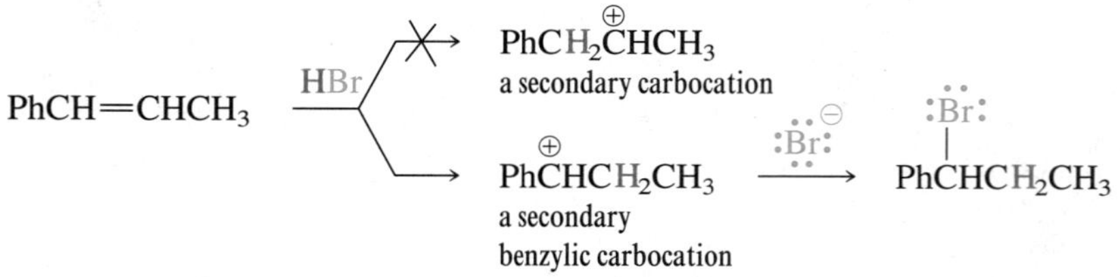

$$PhCH{=}CHCH_3 \xrightarrow{HBr} \begin{cases} \xcancel{\quad} PhCH_2\overset{\oplus}{C}HCH_3 \\ \text{a secondary carbocation} \\ \\ Ph\overset{\oplus}{C}HCH_2CH_3 \\ \text{a secondary} \\ \text{benzylic carbocation} \end{cases}$$

$$Ph\overset{\oplus}{C}HCH_2CH_3 \xrightarrow{\;:\overset{..}{\underset{..}{Br}}:^{\ominus}\;} \begin{array}{c} :\overset{..}{\underset{..}{Br}}: \\ | \\ PhCHCH_2CH_3 \end{array}$$

Because of its resonance stabilization, the secondary benzylic carbocation is more stable than the secondary carbocation so the product has the bromine (nucleophile) bonded to the carbon attached to the phenyl group.

Next, let's consider the stereochemistry of these reactions. Overall, addition reactions are the reverse of the elimination reactions we saw in Chapter 8. As was the case with the eliminations, there are two possible stereochemistries for the addition—syn and anti:

syn addition

anti addition

In a syn addition the electrophile and the nucleophile both add from the same side of the plane of the double bond; in an anti addition they add from opposite sides of this plane.

The mechanism for the addition of the hydrogen halides to alkenes proceeds through a carbocation intermediate. As was the case in the S_N1 reaction, the nucleophile can approach the planar carbocation equally well from either side, so we expect that the products should result from a mixture of syn and anti addition. Indeed, this is often the case. Under some conditions, however, the stereochemisty results from predominant syn addition, while anti addition is the favored pathway under other conditions. This occurs because these reactions are often conducted in nonpolar solvents in which ion pair formation is favored. The details of how this may affect the stereochemistry of these reactions are complex. Fortunately, stereochemistry is not an issue in most of the reactions in which hydrogen halides add, including all the examples presented above, because the carbon to which the proton is adding usually has at least one hydrogen already bonded to it. In such situations, syn addition and anti addition give identical products. Stereochemistry will be more important in some of the other reactions that are discussed later in this chapter.

Because carbocations are intermediates in these reactions, rearrangements can occur. The carbocation formed initially in the example below is secondary. Part of the time (17%), the chloride nucleophile intercepts the carbocation before it has a chance to rearrange. But a majority of the time (83%), the carbocation rearranges to a more stable tertiary carbocation, which, on reaction with chloride ion, produces the rearranged product.

this secondary carbocation reacts with chloride (17%)
and rearranges to the more stable tertiary carbocation (83%)

a tertiary carbocation

The hydrogen halides also add to the triple bond of alkynes. The regiochemistry of the reactions follows Markovnikov's rule. It is usually possible to add to just one of the pi bonds, producing a vinyl halide, or to both of the pi bonds, producing a dihaloalkane. Some examples are provided in the following equations:

$$Ph-C\equiv C-H \; + \; HCl \quad \xrightarrow{\;CH_2Cl_2\;} \quad \underset{Ph}{\overset{Cl}{\diagdown}}C{=}CH_2 \quad (73\%)$$

$$CH_3CH_2CH_2CH_2-C\equiv C-H \; + \; HBr \quad \xrightarrow{\;CH_2Cl_2\;} \quad \underset{CH_3CH_2CH_2CH_2}{\overset{Br}{\diagdown}}C{=}CH_2 \quad (89\%)$$

$$CH_3CH_2-C\equiv C-H \; + \; 2HBr \quad \longrightarrow \quad CH_3CH_2-\overset{\overset{\displaystyle Br}{|}}{\underset{\underset{\displaystyle Br}{|}}{C}}-CH_3$$

Problem

Problem 10.4   Show the products of these reactions.

a) 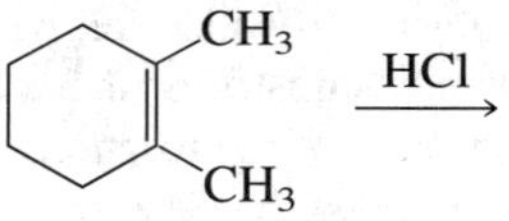 $\xrightarrow{\;HCl\;}$

b) $CH_3\overset{\overset{\displaystyle CH_3}{|}}{C}HCH{=}CH_2 \quad \xrightarrow{\;HBr\;}$

c) $PhC\equiv CH \quad \xrightarrow{\;HBr\;}$

d) $CH_3CH_2CH_2C\equiv CH \quad \xrightarrow{\;2\;HBr\;}$

Solution

a) Because the addition of HCl to an alkene proceeds through a carbocation intermediate, products from both syn and anti addition are usually formed. This means that products are expected with the H and the Cl both *cis* and *trans*.

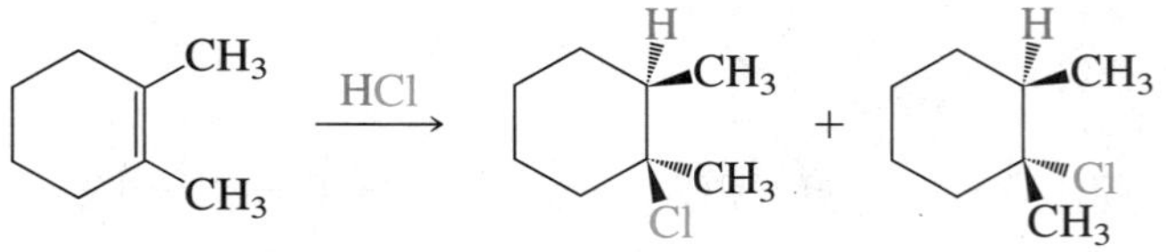

+ enantiomer   + enantiomer

10.4   Addition of Halogens

Both Cl_2 and Br_2 add to carbon–carbon double bonds to produce dihalides as illustrated in the following examples. The other halogens are not commonly used— F_2 because it is too reactive and I_2 because it is not reactive enough. These reactions are usually run in an inert solvent such as CCl_4 , $CHCl_3$, or CH_2Cl_2 .

$$CH_3CH{=}CHCH_2CH_3 + Cl_2 \xrightarrow{CHCl_3} \overset{\displaystyle Cl}{\underset{|}{CH_3CH}}{-}\overset{\displaystyle Cl}{\underset{|}{CHCH_2CH_3}} \quad (81\%)$$

$$PhCH{=}CH{-}\overset{\displaystyle O}{\overset{\|}{C}}OCH_2CH_3 + Br_2 \xrightarrow{CCl_4} \overset{\displaystyle Br}{\underset{|}{PhCH}}{-}\overset{\displaystyle Br}{\underset{|}{CH}}{-}\overset{\displaystyle O}{\overset{\|}{C}}OCH_2CH_3 \quad (85\%)$$

The reaction with bromine is a classical test for the presence of double (or triple) bonds in an unknown compound. In the test, a solution of bromine in a solvent such as CCl_4 or CH_2Cl_2 is added dropwise to a solution of the unknown. The bromine solution has a red-brown color. If the unknown contains carbon–carbon double or triple bonds, the addition reaction is nearly instantaneous. Since the addition products are colorless, the rapid disappearance of the bromine color constitutes a positive test for the presence of unsaturation.

The mechanism of these additions has an interesting variation from the one for the addition of the hydrogen halides. It begins with electrophilic attack of the halogen at the pi electrons of the double bond, as shown in Figure 10.1. As the

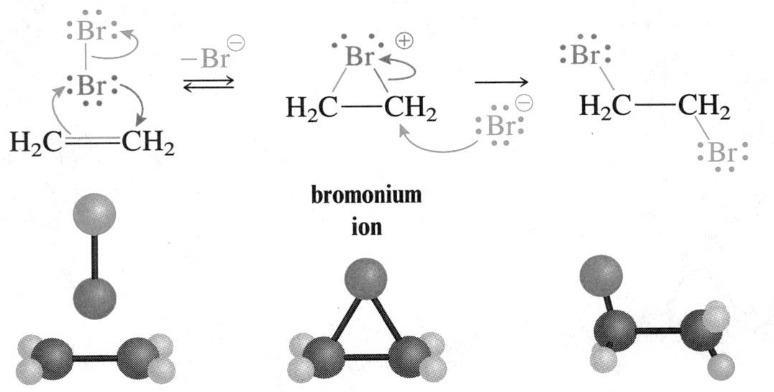

bromonium ion

In this reaction the electrophile is the bromine molecule. One Br leaves as a negative ion. As this Br leaves, the other Br becomes electrophilic and attracts the electrons of the alkene pi bond. If a carbocation were formed at this stage, it would look like this.

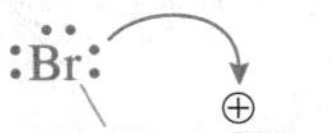

However, in this carbocation, the Br is very close to the positive carbon and can easily use a pair of its unshared electrons to form a bond to the carbon. The resulting species, a bromonium ion, is more stable than the carbocation because the octet rule is satisfied for all of the atoms. The carbocation is not actually formed in the reaction. Instead, the bromonium ion is formed directly. As the bromine–bromine bond is breaking, both carbon–bromine bonds are forming simultaneously.

The bromonium ion resembles a protonated epoxide (see Section 9.11) in both structure and reactions. Therefore when the nucleophilic bromide attacks to open the three-membered ring, it approaches from the side opposite the other bromine. The result is an anti addition of the two bromines to the double bond.

Figure 10.1

Mechanism of the Addition of Bromine to Ethene.

carbene, but rather a **carbenoid,** an organometallic species that reacts like a carbene. This species is generated by reaction of diiodomethane with zinc metal from a special alloy of zinc and copper:

$$CH_2I_2 + :Zn(Cu) \longrightarrow I—CH_2—Zn—I$$

a carbenoid
reacts like $\overset{..}{C}H_2$

The mechanism for the formation of this carbenoid and for its reaction with alkenes need not concern us here. Just remember that it reacts as though it is methylene. The Simmons-Smith reaction is an excellent way to prepare cyclopropane derivatives, as shown in the following examples. Note the stereochemistry in the second equation.

The reactions of carbenes with alkenes, as illustrated in this section, are the best methods for the synthesis of cyclopropanes.

Problem

Problem 10.22 Show the products of these reactions.

ELABORATION

Singlet and Triplet Carbenes

The electronic structure of a carbene is somewhat more complicated than what has been presented so far. A carbene has two bonds and therefore has two atomic orbitals in which to place the two unshared electrons. In a **singlet carbene** these two electrons are in the same AO with opposite spins. The carbon is roughly sp^2 hybridized with the unshared electron pair occupying the third sp^2 hybrid AO and no electrons occupying the remaining p AO. This is the picture for a carbene that we have been using up to this point. The other possible electronic structure is termed a **triplet carbene.** It has the two unshared electrons in different AOs with the same spins. It can be viewed as roughly sp hybridized with the sp AOs used to form the sigma bonds to the hydrogens. Each of the two remaining p AOs contains one electron, and these electrons have the same spin. According to Hund's rule, this minimizes electron–electron repulsion. (This is an oversimplified but adequate picture.) Calculations suggest that the triplet state of methylene is more stable than the singlet state by about 10 kcal/mol (42 kJ/mol).

singlet carbene

triplet carbene

A simplified picture has the carbon sp^2 hybridized with the unshared electrons in an sp^2 AO with opposite spins. (The red arrows are used to indicate the spins of the electrons.)

A simplified picture has the carbon sp hybridized with the unshared electrons in different p AOs with the same spin.

The methods that have been presented so far for the generation of carbenes (decomposition of diazomethane by heat or light and 1,1-eliminations) generate the singlet state of the carbene. This is expected because all the electrons in the reactant are paired; that is, there is the same number of electrons with both spin states. Unless a spin change occurs, the products must also have the same number of electrons with each spin state. Therefore only singlet products are formed:

$$\overset{\ominus}{\ddot{C}}H_2 - \overset{\oplus}{N} \equiv N\colon \quad \xrightarrow[\text{or } \Delta]{h\nu} \quad \ddot{C}H_2 + N_2$$

singlet

The triplet state of the carbene can be produced by a special photochemical reaction using a photosensitizer. While we need not be concerned with all of the details of this process, the photosensitizer absorbs the light and is converted to its triplet excited state. The triplet photosensitizer then reacts with the diazomethane to produce the triplet state of the carbene, nitrogen (singlet state) and the photosensitizer in its singlet ground state. Overall, spin is conserved.

$$\overset{\ominus}{\ddot{C}}H_2 - \overset{\oplus}{N} \equiv N\colon \quad \xrightarrow[\text{photosensitizer}]{h\nu} \quad \ddot{C}H_2 + N_2$$

triplet

Although singlet and triplet carbenes have similar chemical reactions, there are some differences. Recall that the carbene additions to alkenes discussed previously, which we now know involve singlet carbenes, proceed by a stereospecific syn addition:

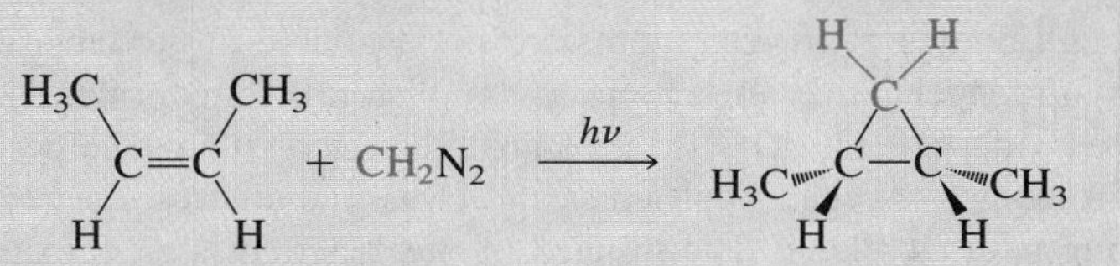

In contrast, the addition of a triplet carbene to an alkene cannot occur in a concerted manner because the numbers of electrons with each spin state are different. Therefore one bond is formed first, leaving one electron on both the other carbon of the double bond and the carbene carbon. Because these electrons have the same spin, one of them must change its spin before the second carbon–carbon bond can form. This spin flip ordinarily takes enough time that rotation about the carbon–carbon bond of the alkene, which was a double bond in the reactant but is now a single bond, can compete. This results in the loss of the stereochemistry that was present in the original alkene and the products of both syn and anti addition are formed. Thus the reaction of (Z)-2-butene with diazomethane in the presence of a photosensitizer gives a mixture of *cis*- and *trans*-1,2-dimethylcyclopropane. A nonstereospecific reaction, like this one, often indicates a nonconcerted mechanism involving an intermediate that can lose stereochemistry:

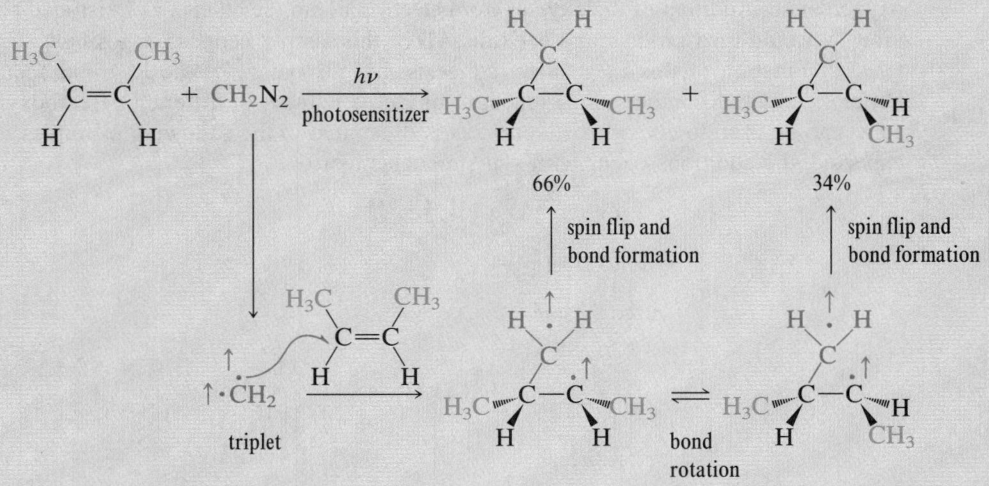

10.10 Epoxidation

An oxygen atom is analogous to a carbene in that it has only six electrons in its valence shell. Like a carbene, it might be expected to act as both an electrophile and a nucleophile in its reaction with an alkene, resulting in the formation of a three-membered ring containing an oxygen—an epoxide.

Although bottles of oxygen atoms are not available in the laboratory, **percarboxylic acids** are reagents that are able to accomplish this transformation. Recall that hydrogen peroxide, H—O—O—H, has an oxygen–oxygen bond. A percarboxylic acid also has an oxygen–oxygen bond and can act as a source of electrophilic oxygen when reacting with an alkene. The product of this reaction is an epoxide.

This epoxidation reaction is concerted. It appears quite complex because of all the bonds that are being made or broken in this one step. It might help to focus on the blue oxygen and imagine the process to occur in a stepwise fashion. A 1,1-elimination of the proton and the gold oxygen would generate a normal carboxylic acid and an oxygen atom. Addition of the oxygen atom electrophile to the alkene, as illustrated previously, would then produce the epoxide. All of this simply happens in a single step.

Examination of this mechanism suggests that the nature of the R group should not make much difference in the reaction. In fact, a number of different percarboxylic acids can be used to epoxidize alkenes, as is illustrated in the following examples. As expected, the additions occur with syn stereochemistry.

Problem

Problem 10.23 Show the products of these reactions.

a) 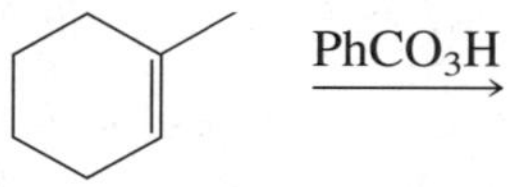
PhCO$_3$H

b)

H$_3$C, CH$_3$, C=C, H, H   → MCPBA →   NaOH / H$_2$O →

10.11 Hydroxylation

There are two reagents that add hydroxy groups to both carbons of a carbon–carbon double bond: osmium tetroxide, OsO$_4$, and potassium permanganate, KMnO$_4$. These operate through similar mechanisms in which one of the oxygens that is bonded to the metal acts as an electrophile and another acts as a nucleophile, as illustrated in the following equation for the reaction of osmium tetroxide:

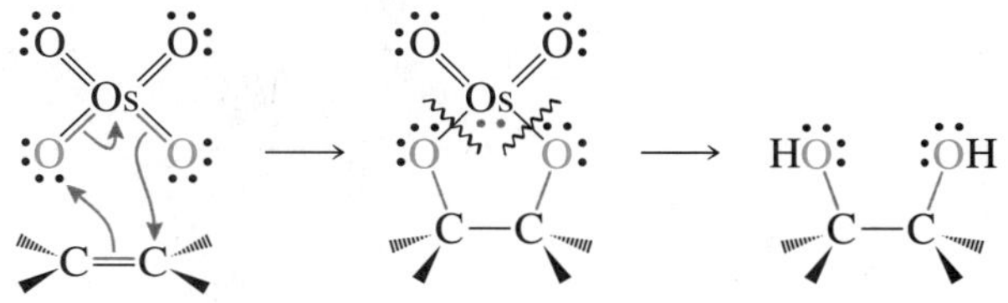

osmate ester   1,2-diol

The cyclic intermediate, called an osmate ester, is not isolated; instead, the osmium–oxygen bonds are cleaved by using a reagent such as sodium sulfite, Na$_2$SO$_3$, resulting in the formation of a 1,2-diol. (The mechanistic details of the cleavage step need not concern us.) Since both the electrophilic and nucleophilic oxygens are attached to the same metal atom, both are delivered from the same side of the plane of the double bond—the reaction is a syn addition.

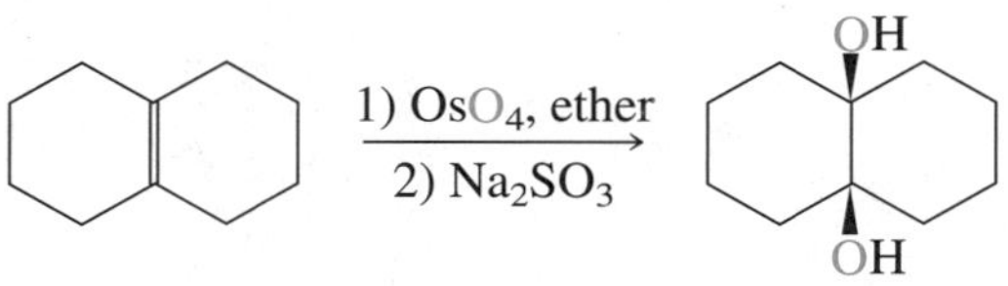

1) OsO$_4$, ether
2) Na$_2$SO$_3$

(81%)

The reaction employing potassium permanganate is conducted in basic aqueous solution. It proceeds by a similar mechanism, and the intermediate manganate ester is cleaved directly under the reaction conditions, resulting in an overall syn addition of two hydroxy groups. The yields in this reaction are often less than 50%, significantly lower than the reaction using osmium tetroxide:

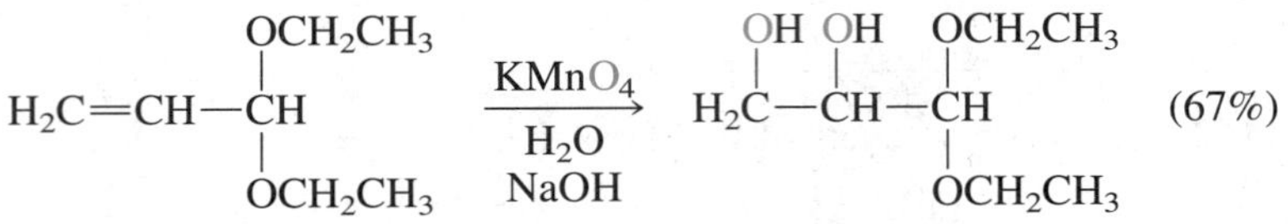

$$H_2C=CH-\overset{\displaystyle OCH_2CH_3}{\underset{\displaystyle OCH_2CH_3}{CH}} \quad \xrightarrow[\substack{H_2O \\ NaOH}]{KMnO_4} \quad H_2C-\overset{\displaystyle OH}{CH}-\overset{\displaystyle OH}{\underset{\displaystyle OCH_2CH_3}{CH}} \overset{\displaystyle OCH_2CH_3}{} \quad (67\%)$$

Whereas hydroxylation using potassium permanganate often gives low yields, the reaction employing osmium tetroxide has its own limitations. The osmium reagent is very expensive and extremely toxic. To help alleviate these problems, methods have been developed that require only a catalytic amount of osmium tetroxide in the presence of some additional oxidizing agent, such as *tert*-butyl peroxide. The reaction proceeds to the cyclic osmate ester as before. The peroxide serves to cleave the ester to the diol and, importantly, oxidizes the osmium back to the tetroxide so that the cycle can be repeated. Therefore only a small amount of osmium tetroxide is needed. The yields are respectable, as shown in the following example. Note the stereochemistry of the product.

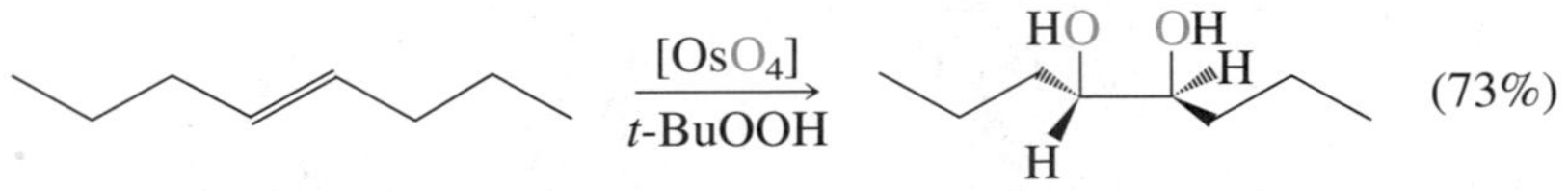

(73%)

Problems

Problem 10.24   Show the products of these reactions.

a) 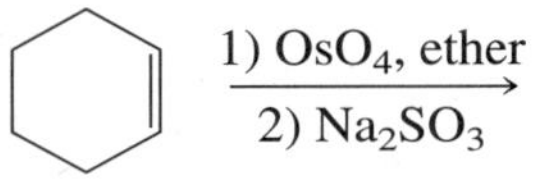
1) OsO$_4$, ether
2) Na$_2$SO$_3$

b) $CH_3CH_2CH=CH_2 \quad \xrightarrow[\substack{H_2O \\ NaOH}]{KMnO_4}$

c) 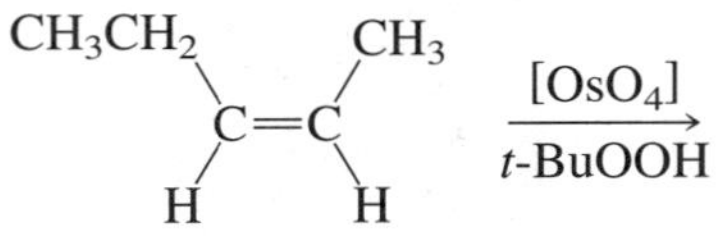
$[OsO_4]$
$t\text{-BuOOH}$

Problem 10.25   Show syntheses of these compounds from (Z)-2-butene.

a)

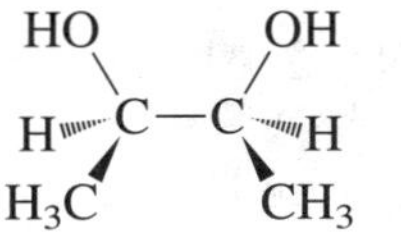

b)

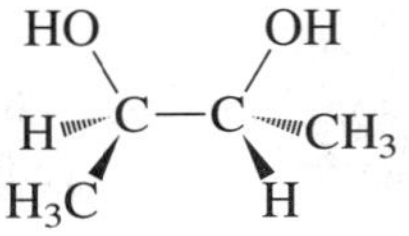

10.12 Ozonolysis

Ozone, O$_3$, is a form of oxygen found in the upper atmosphere. Its presence there is beneficial because it absorbs some harmful ultraviolet radiation, preventing it from reaching the surface of the earth. Ozone is also found at the surface of the earth as a component of smog. Here, its presence is detrimental because of its high reactivity. One of the important reactions of ozone occurs with alkenes and results in the cleavage of the carbon–carbon double bond as illustrated in the following example. The ozone first adds to the double bond. This addition can be viewed as though one oxygen atom acts as an electrophile and the other acts as a nucleophile, although the mechanism is actually an example of a pericyclic reaction (see Chapter 20). The initial product, called a molozonide, spontaneously rearranges to an ozonide. The ozonide is not usually isolated. Instead, it is treated with a reducing agent (dimethyl sulfide works nicely) that cleaves the O—O bond and results in the formation of two carbonyl groups. This sequence results in the cleavage of the carbon–carbon double bond and the formation of two carbon–oxygen double bonds.

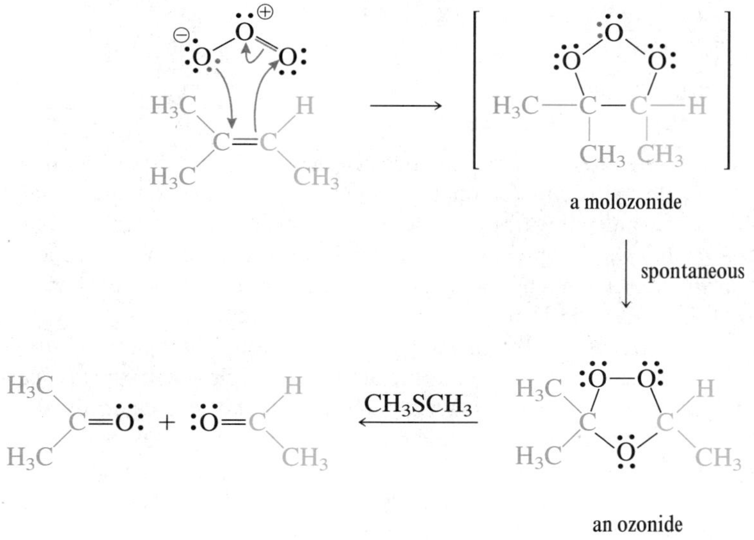

a molozonide

spontaneous

CH_3SCH_3

an ozonide

Some examples are shown in the following equations:

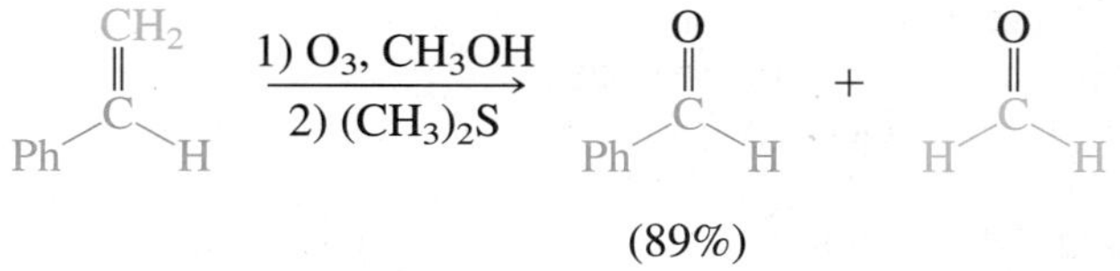

1) O_3, CH_3OH
2) $(CH_3)_2S$

(89%)

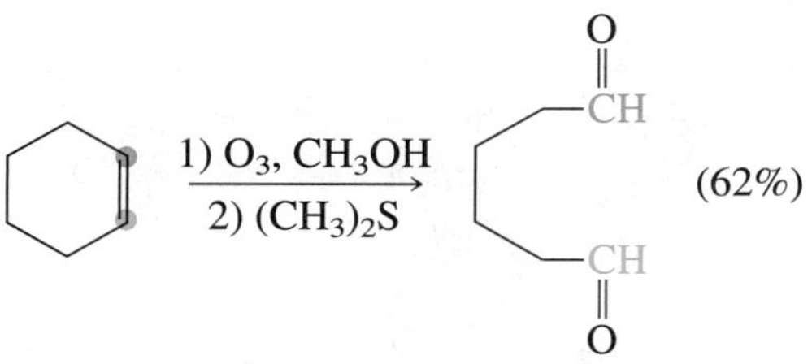

1) O_3, CH_3OH
2) $(CH_3)_2S$

(62%)

Acid anhydrides result from substituting the acyl group of one acid for the hydroxy hydrogen of another. They are called anhydrides because they can be viewed as resulting from the loss of water from two carboxylic acid molecules (removing H from one and OH from the other). Symmetrical anhydrides derived from two molecules of the same carboxylic acid are most often encountered. These are simply named by replacing *acid* in the name of the carboxylic acid with *anhydride*.

$$CH_3\overset{\displaystyle O}{\overset{\|}{C}}-O-\overset{\displaystyle O}{\overset{\|}{C}}CH_3$$

acetic anhydride
ethanoic anhydride

benzoic anhydride

Esters can be viewed as resulting from the combination of a carboxylic acid with an alcohol, using the alkoxy (OR) group of the alcohol to replace the hydroxy (OH) group of the acid. The name therefore must designate both the alcohol part and the acid part of the ester. The name uses two separate words. First the R group of the alcohol is named just like other *groups* we have encountered, using a -yl suffix. This is the first word in the name. Then the acid part is named as usual, and the -ic acid suffix is replaced with the -ate. This is the second word in the name.

$$\overset{\displaystyle O}{\overset{\|}{C}}-OCH_2CH_2CH_3$$

$$\overset{\displaystyle O}{\overset{\|}{C}}-OCH_2\overset{\displaystyle CH_3}{\overset{|}{C}}HCH_2CH_3$$

propyl benzoate
(The red part of this ester is derived from the alcohol [propanol], and the blue part from the carboxylic acid [benzoic acid]. Be careful to correctly identify which part of each ester comes from the acid and which from the alcohol. Note that *propyl* and *benzoate* are separate words.)

(2-methylbutyl) 3-methyl-1-cyclohexanecarboxylate
(The complex group from the alcohol portion of the ester is named in the same manner as the complex groups of alkanes.)

Amides can be viewed as resulting from the replacement of the OH of a carboxylic acid with an NH_2 (primary amide), NHR (secondary amide), or NR_2 (tertiary amide) group. To name an amide, the longest carbon chain having the carbonyl group at one terminus is chosen as the root, as usual. Systematic names are formed by replacing the final -e of the *hydrocarbon name* for this root with -amide. Amide names can also be derived from common names of acids by replacing -ic acid or -oic acid with -amide. Other groups attached to the nitrogen are designated with the prefix *N*-, as was done in the case of amines.

$$\underset{}{CH_3CH_2CH_2CH_2\overset{\overset{O}{\|}}{C}}-NH_2$$

pentanamide
a primary amide
(This name is derived from the five-
carbon hydrocarbon, pentane.)

$$\underset{6\quad5\quad4\quad3\quad2\;1}{CH_2{=}CHCH\overset{\overset{CH_3}{|}}{C}H{=}CH\overset{\overset{O}{\|}}{C}}-NHCH_2CH_3$$

N-ethyl-4-methyl-2,5-hexadienamide
a secondary amide
(Note the N- that is used to show that
the ethyl group is bonded to the nitrogen.)

$$H-\overset{\overset{O}{\|}}{C}-\overset{\overset{CH_3}{|}}{N}-CH_3$$

N,N-dimethylformamide
a tertiary amide
(The common name for the one-
carbon carboxylic acid is
formic acid.)

On first consideration, **nitriles** do not appear to be related to the other carboxylic acid derivatives because they have no acyl group. However, they can be viewed as resulting from the removal of H_2O from a primary amide (loss of both H's from the N and the O from the carbonyl group), and their chemical reactions are related to other carboxylic acid derivatives. Therefore it is convenient to include them with the other carboxylic acid derivatives. They are named in a similar manner to amides, that is, -nitrile is appended to the hydrocarbon name. (Do not forget to count the carbon of the —CN group and to give this carbon the number 1.) Common names are obtained from the common name of the carboxylic acid by replacing the -ic acid or -oic acid with -onitrile. In complex compounds the —CN group can be named as a cyano group.

$$\underset{5\quad4\;\;\;3\;2\quad1}{CH_3C{\equiv}CCH_2C{\equiv}N}$$

3-pentynenitrile
(This is a systematic name. Because
the C of the CN must be given the
number 1, the related hydrocarbon is
3-pentyne.)

benzonitrile
(The related carboxylic acid is benzoic
acid.)

methyl 4-cyanobenzoate
(In a complex compound like this, with
more than one functional group, the
—CN is named as a substituent group, a
cyano group, attached to the root in the
same manner as a halogen.)

Carboxylate salts consist of a carboxylate anion (the anion formed by removal of the proton from the OH of a carboxylic acid) and a cation. They are named in a manner similar to esters, using two words. The first word designates the cation. The second word designates the carboxylate anion, using the -ate suffix, just as is done for esters.

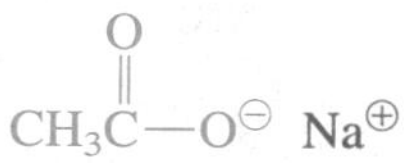

$$CH_3\overset{O}{\overset{||}{C}}-O^{\ominus} \ Na^{\oplus}$$

sodium acetate

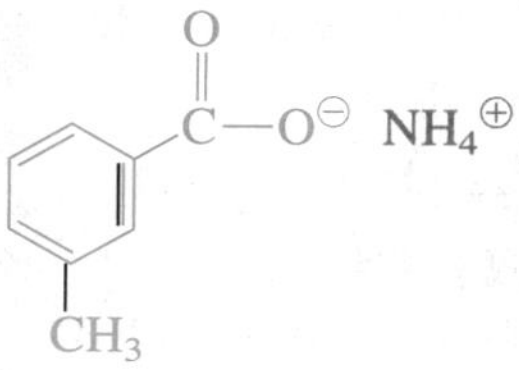

ammonium 3-methylbenzoate

Problems

Problem 11.14 Provide names for these compounds.

a)

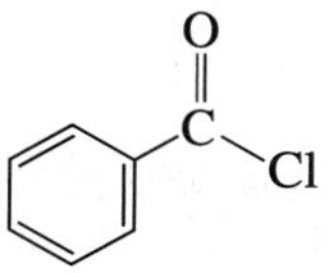

b)  $CH_3CH_2\overset{O}{\overset{||}{C}}O\overset{O}{\overset{||}{C}}CH_2CH_3$

c) $CH_3CH_2\overset{O}{\overset{||}{C}}OCH_3$

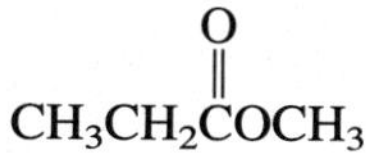

d)

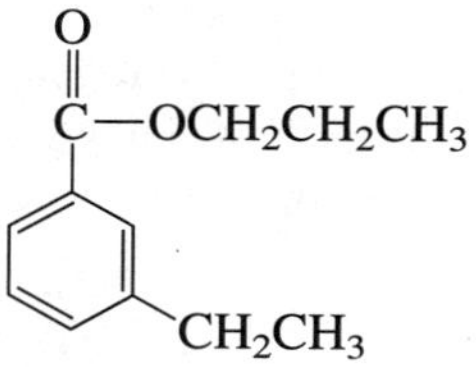

e) 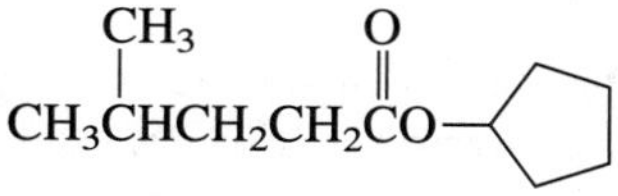 $CH_3\overset{CH_3}{\overset{|}{C}}HCH_2CH_2\overset{O}{\overset{||}{C}}O\text{–}$

f)  $CH_3C{\equiv}CCH_2\overset{O}{\overset{||}{C}}NHCH_3$

g) 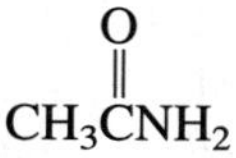 $CH_3\overset{O}{\overset{||}{C}}NH_2$

h)

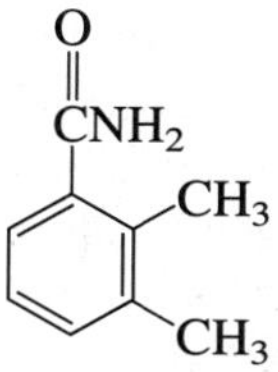

i)

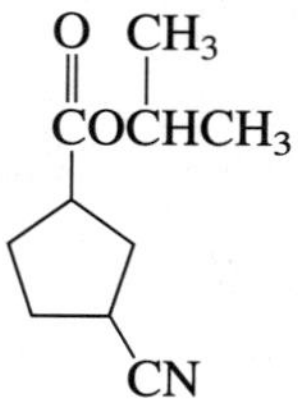

j) $CH_3CH_2CH_2CH_2C{\equiv}N$

k) 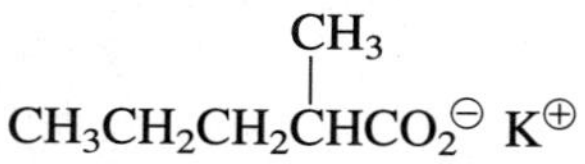 $CH_3CH_2CH_2\overset{CH_3}{\overset{|}{C}}HCO_2^{\ominus} \ K^{\oplus}$

Problem 11.15 Draw structures for these compounds.

   a) propanoyl chloride
   b) *N,N*-dimethylacetamide
   c) pentanoic anhydride
   d) sodium *p*-nitrobenzoate
   e) hexanamide
   f) isopropyl acetate
   g) benzyl benzoate
   h) ethyl cyclopentanecarboxylate
   i) 3-chlorobenzonitrile
   j) 3-methylheptanenitrile

Acid chlorides, acid anhydrides, and esters all contain the carbonyl group but the presence of this polar group has only a small effect on their melting and boiling points. Amides, however, are considerably more polar because of the significant contribution of a charged resonance structure to the resonance hybrid. In addition, primary and secondary amides, with hydrogens bonded to the nitrogen, can also form hydrogen bonds among themselves. For these reasons, amides melt and boil even higher than carboxylic acids of similar molecular mass.

$$R-\overset{\overset{\ddot{O}\cdot}{\|}}{C}-\ddot{N}H_2 \longleftrightarrow R-\overset{\overset{:\ddot{O}:^{\ominus}}{|}}{C}=\overset{\oplus}{N}H_2$$

Table 11.2 lists the melting points and boiling points of a series of compounds of nearly identical molecular mass but containing a variety of different functional groups. Note the increase in the melting point and boiling point that occurs when the polar carbonyl group is introduced. The effect is small (or may be absent) for the ester but is more significant for the aldehyde and ketone, especially on their boiling points. There is an additional increase in the boiling point for the alcohol due to hydrogen bonding and a further increase in the melting and boiling points of the carboxylic acid due to its polar, hydrogen-bonding carboxy group. Finally, the amide has the highest melting and boiling points because of its highly polar nature in combination with its ability to form hydrogen bonds.

Problem

Problem 11.16 Explain which compound has the higher melting point or boiling point.

   a) higher m.p. $CH_3CH_2CH_2\overset{O}{\overset{\|}{C}}NH_2$ or $CH_3\overset{O}{\overset{\|}{C}}N-CH_3$ with CH_3

   b) higher b.p. $CH_3CH_2CH_2\overset{O}{\overset{\|}{C}}OH$ or $CH_3CH_2\overset{O}{\overset{\|}{C}}OCH_3$

Table 11.2 Comparison of the Effect of Functional Groups on Melting and Boiling Points of Compounds of Comparable Molecular Mass.

Structure	Functional Group	m.p. (°C)	b.p. (°C)
$CH_3CH_2CH_2CH_2CH_3$	alkane	−130	36
$CH_3CH_2OCH_2CH_3$	ether	−116	35
$CH_3\overset{\text{O}}{\overset{\|}{C}}OCH_3$	ester	−98	57
$CH_3CH_2CH_2\overset{\text{O}}{\overset{\|}{C}}H$	aldehyde	−99	76
$CH_3CH_2\overset{\text{O}}{\overset{\|}{C}}CH_3$	ketone	−86	80
$CH_3CH_2CH_2CH_2OH$	alcohol	−90	117
$CH_3CH_2\overset{\text{O}}{\overset{\|}{C}}OH$	carboxylic acid	−21	141
$CH_3CH_2\overset{\text{O}}{\overset{\|}{C}}NH_2$	amide	81	213

c) higher b.p.

or

d) higher m.p. $CH_3CH_2CH_2\overset{\text{O}}{\overset{\|}{C}}OH$ or $CH_3CH_2CH_2\overset{\text{O}}{\overset{\|}{C}}NH_2$

In general, acid chlorides and acid anhydrides are too reactive to occur naturally, and nitriles are rare in nature. Esters and amides, on the other hand, are very common. Many esters have pleasant odors, often sweet or fruity, and are responsible for the fragrant odors of fruits and flowers. They are components of many flavorings, both natural and artificial. For example, isopentyl acetate has a strong banana odor, and methyl butanoate is used as an artificial rum flavoring. Typical fats are triesters formed from long-chain "fatty" acids and the triol glycerol.

$$CH_3\overset{\displaystyle O}{\overset{\|}{C}}-OCH_2CH_2\overset{\displaystyle CH_3}{\overset{|}{C}}HCH_3$$

isopentyl acetate

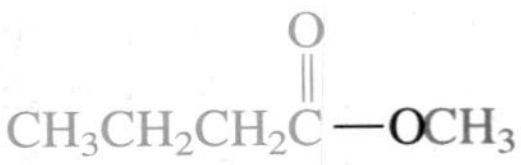

$$CH_3CH_2CH_2\overset{\displaystyle O}{\overset{\|}{C}}-OCH_3$$

methyl butanoate

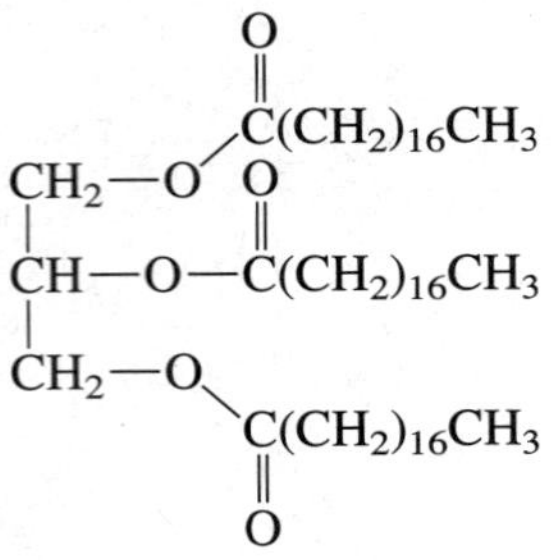

a fat

$$\begin{array}{l} CH_2OH \\ | \\ CHOH \\ | \\ CH_2OH \end{array}$$

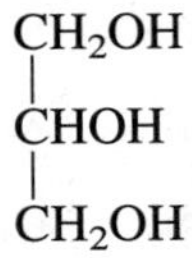

glycerol

Amides often have pronounced physiological activity; acetaminophen, a common pain reliever, and the diethyl amide of lysergic acid (LSD), a hallucinogen, are two examples. The important peptide bond of proteins is actually an amide bond. Dimethylformamide (DMF) and dimethylacetamide (DMA) are important solvents in the organic laboratory. They are highly polar and are capable of dissolving many ionic reagents in addition to less polar organic compounds. They are fairly unreactive because the bond between the carbonyl group and the nitrogen of an amide is reasonably strong, and they do not have hydrogens on the nitrogens that might react as acids. They are especially good solvents for S_N2 reactions because they are aprotic.

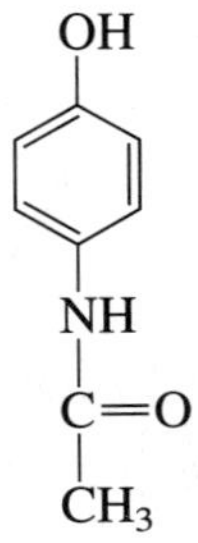

acetaminophen

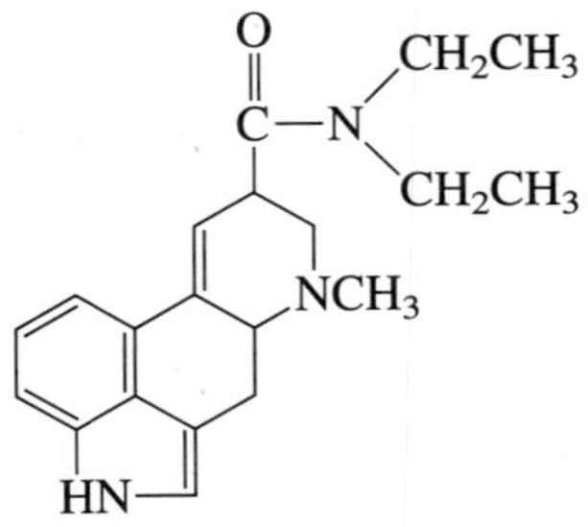

lysergic acid diethylamide
(LSD)

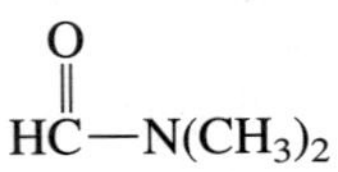

$$HC\overset{\displaystyle O}{\overset{\|}{}}-N(CH_3)_2$$

dimethylformamide
(DMF)

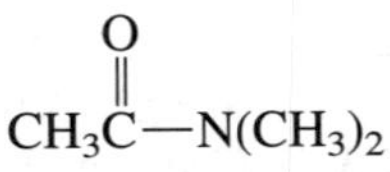

$$CH_3C\overset{\displaystyle O}{\overset{\|}{}}-N(CH_3)_2$$

dimethylacetamide
(DMA)

If you have spent very much time in a laboratory where organic chemicals are used, you are well aware that many organic compounds have very powerful odors. In some cases these odors are disagreeable, but others are pleasant. This is especially true of esters, which often have fruity odors, and aldehydes and ketones, many of which have floral odors. In fact, aldehydes, ketones, and esters with relatively simple structures are major components of numerous natural scents and flavors. These natural materials are usually extremely complex mixtures, sometimes containing hundreds of compounds. Artificial scents and flavors usually contain fewer components, but many still have complex recipes consisting of dozens of ingredients. Isopentyl acetate (banana) and methyl butanoate (rum) are examples of fragrant esters that have already been mentioned (see page 481). Other examples include isopentenyl acetate, the flavoring used in Juicy Fruit gum; ethyl phenylacetate, which has a honey odor; methyl salicylate, a major component of oil of wintergreen; and benzyl acetate, which composes over 60% of jasmine oil.

Fragrant Organic Compounds

isopentenyl acetate
(Juicy Fruit gum)

ethyl phenylacetate
(honey)

methyl salicylate
(wintergreen)

benzyl acetate
(jasmine)

Although many esters are used in artificial flavorings, they are less often employed as ingredients of perfumes because they are slowly hydrolyzed to a carboxylic acid and an alcohol on the skin, as shown in the following equation:

$$R-\overset{\displaystyle O}{\overset{\|}{C}}-O-R' \ + \ H_2O \ \longrightarrow \ R-\overset{\displaystyle O}{\overset{\|}{C}}-O-H \ + \ H-O-R'$$

This is quite undesirable in a perfume because many carboxylic acids have objectionable odors. For example, the sharp, penetrating odor of vinegar is due to acetic acid, butanoic acid (butyric acid) smells like rancid butter, and 2-methylpropanoic acid (isobutyric acid) is a component of sweat. A common name for hexanoic acid is caproic acid, derived from the Latin word *caper*, which means goat. If you have ever been around goats, perhaps you can imagine the odor of hexanoic acid.

ELABORATION

$$CH_3CH_2CH_2COH$$

butanoic acid
butyric acid
(rancid butter)

$$CH_3CH{-}COH$$ (with CH_3 substituent)

2-methylpropanoic acid
isobutyric acid
(sweat)

$$CH_3CH_2CH_2CH_2CH_2COH$$

hexanoic acid
caproic acid
(goat)

Aldehydes and ketones are also important components of many fragrances and flavors. For example, butanal (butyraldehyde) is used to impart a buttery flavor to margarine and other foods. Since aldehydes are slowly oxidized to carboxylic acids by the oxygen of air, it is readily apparent how the odor of rancid butter arises. Coumarin has an odor that is often described as "new-mown hay" or "woody" and is used in men's toiletries. Along with vanillin (see page 472), it is a component of natural vanilla. A combination of these two compounds, prepared synthetically in the laboratory, is used in artificial vanilla, and the flavoring in cream soda consists of coumarin and vanillin. Although α-pentylcinnamaldehyde does not occur naturally, it has been found to have a powerful jasmine odor and is used in many perfumes and soaps. α-Ionone is a naturally occurring ketone with an odor resembling violets. Many large-ring ketones have a musky odor and are prized ingredients in perfumes. Muscone (see page 472), which was first isolated from the musk deer, has a 15-membered ring, and civetone, from the civet cat, has a 17-membered ring. These compounds were extremely expensive when they could only be obtained from natural sources. Once their structures were determined, they were prepared in the laboratory. Although the preparation of such large rings is difficult, these synthetic materials are still considerably less expensive than their natural counterparts. Other fragrant aldehydes and ketones are mentioned on page 472.

$$CH_3CH_2CH_2CH$$

butanal
butyraldehyde
(buttery)

coumarin
(new-mown hay)

α-pentylcinnamaldehyde
(jasmine)

α-ionone
(violet)

civetone
(musky)

11.7   Sulfur and Phosphorus Compounds

Sulfur occurs directly beneath oxygen in the periodic table, and, like oxygen, it often exhibits a valence of two. Therefore sulfur analogs of alcohols and ethers are often encountered. However, because sulfur is in the third period of the periodic table, it can also have a higher valence. Structures with four or six bonds to a sulfur are common. In organic chemistry the most important of these "expanded valence" compounds have the sulfur bonded to one or two extra oxygens.

Similarly, phosphorus occurs directly beneath nitrogen in the periodic table and therefore often exhibits a valence of 3. Again, structures with an expanded valence, having five bonds to the phosphorus, are common, especially when the extra bonds are to oxygen. This book is not concerned with all the possible sulfur and phosphorus compounds, nor does it spend much time on their nomenclature. Instead, it concentrates on those of most importance in organic chemistry and biochemistry. Let's begin with a discussion of some common sulfur compounds.

Sulfur analogs of alcohols are called **thiols** or **mercaptans.** They are named in the same general manner as alcohols but with the suffix -thiol added to the name of the hydrocarbon. (In chemistry, thi- or thio-, from the Greek word for sulfur, *theion*, is used to indicate a compound that contains sulfur.) Thus the sulfur analog of ethanol, CH_3CH_2SH, is named ethanethiol. Note that the "e" at the end of the hydrocarbon name is retained to aid in pronunciation. Sulfur analogs of ethers are called **sulfides.** They are named in the same general manner as ethers but with sulfide replacing ether in the name. Some examples of thiols and sulfides are the following:

$$\underset{\text{3-methyl-1-butanethiol}}{CH_3\overset{\displaystyle CH_3}{\overset{\displaystyle |}{C}}HCH_2CH_2{-}SH} \qquad \underset{\text{2-butene-1-thiol}}{CH_3CH{=}CHCH_2{-}SH} \qquad \underset{\text{dimethyl sulfide}}{CH_3{-}S{-}CH_3}$$

A characteristic of organic sulfur compounds, especially volatile (low molecular mass) thiols, is their disagreeable odors. For example, 3-methyl-1-butanethiol and 2-butene-1-thiol are ingredients of a skunk's "perfume," and methanethiol or ethanethiol is usually added, in small amounts, to natural gas, which is odorless by itself, so that leaks can be readily detected. The chemical properties of thiols and sulfides differ from those of alcohols and ethers in that thiols are somewhat stronger acids than alcohols and the sulfur atoms of these compounds are considerably more nucleophilic than the oxygen of their analogs. They are excellent nucleophiles in substitution reactions.

Other sulfur compounds of importance in organic chemistry have additional oxygens bonded to the sulfur, as in the **sulfoxide** and **sulfone** functional groups. (You may encounter sulfoxides written as either of their resonance structures.) The chemistry of these compounds is not covered in detail in this text. It is worth noting, however, that dimethylsulfoxide (DMSO) is an important solvent that is used in many organic reactions. It is quite polar, so it will dissolve both organic compounds and inorganic reagents, and it is fairly unreactive, so it will not interfere with many reactions. It is also aprotic.

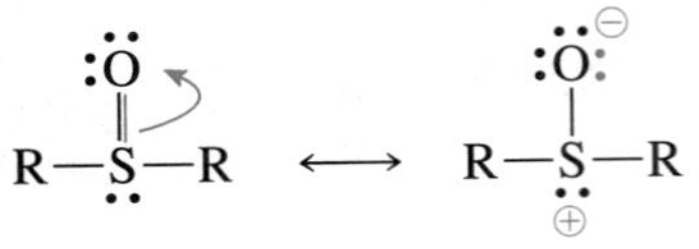

a sulfoxide

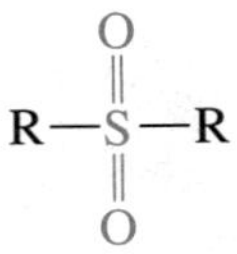

a sulfone 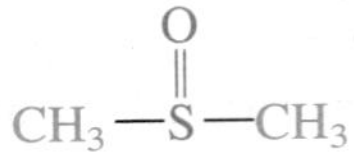   dimethylsulfoxide
(DMSO)

Another important group of sulfur-containing compounds can be viewed as being derived from sulfuric acid. Replacing one of the OH groups of sulfuric acid with a carbon group results in a new functional group called a **sulfonic acid.** Sulfonic acids are strong acids, comparable to sulfuric acid in strength. *p*-Toluenesulfonic acid is often used when a strong acid that is soluble in organic solvents is needed. The chemistry of sulfonic acids has some similarities to that of carboxylic acids. Thus derivatives of sulfonic acid, like derivatives of carboxylic acids, include sulfonyl chlorides, sulfonate esters, and sulfonamides. As shown in the examples below, nomenclature of these compounds attaches -sulfon- to the hydrocarbon name and adds the same final suffixes as are used for the carboxylic acid derivatives.

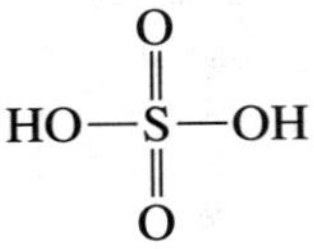

sulfuric acid

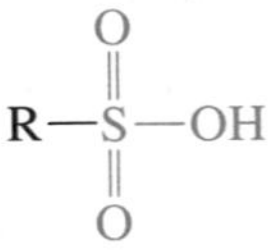

a sulfonic acid

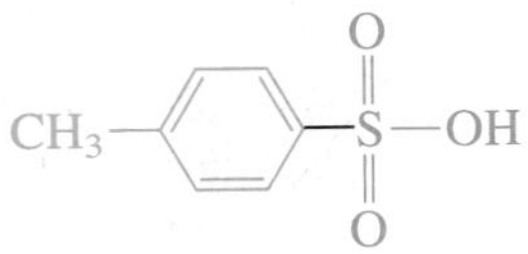

p-toluenesulfonic acid
(The suffix -sulfonic acid is added to the hydrocarbon name.)

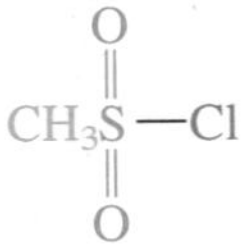

methanesulfonyl chloride
a sulfonyl chloride

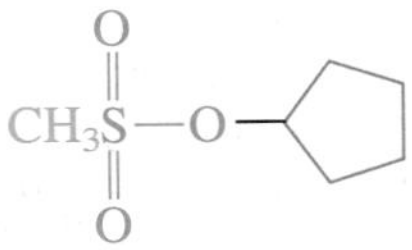

cyclopentyl methanesulfonate
a sulfonate ester

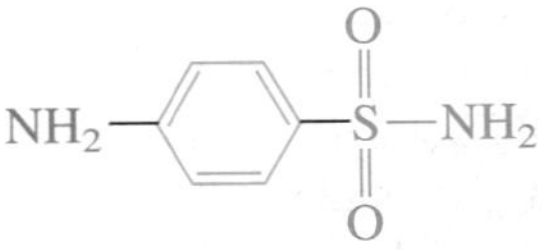

p-aminobenzenesulfonamide
a sulfonamide
(This is better known as sulfanilamide, a sulfa drug.)

The phosphorus analogs of amines are called **phosphines.** The parent, PH_3, is called phosphine and is a toxic gas with a characteristic, unpleasant odor. Triphenylphosphine is a good nucleophile that is employed in certain organic reactions. Esters derived from phosphoric acid play an important role in living organisms. One, two, or all three of the OH groups of phosphoric acid can be replaced with OR groups to form various phosphate esters. The anions produced by the ionization of dialkyl hydrogen phosphates in water by an acid–base reaction are especially important in biological systems. This group provides the backbone for DNA and RNA. As another example, phosphatidyl-choline is a phospholipid that is an important constituent of cell membranes.

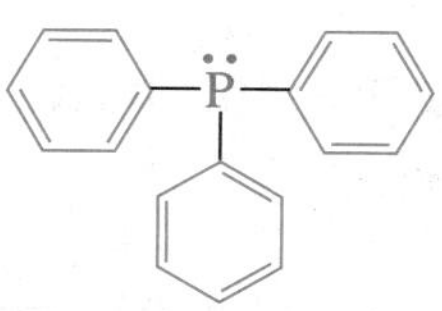

triphenylphosphine

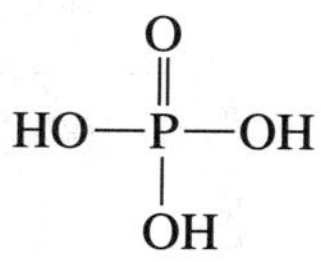

phosphoric acid

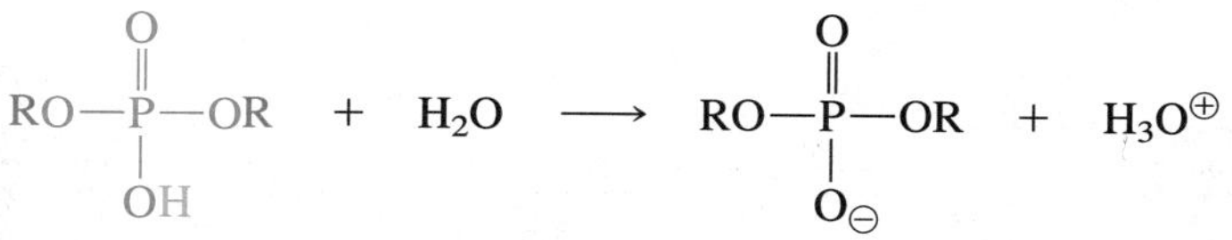

a dialkyl hydrogen phosphate

$$RO-\overset{\overset{\displaystyle O}{\|}}{\underset{\underset{\displaystyle OH}{|}}{P}}-OR \ + \ H_2O \ \longrightarrow \ RO-\overset{\overset{\displaystyle O}{\|}}{\underset{\underset{\displaystyle O^{\ominus}}{|}}{P}}-OR \ + \ H_3O^{\oplus}$$

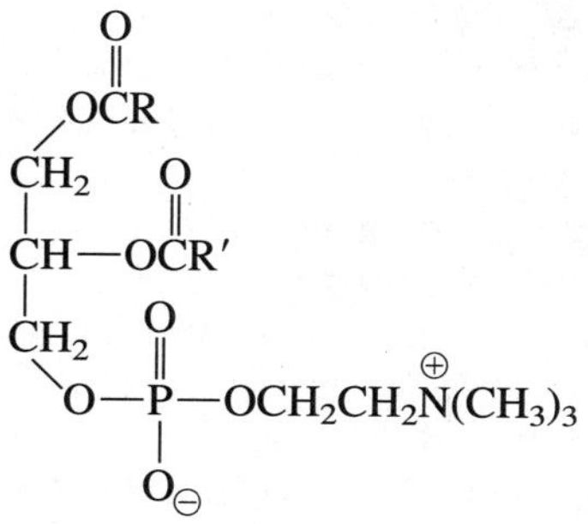

phosphatidylcholine

Problem

Problem 11.17   Provide names for these compounds.

a)

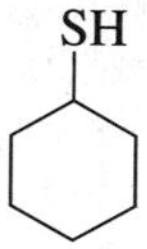

b) $CH_3CH_2CH_2SCH_2CH_2CH_3$

c) 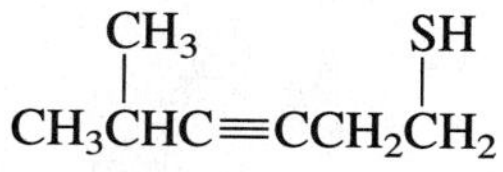 $CH_3\overset{\overset{\displaystyle CH_3}{|}}{CH}C\equiv CCH_2\overset{\overset{\displaystyle SH}{|}}{CH_2}$

d) 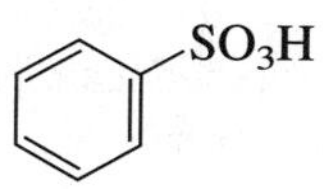   —SO$_3$H

e) 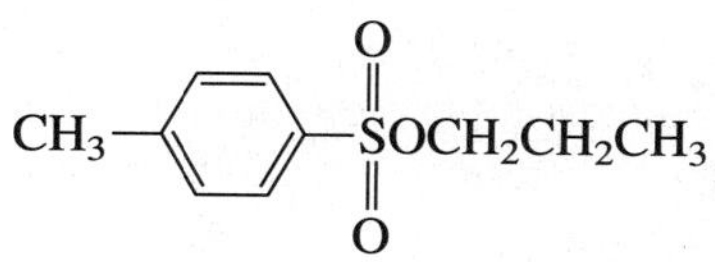 CH_3—⟨ ⟩—$\overset{\overset{\displaystyle O}{\|}}{\underset{\underset{\displaystyle O}{\|}}{S}}OCH_2CH_2CH_3$

Problem 11.18 Draw structures for these compounds.

  a) 2-butanethiol

  b) benzenethiol

  c) isopropyl methanesulfonate

  d) *p*-bromobenzenesulfonic acid

  e) phenyl trichloromethyl sulfide

11.8   Nomenclature of Compounds with Several Functional Groups

Most of the compounds that you have encountered in this text so far have been fairly simple. In addition to any carbon–carbon double or triple bonds, they have con-

Medicinal Use of DMSO

Dimethylsulfoxide (DMSO) is an important solvent in the organic chemistry laboratory. Unlike many organic solvents, it is miscible with water and with many organic solvents. It is very polar, so it dissolves a wide variety of compounds. And, as discussed in Chapter 9, it is aprotic, so it especially useful in S_N2 reactions. DMSO is only weakly acidic, so it is a suitable solvent for the preparation and use of many strongly basic reagents.

In addition, DMSO has a very interesting and controversial history in the medicinal area. When applied to the skin, it penetrates rapidly and can be detected in the blood within five minutes. In addition, it can be tasted soon after a topical application. It is metabolized to various sulfides that cause a distinctive garlicky odor to appear on the breath. It has local analgesic and anti-inflammatory properties, and because of its great penetrating power, it has been used to treat musculoskeletal injuries. It can also be used to transport drugs through the skin.

DMSO was approved for clinical trials by the FDA in 1964. It was being tested for a wide variety of applications, ranging from sprains to cancer therapy, when a report appeared describing changes in the lenses of the eyes of some animals that were given high doses of this chemical. The FDA immediately banned further trials. (This was shortly after the thalidomide catastrophe.) Further studies have shown DMSO to be quite safe, but its effectiveness for the treatment of any condition has seldom been conclusively demonstrated. One problem in testing DMSO is that the traditional double-blind studies are not possible because the patient certainly knows whether a placebo or the real material has been applied. Its only approved use in humans is for the treatment of certain types of bladder infections, although it is used in veterinary medicine as a topical anti-inflammatory agent in the treatment of horses.

tained only one functional group and could be named by using suffixes such as -ynol or -dienone. However, for a compound that contains more than one heteroatom functional group, only one of these functional groups can be designated in the suffix. For example, consider the following compound, which has both alcohol and ketone functional groups.

$$CH_3\overset{OH}{\underset{5}{\overset{|}{C}}}H\underset{4}{C}H_2\overset{O}{\underset{3\;2}{\overset{\|}{C}}}CH_2CH_3$$
$$6 \quad 5 \quad 4 \quad 3\;2 \quad 1$$

5-hydroxy-3-hexanone

Because both functional groups cannot be denoted in the suffix, one must be chosen as higher priority and used to control both the numbering and the suffix. A prefix is then used to indicate the presence of the lower-priority functional group on the main chain. In the case of the compound above, the ketone group has a higher priority than the alcohol group, so an -one suffix is used. Numbering begins with the right carbon so that the carbon of the carbonyl group gets the lower number. The group name for OH is hydroxy. Therefore the compound is named as 5-hydroxy-3-hexanone. Table 11.3 lists the priorities and group names for selected other functional groups. Some examples of naming other complex molecules follow:

$$CH_3\overset{O}{\overset{\|}{C}}CH_2CH_2\overset{O}{\overset{\|}{C}}CH_3$$
$$1 \quad 2\;3 \quad 4 \quad 5\;6$$

2,5-hexanedione
(The presence of several identical functional groups can be denoted with di-, tri-, etc.)

$$CH_3\overset{O}{\overset{\|}{C}}CH_2\overset{O}{\overset{\|}{C}}OCH_2CH_3$$
$$4 \quad 3\;2 \quad 1$$

ethyl 3-oxobutanoate
(The ester group has higher priority, so it determines the numbering and the suffix. The ketone group is designated by the prefix oxo-. In common nomenclature the $CH_3C{=}O$ group is called acetyl or aceto. The common name for this compound is ethyl acetoacetate.)

$$CH_3\overset{O}{\overset{\|}{C}}-$$

the acetyl or aceto group

$$CH_3\overset{OH}{\underset{6}{\overset{|}{C}}}HCH{=}CH\overset{CN}{\underset{3}{\overset{|}{C}}}HCH_2\overset{O}{\overset{\|}{C}}H$$
$$7 \quad 6 \quad 5 \quad 4 \quad 3 \quad 2 \quad 1$$

3-cyano-6-hydroxy-4-heptenal
(The aldehyde is the highest-priority functional group, so the OH and the CN must be denoted by their group prefixes.)

3-amino-4-nitrobenzoic acid
(The acid group has the highest priority.)

Table 11.3 Order of Priority for Selected Functional Groups.

Functional Group	Group Prefix	
carboxylic acid		highest priority
ester		
acid chloride		
amide		
aldehyde	oxo-	
nitrile	cyano-	
ketone	oxo-	
alcohol	hydroxy-	
amine	amino-	
ether	alkoxy-	lowest priority
halogen	fluoro-, chloro-	
	bromo-, iodo-	
—NO_2	nitro-	

Note that the ether, halogen, and —NO_2 groups are always denoted by prefixes in systematic nomenclature.

Problems

Problem 11.19 Provide names for these compounds.

a)

b)

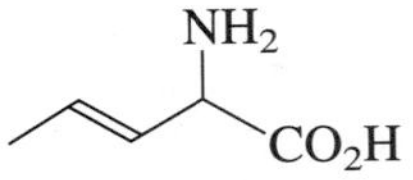

c) 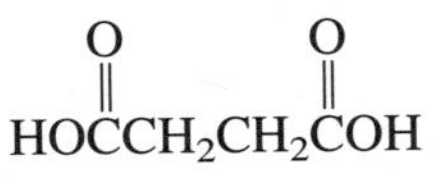 $HOCCH_2CH_2COH$ (with two O groups)

d)

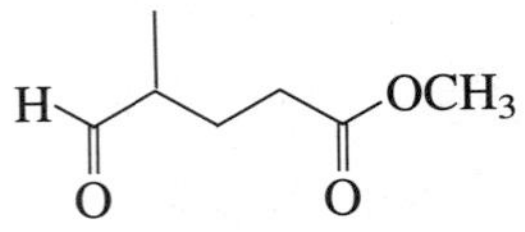

e)

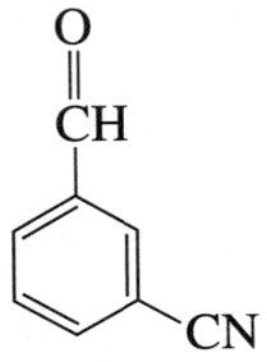

f)

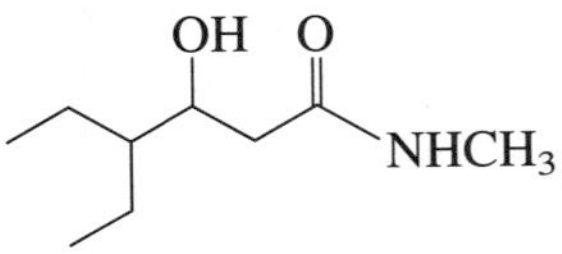

g) 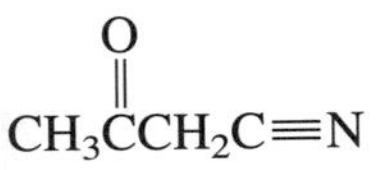 $CH_3CCH_2C{\equiv}N$

h)

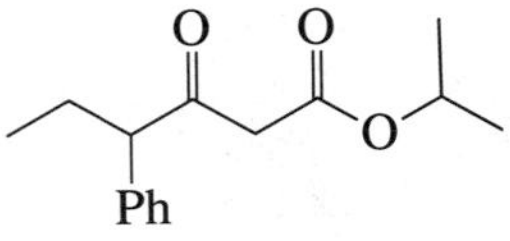

Problem 11.20   Draw structures for these compounds.

    a)   1,6-hexanedioic acid

    b)   ethyl 2-ethyl-2-hydroxybutanoate

    c)   2-amino-3-cyclohexyl-1-propanol

    d)   *tert*-butyl 2-hydroxy-5-octenoate

    e)   *N,N*,3-trimethyl-2-oxobutanamide

As you can see, organic nomenclature is a very complex subject. There are many more rules and examples that are beyond the scope of this book. You should not expect to be able to name every compound that you might encounter. However, based on the rules and examples given here and in Chapter 5, your knowledge of nomenclature should be sufficient to allow you to read and discuss organic chemistry.

Finally, you should realize that organic nomenclature is not an exact science. In complicated molecules there will often be ambiguities, and more than one correct systematic name may result. In addition, many complex molecules are best identified by a short common name rather than by a purely systematic name that is unmanageably long or complicated. As an example, the systematic name for morphine would be so complex that it would convey little immediate information about the compound to most chemists. In practice, therefore, common names continued to be coined for many molecules. For example, the systematic name pentacyclo[4.2.0.0^{2,5}.0^{3,8}.0^{4,7}]octane must be carefully analyzed by the average organic chemist, probably using pencil and paper, before its structure can be recognized. However, the common name given to this compound when it was first synthesized is very descriptive and immediately summons a mental picture for the compound. It is best known as cubane!

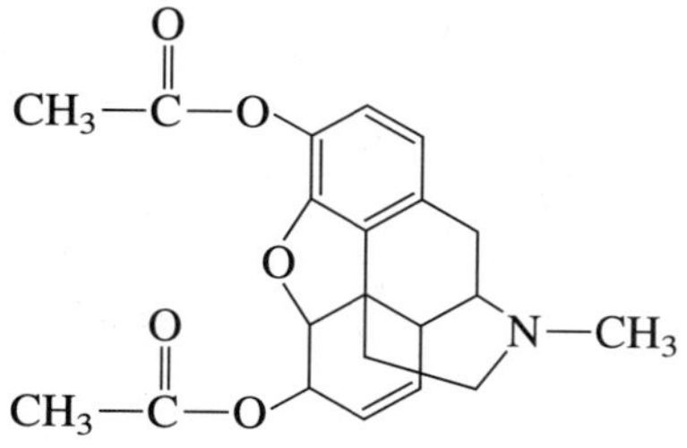

morphine 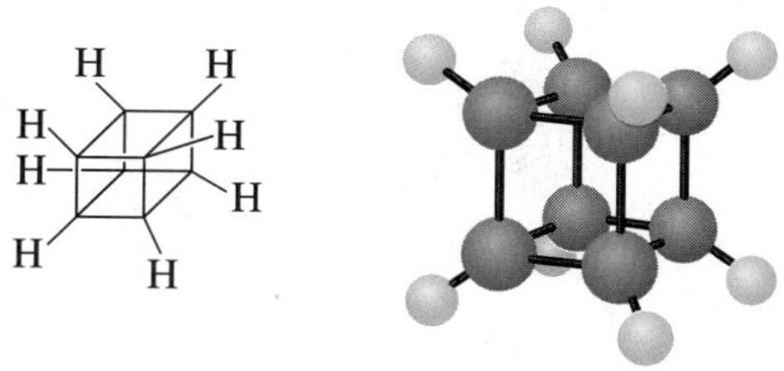   cubane

11.9   Summary

After completing this chapter, you should be able to:

1. Name an aromatic compound, a phenol, an aldehyde, a ketone, a carboxylic acid, an acid chloride, an anhydride, an ester, an amide, a nitrile, and a carboxylic acid salt.

2. Draw the structure of a compound containing one of these functional groups when the name is provided.

3. Recognize the common functional groups that contain sulfur or phosphorus.

4. Name a compound containing more than one functional group or draw the structure of such a compound when the name is provided.

5. Understand how the physical properties of these compounds depend on the functional group that is present.

End-of-Chapter Problems

11.21 Provide names for these compounds.

a)

b) 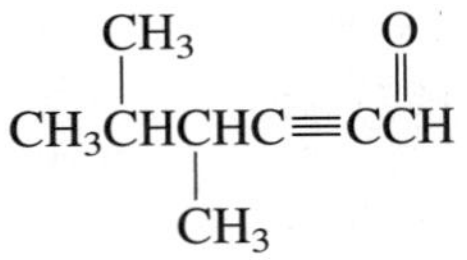 $CH_3CHCHCH C\equiv CCH$ with CH_3 and CH_3 substituents, O

c)

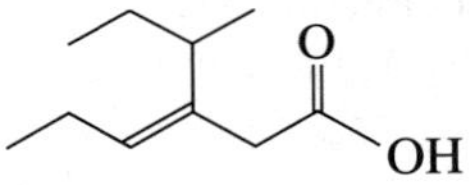

d)

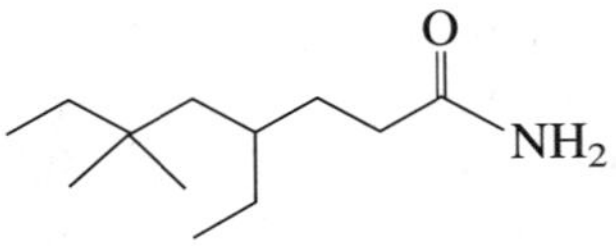

e)

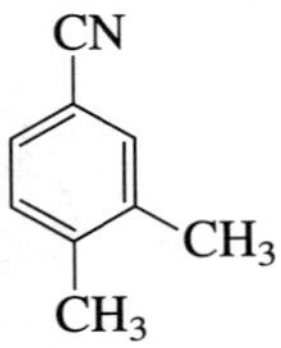

f) 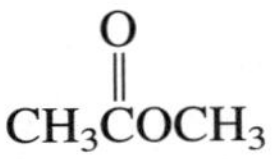 CH_3COCH_3 with O

g)

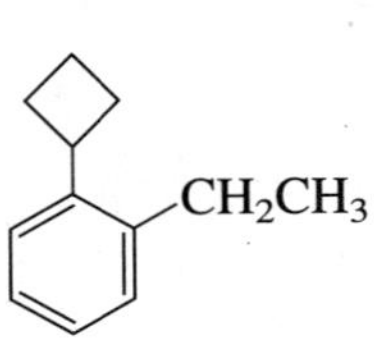

h)

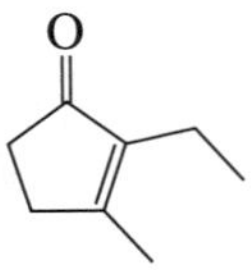

i)

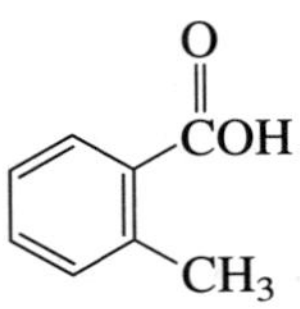

j) 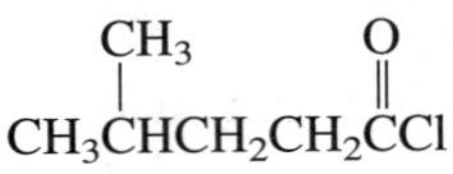 $CH_3CHCH_2CH_2CCl$ with CH_3 and O

k) 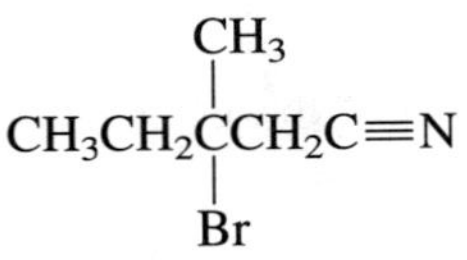 $CH_3CH_2CCH_2C\equiv N$ with CH_3 and Br

l)

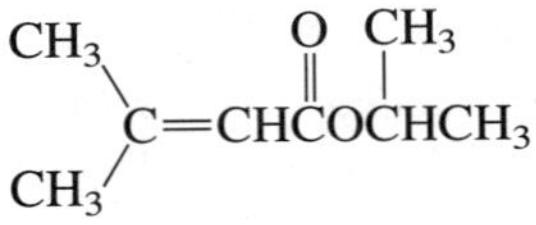

m)

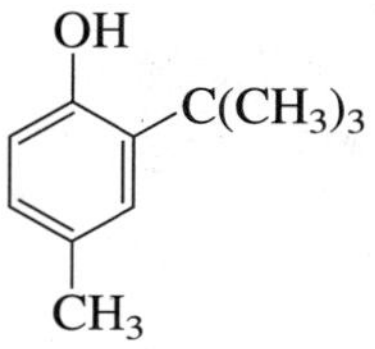

n)

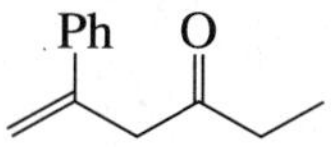

o)

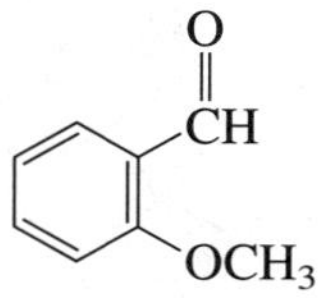

p)

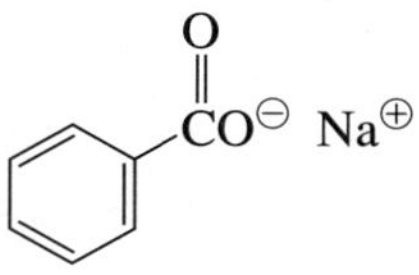

q)

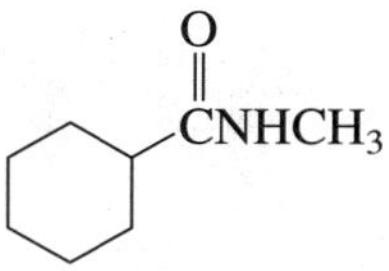

r)

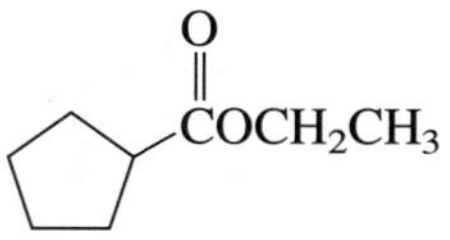

11.22 Draw structures for these compounds.

a) 3-methoxybenzoic acid

b) *p*-phenylphenol

c) 2-methylbutyl *p*-toluenesulfonate

d) 4-cyano-3,5-dimethyloctanal

e) sodium acetate

f) 2,4-dinitrotoluene

g) ethyl 3-hydroxy-4-oxohexanoate

h) acetyl chloride

i) ethyl 2-chlorobenzoate

j) phenyl vinyl sulfide

k) 1,2,4,5-tetramethylbenzene

l) 2,4-cyclooctadienone

m) acetonitrile

n) propyl 3-isopropyl-4-heptenoate

o) *N*-methylpentanamide

p) 2,5-hexanedione

q) 4-(1-methylpropyl)benzaldehyde

r) butanoic anhydride

s) 4-hydroxyheptanoic acid

t) 2,2-diethylcyclobutanecarboxylic acid

11.23 Provide names for these compounds.

a)

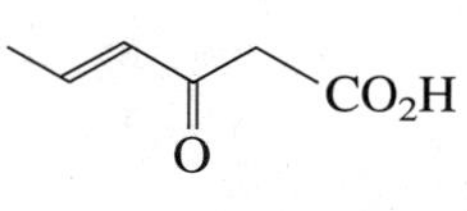

b)

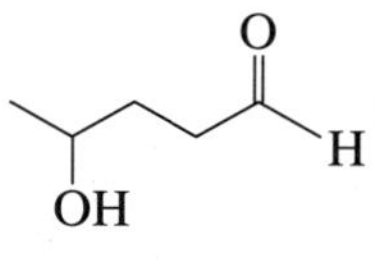

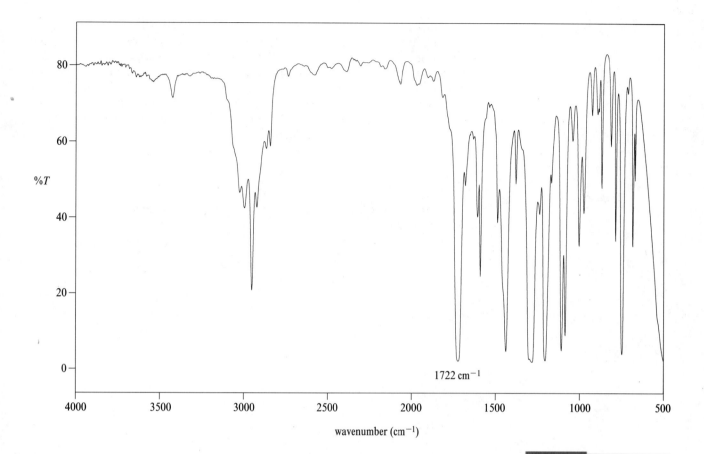

%T

80

60

40

20

0

1722 cm⁻¹

4000   3500   3000   2500   2000   1500   1000   500

wavenumber (cm⁻¹)

Figure 12.23

The Infrared Spectrum of an Unknown Compound.

appear to be an amide (no N—H, carbonyl absorption too high, no extra bands at slightly lower wavenumbers than the carbonyl band), an anhydride (absence of a second carbonyl band), or an acyl chloride (carbonyl position too low). This leaves a ketone or an ester as possibilities. The strong absorption at 1282 cm⁻¹ suggests that the unknown is an ester. The bands at 1607, 1591, 1489, and 1437 cm⁻¹ along with the absorptions at 3100–3000 and 746 cm⁻¹ suggest the presence of an aromatic ring. The carbonyl of the ester occurs at slightly lower wavenumbers (1722 cm⁻¹) than the usual position (1740 cm⁻¹), indicating that it might be conjugated. Therefore we conclude that the unknown is probably an ester, that it may have an aromatic ring, and that the ester may be conjugated (with the aromatic ring?). However, these conclusions must be considered tentative until confirming evidence is obtained from other sources. The unknown is actually methyl 3-methylbenzoate:

methyl 3-methylbenzoate

The exact identity of an unknown cannot be established only on the basis of its IR spectrum (unless, of course, the spectrum of a known exactly matches the spectrum of the unknown). It is possible to determine the functional groups that are present in the unknown, but the IR spectrum provides little information about the hydrocarbon part of the compound. However, this is exactly the information supplied by the nuclear magnetic resonance spectrum. That is why the combination of these two types of spectroscopy is so valuable to the organic chemist. Let's now discuss nuclear magnetic resonance spectroscopy.

Problems

Problem 12.10   Predict the positions of the major absorption bands in the IR spectra of these compounds.

a) $CH_3CH{=}CHCCH_3$ (with C=O)

b)   benzene ring with CH_2NH_2

c) CH_3O substituted cyclohexene

d) $CH_3CH_2CH_2OH$

e) $CH_2CH_2C{\equiv}C{-}H$ with NO_2

f)   benzaldehyde with C(=O)H and CH_3

Problem 12.11   Explain how IR spectroscopy could be used to distinguish between these compounds.

a) $CH_3CH_2CH{=}CH_2$ and $CH_3CH_2C{\equiv}CH$

b) $CH_3{-}C(CH_3)(CH_3){-}OH$ and   benzene ring with CH_2OH

c)   benzaldehyde (CHO)   and   acetophenone (CCH_3, C=O)

d) $CH_3CH_2CH_2CH_2NH_2$ and $CH_3CH_2NHCH_2CH_3$

Problem 12.12   Explain which functional groups are present in these compounds on the basis of their IR spectra.

a)

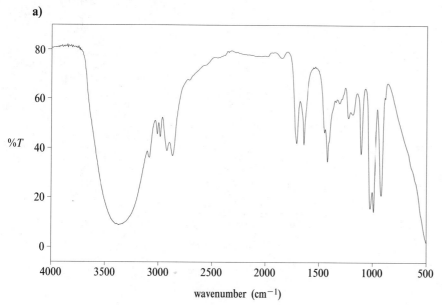

Solution

a)   The broad absorption centered near 3300 cm^{-1} indicates the presence of a hydroxy group. The absorption at 3005 cm^{-1} suggests the presence of H's bonded to sp^2 hybridized C's. (Note that you are not expected to read peak positions this exactly from any of these spectra.) This is supported by the absorption for a C=C at 1646 cm^{-1}. The absorptions in the region of 3000–2850 cm^{-1} indicate the presence of H's bonded to sp^3-hybridized C's. Although the compound has a C=C, there is no indication of the presence of an aromatic ring due to the absence of the four bands in the 1600–1450 cm^{-1} region and the absence of bands in the 900–675 cm^{-1} region.

In summary, the structural features that can be identified from the IR spectrum are as follows:

$$O-H \qquad C=C\diagup^{H} \qquad -\overset{|}{\underset{|}{C}}-H$$

(This is the spectrum of 2-propen-1-ol or allyl alcohol. The structure cannot be determined only from this IR, but the conclusions reached above are consistent with this structure.)

$$\overset{\displaystyle OH}{\underset{\displaystyle |}{CH_2=CH-CH_2}}$$

2-propen-1-ol (allyl alcohol)

a)

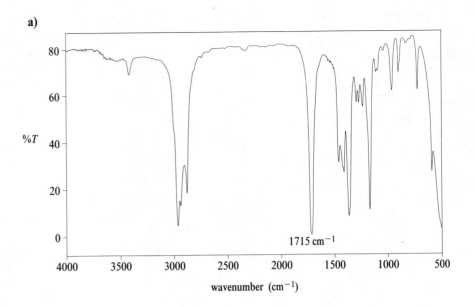

b)

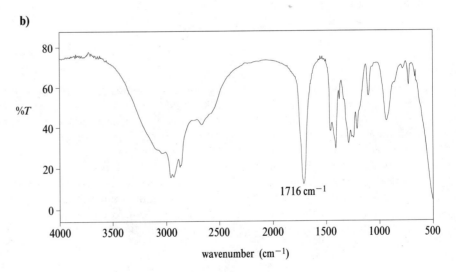

c)

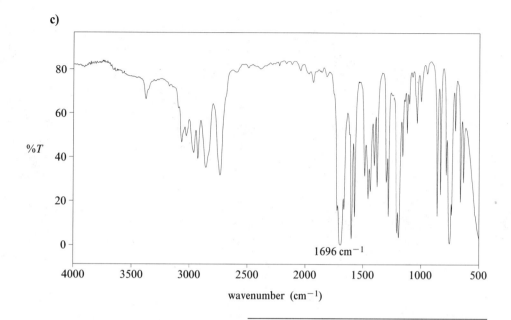

1696 cm^{-1}

wavenumber (cm^{-1})

The amount of carbon dioxide in the atmosphere has been steadily increasing since the Industrial Revolution because of the burning of large amounts of organic carbon from natural gas, oil, coal, trees, and other biomass. Recently, there has been considerable discussion concerning the environmental effects of this increase. Some scientists believe that there will be a significant increase in the average global temperature as a result of a "greenhouse effect," in which the increase in carbon dioxide traps more heat on the earth, preventing it from escaping into space.

Ultraviolet and visible radiation from the sun brings considerable energy to the surface of the earth. This energy is absorbed and converted to vibrational, translational, and even chemical energy. Part of the vibrational energy is reemitted as infrared radiation. If the atmosphere is transparent to this IR radiation, then it escapes into space, a net cooling process. But if there are components of the atmosphere that absorb it, then it is trapped and contributes to the warming of the earth.

The figure shows an IR spectrum of normal air. The two major components of air, N_2 and O_2, do not absorb IR radiation because their dipole moments do not change on vibration. Water vapor has strong IR absorptions from 3400 to 4000 cm^{-1} (O—H stretches) and also from 1200 to 1800 cm^{-1} (bends). Air is transparent to much of the rest of IR radiation, except for the strong absorption of carbon dioxide from 2300 to 2400 cm^{-1}.

As the concentration of carbon dioxide in the atmosphere increases because of human activity, more IR radiation is trapped. However, because the climate of the earth is extremely complex and dependent on numerous interrelated factors, it is very difficult to predict the overall effect on the temperature of the earth. Some

The Greenhouse Effect

ELABORATION

models predict an increase in the average temperature of several degrees, which would cause severe changes in climate throughout the world; other models predict much less significant changes.

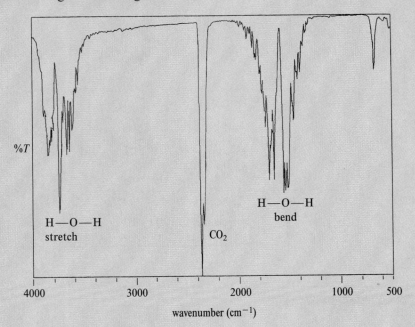

Problem

Problem 12.13   Significant amounts of methane are produced naturally by processes such as the decay of plants. Methane is also produced by the digestive process of ruminants such as cows. Comment on the potential of methane to act as a greenhouse gas.

12.12   Nuclear Magnetic Resonance Spectroscopy

The nuclei of certain isotopes of some elements have two (or more) energy states available when they are in a magnetic field. The transitions between these energy states can be investigated by using the technique of nuclear magnetic resonance (NMR) spectroscopy. Although there are many nuclei that exhibit this phenomenon, the two that are of most use to organic chemists are the hydrogen nucleus (a proton, ^{1}H) and the nucleus of the isotope of carbon with an atomic mass number of 13 (^{13}C). First, proton magnetic resonance (^{1}H-NMR) spectroscopy is discussed. Later sections deal with ^{13}C-NMR.

Proton magnetic resonance spectroscopy provides information about the relative numbers of different kinds of hydrogens in the compound, the nature of the carbons bonded to them, and which hydrogens are nearby. From this information it is possi-

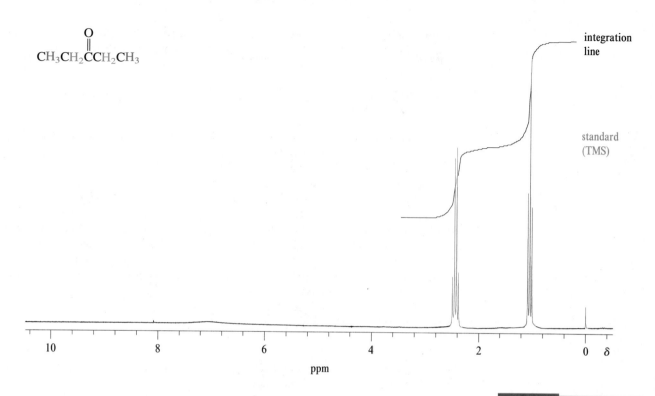

Figure 12.24

The ¹H-NMR Spectrum of
3-Pentanone.

ble to get a good idea about the structure of the hydrocarbon part of the compound. In combination with the knowledge of the functional group obtained from the IR spectrum, the NMR spectrum often enables the structure of a compound to be assigned with certainty.

Before we look at the theory behind the NMR technique, it is useful to see a sample spectrum and find out what kind of information it provides. Figure 12.24 shows the ¹H-NMR spectrum of 3-pentanone. Note that the absorption peaks extend up from the base line at the bottom of the spectrum, in contrast to IR spectra.

Three types of information are present in an NMR spectrum:

1. The position on the *x*-axis. Called the **chemical shift,** this provides information about the carbon (or other atom) to which the hydrogen is attached.

2. The number of peaks in each group. Called the **multiplicity,** this provides information about the other hydrogens that are near the hydrogen(s) that produces the peaks.

3. The **integral.** The area under a group of peaks is proportional to the number of hydrogens that produce that group of peaks.

Many NMR spectrometers provide the integral in the form of a line drawn on the spectrum that increases in height in proportion to the total area of the peaks. The height of each "step" is proportional to the number of hydrogens that produce the peaks under that step. The heights must be put into small whole number ratios to provide the relative numbers of each type of hydrogen. To simplify interpretation, the spectra in this book have already been integrated, and the relative numbers of hydrogens for each group of peaks are provided. However, remember that the actual numbers of hydrogens can be multiples of the numbers provided by the integrals.

12.13   Theory of ^{1}H-NMR

The following discussion of the theory behind ^{1}H-NMR is adequate for our needs. The charge of some nuclei, including ^{1}H, "spins" on the nuclear axis. This spinning charge generates a small magnetic field. For the purposes of further discussion the nucleus can be considered as a small bar magnet.

The hydrogen nucleus has two possible spin states that have identical energies under normal circumstances. However, if the atoms are placed in an external magnetic field—that is, between the poles of a large magnet in the laboratory—then the spin states have different energies. As shown in Figure 12.25, one state has its magnetic field oriented in the same direction as the external magnetic field, B_0, and is lower in energy than the other state that has its field oriented in opposition to the external field. With two states of different energy, spectroscopy can be done.

The difference in energy between the two states is related to the strength of the external magnetic field by the equation

$$\Delta E = h\gamma B_0/2\pi$$

where B_0 is the strength of the external magnetic field and γ is the magnetogyric ratio, which differs for each kind of atomic nucleus.

The magnetogyric ratio is extremely small. Therefore ΔE is very small even when very large magnets are used. For example, early NMR spectrometers employed a 14,000-gauss magnet. (The magnetic field of the earth is about 0.5 gauss, whereas small magnets, such as those used to hold notes on a refrigerator, have fields of hundreds of gauss.) For these instruments, ΔE is about 10^{-6} kcal/mol, requiring radiation with a frequency of 60×10^6 s^{-1} or 60 MHz. Because better spectra are obtained

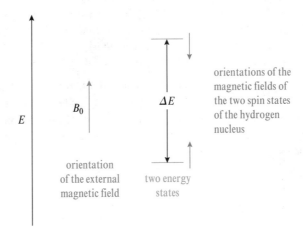

Figure 12.25

Energy Level Diagram for the Two Spin States of a Hydrogen Nucleus in a Magnetic Field.

with larger ΔE's, many current NMR instruments use superconducting magnets, cooled with liquid helium, with fields that are substantially larger than those just described. Instruments that operate at 200 MHz (46,700-gauss magnet) and 300 MHz (70,000-gauss magnet) are relatively common, and some operating at 500 or 600 MHz, although quite expensive, are also available. The spectra in this book were obtained on an instrument operating at 200 MHz.

Other nuclei, such as ^{13}C, ^{19}F, ^{2}H, and ^{31}P also have nuclear spins and can be studied by NMR techniques. However, since γ is different for these nuclei, they appear in a very different region of the spectrum from hydrogen and are not seen in a ^{1}H-NMR spectrum. Both ^{12}C and ^{16}O, which are very common in organic compounds, do not have nuclear spin and therefore have no NMR absorptions.

To obtain a NMR spectrum, the sample is placed in a strong magnetic field in the spectrometer. One type of instrument irradiates the sample with electromagnetic radiation of fixed frequency while the magnetic field strength is varied. When the magnetic field strength is such that the difference between the energy states of the hydrogen matches the energy of the radiation, the hydrogen absorbs the energy of the radiation and is said to be in resonance.

If all of the hydrogen nuclei in a compound absorbed at identical magnetic field strengths, the NMR technique would not be very useful. However, the exact field required for resonance depends on the local environment around the hydrogen. The electrons in the molecule circulate and create magnetic fields that oppose the external magnetic field. Therefore the magnetic fields due to the electrons partially screen the hydrogen from the external magnetic field. Because the electron density varies throughout a molecule, different hydrogens require different field strengths to absorb the fixed frequency radiation. The variation is quite small, about 2500 Hz on an instrument that operates at 200 MHz, or about 0.001%, so an NMR spectrum must be obtained at very high resolution to be able to detect these differences.

12.14 The Chemical Shift

As was discussed above, the chemical shift of a hydrogen signal—that is, the field required for the hydrogen to be in resonance—varies slightly with the chemical environment of the hydrogen. To measure chemical shifts, a small amount of a reference compound, usually tetramethylsilane (TMS), is added to the sample. The separation, in hertz, between the peak of interest and the peak due to TMS is measured. TMS is chosen as the reference because it has only one NMR peak and this peak occurs in a region of the spectrum where it does not usually overlap with other absorptions. Figure 12.26 illustrates the use of TMS as a reference compound in the spectrum of acetone.

$$
\begin{array}{c}
CH_3 \\
| \\
H_3C-Si-CH_3 \\
| \\
CH_3
\end{array}
$$

tetramethylsilane

In the plot of a typical NMR spectrum the field strength increases from left to right. The protons of TMS absorb at higher field than most other protons, so the TMS signal occurs at the right edge of the spectrum, as can be seen in Figure 12.26. The signals for most other types of hydrogens appear to the left of the TMS peak. Left on an NMR spectrum is termed the **downfield** direction; right is termed the **upfield** direction. The absorption for acetone occurs 436 Hz downfield from TMS in this spectrum.

Recall that the energy separation of the two nuclear spin states of the hydrogen is directly proportional to the magnetic field strength, B_0, of the NMR instrument. This means that the chemical shift, in hertz, also is directly proportional to the magnetic field strength. On an instrument with a 46,700-G magnet, which operates at a frequency of 200 MHz, the peak for acetone occurs 436 Hz downfield from TMS. On a spectrometer with a magnet that is 1.5 times stronger, which operates at a frequency of 300 MHz, the peak for acetone occurs 654 Hz downfield from TMS, farther downfield by a factor of 1.5. Chemical shifts that do not depend on the particular instrument that is used to acquire the spectrum are obtained by dividing the chemical shift, in hertz, by the operating frequency of the instrument. The result is multiplied by 10^6 to get a number of a more convenient magnitude. The resulting values for chemical shifts, called **parts per million** (ppm) or δ, do not depend on the operating frequency of the instrument.

$$\delta = \frac{\text{observed position of peak (Hz)}}{\text{operating frequency of instrument (Hz)}} \times 10^6$$

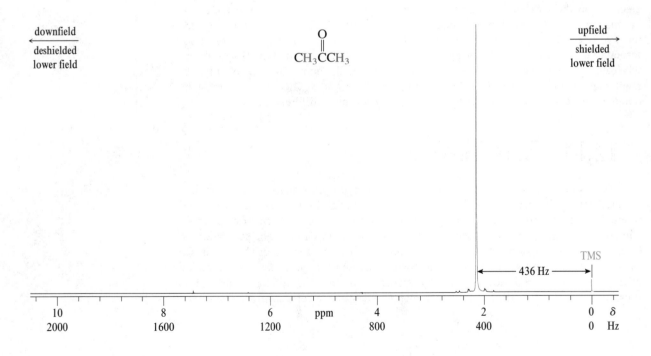

Figure 12.26

The ^{1}H-NMR Spectrum of Acetone at 200 MHz.

Figure 12.32 provides another example. An interesting feature of this case is the multiplicity of the signal for the center CH_2 group. These hydrogens are coupled to two hydrogens on one side and three different hydrogens on the other side and could appear as $3 \times 4 = 12$ lines. However, because the coupling constants are very similar, the signal behaves as though there were five identical hydrogens doing the splitting, and it appears as a sextet.

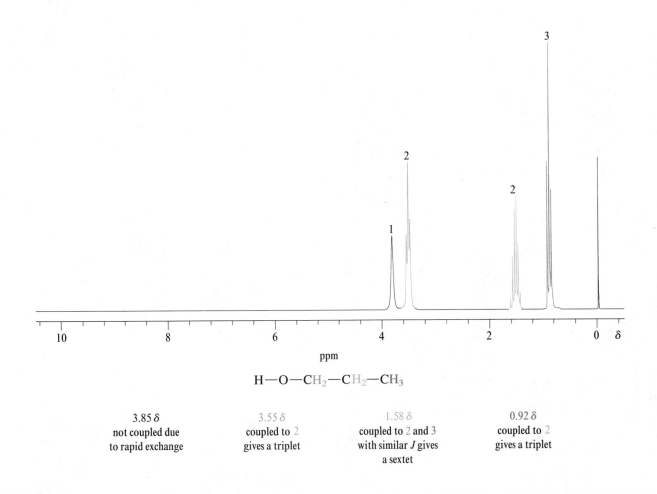

$$H-O-CH_2-CH_2-CH_3$$

3.85 δ	3.55 δ	1.58 δ	0.92 δ
not coupled due to rapid exchange	coupled to 2 gives a triplet	coupled to 2 and 3 with similar *J* gives a sextet	coupled to 2 gives a triplet

Figure 12.32

The ¹H-NMR Spectrum of an Unknown Compound, C_3H_8O.

This compound shows an OH in its IR spectrum. Its DU equals 0. Examination of its NMR spectrum indicates only alkyl type H's. The integral provides the actual number of H's in this case, since they total eight. The broad singlet due to 1 H at 3.85 δ is probably due to the hydroxy H, which is not coupled due to rapid chemical exchange. The 2 H's at 3.55 δ appear as a triplet and must be coupled to 2 H's—the 2 H's at 1.58 δ. The 3 hydrogens at 0.92 δ appear as a triplet and must be coupled to the 2 H's at 1.58 δ also. If this is the case, then the 2 H's at 1.58 δ are coupled to $3 + 2 = 5$ H's and should appear as six lines, a sextet. This information allows the fragment $CH_2CH_2CH_3$ to be written. Combining this with the HO indicates that the unknown is 1-propanol. The calculated chemical shifts (see page 550) agree well with those in the spectrum.

Problem

Problem 12.21   Determine the structures of these compounds from their ^{1}H-NMR spectra.

a)   The formula is C_3H_6O.

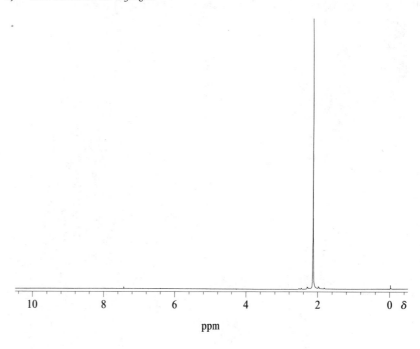

b)   The formula is C_8H_{10}.

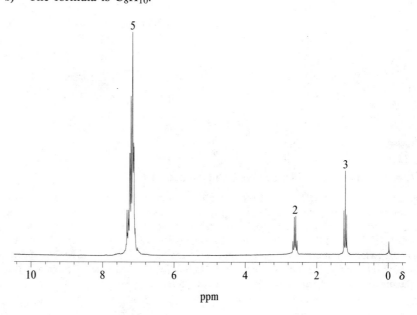

c) The formula is C_2H_4O.

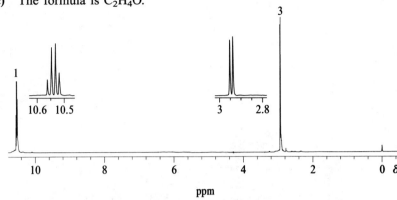

d) The formula is C_6H_{12}.

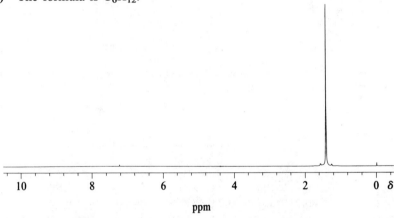

e) The formula is $C_5H_{10}O_2$.

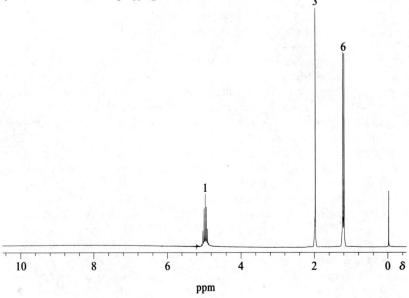

f)   The formula is C_4H_9Cl.

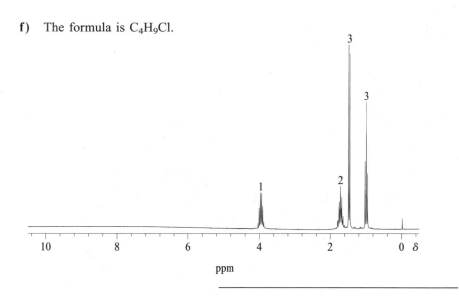

Problems that use both IR and NMR spectra to determine the identity of unknowns are provided in Section 12.21.

Magnetic Resonance Imaging

Physicians are always looking for methods for viewing the internal organs of the human body without invasive techniques such as surgery. One method that has found considerable use is computerized tomography (CT), also known as computer-assisted tomography (CAT). In a CAT scan, X-rays are used to generate the images that are collected and processed by computer. X-rays interact more strongly with atoms of larger atomic mass, so imaging agents must often be administered to the patient to enhance the pictures of soft tissues, which are composed primarily of C, H, N, and O. Another potential disadvantage of the CAT technique is the high energy of the radiation that is used. X-rays are called ionizing radiation because they have enough energy to eject electrons from the orbitals of atoms. Chemical reactions caused by the resulting ions can cause damage to living tissue. Although ionizing radiation techniques pose little hazard if done properly, techniques using less energetic radiation are desirable.

Magnetic resonance imaging (MRI) is a newer technique based on the same principles as ^{1}H-NMR. The patient is placed within the field of a huge magnet and radio frequency radiation is used to excite hydrogen nuclei to their higher-energy spin state. (The low energy of this radiation poses no danger to the patient.) The magnetic field of an MRI instrument is not nearly as uniform as that of an NMR spectrometer, so the signal for the protons is a very broad peak rather than the individual multiplets that we have seen in an NMR spectrum. The instrument detects differences in the intensity of the proton signal. The intensity depends on the concentration of hydrogens in the small area being sampled and on the "relaxation times," that is, the time that it takes a hydrogen in the higher-energy spin state to return to the ground state. Both of these factors cause different environments, such as fluids, tissues, and even diseased tissues, to produce signals of different intensities. The data are gathered by a computer, which produces a map or picture of the intensities. Because MRI is looking at hydrogens, it gives a particularly good image of soft tissue and therefore complements CAT. The figure shows an MRI image of a human skull.

Actually, this technique should be called nuclear magnetic resonance imaging. Considering the poor image of anything "nuclear" with the general public, it is not surprising that the medical community has decided to drop this term from the name.

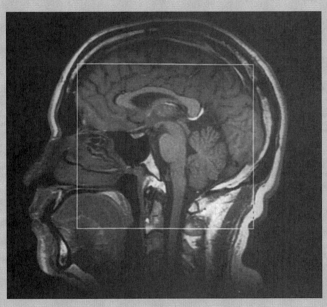

12.20 Carbon-13 Magnetic Resonance Spectroscopy

The isotope of carbon with seven neutrons, ^{13}C, composes about 1.1% of carbon atoms. It is similar to hydrogen in that it has two nuclear spin states of different energy when it is in an external magnetic field. The spectroscopy that is done using this nucleus, ^{13}C-NMR, provides direct information about the carbon chains in the compound, information that is often complementary to that obtained from ^{1}H-NMR spectroscopy.

Figure 12.33 shows the ^{13}C-NMR spectrum of 3-buten-2-one. In contrast to ^{1}H-NMR, the peak for each carbon appears as a sharp singlet. Chemical shifts are measured by using the carbons of TMS as a standard. The chemical shift range is much larger here—peaks appear as far as 240 ppm downfield from TMS. Therefore overlap of peaks resulting from different carbons occurs much less often than over-

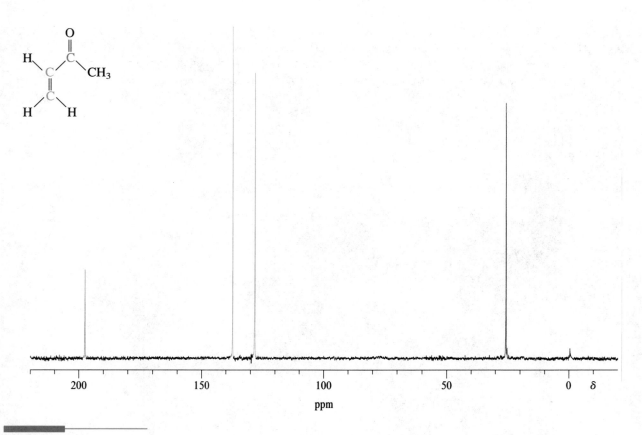

Figure 12.33

The ^{13}C-NMR Spectrum of 3-Buten-2-one.

lap of hydrogen peaks. It is usually possible to count all of the different types of carbons in a compound by examination of its ^{13}C-NMR spectrum.

Problem

Problem 12.22　How many different absorption bands would appear in the ^{13}C-NMR spectra of these compounds?

a)

b)

c)　$CH_3CH_2CH_2CH_2CH_3$

d)　$CH_3CCH_2CH_2CH_2CH_3$

e)

Solution

a) The compound has a plane of symmetry, so some carbons are chemically equivalent to others. There are five different types of carbons, C-7, C-1, C-2 = C-6, C-3 = C-5, and C-4, so there are five absorptions in the ^{13}C-NMR spectrum.

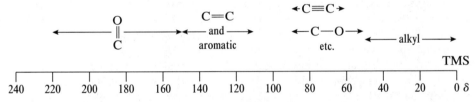

The factors that control the chemical shifts of hydrogens, such as the electronegativities of nearby atoms, have similar effects on the chemical shifts of carbon signals. Therefore the chemical shifts for carbons parallel the shifts for hydrogens, although they are approximately 20 times larger in the case of carbon. Alkyl carbons appear in the most upfield positions, carbons attached to an electronegative element such as oxygen are shifted downfield, the carbons of an aromatic ring appear farther downfield, and so forth. The carbons of carbonyl groups are easily recognized because they appear farthest downfield. The following diagram provides approximate chemical shifts for the various types of carbons that are encountered in organic compounds:

Note the similarity of this diagram to that for proton chemical shifts on page 548. Numerous tables, empirical equations, and even computer programs are available that enable the chemical shifts of the carbons of most compounds to be predicted rather accurately. However, it is possible to assign the carbons responsible for the peaks in many spectra based only on the limited information presented here. For example, the peak at 198.0 δ in Figure 12.3 is assigned to the carbonyl carbon, and the peak at 26.1 δ is due to the methyl carbon. The two alkene carbons appear at 137.5 and 128.5 δ.

Problem

Problem 12.23   Assign the absorptions in the ^{13}C-NMR spectra of these compounds to the appropriate carbons.

a)   1-butanol; absorptions at 61.4, 35.0, 19.1, and 13.6 δ.

b)   cyclohexanone; absorptions at 209.7, 41.9, 26.6, and 24.6 δ.

Let's now deal with the issue of spin coupling in ^{13}C-NMR spectra. No ^{13}C—^{13}C coupling is observed because of the low natural abundance of this isotope. A particular ^{13}C has another ^{13}C adjacent to it in only 1% of the situations, so the split peaks are very small in comparison to the unsplit peaks. However, coupling between carbon and hydrogen is strong and occurs even when they are separated by several intervening bonds. Because the resulting spectra are often quite complex and difficult to interpret, C—H coupling is removed by a technique called **broadband decoupling.** In this technique, as the carbon spectrum is being obtained, the sample is simultaneously irradiated with a band of radio frequency radiation that excites all of the hydrogens. This causes each of the hydrogens to flip rapidly between its two spin states, so its two magnetic field orientations average to zero. No coupling occurs with the carbon and each peak appears as a singlet.

Several techniques have been developed that enable the number of hydrogens attached to the carbon to be determined. An older technique, called **off-resonance decoupling,** allows hydrogens and carbons that are directly bonded to couple, but removes any longer-range coupling. In an off-resonance decoupled spectrum a CH_3 appears as a quartet, a CH_2 appears as a triplet, a CH appears as a doublet, and a C that has no hydrogens bonded to it appears as a singlet. A newer and more convenient technique, called **DEPT-NMR** (distortionless enhancement by polarization transfer), also allows the determination of the number of hydrogens attached to each carbon. In a DEPT experiment, three spectra are obtained. One is a normal broadband decoupled spectrum. Another spectrum (DEPT 90° spectrum) is obtained under special conditions in which only carbons bonded to a single hydrogen (CH's) appear. A third spectrum (DEPT 135° spectrum) is obtained under conditions in which CH's and CH_3's appear as positive absorptions and CH_2's appear as negative absorptions. By combining the information in these spectra, each peak can be assigned as resulting from a CH_3, CH_2, CH, or C group. Thus signals that appear only in the broadband decoupled spectrum are due to C's with no attached H's. Signals that appear in the DEPT 90° spectrum are due to CH groups. Signals that appear as negative peaks in the DEPT 135° spectrum are due to CH_2 groups, and signals that appear as positive peaks in the DEPT 135° spectrum but are absent from the DEPT 90° spectrum are due to CH_3 groups. DEPT and broadband decoupled spectra of ethyl 2-propenoate are shown in Figure 12.34. The remaining ^{13}C-NMR spectra in this book are broadband decoupled spectra with the information obtained from the DEPT spectra indicated above each peak as C, CH, CH_2, or CH_3.

As you examine more ^{13}C-NMR spectra, you will find that the heights (which are proportional to the areas) of the peaks often do not directly correspond to the number of carbons responsible for those peaks. In addition to the number of carbons, there are other factors that affect the areas of the peaks. Although it is possible to experimentally obtain an accurate integral, the process is time consuming and is not usually done.

In summary, considerable information is available from a ^{13}C-NMR spectrum. First, the number of different carbons can be counted, providing information about the symmetry of the molecule. Second, the chemical environment of the carbons can be deduced from their chemical shifts. Third, information from the DEPT spectra tells how many hydrogens are bonded to each carbon. Some of this same information is provided by the ^{1}H-NMR spectrum. However, the two types of spectra often provide complementary information and help to solidify deductions made with one alone. ^{13}C-NMR is especially useful when the ^{1}H-NMR spectrum is too complex for ready interpretation.

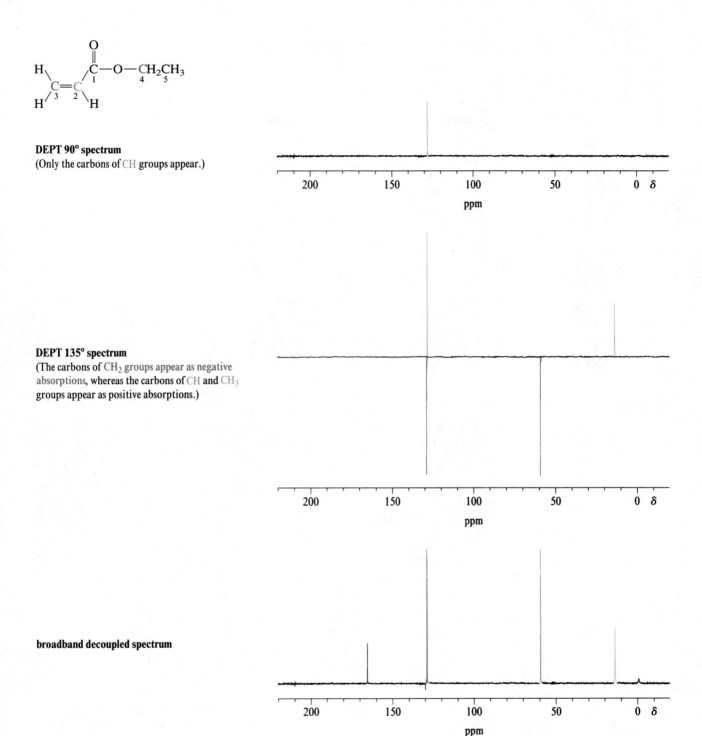

DEPT 90° spectrum
(Only the carbons of CH groups appear.)

DEPT 135° spectrum
(The carbons of CH_2 groups appear as negative absorptions, whereas the carbons of CH and CH_3 groups appear as positive absorptions.)

broadband decoupled spectrum

Figure 12.34

Broadband Decoupled and DEPT ^{13}C-NMR Spectra of Ethyl 2-Propenoate.

The peaks are assigned as follows: the peak at 166 δ is due to a C with no attached H's (C-1) because it appears only in the broadband decoupled spectrum; the peak at 128.7 δ is due to an alkene CH_2 (C-3) because it appears as a negative peak in the DEPT 135° spectrum; the peak at 129.7 δ is due to an alkene CH (C-2) because it appears in the DEPT 90° spectrum; the peak at 60 δ is due to a CH^2 (C-4) because it appears as a negative peak in the DEPT 135° spectrum; and the peak at 14 δ is due to a CH_3 (C-5) because it appears as a positive peak in the DEPT 135° spectrum but does not appear in the DEPT 90° spectrum.

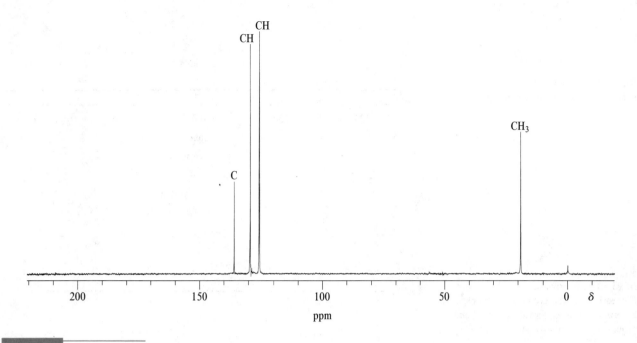

Figure 12.35

The ^{13}C-NMR Spectrum of C_8H_{10}.

Let's look at some examples of structure determination using ^{13}C-NMR spectroscopy. Figure 12.35 shows the spectrum of C_8H_{10}. First, calculation indicates that the DU is 4. The compound has some symmetry, since the spectrum shows the presence of only four different types of carbons. (There are probably two of each type, although other combinations are possible.) Examination of the chemical shifts shows one alkyl type and three alkene/aromatic types. A benzene ring is consistent with the DU of 4. The signal at 19.5 δ is due to a methyl group. The signals at 125.9 and 129.7 δ are due to carbons bonded to one hydrogen, and the signal at 136.2 δ is due to a carbon that is not bonded to any hydrogens. If there are indeed two of each of these types of carbons, then the fragments 2 CH_3, 2 CH, 2 CH, and 2 C can readily be assembled to form dimethylbenzene (xylene). However, there are three isomers of xylene: *ortho*, *meta*, and *para*.

4 ^{13}C-NMR signals	5 ^{13}C-NMR signals	3 ^{13}C-NMR signals

These isomers can be distinguished on the basis of their ^{13}C-NMR spectra because they have different symmetries. The *ortho* isomer has four different carbons, the *meta* isomer has five, and the *para* isomer has only three. The unknown must be *ortho*-xylene. Note that this compound could be identified as one of the xylene isomers on the basis of its ^{1}H-NMR spectrum, but it would be difficult to establish which isomer it is from just that information.

12.39 Suggest a structure for the compound with the formula $C_6H_{10}O_2$ that has the following IR, ^{1}H-NMR, and ^{13}C-NMR spectra:

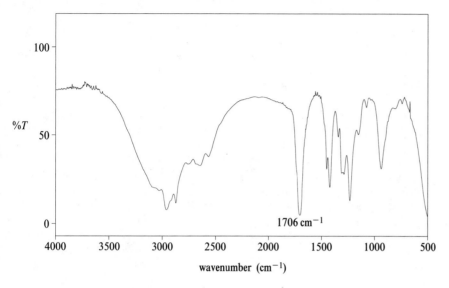

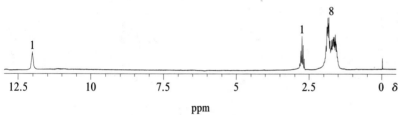

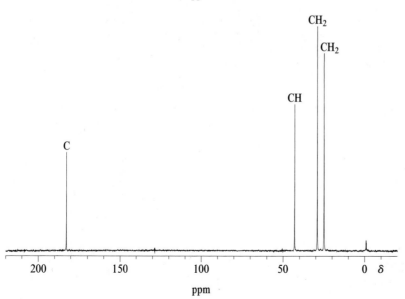

12.40 Suggest a structure for the compound with the formula $C_6H_{12}O_2$ that has the following IR, ^{1}H-NMR, and ^{13}C-NMR spectra:

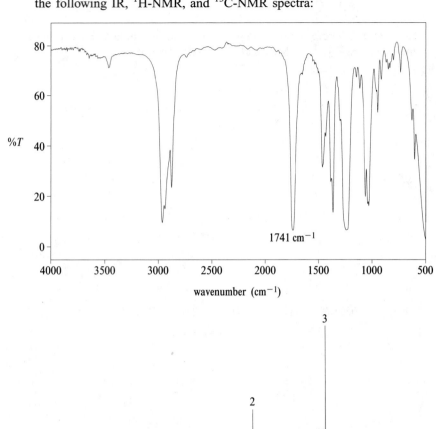

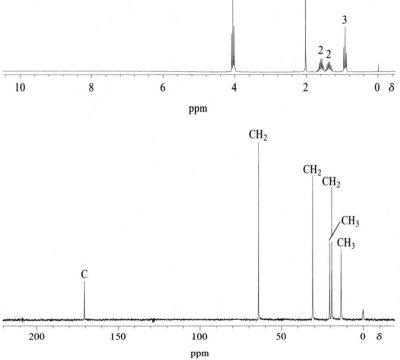

12.41   Suggest a structure for the compound with the formula $C_8H_{18}O$ that has the following IR, ^{1}H-NMR, and ^{13}C-NMR spectra:

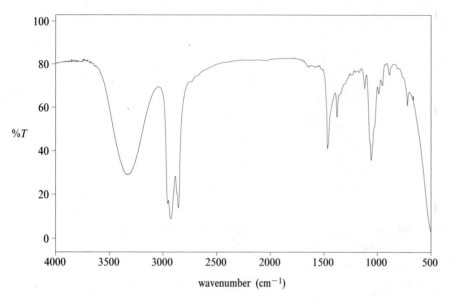

wavenumber (cm^{-1})

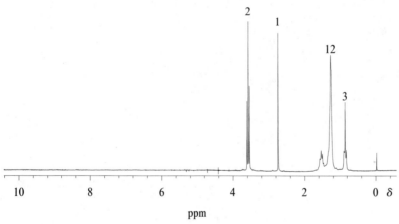

ppm

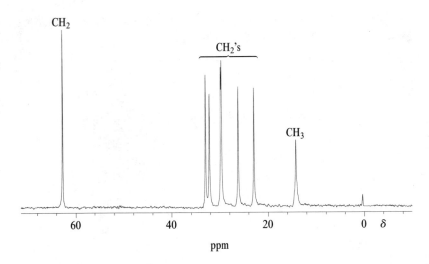

12.42 Suggest a structure for the compound with the formula $C_7H_{12}O$ that has the following IR, ^{1}H-NMR, and ^{13}C-NMR spectra:

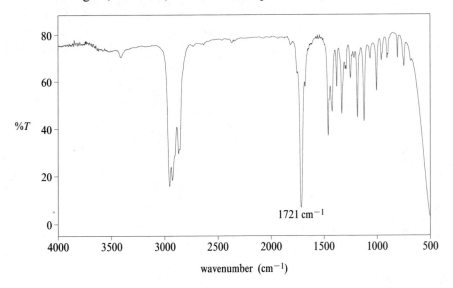

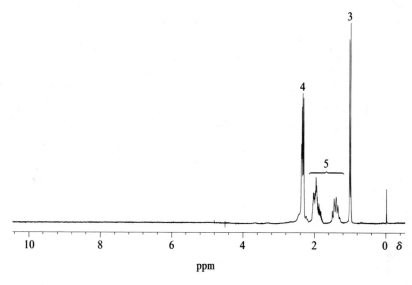

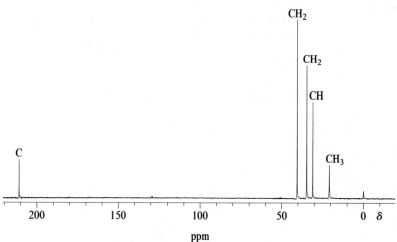

12.43 An unknown compound, **A** (C_7H_{10}), shows four absorptions in its ^{13}C-NMR spectrum, at 22 (CH_2), 24 (CH_2), 124 (CH), and 126 (CH) δ. On reaction with excess H_2 and a Pt catalyst, **A** produces **B** (C_7H_{14}). **B** shows a single peak at 28.4 δ in its ^{13}C-NMR spectrum. Ozonolysis of **A** gives $C_2H_2O_2$ and $C_5H_8O_2$. Suggest structures for **A** and **B**.

12.44 The product of the following reaction has a broad absorption at 3330 cm^{-1} in its IR spectrum. Its ^{13}C-NMR spectrum shows absorptions at 70 (C), 34 (CH_2), 30 (CH_3), and 15 (CH_3). Suggest a structure for this compound.

$$\underset{\displaystyle CH_3CHCH=CH_2}{\overset{\displaystyle CH_3}{|}} \quad \xrightarrow[\displaystyle H_2SO_4]{\displaystyle H_2O}$$

12.45 Explain how many absorptions appear in the ^{1}H-NMR spectrum of this compound.

Structure Determination by Spectroscopy II: Ultraviolet-Visible Spectroscopy and Mass Spectrometry

13.1 Introduction

In addition to IR and NMR spectroscopy there are many other instrumental techniques that are useful to the organic chemist. Two of these, ultraviolet-visible spectroscopy and mass spectrometry, are discussed in this chapter. Ultraviolet-visible spectroscopy is presented first. The use of this technique to obtain information about the conjugated part of a molecule is described. Then mass spectrometry is discussed. This technique provides the molecular mass and formula for a compound. In addition, the use of the mass spectrum to provide structural information about the compound under investigation is presented.

13.2 Ultraviolet-Visible Spectroscopy

As was discussed in Chapter 12, the ultraviolet (UV) region of the spectrum covers the range from 200 to 400 nm, and the visible region covers the range from 400 to 800 nm. The amount of energy available in this radiation, ranging from 143 kcal/mol (600 kJ/mol) to 36 kcal/mol (150 kJ/mol), is enough to cause an **electronic transition** in a molecule, that is, to excite an electron from an occupied MO to an antibonding MO.

Figure 13.1 shows a general diagram for such an electronic transition. The lowest-energy electron arrangement, the ground electronic state, is illustrated on the left side of the diagram. Only the highest-energy orbital that is occupied by electrons (the **highest occupied MO,** or **HOMO**) and the lowest-energy empty MO (**lowest unoccupied MO,** or **LUMO**) are shown in this diagram. In general there are other occupied MOs at lower energies than the HOMO and other unoccupied MOs at higher energies than the LUMO (see Section 3.10). The HOMO may be a bonding or a nonbonding MO; the LUMO is usually an antibonding MO. When the energy of the light

Figure 13.1

General Diagram for an Electronic Transition.

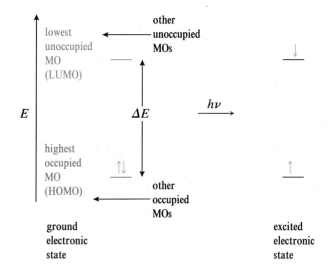

matches the energy difference between the HOMO and the LUMO, then the light is absorbed and an electron is promoted to the LUMO, producing the excited state. The electronic transition shown in Figure 13.1 is the lowest-energy one. Other transitions, resulting from the excitation of an electron from the HOMO to higher-energy unoccupied MOs or from the excitation of an electron from lower-energy occupied MOs to any of the unoccupied MOs, are also possible. Of course, all of these higher-energy transitions result in the absorption of shorter wavelengths of light.

Figure 13.2 shows the UV spectrum of 2,5-dimethyl-2,4-hexadiene. The wavelength of the light, in nanometers, is plotted along the x-axis, and the absorption band comes up from this axis, as was the case for NMR spectra. The wavelength range for most UV spectra begins at 200 nm because the O_2 of air and quartz glass absorb light with wavelengths shorter than this. Most absorption bands due to electronic transitions are broad and rather featureless like this one because each electronic energy level has numerous vibrational and rotational sublevels (see Figure 12.2).

The amount of light absorbed by the sample is plotted along the y-axis as the **absorbance** (A), which is defined as

$$A = \log\left(\frac{I_0}{I}\right)$$

where I_0 is the intensity of the light striking the sample and I is the intensity of the light emerging from the sample. Absorbance is directly proportional to the concentration of the sample and the path length through the sample. The equation expressing this proportionality, known as the Lambert-Beer law, is

$$A = \epsilon c l$$

where ϵ is the **molar absorptivity** or the **molar extinction coefficient,** in units of $\mathbf{M}^{-1}\ \mathrm{cm}^{-1}$, c is the concentration of the compound in moles per liter, and l is the path length in centimeters. (The Lambert-Beer law is very useful for quantitative analysis. You probably used this equation in the general chemistry laboratory to measure the concentration of a dissolved solute that absorbed UV or visible light.)

For a particular compound, the wavelength at the maximum of the absorption band (λ_{max}) and the extinction coefficient at the maximum (ϵ_{max}) are characteristic constants for that compound and are often listed in reference books along with the melting point, boiling point, and other physical constants of the compound. The solvent in which the sample is dissolved is also reported because λ_{max} and ϵ_{max} vary slightly with the solvent. 2,5-Dimethyl-2,4-hexadiene, whose UV spectrum is shown in Figure 13.2, has $\lambda_{max} = 242$ nm and $\epsilon_{max} = 13,100$ in methanol as solvent.

Although the λ_{max} and ϵ_{max} can be used like a melting point to aid in the identification of a compound, they also provide information about the energy separation between the MOs of the compound. Let's see how to use this to obtain information about the structure of the compound.

Problems

Problem 13.1   Anthracene has $\epsilon = 1.80 \times 10^5$ $\mathbf{M}^{-1}$ $\mathbf{cm}^{-1}$ at $\lambda_{max} = 256$ nm. Calculate the absorbance of a 1.94×10^{-6} $\mathbf{M}$ solution of anthracene in a 1 cm cell.

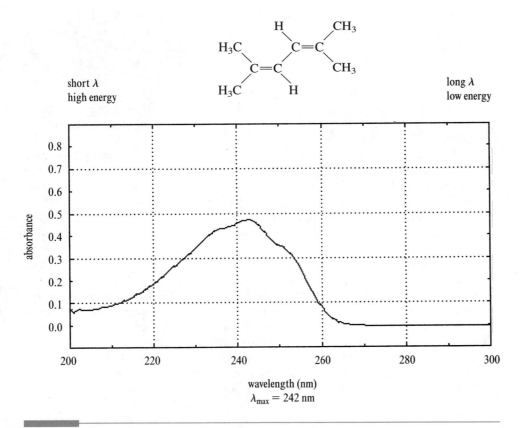

Figure 13.2

The Ultraviolet Spectrum of 2,5-Dimethyl-2,4-hexadiene.

Problem 13.2   A solution of 0.0014 g of benzophenone in 1 L of ethanol has $A = 0.153$ (1 cm cell) at $\lambda_{max} = 252$ nm. Calculate the molar absorptivity of benzophenone.

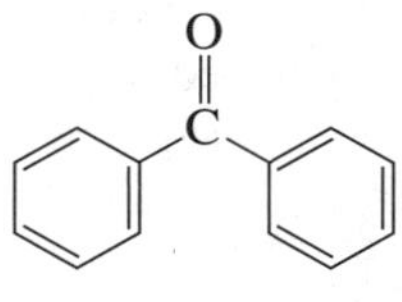

benzophenone

Problem 13.3   *trans*-1-Phenyl-1,3-butadiene has $\lambda_{max} = 280$ ($\epsilon = 27,000$). Calculate the concentration of a solution that has $A = 0.643$ at 280 nm in a 1 cm cell.

13.3   Types of Electronic Transitions

In a simple alkane such as ethane, the HOMO is a sigma bonding MO (it is not really important whether it is the σ_{CC} or a σ_{CH} MO) and the LUMO is a sigma antibonding MO (see Figure 3.23). The lowest-energy electronic transition that can occur involves the excitation of an electron from a bonding to an antibonding MO and is termed a $\sigma \longrightarrow \sigma^*$ transition. This transition occurs at high energy in ethane ($\lambda_{max} = 135$ nm) and other alkanes because sigma bonds are strong and the energy separation between bonding and antibonding MOs is large. Therefore alkanes are transparent in the accessible UV region above 200 nm and are often used as solvents for obtaining UV spectra of other compounds.

If a compound is to absorb in the region above 200 nm, the energy separation between its HOMO and LUMO must be smaller than in the case of alkanes. The bonding and antibonding MOs for pi bonds, which are weaker than sigma bonds, are closer together in energy (see Figure 3.24). While the $\pi \longrightarrow \pi^*$ transition for ethene ($\lambda_{max} = 165$ nm, $\epsilon_{max} = 10,000$) still does not occur in the accessible UV region, it does occur at lower energy and longer wavelength than the $\sigma \longrightarrow \sigma^*$ transition of ethane.

Delocalized pi MOs must be used to describe compounds with conjugated pi bonds. This topic is covered in more detail in Chapters 17 and 20. For now it is enough to note that the energy of the highest-bonding pi MO increases and the energy of the lowest-antibonding pi MO decreases as the number of conjugated pi bonds in the compound increases. Figure 13.3 illustrates this for the series ethene, 1,3-butadiene, and 1,3,5-hexatriene. As would be expected based on this figure, λ_{max} for the $\pi \longrightarrow \pi^*$ transition increases as the number of conjugated pi bonds increases and moves into the accessible UV region for dienes and trienes. As the **chromophore** (the part of the molecule that is responsible for the absorption of ultraviolet or visible light) becomes more conjugated, the absorption maximum moves to even longer wavelengths. Note also that the molar absorptivities for these $\pi \longrightarrow \pi^*$ transitions are all quite large.

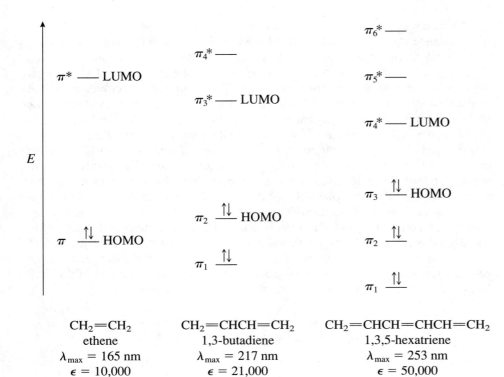

$$E$$

$\pi^* $ —— LUMO	$\pi_4^* $ ——	$\pi_6^* $ ——
	$\pi_3^* $ —— LUMO	$\pi_5^* $ ——
		$\pi_4^* $ —— LUMO
		$\pi_3 \underline{\uparrow\downarrow}$ HOMO
	$\pi_2 \underline{\uparrow\downarrow}$ HOMO	$\pi_2 \underline{\uparrow\downarrow}$
$\pi \underline{\uparrow\downarrow}$ HOMO	$\pi_1 \underline{\uparrow\downarrow}$	$\pi_1 \underline{\uparrow\downarrow}$

$CH_2{=}CH_2$
ethene
$\lambda_{max} = 165$ nm
$\epsilon = 10{,}000$

$CH_2{=}CHCH{=}CH_2$
1,3-butadiene
$\lambda_{max} = 217$ nm
$\epsilon = 21{,}000$

$CH_2{=}CHCH{=}CHCH{=}CH_2$
1,3,5-hexatriene
$\lambda_{max} = 253$ nm
$\epsilon = 50{,}000$

Figure 13.3

Energies of the Pi MOs of Ethene, 1,3-Butadiene, and 1,3,5-Hexatriene.

In the case of an aldehyde or a ketone, the chromophore is the carbonyl group. The $\pi \longrightarrow \pi^*$ transition for acetone, like that for ethene, occurs below 200 nm ($\lambda_{max} = 188$ nm) because there is no conjugation. However, acetone has electrons in nonbonding orbitals that are higher in energy than the pi electrons. The longest-wavelength absorption for acetone is due to an $n \longrightarrow \pi^*$ transition ($\lambda_{max} = 279$ nm, $\epsilon_{max} = 15$). The molar absorptivity for the $n \longrightarrow \pi^*$ transition is much smaller than that for a typical $\pi \longrightarrow \pi^*$ transition because the n orbital and the π^* occupy different regions of space, resulting in a less probable transition.

Another chromophore that is commonly encountered in organic compounds is the benzene ring. Benzene itself has $\pi \longrightarrow \pi^*$ transitions with $\lambda_{max} = 204$ nm ($\epsilon_{max} = 7{,}900$) and 256 nm ($\epsilon_{max} = 200$).

Problems

Problem 13.4   Nitromethane has $\lambda_{max} = 275$ nm ($\epsilon = 15$). What kind of transition is responsible for this absorption?

Problem 13.5   3-Buten-2-one has $\lambda_{max} = 213$ nm ($\epsilon = 7080$) and $\lambda_{max} = 320$ nm ($\epsilon = 21$). What kind of transition is responsible for each of these absorptions?

For a compound to be colored, it must absorb light in the visible region. The color of the compound is complementary to the color of the light that is absorbed (see Table 13.1). Most simple organic compounds are colorless because the amount of energy separating their HOMOs and LUMOs is too large for the absorption of visible light to occur. In the case of some compounds with λ_{max} in the ultraviolet region, the tail of the absorption band extends into the visible region. Such compounds absorb some of the violet light and appear yellow. 1,2-Benzanthracene (λ_{max} = 386 nm) is an example of such a yellow compound. Compounds with even more extensive conjugation absorb visible light and appear colored. β-Carotene, which absorbs at 452 nm, is an example of a compound in which a long series of conjugated double bonds results in the absorption of visible light. Because it absorbs blue light, β-carotene appears orange. It is the pigment responsible for the color of carrots. Indigo absorbs yellow-orange light at 600 nm. It appears blue and is used to dye denim for blue jeans. Azo dyes, which constitute over 60% of the manufactured dyes, contain a nitrogen–nitrogen double bond. Usually they have an aromatic group bonded to each nitrogen of the azo group. The wavelength of the absorption maximum of the dye, and thus its color, depends on the substituents on the aromatic rings. An example is FD&C Red No. 40, a red azo dye (λ_{max} = 508 nm) that is used in food coloring.

β-carotene λ_{max} = 452 nm

Table 13.1 Relationship Between Color Absorbed and Color Observed.

Wavelength Absorbed (nm)	Color Absorbed	Color Observed
400	violet	yellow-green
425	blue-violet	yellow
450	blue	orange
490	blue-green	red
510	green	purple
530	yellow-green	violet
550	yellow	blue-violet
590	orange	blue
640	red	blue-green
730	purple	green

m/z 100   *m/z* 58   mass 42

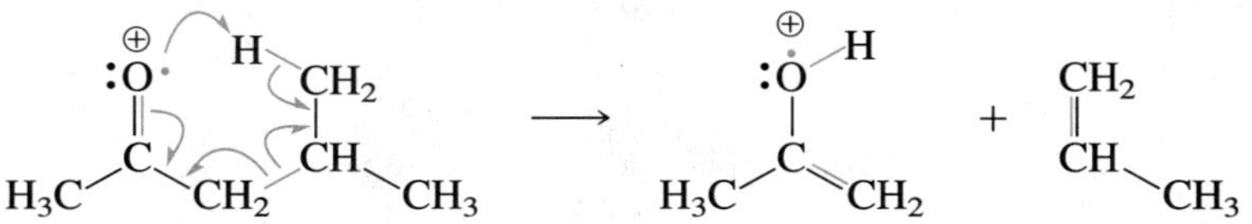

These fragmentations serve to illustrate many of the major types. The driving force behind all of them is the formation of stable cations and radicals. Fragmentations of functional groups that have not been covered here are often similar to those above. Although this has been only a very brief introduction to mass spectrometry, the power and utility of this technique should be apparent.

Problem

Problem 13.14   Show equations to account for the major fragment ions that occur at the indicated *m/z* for these compounds.

a) 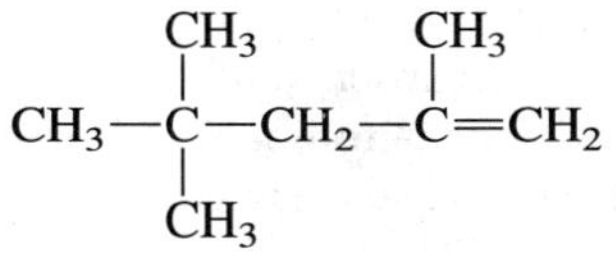
CH_3—$\overset{\underset{|}{CH_3}}{\underset{|}{C}}$—$CH_2$—$\overset{\underset{}{CH_3}}{\underset{|}{C}}$=$CH_2$ *m/z* 97, 57

b) 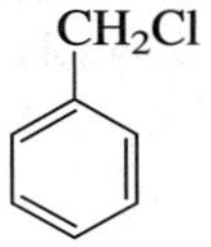
m/z 91

c) $CH_3CH_2CH_2CH_2OH$ *m/z* 31

d) 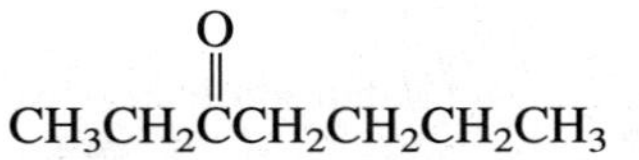 $CH_3CH_2\overset{\overset{O}{\|}}{C}CH_2CH_2CH_2CH_3$ *m/z* 85, 72, 57

Solution

a)   The molecular ion for this compound (C_8H_{16}) has *m/z* 112. The peak at *m/z* 97 results from the loss of a fragment of mass 15, which must be a CH_3 group. This fragmentation is important because it produces a tertiary carbocation.

m/z 112   mass 15   *m/z* 97

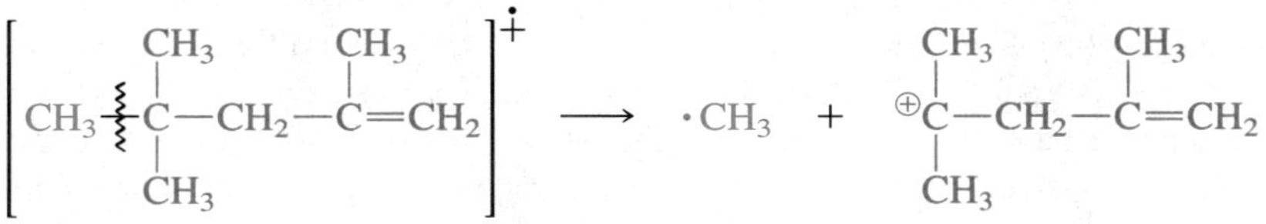

The peak at *m/z* 57 results from the fragmentation to produce a tertiary carbocation and an allylic radical according to the following equation. (As might be expected, the peak at *m/z* 57 is the base ion in the mass spectrum of this compound.)

$$
\begin{bmatrix}
& \overset{\text{CH}_3}{\underset{\text{CH}_3}{|}} & \overset{\text{CH}_3}{|} \\
\text{CH}_3-\text{C}-\text{CH}_2-\text{C}=\text{CH}_2
\end{bmatrix}^{+\cdot}
\longrightarrow
\text{CH}_3-\overset{\text{CH}_3}{\underset{\text{CH}_3}{\overset{|}{\underset{|}{\text{C}}}}}{}^{\oplus} \quad + \quad \overset{\cdot}{\text{CH}_2}-\overset{\text{CH}_3}{\overset{|}{\text{C}}}=\text{CH}_2
$$

m/z 112   *m/z* 57   mass 55

13.8   Summary

UV-visible spectroscopy and mass spectrometry provide two additional techniques that can be quite useful for structural determination under certain circumstances. From the UV-visible spectrum the presence or absence of a conjugated chromophore can be determined. Often, the structure of the chromophore is suggested by simple empirical calculations or by comparison of the spectrum with those of model compounds.

Mass spectrometry provides the molecular mass of the compound, and the formula can be obtained from a high-resolution mass spectrum. Considerable structural information is available from fragmentation of the molecular ion.

ELABORATION

Gas Chromatography and Mass Spectrometry

The techniques of gas chromatography (GC) and mass spectrometry (MS) have been combined into a very powerful instrument, a GC/MS, for the identification of organic compounds. In this technique a small sample, containing a mixture of compounds, is injected onto the gas chromatograph, which separates the compounds. As the individual components elute from the column, they are passed directly into a mass spectrometer. The mass spectrum can be obtained without ever isolating the pure component.

Gas chromatography is an extremely powerful separation method. It requires only a very small amount of material and is able to separate even a quite complex mixture into its numerous components. The time that it takes for a particular compound to elute from the column under a given set of conditions, called the retention time, is characteristic for that compound. However, the identity of a component cannot be established solely on the basis of its retention time because another compound may coincidentally have an identical retention time. Whenever an analysis has a potential for legal or disciplinary consequences, the identity of the compound must be established by some other technique. Mass spectrometry is an ideal partner because it can establish the identity of a compound with certainty by using only the small amount of sample obtained from the GC.

After completing this chapter you should be able to:

1. Determine whether a compound will absorb light in the ultraviolet or visible region.

2. Identify the chromophore and type of transition responsible for absorption of UV-visible radiation.

3. Use a high-resolution mass spectrum to determine the formula of a compound.

4. Determine whether sulfur, chlorine, bromine, or nitrogen is present in a compound by examination of the M, M + 1, and M + 2 peaks in its mass spectrum.

5. Explain the major fragmentation pathways for compounds containing some of the simple functional groups.

End-of-Chapter Problems

13.15 A student wishes to record the UV spectrum of *trans*-stilbene, which has λ_{max} = 308 nm (ϵ = 25,000). What concentration should be prepared if the desired absorbance is 0.5 at the maximum?

Extensive drug screening is done at many athletic events, such as the Olympic Games. Usually, separate analyses, using different extraction procedures, are done for stimulants, narcotics, anabolic steroids, diuretics, and peptide hormones. In the analysis for stimulants, which are amines such as amphetamine and cocaine, a 5 mL urine sample is first made basic with KOH to ensure that the amines are present as the neutral molecules rather than as salts. The free amines are then extracted from the sample with diethyl ether. To save time and expense, the sample is first analyzed by gas chromatography only. If a peak appears with the retention time of one of the proscribed stimulants, then the sample is reanalyzed by GC/MS to confirm the identity of the suspected compound.

amphetamine

cocaine

13.16 Indicate the types of transitions responsible for the absorptions of these compounds.

a)

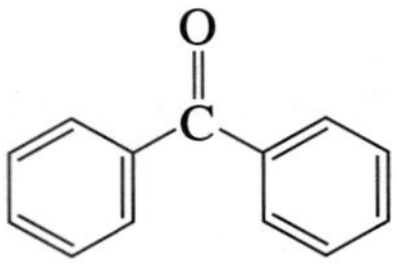

λ_{max} = 252 nm (ϵ = 20,000)
λ_{max} = 325 nm (ϵ = 180)

b)

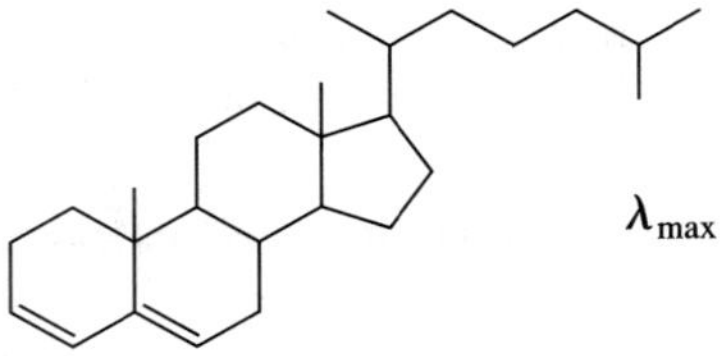

λ_{max} = 235 nm (ϵ = 19,000)

c)

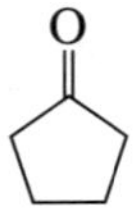

λ_{max} = 299 nm (ϵ = 20)

d) $CH_3C\equiv C-C\equiv CCH_3$ λ_{max} = 227 nm (ϵ = 360)

13.17 Which of these compounds are expected to have an absorption maximum in the region of 200–400 nm in their UV spectra?

a) $CH_3\overset{\overset{\displaystyle O}{\|}}{C}CH_2CH_3$

b)

c)

d) 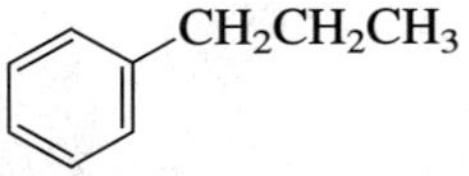 $CH_2CH_2CH_3$

e)

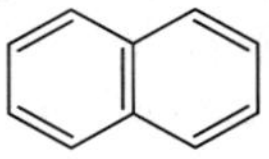

f) $CH_3CH_2OCH_2CH_3$

g) 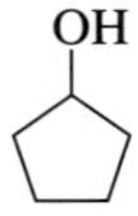   OH

h)

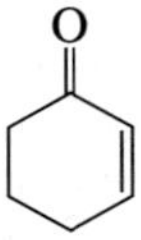

13.18 Predict the major fragments and their *m/z* that would appear in the mass spectra of these compounds.

a) $CH_3CH_2CH_2CH_2CH_2CH_3$

b) $CH_3CH_2CH_2\overset{\overset{\displaystyle CH_3}{|}}{C}HCH_2CH_3$

c)   H$_3$C—C—OH
        with CH$_3$ above and CH$_3$ below the central C

d)   (benzene ring) with CH$_2$CH$_2$CH$_3$ and H$_3$C substituents

e)   CH$_3$CH$_2$CCH$_2$CH$_2$CH$_3$
        with CH$_3$ above and CH$_3$ below

f)   CH$_3$CCH$_2$CH$_3$
        with OH above and CH$_3$ below

g)   CH$_3$CCH$_2$CH$_2$CH$_2$CH$_3$
        with O (double bond) above the second C

h)   CH$_3$CH$_2$CH$_2$CHCH$_3$
        with OH above the fourth C

i)   CH$_3$CH=CHCH$_2$CH$_2$CH$_3$

13.19 Explain how mass spectrometry could be used to distinguish between these compounds.

a)   CH$_3$CH$_2$CH$_2$CH$_3$ and   CH$_3$CHCH$_3$
        with CH$_3$ above the middle C

b)   CH$_3$CCH$_2$CHCH$_2$CH$_3$ and   CH$_3$CCHCH$_2$CH$_2$CH$_3$
        with O (double bonds) and CH$_3$ substituents

c)   CH$_3$CCH$_2$CH$_2$CH$_3$ and   CH$_3$CH$_2$CCH$_2$CH$_3$
        with O (double bonds) above

13.20 Explain why neopentane shows no molecular ion in its mass spectrum. Predict the structure and *m/z* for the base ion in its mass spectrum.

H$_3$C—C—CH$_3$
    with CH$_3$ above and CH$_3$ below the central C

neopentane

13.21 Explain how the peaks at *m/z* 115, 101, and 73 arise in the mass spectrum of 3-methyl-3-heptanol.

13.22 The mass spectra of 3-methyl-2-pentanone and 4-methyl-2-pentanone are as follows. Explain which spectrum goes with which compound. What is the structure of the ion responsible for the peak at *m/z* 43 in each spectrum?

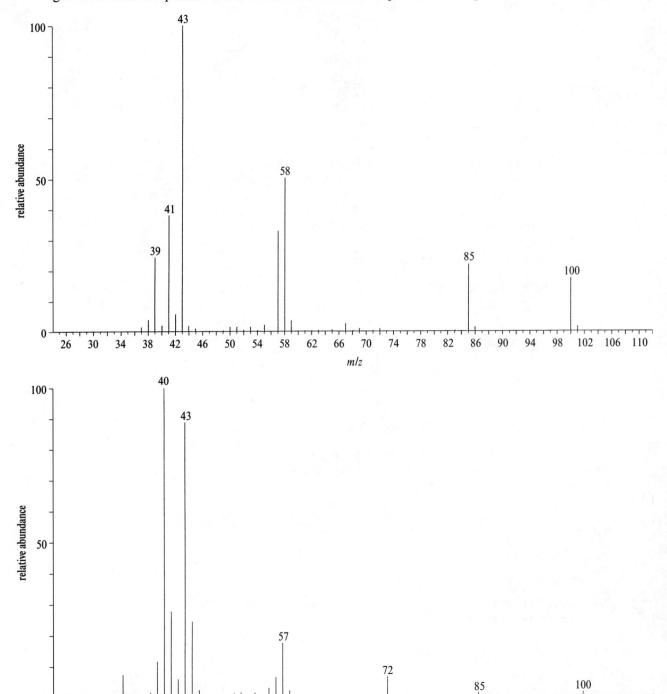

It is interesting to note that an alkene can be prepared by two different Wittig pathways, depending on which of the doubly bonded carbons was originally the carbon of the carbonyl group and which was the carbon of the ylide. Thus the synthesis of β-carotene has also been accomplished by using the reaction of a diylide with two equivalents of an aldehyde, as illustrated in the following equations:

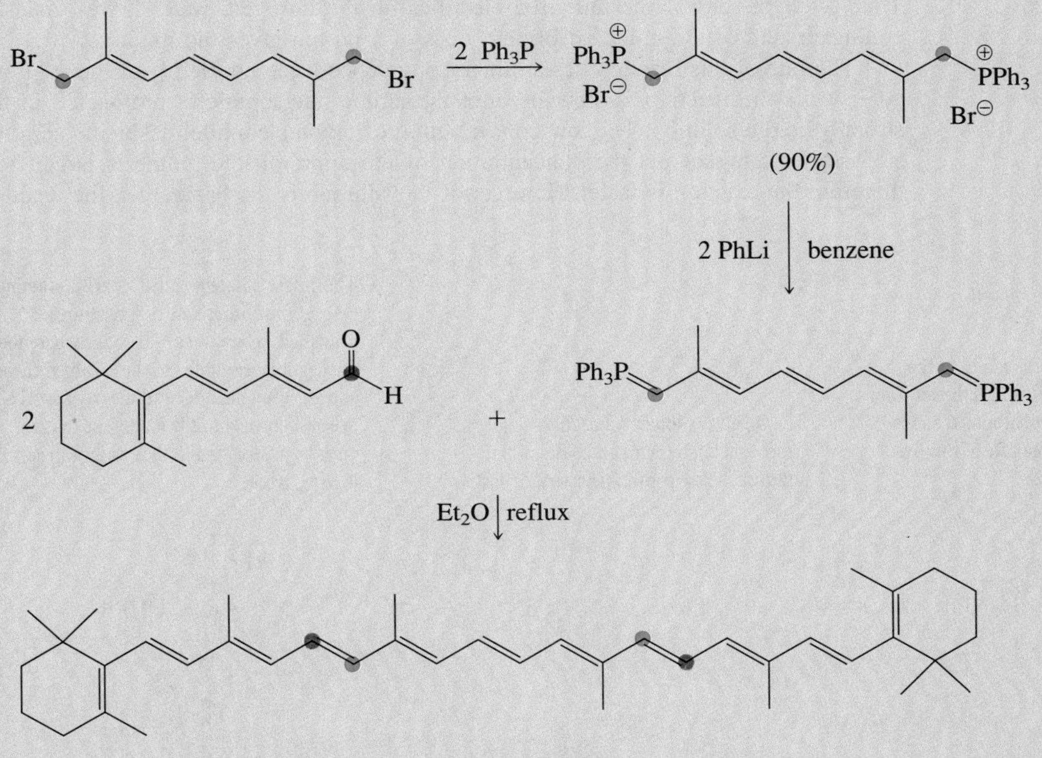

(90%)

2 PhLi | benzene

2

+

Et$_2$O | reflux

β-carotene (51%)

14.9   Addition of Nitrogen Nucleophiles

Amines add to the carbonyl groups of aldehydes or ketones to produce compounds containing carbon–nitrogen double bonds. These nitrogen analogs of aldehydes and ketones are called **imines.**

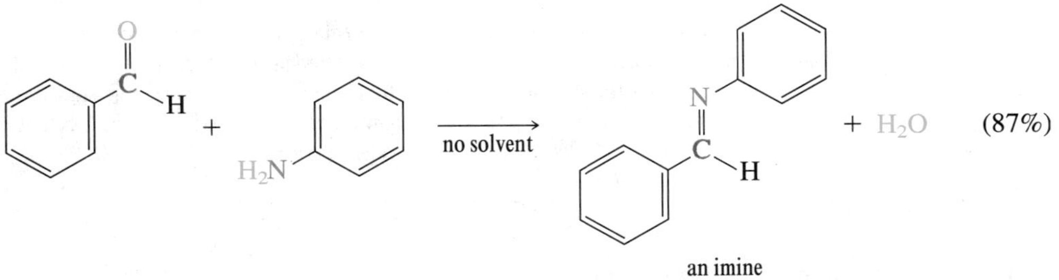

+ H$_2$O   (87%)

an imine

The reaction proceeds according to the mechanism for basic conditions to form the addition product, called a carbinolamine (see Figure 14.3). Because the nitrogen atom of the carbinolamine has another pair of electrons that can be used for the formation of a bond to the electrophilic carbon, the reaction does not stop at the addition stage. Instead, it proceeds to the addition–elimination product by the loss of water to form a carbon–nitrogen double bond. First the oxygen of the carbinolamine is protonated. Then, with the aid of the unshared electrons on the nitrogen, water leaves, and the conjugate acid of the imine is formed. Loss of a proton gives the imine.

The rate of this reaction has an interesting dependence on the pH of the solution. At low pH the reaction is slow because the amine nucleophile is protonated in the strongly acidic solution. The low concentration of the nucleophile makes the addition step slow. At higher pH the concentration of the unprotonated amine is larger, and therefore the reaction is faster. However, if the solution is too basic, then the reaction

The addition step occurs according to the mechanism for basic conditions. If the solution is too acidic, the amine is protonated and is no longer nucleophilic. The rate of this first step then becomes very slow.

The O gets protonated and the N gets deprotonated by acids and bases in the solution. These proton transfers are fast.

This intermediate is called a **carbinolamine.** (Most of the other addition reactions presented so far stop at this stage where the nucleophile has added to the carbonyl carbon and a proton has added to the oxygen.) To proceed onward to the imine, the oxygen must be protonated so that water can act as a leaving group.

The product of the elimination step is a protonated imine, the conjugate acid of the final product. The proton is transferred to a solvent molecule in an acid–base reaction to produce the imine.

If the solution is too basic, the concentration of the protonated carbinolamine is low, and this elimination step becomes very slow. Note how the unshared electrons on the N help the water to leave.

Figure 14.3

Mechanism for the Addition of an Amine to an Aldehyde to Form an Imine.

is again slow because the concentration of the protonated carbinolamine is low, and therefore the elimination of water is slow. The maximum reaction rate occurs in the pH range of 4 to 6.

Removal of Water

Whenever a reaction equilibrium does not strongly favor the products, it is necessary to drive the equilibrium toward the products if the reaction is to be synthetically useful. According to Le Chatelier's principle, this can be accomplished either by using an excess of one of the reagents, if one is readily available and inexpensive, or by removing one of the products. Several reactions with unfavorable equilibria, such as the formation of imines and the formation of acetals (see Section 14.10), have water as one of the products. The equilibria in such reactions are often driven to the products by removal of water.

The most common way to accomplish this employs a Dean-Stark water separator. The reaction is conducted at reflux in a solvent such as benzene or toluene. The water is carried off with the solvent vapors because these two solvents form an azeotrope with water. When the vapor condenses, the water, which is not soluble in the solvent, separates. The design of the apparatus allows the less dense organic solvent to flow back into the reaction flask while the water is retained in the separator.

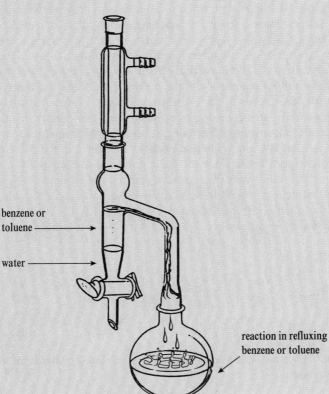

benzene or toluene

water

reaction in refluxing benzene or toluene

Because a carbon–oxygen double bond is considerably stronger than a carbon–nitrogen double bond, the equilibrium in these reactions often favors the carbonyl compound rather than the imine. In such cases it is necessary to drive the equilibrium to the product. This is usually accomplished by removing the water as it is formed. Some additional examples of imine formation are provided in the following equations:

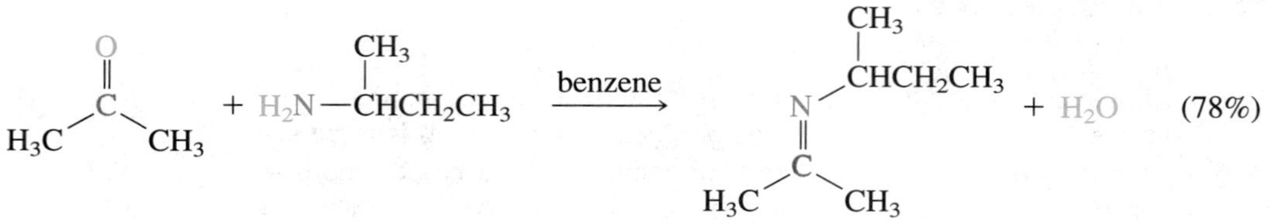

The product in the following reaction results from the formation of two carbon–nitrogen double bonds. In this case the equilibrium favors the product because of the additional resonance stabilization provided by the new aromatic ring.

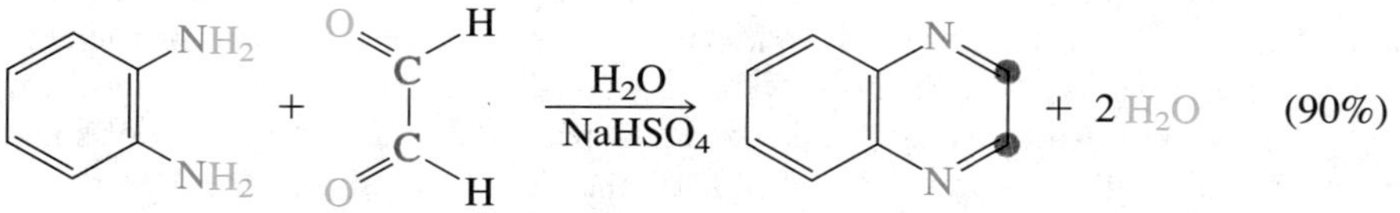

In the days before the advent of spectroscopic techniques, reactions that form carbon–nitrogen double bonds were often used in determining the structure of unidentified aldehydes and ketones. After the functional group of an unknown compound was determined and its possible identity was narrowed to a few choices, the final step in the structure determination was often the conversion of the unknown to a solid derivative using a standard chemical reaction of that functional group. The melting point of the derivative was then compared with the melting points that had been reported in the literature for the derivatives of the possible candidates. A match between the melting points was considered strong evidence in establishing the identity of the unknown.

In the case of aldehydes and ketones, several C=N forming reactions were used to make derivatives. The most common of these are shown in the following equations. Note that each reagent has an electronegative group substituted on the NH_2. This helps to shift the equilibrium toward the product and makes it more likely to be a solid. Tables of the melting points of these derivatives for common aldehydes and ketones can be found in many reference books.

The reaction of an aldehyde or a ketone with **hydroxylamine** produces an **oxime** derivative:

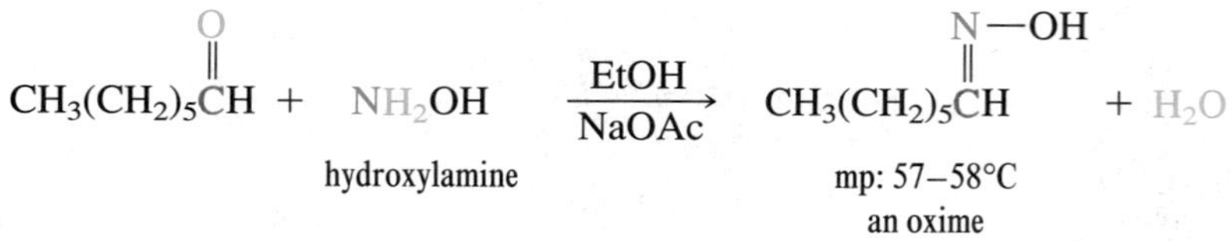

The reaction of an aldehyde or a ketone with **semicarbazide** produces a **semicarbazone** derivative:

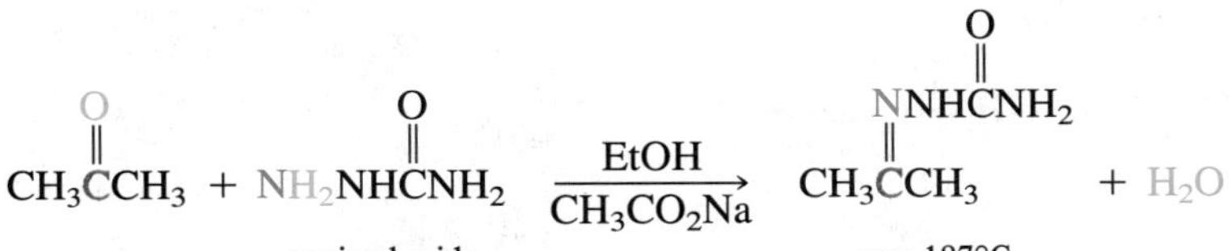

$$CH_3CCH_3 + NH_2NHCNH_2 \xrightarrow[CH_3CO_2Na]{EtOH} CH_3CCH_3 + H_2O$$

semicarbazide

mp: 187°C
a semicarbazone

The reaction of an aldehyde or a ketone with **phenylhydrazine** produces a **phenylhydrazone** derivative:

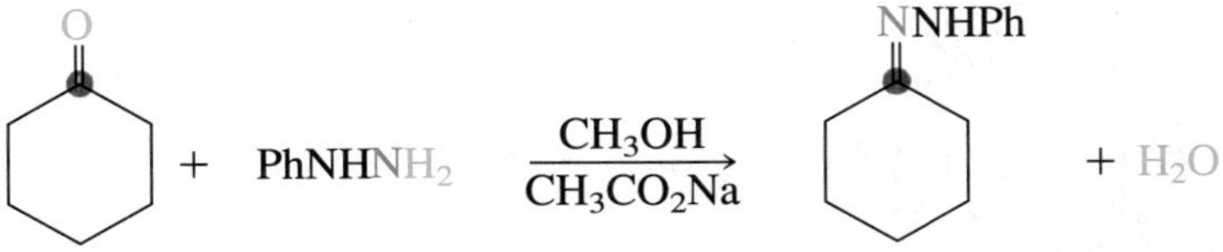

phenylhydrazine

mp: 77°C
a phenylhydrazone

The reaction of an aldehyde or a ketone with **2,4-dinitrophenylhydrazine** produces a **2,4-dinitrophenylhydrazone** derivative. This reaction is also used as a test for the presence of an aldehyde or ketone. A drop or two of a suspected aldehyde or ketone is added to a solution of 2,4-dinitrophenylhydrazine in ethanol and water. The formation of a precipitate, usually orange or red, of the derivative indicates that the unknown is an aldehyde or ketone:

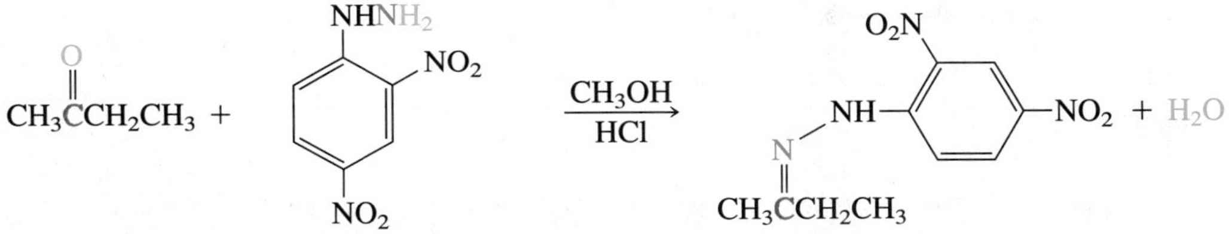

2,4-dinitrophenylhydrazine

mp: 117°C
a 2,4-dinitrophenylhydrazone

Problem

Problem 14.16 Show the products of these reactions.

a)

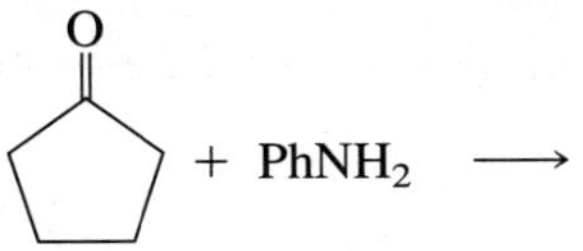

+ PhNH₂ ⟶

b)

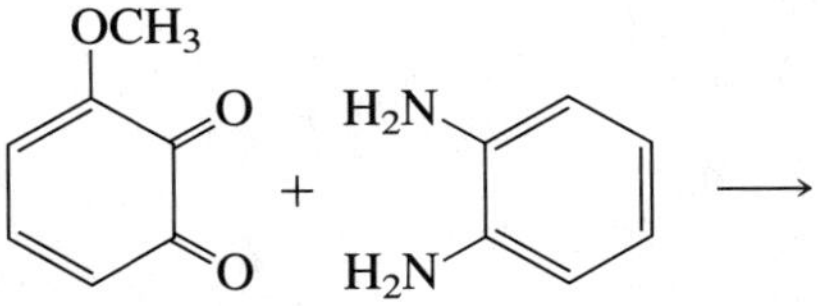

c) $CH_3\overset{O}{\overset{\|}{C}}CH_2CH_3 + NH_2OH \longrightarrow$

d) [benzaldehyde structure] $+ NH_2NH\overset{O}{\overset{\|}{C}}NH_2 \longrightarrow$

e) [cycloheptanone structure] $+$ [2,4-dinitrophenylhydrazine structure with NHNH$_2$, NO$_2$, NO$_2$] $\longrightarrow$

So far, all of the examples have involved primary amines. The reaction of ammonia with aldehydes and ketones also forms imines, but the products are unstable and cannot usually be isolated. If a secondary amine is used, an enamine, rather than an imine, is formed. An **enamine** has an amino group bonded to one of the carbons of a carbon–carbon double bond. It is related to the imine in the same manner as an enol is related to a ketone (see Section 10.7). The mechanism for its formation can be outlined as follows:

$$CH_3-\overset{\overset{..}{\overset{\displaystyle O:}{\|}}}{C}-CH_3 + \overset{\frown}{\underset{\underset{H}{|}}{N}:} \xrightarrow{\text{H—A}} CH_3-\overset{\overset{\oplus}{N}}{\overset{\|}{C}}-CH_2 \rightleftharpoons CH_3-\overset{\overset{..}{N}}{C}=CH_2$$

an enamine

The reaction of an aldehyde or a ketone with a secondary amine follows exactly the same mechanism as the reaction with a primary amine (see Figure 14.3) until the final step. Unlike the case with a primary amine, the nitrogen of the iminium ion does not have a proton that can be removed to produce a stable imine. Therefore a proton is removed from an adjacent carbon, resulting in the formation of an enamine. Enamine formation is illustrated in the following equations. In each case the equilibrium is driven toward the products by removal of water.

[cyclohexanone] $+$ [morpholine] $\xrightarrow[\text{toluene reflux}]{\text{TsOH}}$ [enamine product] $+ H_2O$ (80%)

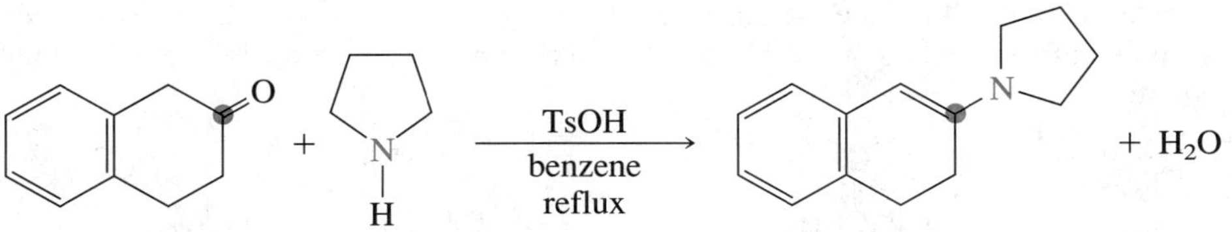

Problems

Problem 14.17 Show all of the steps in the mechanism for this reaction:

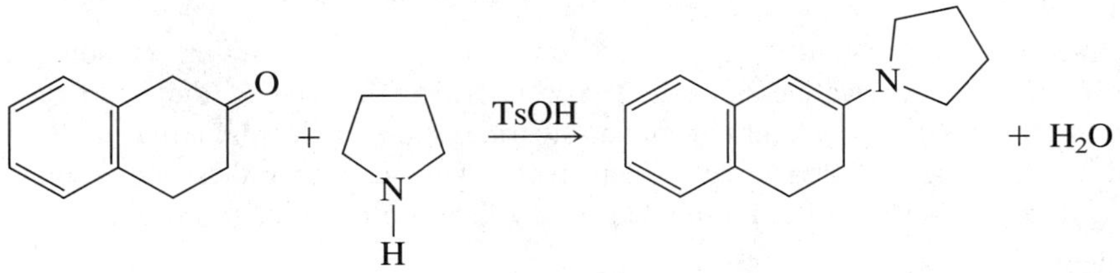

Problem 14.18 Explain why the reaction in problem 14.17 produces the enamine shown in that equation rather than this enamine.

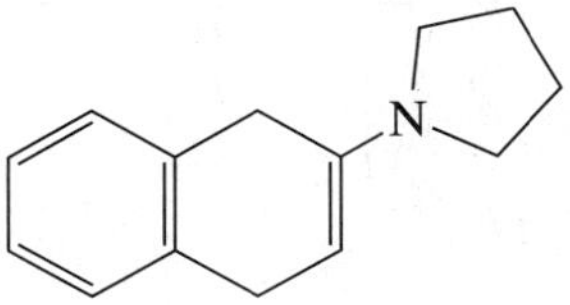

not formed

Problem 14.19 Show the products of these reactions.

a)

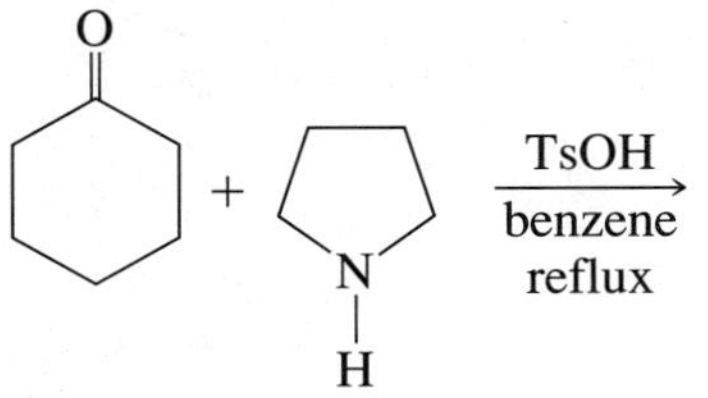

b) 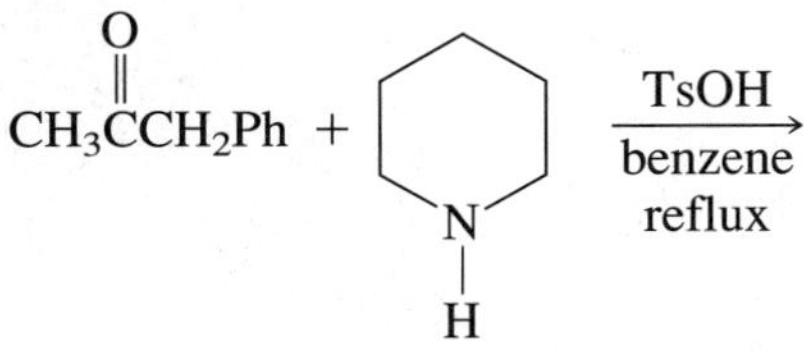 CH_3CCH_2Ph +

Imines are also intermediates in a useful process called reductive amination. In this reaction an aldehyde or a ketone is reacted with an amine to form an imine. A reducing agent, such as hydrogen and a catalyst, is also present in the reaction mixture. The reducing agent chosen is one that will not reduce the carbonyl compound but will reduce the more reactive imine. The imine is reduced as rapidly as it is formed and is not isolated. The reaction is illustrated in the following equation:

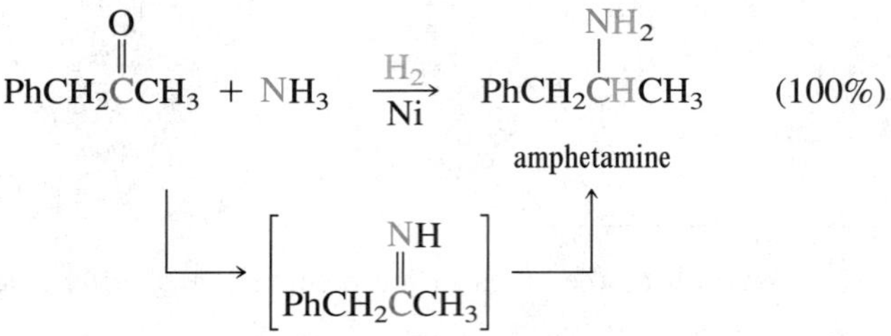

$$PhCH_2\overset{\displaystyle O}{\overset{\|}{C}}CH_3 \;+\; NH_3 \;\xrightarrow[\text{Ni}]{H_2}\; PhCH_2\overset{\displaystyle NH_2}{\underset{}{\overset{|}{C}H}}CH_3 \quad (100\%)$$

amphetamine

$$\longrightarrow \left[PhCH_2\overset{\displaystyle NH}{\overset{\|}{C}}CH_3 \right] \longrightarrow$$

Another reducing agent that can be employed in this reaction is sodium cyanoborohydride, a derivative of sodium borohydride with one of the hydrogens replaced by a cyano group. Sodium cyanoborohydride is less nucleophilic than sodium borohydride and does not react with aldehydes or ketones under these conditions. However, it does react with the protonated form of the imine, which is considerably more electrophilic:

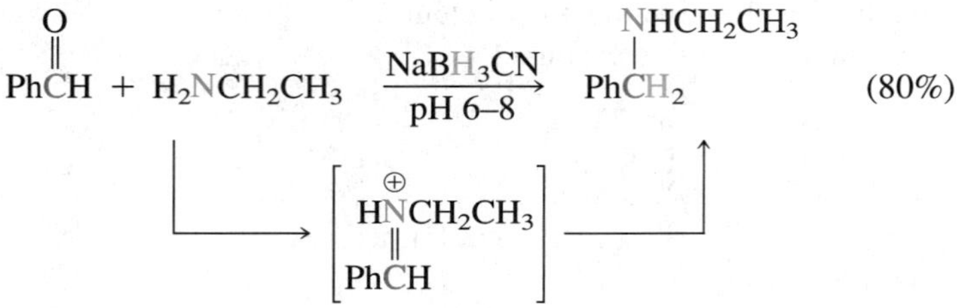

$$Ph\overset{\displaystyle O}{\overset{\|}{C}}H \;+\; H_2NCH_2CH_3 \;\xrightarrow[\text{pH 6–8}]{NaBH_3CN}\; PhCH_2\overset{\displaystyle NHCH_2CH_3}{\underset{}{|}} \quad (80\%)$$

$$\longrightarrow \left[Ph\overset{\displaystyle \overset{\oplus}{H}NCH_2CH_3}{\overset{\|}{C}H} \right] \longrightarrow$$

protonated imine

$$(71\%)$$

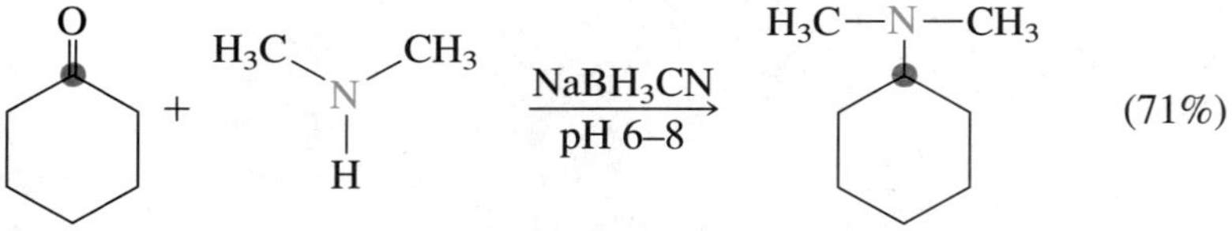

Problem

Problem 14.20 Show the products of these reactions.

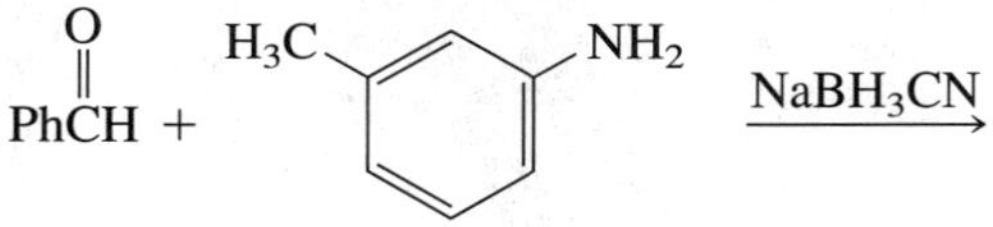

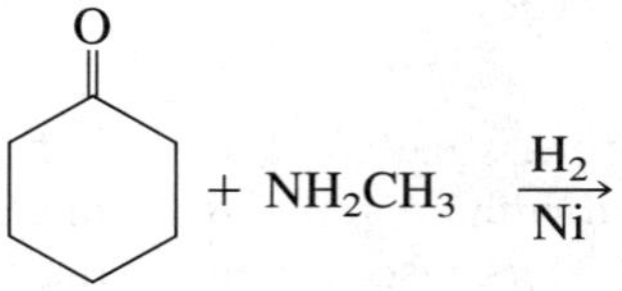

Meth-amphetamine

Amphetamines have been abused in the United States for many years. In the 1950s the drug was taken in pill form, known as "uppers," to decrease appetite and provide energy and euphoria. In the 1960s, methamphetamine, known as "speed," was taken by injection. This dangerous abuse led to the saying "Speed kills." Even though the dangers of its use are well known, snorting and injecting methamphetamine continue today.

Most illicit drugs are isolated from natural sources because their structures are too complex to allow easy synthesis in the laboratory, although some simple chemical modifications may be conducted such as in the conversion of morphine to heroin. However, methamphetamine has a simple enough structure that it can be synthesized in clandestine laboratories by someone with only rudimentary skills in organic chemistry. A common synthesis of so-called "crank" involves the reductive amination of phenylacetone with methylamine as shown in the following equation:

phenylacetone   methamphetamine

To help control illegal synthesis of methamphetamine, phenylacetone is a controlled substance; that is, buyers must have a license to purchase it.

14.10 Addition of Alcohols

Aldehydes and ketones add two equivalents of alcohols to form **acetals.** (The term *ketal,* which was formerly used to describe the product formed from a ketone may still be encountered.)

an acetal

The mechanism for this reaction, shown in Figure 14.4, is as long and complex as any that you will encounter in this text. You can make the task of learning this mechanism much easier by recognizing its similarities to other mechanisms in this chapter and the similarities among the steps within this mechanism. The initial addition follows the mechanism for acidic conditions. Steps 1, 2, and 3 are nearly identical to the mechanism for acid-catalyzed hydration (Section 14.4) with the exception that the nucleophile is the oxygen of methanol rather than the oxygen of water. The addition product, a **hemiacetal,** is very similar to the hydrate or the carbinolamine. Steps 4 and 5 are very similar to steps in imine formation (Figure 14.3). First the oxygen is protonated, and then water leaves with the help of the unshared electrons on the other oxygen. However, in contrast to imine formation, the product of step 5 cannot be stabilized by the loss of a proton. Instead, the oxygen of a second molecule of alcohol acts as a nucleophile.

ELABORATION

Imines in Living Organisms

Imines are readily formed from amines and aldehydes or ketones under physiological conditions. In addition, they are readily hydrolyzed back to the amine and the carbonyl compound. These properties provide a common way in biochemical reactions to temporarily link an aldehyde or a ketone to a protein by reacting the carbonyl with a free amino group of the protein to form an imine.

One example of this process is found in the chemistry of vision. In the rods and cones of the eye the aldehyde 11-*cis*-retinal forms an imine by reaction with an amino group of the protein opsin. Studies have shown that the nitrogen of the imine of the product, called rhodopsin, is protonated.

11-*cis*-retinal

$+ H_2N$—(opsin)

rhodopsin

When rhodopsin absorbs light in the vision process, the *cis* double bond between carbons 11 and 12 isomerizes to a *trans* double bond. This isomerization triggers a nerve impulse telling the brain that light has been absorbed by the eye. The imine of the isomerized product is unstable and is hydrolyzed to opsin and the all-*trans* form of retinal (also known as vitamin A aldehyde). All-*trans* retinal is converted back to 11-*cis*-retinal by enzymes so it can be used again in rhodopsin formation.

Imine formation is also important in the enzymatic decarboxylation of acetoacetate anion to form acetone, which occurs during the metabolism of glucose. Initial formation of a protonated imine facilitates the loss of carbon dioxide to form an enamine. Hydrolysis of the enamine produces acetone and regenerates the enzyme catalyst:

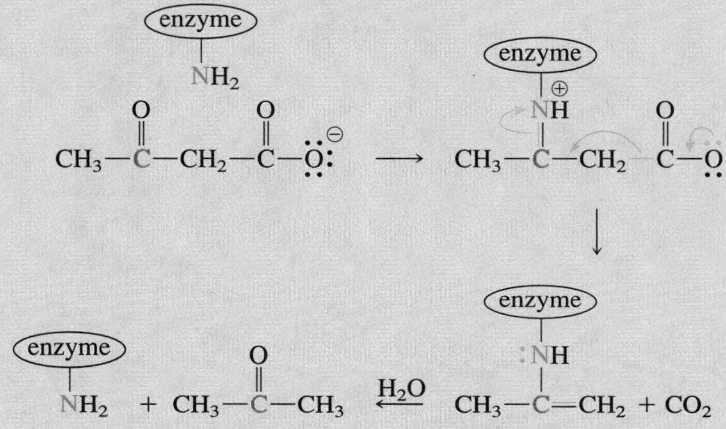

Many enzymes require the presence of an additional compound, called a coenzyme, to carry out their catalytic functions. The coenzyme often bonds to the substrate, modifying its structure so that the enzyme can more easily accomplish its catalytic reaction. Pyridoxal, also known as vitamin B_6, acts as a coenzyme by forming an imine between its aldehyde group and an amine group of the substrate. (Some of the reactions that are catalyzed by pyridoxal are described in the elaboration on page 1160.) In these biological reactions it is important that the equilibrium constant for imine formation not be too large so that the carbon–nitrogen double bond can be cleaved by hydrolysis when the reaction is finished, thus completing the catalytic cycle. The compounds that are used to form derivatives of carbonyl compounds—hydroxylamine, hydrazine, phenylhydrazine, and semicarbazide—are poisonous because they form imine derivatives of pyridoxal. The equilibrium constant for the formation of these derivatives is so large that there is not enough free pyridoxal available to carry out its catalytic functions.

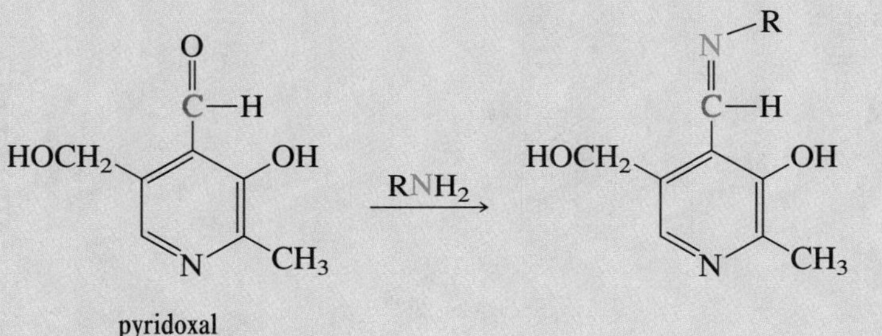

pyridoxal
(vitamin B_6)

The reaction follows the mechanism for acidic conditions. First, the carbonyl oxygen is protonated, making the carbon more electrophilic. (HA represents an acid in the solution.)

Then the oxygen of an alcohol molecule acts as a nucleophile.

Next a proton is transferred to some base. This could be the conjugate base of the acid or even CH_3OH.

This addition product is called a hemiacetal. It resembles a carbinolamine, and the next two steps are similar to those in imine formation: The oxygen is protonated and water leaves. Step 4 and the reverse of step 3 are nearly identical. They differ only in which oxygen is protonated.

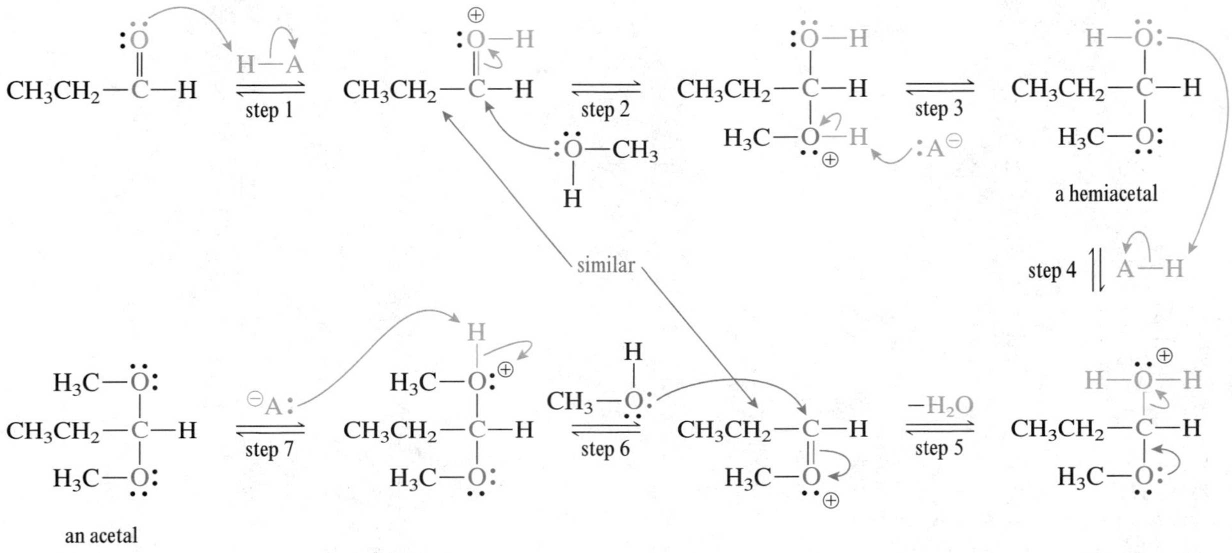

This intermediate resembles the product of step 2 but with a methyl replacing one hydrogen. Transfer of a proton to some base in the solution produces the final product, an acetal.

This intermediate resembles the protonated aldehyde above. It reacts in a similar fashion: Steps 6 and 7 are quite similar to steps 2 and 3.

The electron pair on the ether oxygen helps water leave.

Figure 14.4

Mechanism for the Formation of an Acetal.

Like hydrates, hemiacetals are not favored at equilibrium and, in general, cannot be isolated. The equilibrium is shifted in their favor by the inductive effects of electron-withdrawing groups, similar to the case of hydrates. In addition, the equilibrium is shifted in their favor if the alcohol nucleophile is part of the same molecule as in the following example:

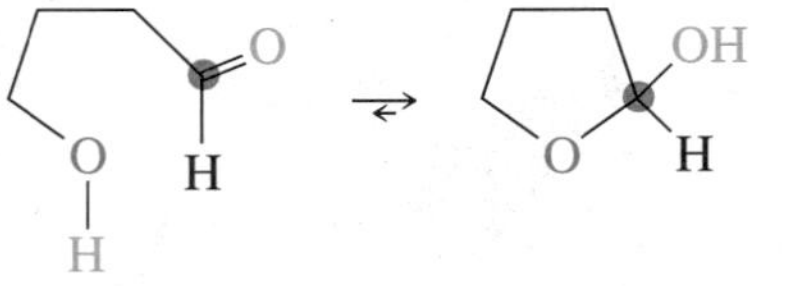

11.4%   88.6%
                a cyclic hemiacetal

In this example the oxygen of the hydroxy group acts as an intramolecular nucleophile. Recall from Section 7.14 that intramolecular reactions are favored by entropy.

15.4 Preparation of Anhydrides

An anhydride can be prepared by the reaction of a carboxylic acid, or its conjugate base, with an acyl chloride as illustrated in the following equation:

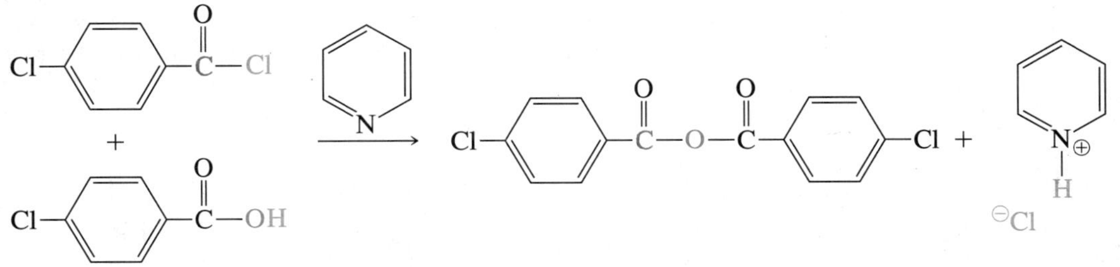

(98%)

Pyridine is often added to these reactions, or used as a solvent, to react with the HCl that is produced and prevent the reaction mixture from becoming strongly acidic.

Anhydrides can also be prepared by the reaction of a carboxylic acid with another anhydride, usually acetic anhydride. A mixed anhydride is an intermediate in this reaction:

$$2\ \text{Ph}\overset{\text{O}}{\overset{\|}{\text{C}}}\text{OH} + \text{CH}_3\overset{\text{O}}{\overset{\|}{\text{C}}}\text{O}\overset{\text{O}}{\overset{\|}{\text{C}}}\text{CH}_3 \rightleftharpoons \text{Ph}\overset{\text{O}}{\overset{\|}{\text{C}}}\text{O}\overset{\text{O}}{\overset{\|}{\text{C}}}\text{Ph} + 2\ \text{CH}_3\overset{\text{O}}{\overset{\|}{\text{C}}}\text{OH} \qquad (74\%)$$

$$\longrightarrow \text{Ph}\overset{\text{O}}{\overset{\|}{\text{C}}}\text{O}\overset{\text{O}}{\overset{\|}{\text{C}}}\text{CH}_3 \longrightarrow$$

To make this preparation useful, the equilibrium is driven to the right by removal of the acetic acid by careful distillation. This is often the method of choice for the preparation of cyclic anhydrides, in which the equilibrium is favored by entropy:

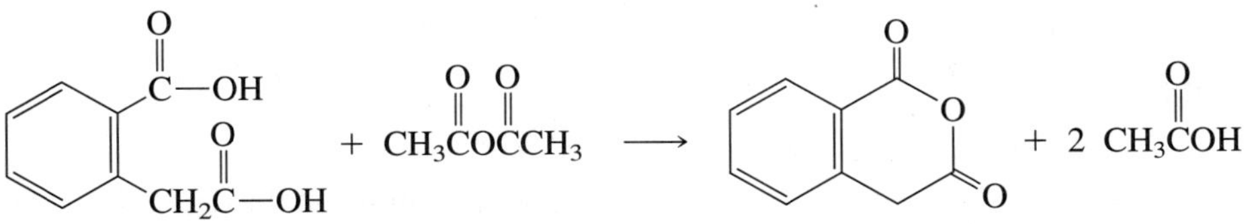

(88%)

Problem

Problem 15.6 Show the products of these reactions.

a) $\text{CH}_3\text{CH}_2\overset{\text{O}}{\overset{\|}{\text{C}}}\text{Cl} + \text{CH}_3\text{CH}_2\overset{\text{O}}{\overset{\|}{\text{C}}}\text{OH} \xrightarrow{\text{pyridine}}$

b) 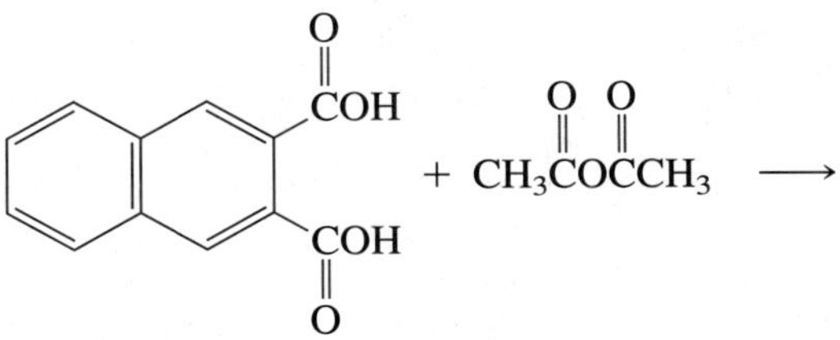

15.5   Preparation of Esters

Esters are readily prepared by reaction of an alcohol with either an acyl chloride or an anhydride. Because it is more easily prepared from the acid, the acyl chloride is commonly employed. Again, a base, such as pyridine, is often added to react with the HCl that is produced. Acetic anhydride, which is commercially available, is often used for the preparation of acetate esters. Several examples are provided in the following equations:

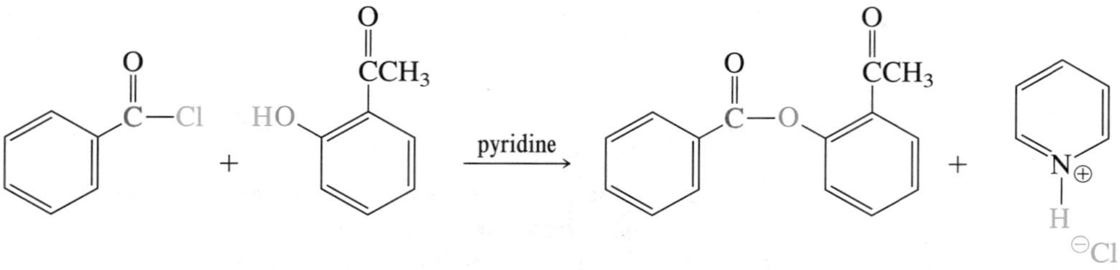

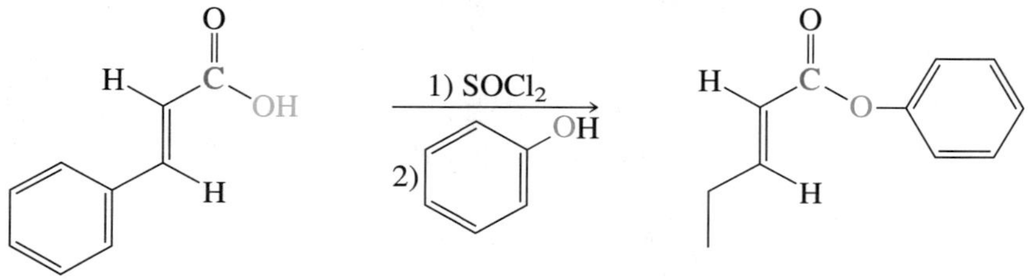

salicylic acid   acetylsalicylic acid
                                                              (aspirin)

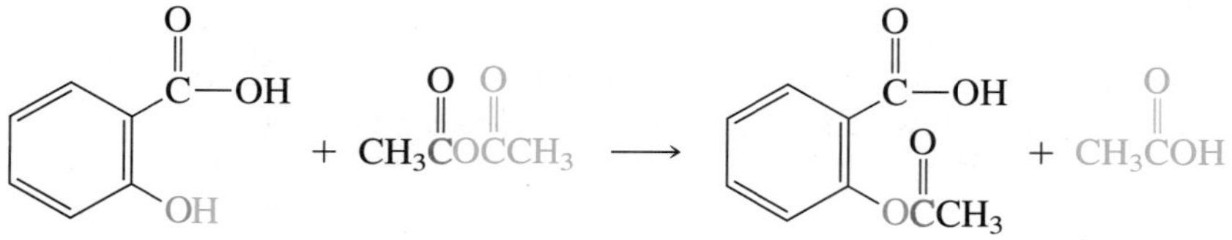

It is also very common to prepare esters directly from the carboxylic acid without passing through the acyl chloride or the anhydride, as shown in the following example:

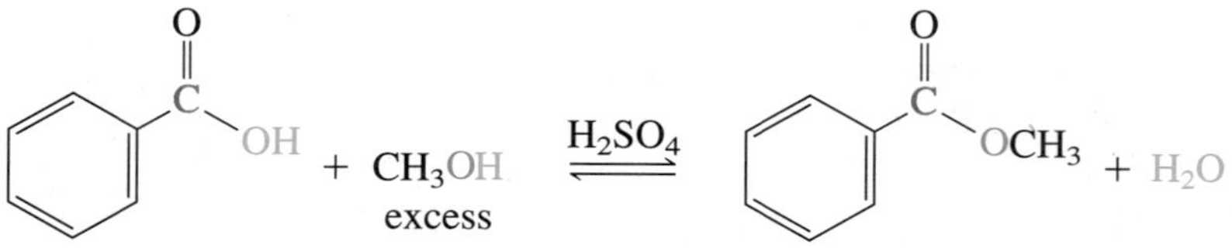

This reaction, known as **Fischer esterification,** requires the presence of an acid catalyst. Since the carboxylic acid and the ester have similar reactivities, the reaction is useful only if a method can be found to drive the equilibrium in the direction of the desired product—the ester. In accord with Le Chatelier's principle, this is accomplished by using an excess of one of the reactants or by removing one of the products. An excess of the alcohol is used if it is readily available, as is the case for methanol or ethanol. Or water can be removed by azeotropic distillation with a solvent such as toluene, as described in the elaboration on page 661.

The mechanism for the Fischer esterification is shown in Figure 15.3. Sulfuric acid, hydrochloric acid, or *p*-toluenesulfonic acid is most often used as a catalyst. The mechanism will be easier to remember if you note the similarities to other acid-catalyzed mechanisms, such as the one for the formation of acetals in Figure 14.4. Also note that the steps leading from the tetrahedral intermediate to the carboxylic acid and alcohol starting materials and to the ester and water products are very similar.

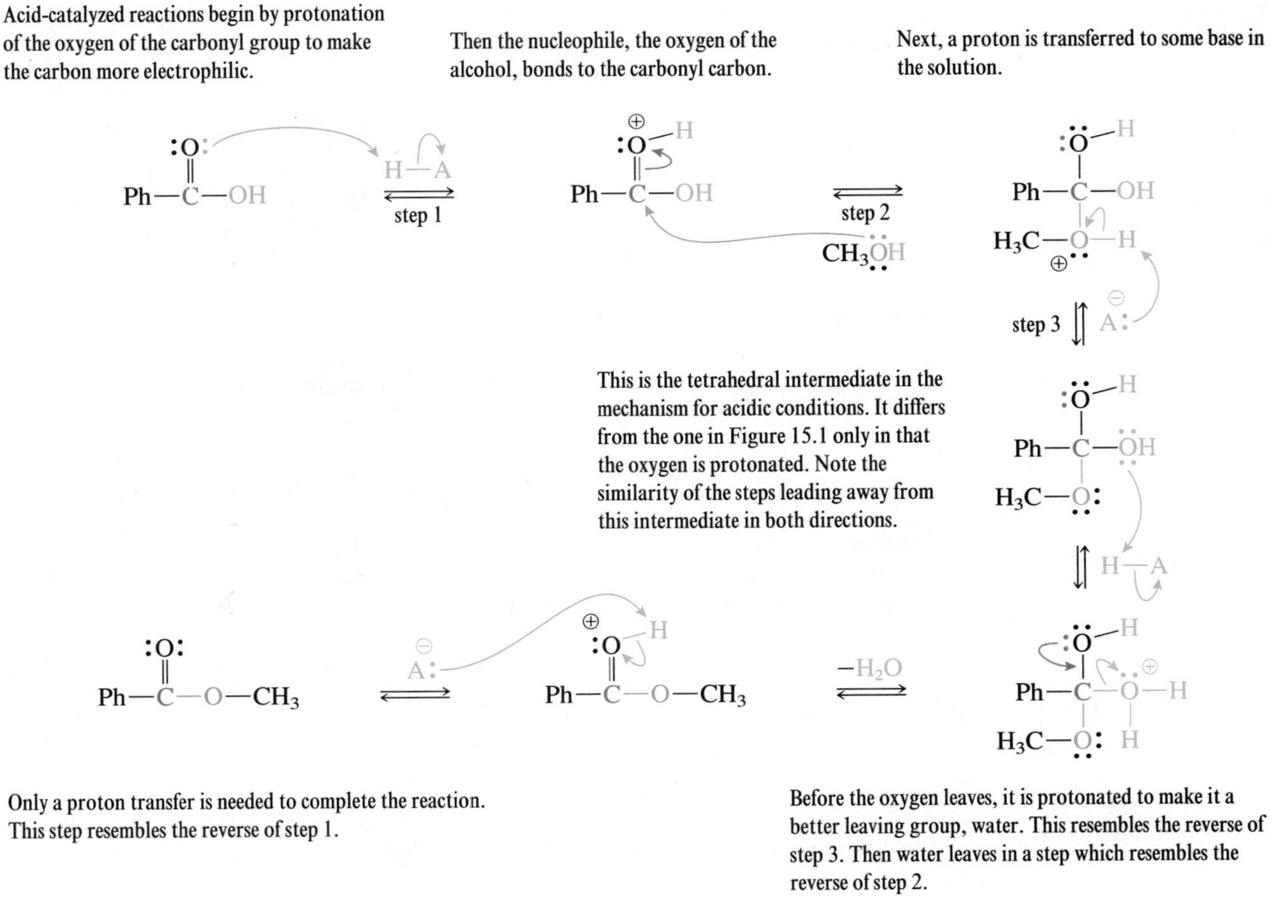

Acid-catalyzed reactions begin by protonation of the oxygen of the carbonyl group to make the carbon more electrophilic.

Then the nucleophile, the oxygen of the alcohol, bonds to the carbonyl carbon.

Next, a proton is transferred to some base in the solution.

This is the tetrahedral intermediate in the mechanism for acidic conditions. It differs from the one in Figure 15.1 only in that the oxygen is protonated. Note the similarity of the steps leading away from this intermediate in both directions.

Only a proton transfer is needed to complete the reaction. This step resembles the reverse of step 1.

Before the oxygen leaves, it is protonated to make it a better leaving group, water. This resembles the reverse of step 3. Then water leaves in a step which resembles the reverse of step 2.

Figure 15.3

Mechanism for Fischer Esterification.

Other examples of the Fischer esterification are provided by the following equations:

$$CH_3CH_2CH_2CH_2\overset{O}{\overset{\|}{C}}OH \; + \; HOCH_2CH_2CH_3 \underset{excess}{\overset{H_2SO_4}{\rightleftharpoons}} \; CH_3CH_2CH_2CH_2\overset{O}{\overset{\|}{C}}OCH_2CH_2CH_3 \; + \; H_2O \quad (78\%)$$

$$+ \; CH_3CH_2OH \underset{toluene}{\overset{TsOH}{\longrightarrow}} \qquad\qquad (97\%)$$

Lactones are cyclic esters. Their formation is very favorable when a hydroxy group and a carboxylic acid in the same molecule can react to form a five- or six-membered ring:

a lactone

Problems

Problem 15.7 Show all of the steps in the mechanism for this reaction:

$$CH_3\overset{O}{\overset{\|}{C}}Cl \; + \; CH_3OH \; \longrightarrow \; CH_3\overset{O}{\overset{\|}{C}}OCH_3 \; + \; HCl$$

Problem 15.8 Salicylic acid has two nucleophilic sites, the oxygen of the phenol and the oxygen of the carboxylic acid. Explain why its reaction with acetic anhydride produces aspirin rather than this compound:

Problem 15.9 Show the products of these reactions.

b) C_6H_5COOH + $CH_3CH_2CH_2OH$ $\xrightarrow{H_2SO_4}$

c) CH_3COCCH_3 + CH_3CHCH_2OH (with CH_3 substituent) $\longrightarrow$

d) $HO\text{-}CH_2CH_2CH_2CH_2\text{-}COOH$ $\xrightarrow{H_2SO_4}$

e) C_6H_5COOH $\xrightarrow{SOCl_2}$ $\qquad$ $\xrightarrow{CH_3OH}$

f) cyclopentane-COOH + CH_3OH $\xrightarrow{TsOH}$

15.6 Preparation of Carboxylic Acids

Carboxylic acids commonly are employed as the starting materials for the preparation of the other acid derivatives. However, any of the acid derivatives can be hydrolyzed to the carboxylic acid by reaction with water under the appropriate conditions. Acid or base catalysis is necessary for the less reactive derivatives.

Acyl chlorides and anhydrides must be protected from water because they react readily, often vigorously, with water to produce carboxylic acids. This reaction is not of much synthetic usefulness because the acyl chloride or anhydride is usually prepared from the acid. However, the hydrolysis reaction is occasionally used for the preparation of a carboxylic acid if the acyl chloride or anhydride is available from some other source. The following equation provides an example:

$$\text{(methylmaleic anhydride)} + H_2O \longrightarrow \text{(dicarboxylic acid)} \quad (94\%)$$

Esters can be hydrolyzed to carboxylic acids under either acidic or basic conditions. Under acidic conditions the mechanism is the exact reverse of the Fischer esterification mechanism shown in Figure 15.3. Again, because the acid and the ester have comparable reactivities, some method must be used to drive the equilibrium toward the desired product—the acid in this case. This can be accomplished by using water as the solvent, providing a large excess of this reagent that, by Le Chatelier's principle, shifts the equilibrium toward the carboxylic acid.

Problem

Problem 15.10   Show all of the steps in the mechanism for this reaction:

$$CH_3\overset{O}{\overset{\|}{C}}OCH_2CH_3 + H_2O \; \underset{\longleftarrow}{\overset{H_2SO_4}{\rightleftharpoons}} \; CH_3\overset{O}{\overset{\|}{C}}OH + CH_3CH_2OH$$

It is more common to hydrolyze esters under basic conditions because the equilibrium is favorable. The mechanism for this process, called **saponification,** is presented in Figure 15.4. Note that the production of the conjugate base of the carboxylic acid, the carboxylate anion, which is at the bottom of the reactivity scale, drives the equilibrium in the desired direction. To isolate the carboxylic acid, the solution must be acidified after the hydrolysis is complete. Some examples are provided in the following equations. We saw another example of this hydrolysis reaction in Chapter 9, where it was used in the preparation of alcohols by the acetate method (see Section 9.3 and Figure 9.1 on page 348).

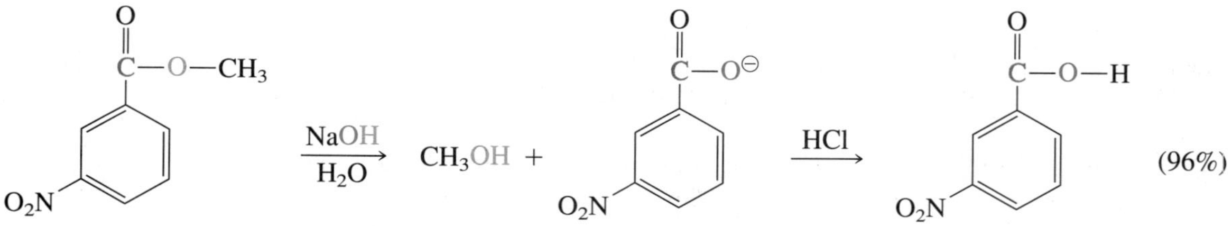

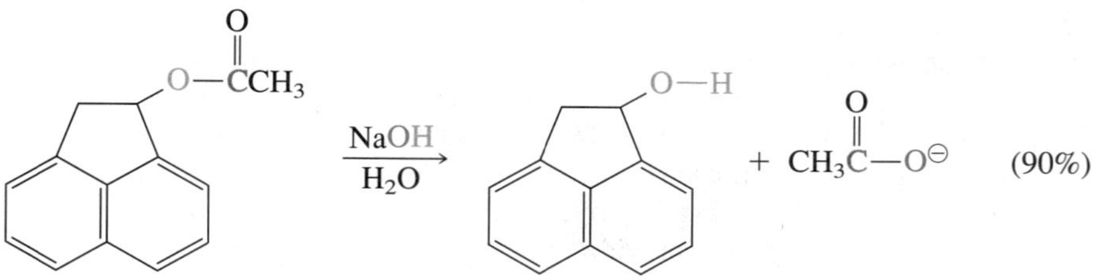

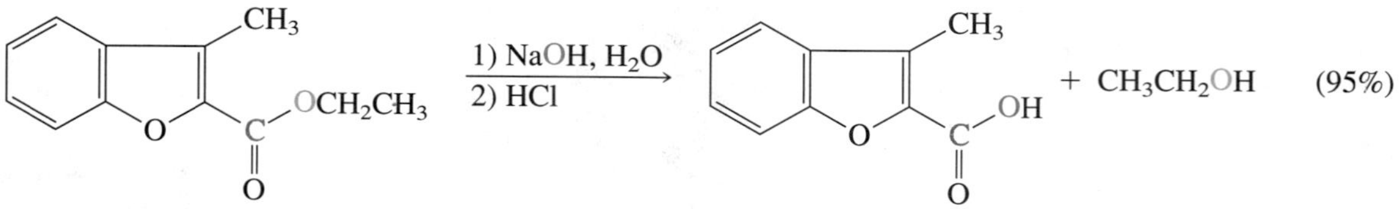

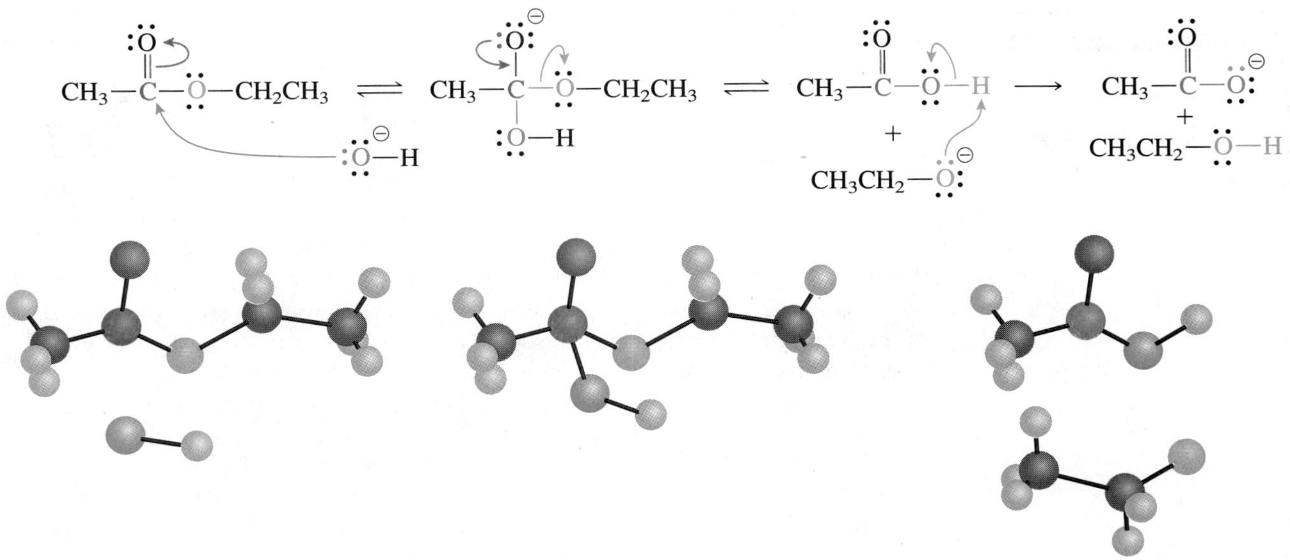

This step is the same as the first step of Figure 15.1. The nucleophile, hydroxide ion in this case, bonds to the carbonyl carbon, displacing the pi electrons onto the oxygen.

Next, the leaving group, ethoxide ion, departs as an electron pair from the negative oxygen reforms the double bond of the carbonyl group.

Ethoxide ion, a strong base, removes a proton from acetic acid. The formation of the weak base, acetate ion, in this step drives the equilibrium to the final products, the alcohol and the carboxylate anion.

Figure 15.4

Mechanism for the Base-Catalyzed Hydrolysis (Saponification) of an Ester.

Amides are less reactive than esters, and their hydrolysis often requires vigorous heating in either aqueous acid or base. The mechanism for acidic conditions is quite similar to the reverse of the Fischer esterification mechanism shown in Figure 15.3. The mechanism for basic conditions is related to that depicted in Figure 15.4 for ester saponification and a similar tetrahedral intermediate is formed in the first step. However, $NH_2^{\ominus}$ is a strong base and a poor leaving group, so the nitrogen must usually be protonated before it can leave. The exact timing of the various proton transfers that occur in this mechanism is difficult to establish and depends on both the structure of the amide and the reaction conditions. One possible mechanism is shown in Figure 15.5.

Under acidic conditions the equilibrium for the hydrolysis of an amide is driven toward the products by the protonation of the ammonia or amine that is formed. Under basic conditions the equilibrium is driven toward the products by the formation of the carboxylate anion, which is at the bottom of the reactivity scale. The pH of the final solution may need to be adjusted, depending on which product is to be isolated. If the carboxylic acid is desired, the final solution must be acidic, while isolation of the amine requires that the solution be basic. Several examples are shown in the following equations. Also, note that the last step of the Gabriel amine synthesis, the

The Preparation of Soap

The preparation of a solution of soap by the reaction of fat with water in the presence of base was probably one of the earliest chemical processes discovered by humans. Although the details of this discovery are lost in antiquity, we can imagine early humans finding that water that had been in contact with wood ashes from the campfire could be used to remove grease from hands and other objects and that this water became a more effective cleaning agent as it was used. The water leaches some alkaline compounds from the ashes, and this basic water hydrolyzes the esters of the fat or grease to alcohols and soap. This is why the hydrolysis of esters under basic conditions is called saponification (the Latin word for soap is *sapo*).

Fats are triesters formed from a triol, glycerol, and fatty acids, which are carboxylic acids with long, unbranched alkyl chains. These alkyl chains may be saturated, or they may have one or more double bonds. Usually, the three carboxylic acids of a fat molecule are not the same. The following equation illustrates the hydrolysis of a representative fat to glycerol and the conjugate bases of stearic acid, palmitoleic acid and linoleic acid. Addition of NaCl causes the fatty acid salts to precipitate. The resulting solid is formed into bars of soap.

$$
\begin{array}{l}
\text{CH}_2\text{—O—C(=O)—(CH}_2)_{16}\text{CH}_3 \\
\text{CH—O—C(=O)—(CH}_2)_7\text{CH=CH(CH}_2)_5\text{CH}_3 \\
\text{CH}_2\text{—O—C(=O)—(CH}_2)_7\text{CH=CHCH}_2\text{CH=CH(CH}_2)_4\text{CH}_3
\end{array}
$$

a representative fat

$$\text{H}_2\text{O} \downarrow \text{NaOH}$$

$$\text{Na}^{\oplus}\ ^{\ominus}\text{O—C(=O)—(CH}_2)_{16}\text{CH}_3 \quad \text{conjugate base of stearic acid}$$

$$
\begin{array}{l}
\text{CH}_2\text{—OH} \\
\text{CH—OH} + \text{Na}^{\oplus}\ ^{\ominus}\text{O—C(=O)—(CH}_2)_7\text{CH=CH(CH}_2)_5\text{CH}_3 \quad \text{conjugate base of palmitoleic acid} \\
\text{CH}_2\text{—OH}
\end{array}
$$

glycerol $\text{Na}^{\oplus}\ ^{\ominus}\text{O—C(=O)—(CH}_2)_7\text{CH=CHCH}_2\text{CH=CH(CH}_2)_4\text{CH}_3$ conjugate base of linoleic acid

The cleaning action of soap is due to the dual nature of the conjugate base of the fatty acid molecule. On one end is the ionic carboxylate anion group, while the rest of the molecule consists of a nonpolar hydrocarbon chain. The ionic part, called the head, is attracted to a polar solvent such as water and is hydrophilic while the long hydrocarbon tail is hydrophobic. In water, soap molecules tend to group together in clusters called micelles with their ionic heads oriented toward the water molecules and their hydrocarbon tails in the interior of the cluster so that unfavorable interactions with water are avoided. Nonpolar grease and fat molecules hydrolysis of the phthalimide (see Section 9.7 and Figure 9.5 on page 364), is an amide hydrolysis.

$$\text{(amide)} \xrightarrow[\text{H}_2\text{O}]{\text{NaOH}} \text{(carboxylate)} + \text{NH}_3 \xrightarrow{\text{HCl}} \text{(carboxylic acid)} + \text{NH}_4^{\oplus} \text{ Cl}^{\ominus} \quad (98\%)$$

dissolve in the interior of the micelle. The ionic heads keep the micelle in solution and allow water to wash it away, along with the grease.

ionic head   hydrocarbon tail

schematic representation
of a fatty acid anion

schematic representation
of a soap micelle

Problem

Problem 15.11   Show the products that are formed when this fat is saponified:

$$\begin{array}{l}
\text{CH}_2\text{—OC(CH}_2)_7\text{CH}=\text{CH(CH}_2)_7\text{CH}_3 \\
\text{CH—OC(CH}_2)_{14}\text{CH}_3 \\
\text{CH}_2\text{—OC(CH}_2)_{12}\text{CH}_3
\end{array}$$

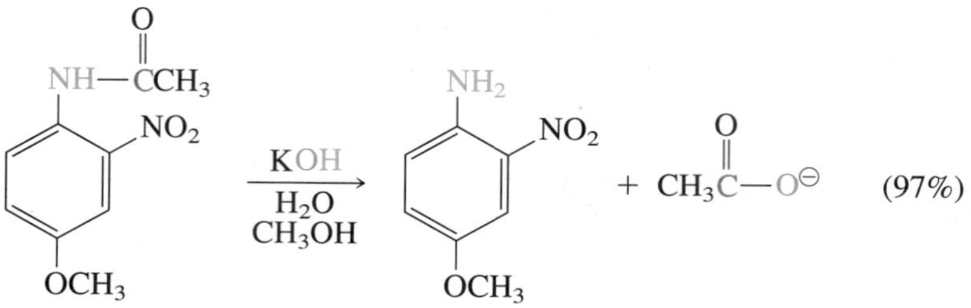

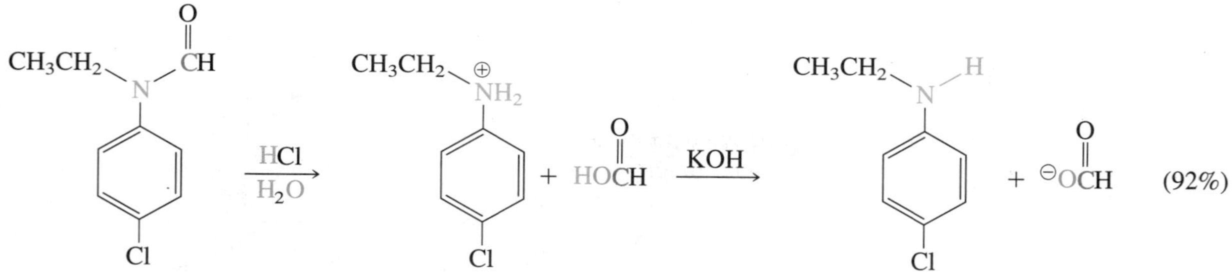

$$CH_3C-O-CH_2CH_2CH_2CH_2CH_3 + (97\%)$$

Establishing the Mechanism of Saponification

As we have seen before, a variety of experiments are used to support or disprove a mechanism that has been postulated for a particular reaction. The mechanism for the basic hydrolysis of esters, shown in Figure 15.4, involves cleavage of the bond between the ether oxygen of the ester and the carbonyl carbon rather than the cleavage of the bond between this oxygen and the carbon of the alcohol part of the ester.

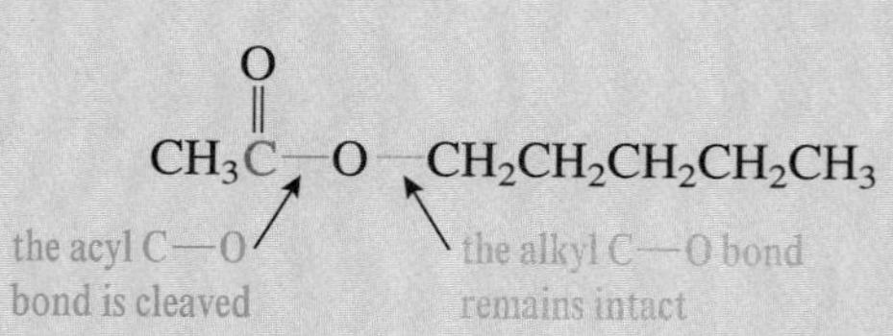

the acyl C—O bond is cleaved   the alkyl C—O bond remains intact

Proof that it is indeed this bond that is cleaved was provided by conducting the hydrolysis in water with an enriched content of ^{18}O. Pentyl acetate was hydrolyzed in this isotopically enriched water. The 1-pentanol product was shown by mass spectrometry to contain only the normal amount of ^{18}O, demonstrating that the alkyl C—O bond was not broken during the reaction.

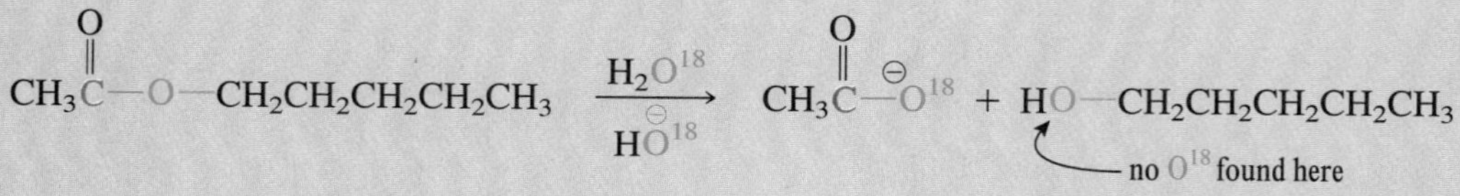

$$CH_3C-O-CH_2CH_2CH_2CH_2CH_3 \xrightarrow[\text{HO}^{18}]{\text{H}_2\text{O}^{18}} CH_3C-O^{18} + HO-CH_2CH_2CH_2CH_2CH_3$$

no O^{18} found here

ELABORATION

The mechanism begins with a hydroxide ion nucleophile bonding to the carbon of the carbonyl group, exactly as the mechanism for the saponification of an ester begins.

At this point, this mechanism deviates slightly from the ester saponification mechanism. Because amide ion (NH_2^-) is a poor leaving group, the nitrogen is first protonated by the solvent to convert it to a better leaving group.

In this step, ammonia leaves while an electron pair from the negative oxygen reforms the double bond of the carbonyl group.

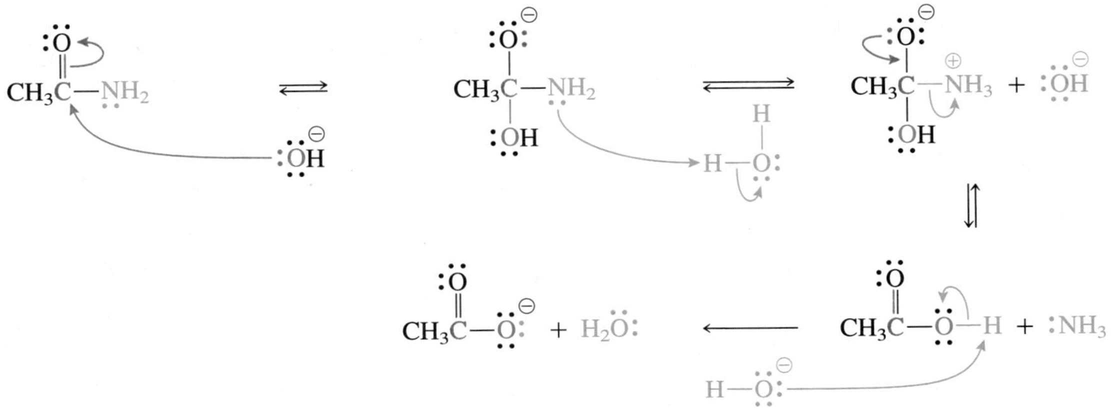

A proton transfer from the carboxylic acid to a hydroxide ion completes the process.

Figure 15.5

A Possible Mechanism for the Hydrolysis of an Amide for Basic Conditions.

Problem

Problem 15.12   Show all of the steps in the mechanism for this reaction:

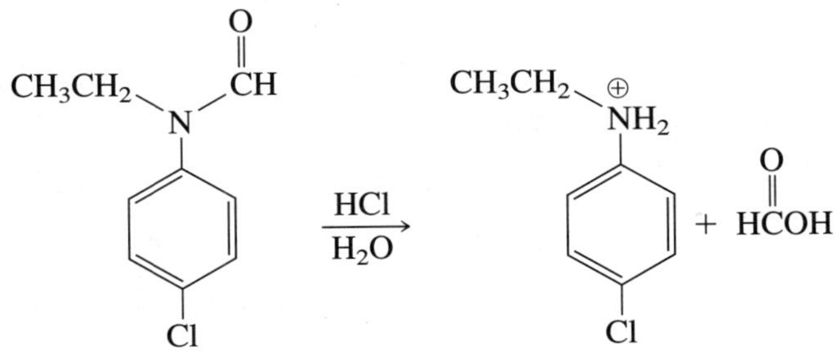

Solution

This acid-catalyzed mechanism resembles the reverse of the mechanism for Fischer esterification:

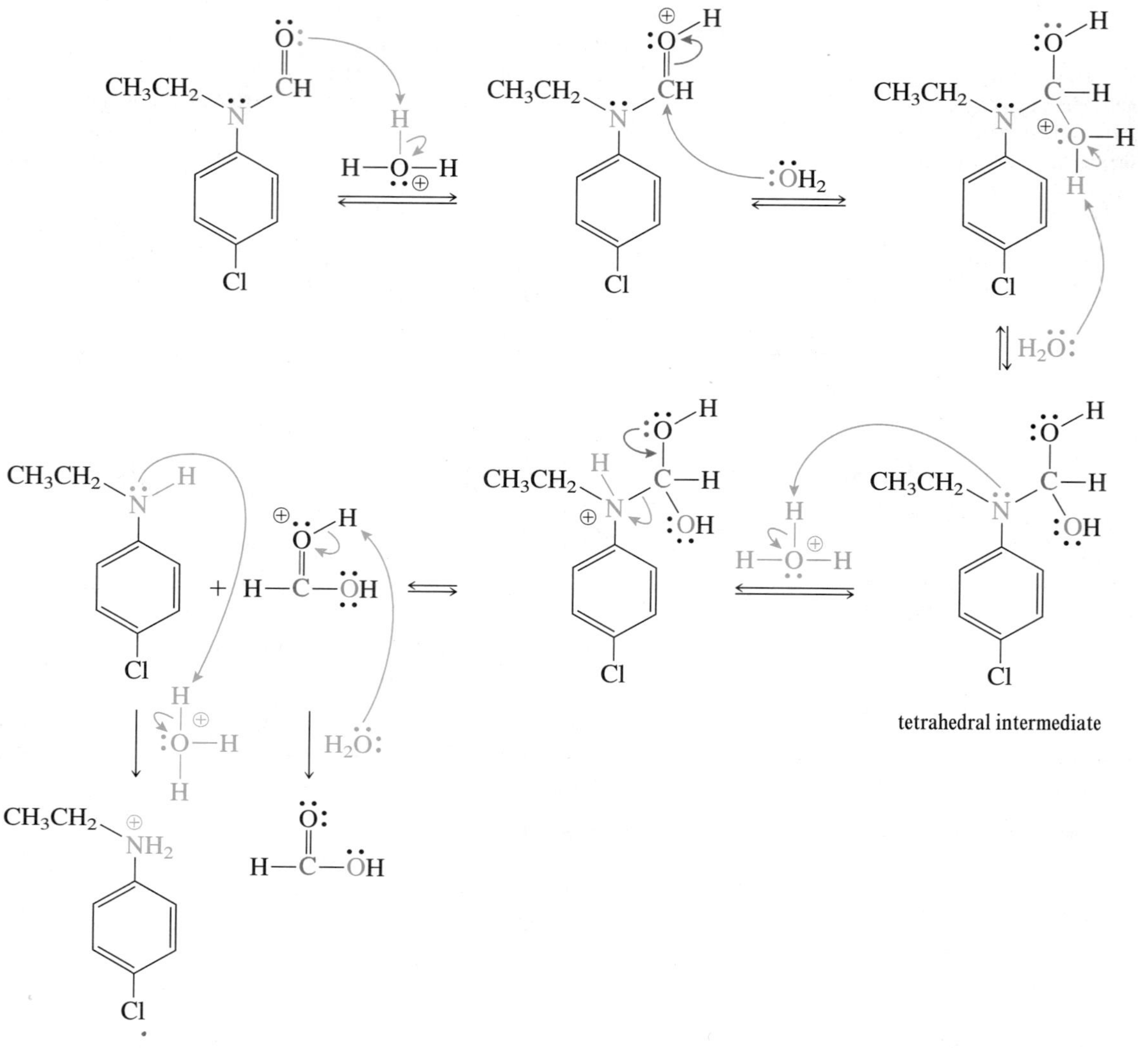

tetrahedral intermediate

Nitriles are often considered as derivatives of carboxylic acids because they can be hydrolyzed to the acid under acidic or basic conditions, as illustrated in the following equation:

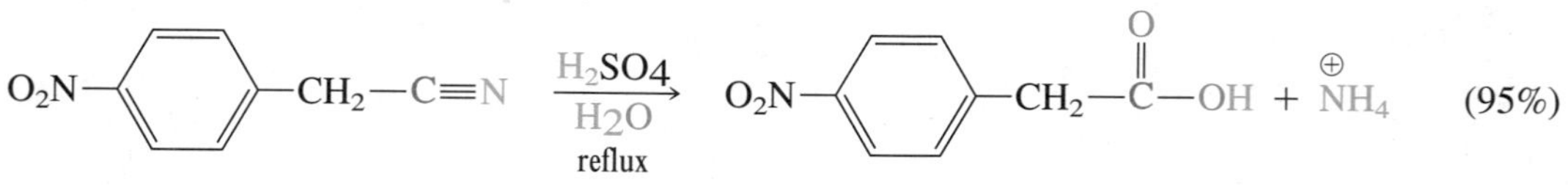

$$O_2N-\!\!\bigcirc\!\!-CH_2-C\equiv N \xrightarrow[\substack{H_2O \\ \text{reflux}}]{H_2SO_4} O_2N-\!\!\bigcirc\!\!-CH_2-\overset{\displaystyle O}{\overset{\|}{C}}-OH + \overset{\oplus}{N}H_4 \quad (95\%)$$

The amide is an intermediate in the hydrolysis, and because it is less reactive than the nitrile, the reaction can often be stopped at the amide stage, if so desired, by using milder reaction conditions, such as shorter reaction times, lower temperatures or weaker base:

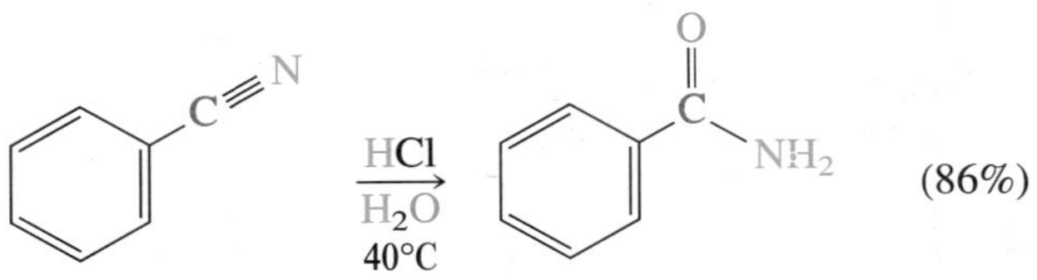

$$\text{C}_6\text{H}_5\text{--C}\equiv\text{N} \xrightarrow[\substack{H_2O \\ 40°C}]{HCl} \text{C}_6\text{H}_5\text{--C(=O)--NH}_2 \quad (86\%)$$

Many of the reactions of a carbon–nitrogen triple bond resemble those of a carbon–oxygen double bond, and the mechanisms have many similarities also. The mechanism for the hydrolysis of a nitrile to an amide for basic conditions is shown in Figure 15.6.

The use of cyanide ion as a nucleophile in an S_N2 reaction (see Section 9.9, page 367), followed by hydrolysis of the product nitrile provides a useful preparation of carboxylic acids that contain one more carbon than the starting compound:

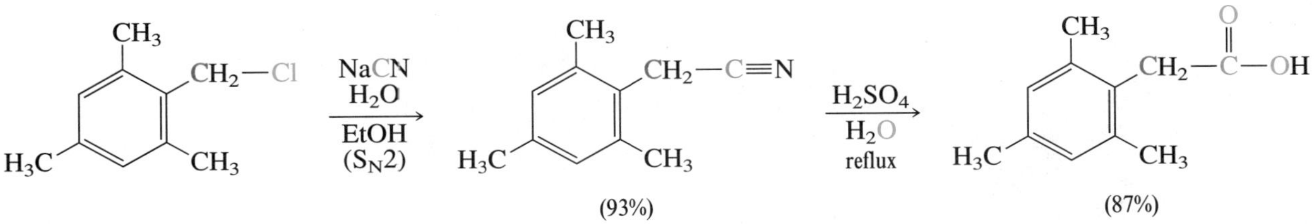

$$\text{ArCH}_2\text{--Cl} \xrightarrow[\substack{H_2O \\ EtOH \\ (S_N2)}]{NaCN} \text{ArCH}_2\text{--C}\equiv\text{N} \; (93\%) \xrightarrow[\substack{H_2O \\ reflux}]{H_2SO_4} \text{ArCH}_2\text{--C(=O)--OH} \; (87\%)$$

The cyano group resembles a carbonyl group in many of its reactions. This mechanism begins by the nucleophile, hydroxide ion, bonding to the electrophilic carbon of the nitrile. One pair of pi electrons is displaced onto the nitrogen.

Next, the negative nitrogen is protonated by a water molecule.

This compound is a tautomer of an amide. Tautomerization occurs in the same manner as was the case for the conversion of an enol to its carbonyl tautomer. First, a proton on the oxygen is removed by a base in the solution.

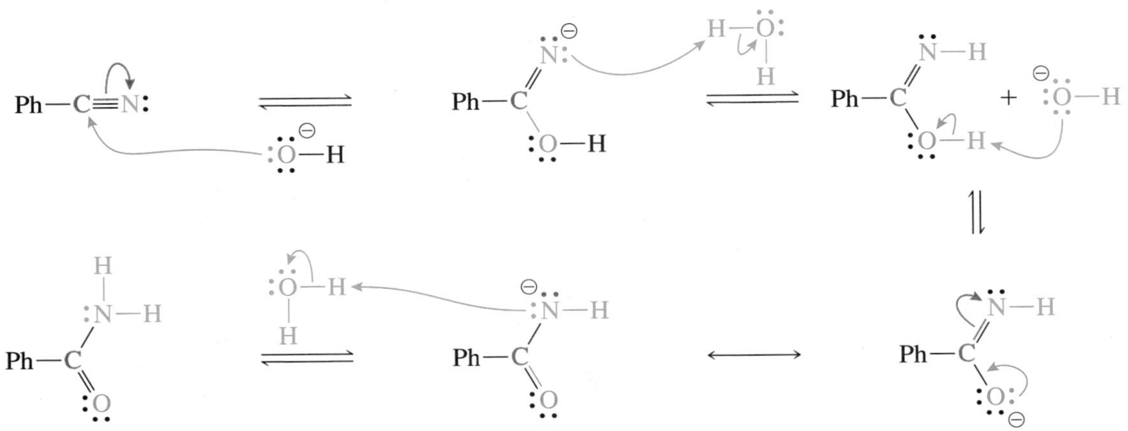

Then reprotonation occurs on the nitrogen to produce the amide. As was the case with the carbonyl–enol tautomerization, the stability of the carbon–oxygen double bond causes the amide tautomer to be favored at equilibrium.

Figure 15.6

Mechanism for the Hydrolysis of a Nitrile to an Amide for Basic Conditions.

example, the first step in the metabolism of glucose is formation of a phosphate ester at the primary hydroxy group to produce glucose 6-phosphate:

glucose   glucose 6-phosphate $\Delta G° = 3.3$ kcal/mol (13.8 kJ/mol)

The free energy for this reaction is 3.3 kcal/mol (13.8 kJ/mol). Recall that a positive free energy means that the equilibrium favors the reactants. Therefore this step in glucose metabolism is unfavorable.

However, glucose reacts readily with ATP to produce glucose 6-phosphate and ADP, as shown in the following equation:

glucose   ATP

glucose 6-phosphate   ADP

The free energy change for this reaction can be calculated by adding that for the hydrolysis of ATP to that for the phosphorylation of glucose, as shown in the following equations:

$$ATP + 2\ H_2O \longrightarrow ADP + phosphate + H_3O^{\oplus} \qquad \Delta G° = -7.3 \text{ kcal/mol}$$
$$(-30.5 \text{ kJ/mol})$$

$$glucose + phosphate \longrightarrow glucose\ 6\text{-phosphate} + H_2O \qquad \Delta G° = +3.3 \text{ kcal/mol}$$
$$(+13.8 \text{ kJ/mol})$$

$$ATP + glucose + H_2O \longrightarrow glucose\ 6\text{-phosphate} + ADP + H_3O^{\oplus} \qquad \Delta G° = -4.0 \text{ kcal/mol}$$
$$(-16.7 \text{ kJ/mol})$$

The free energy change for this reaction is −4.0 kcal/mol (−16.7 kJ/mol). The coupling of the very favorable hydrolysis of ATP and the unfavorable phosphorylation of glucose has resulted in a reaction that has a favorable free energy change. The energy of the ATP is used to shift the equilibrium toward the products, making the formation of glucose 6-phosphate favorable. The situation is analogous to the reaction of a carboxylic acid anhydride with an alcohol to form an ester. In each case the high-energy anhydride is used to drive the equilibrium for the formation of the ester completely to the right.

How does nature prepare a high-energy compound such as ATP? This is accomplished by coupling ATP formation with one of the even more favorable reactions that occur during photosynthesis, aerobic metabolism, or fermentation. For example, phosphoenolpyruvate is a high-energy compound that is an intermediate formed in the metabolism of glucose. The energy of this compound is high because it is a phosphate ester of an enol, the unstable tautomer of a ketone (see Section 10.7). Its hydrolysis to phosphate and pyruvate has a very favorable free energy change of −14.8 kcal/mol (−61.9 kJ/mol). When this reaction is coupled with the conversion of ADP to ATP, the overall process has a favorable free energy of −7.5 kcal/mol (−31.4 kJ/mol).

$$CH_2{=}C \begin{array}{c} CO_2^{\ominus} \\ \\ OPO_3^{2-} \end{array} + H_2O \longrightarrow CH_3\overset{O}{\overset{\|}{C}}CO_2^- + HPO_4^{2\ominus} \qquad \Delta G^\circ = -14.8 \text{ kcal/mol}$$
$$(-61.9 \text{ kJ/mol})$$

phosphoenolpyruvate   pyruvate

$$ADP + HPO_4^{2-} \longrightarrow ATP + H_2O \qquad \Delta G^\circ = +7.3 \text{ kcal/mol}$$
$$(+30.5 \text{ kJ/mol})$$

$$\text{phosphoenolpyruvate} + ADP \longrightarrow \text{pyruvate} + ATP \qquad \Delta G^\circ = -7.5 \text{ kcal/mol}$$
$$(-31.4 \text{ kJ/mol})$$

ATP is continuously synthesized in biological systems. It is consumed in a variety of reactions within 1 minute of its formation. The amount that is made and used is enormous. A resting human synthesizes and consumes about 40 kg of ATP per day. During periods of strenuous activity, up to 0.5 kg per minute can be used.

15.13 Summary

The reactions in this chapter follow the same general mechanism. First, the nucleophile bonds to the carbonyl carbon, displacing the pi electrons of the carbon–oxygen double bond onto the oxygen and forming the tetrahedral intermediate. In the second step the unshared electrons on the oxygen reform the pi bond as the leaving group leaves:

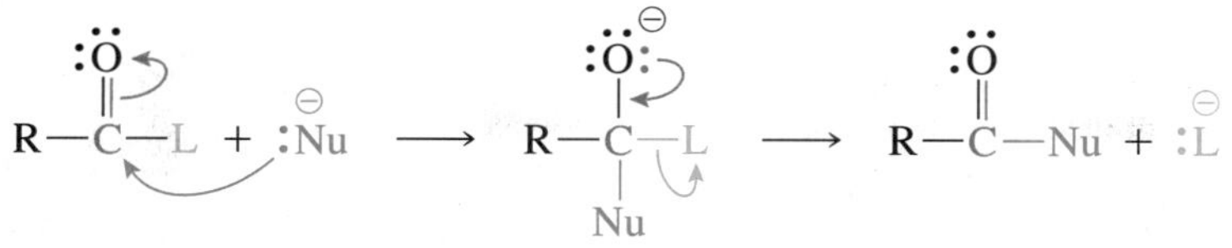

tetrahedral
intermediate

The mechanism for acidic conditions is very similar, except that the carbonyl oxygen and/or the leaving group is protonated. The reactivity of the carboxylic acid derivative is affected by the following:

Resonance effects: Electron donors slow the reaction; withdrawers accelerate it.

Inductive effects: Electron withdrawers accelerate the reaction.

Steric effects: Steric hindrance slows the approach of the nucleophile.

The overall reactivity order for the carboxylic acid derivatives (see Table 15.1) is as follows:

most reactive ... least reactive

$$\mathrm{RC(=O)Cl} > \mathrm{RC(=O)OC(=O)R} > \mathrm{RC(=O)OR'}\ \&\ \mathrm{RC(=O)OH} > \mathrm{RC(=O)NH_2} > \mathrm{RC(=O)O^{\ominus}}$$

Table 15.2 provides a summary of the reactions presented in this chapter.

Problem

Problem 15.27

a) $2\ CH_3CH_2OH + Cl{-}C(=O){-}Cl \longrightarrow$

b) $CH_3CH_2C(=O)OH + SOCl_2 \longrightarrow$

c) 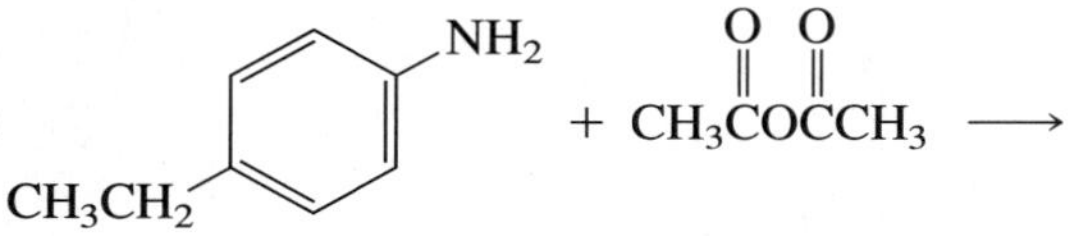

d) $PhCH_2C(=O)OEt + 2\ CH_3CH_2MgI \longrightarrow \xrightarrow[H_2O]{NH_4Cl}$

e) $PhCH_2C(=O)OH \xrightarrow{SOCl_2} \xrightarrow{PhCH_2C(=O)O^{\ominus}}$

f) 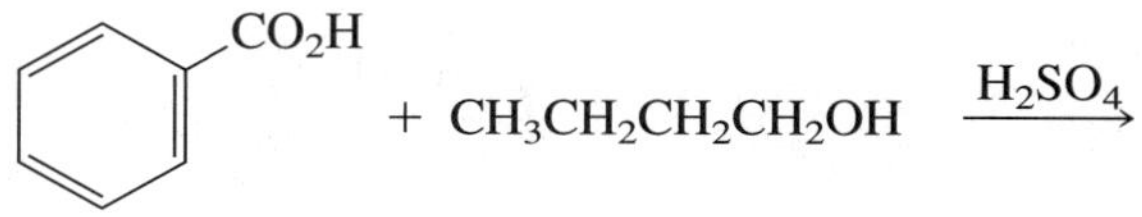

g) 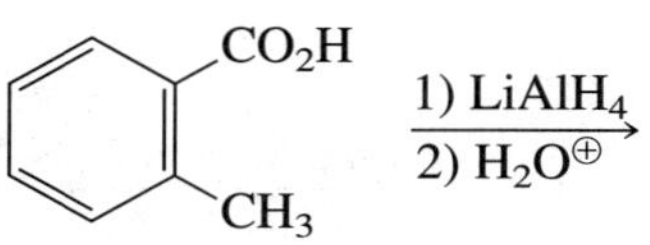

h) $CH_3CH_2CH_2CH_2C(=O)Cl + (CH_3CH_2CH_2CH_2)_2CuLi \longrightarrow$

i) $CH_3CH_2CH_2C{\equiv}N \xrightarrow[2)\ H_2O]{1)\ LiAlH_4}$

j) $CH_2{=}CHC(=O)Cl + CH_3CH_2OH \longrightarrow$

k) $\underset{\displaystyle CH_3}{CH_3CHCO_2H}$ $\xrightarrow{SOCl_2}$ $\xrightarrow[H_2O]{NH_3}$

l) $CH_3(CH_2)_{10}CH_2OH$ + $ClSO_2-\!\!\!\!\bigcirc\!\!\!\!-CH_3$ $\xrightarrow{pyridine}$

m) $NCCH_2CH_2CH_2CN$ $\xrightarrow[\substack{H_2O \\ \Delta}]{HCl}$

n) $H_3C-\!\!\!\!\bigcirc\!\!\!\!-CH_2\overset{\displaystyle O}{\overset{\|}{C}}OCH_3$ $\xrightarrow[2)\ H_2O^{\oplus}]{1)\ LiAlH_4}$

o) (4-methoxyphenyl $\overset{O}{\overset{\|}{C}}OCH_2CH_3$, OCH$_3$ para) $\xrightarrow[2)\ H_3O^{\oplus}]{1)\ DIBAH,\ -78°C}$

p) (naphthalene-$C\!\!\equiv\!\!N$) $\xrightarrow{CH_3MgI}$ $\xrightarrow{H_3O^{\oplus}}$

q) $\underset{\displaystyle OH}{PhCH}-\underset{\displaystyle O}{\overset{\|}{C}}Ph$ + $CH_3\overset{O}{\overset{\|}{C}}O\overset{O}{\overset{\|}{C}}CH_3$ $\longrightarrow$

r) (4-methylphenyl, $NHCCH_3$, $\overset{O}{\|}$) $\xrightarrow[\substack{H_2O \\ \Delta}]{HCl}$

s) (2-methylphenyl $CH_2O\overset{O}{\overset{\|}{C}}CH_3$) $\xrightarrow[H_2O]{NaOH}$

t) $HO\overset{O}{\overset{\|}{C}}-\overset{O}{\overset{\|}{C}}OH$ + $\underset{\displaystyle CH_3OH}{excess}$ $\xrightarrow{H_2SO_4}$

u) (cyclohexyl $\overset{O}{\overset{\|}{C}}N(CH_3)_2$) $\xrightarrow[2)\ H_2O]{1)\ LiAlH_4}$

v) $Cl\overset{O}{\overset{\|}{C}}-\!\!\!\!\bigcirc\!\!\!\!-\overset{O}{\overset{\|}{C}}Cl$ $\xrightarrow[-78°C]{2\ LiAlH(Ot\text{-}Bu)_3}$

Table 15.2 Nucleophilic Substitution Reactions at Carbonyl Carbons.

Reaction	Comment
$$\underset{\text{(PCl}_3\text{ or PCl}_5)}{\text{RCOH} + \text{SOCl}_2 \longrightarrow \text{RCCl}}$$	Preparation of acyl chlorides. Acyl chlorides are commonly used to prepare other carboxylic acid derivatives.
$$\text{RCOH} + \text{RCCl} \longrightarrow \text{RCOCR}$$	Preparation of anhydrides.
$$2\,\text{RCOH} + \text{CH}_3\text{COCCH}_3 \longrightarrow \text{RCOCR} + 2\,\text{CH}_3\text{COH}$$	Preparation of anhydrides by exchange.
$$\text{RCCl} + \text{R'OH} \longrightarrow \text{RCOR'}$$ or $$\text{RCOCR}$$	Preparation of esters.
$$\text{RCOH} + \text{R'OH} \longrightarrow \text{RCOR'}$$	Preparation of esters by Fischer esterification. The equilibrium must be driven to favor the ester.
$$\text{RCCl} + \text{H}_2\text{O} \longrightarrow \text{RCOH}$$ or $$\text{RCOCR}$$	Hydrolysis of acyl chlorides and anhydrides. These derivatives must be protected from water to avoid these reactions.
$$\text{RCOR'} + \text{H}_2\text{O} \xrightarrow[\substack{\text{or} \\ {}^{\ominus}\text{OH}}]{\text{H}^{\oplus}} \text{RCOH} + \text{R'OH}$$	Hydrolysis of esters. Base is most commonly used in process known as saponification.
$$\text{RCNH}_2 + \text{H}_2\text{O} \xrightarrow[\substack{\text{or} \\ {}^{\ominus}\text{OH}}]{\text{H}^{\oplus}} \text{RCOH} + \text{NH}_3$$	Hydrolysis of amides. This reaction can be accomplished by using either acid or base catalysis.
$$\text{RC}{\equiv}\text{N} + \text{H}_2\text{O} \xrightarrow[\substack{\text{or} \\ \text{OH}^{\ominus}}]{\text{H}^{\oplus}} \text{RCNH}_2 \longrightarrow \text{RCOH}$$	Hydrolysis of nitriles. This reaction can be stopped at the amide or carried to the carboxylic acid.
$$\text{R}{-}\text{L} + {}^{\ominus}\text{CN} \longrightarrow \text{RCN}$$	The preparation of nitriles by S$_N$2 reactions combined with hydrolysis of nitriles provides a carboxylic acid preparation.

Table 15.2   *Continued*

Reaction	Comment
$$\underset{\text{or}}{\overset{\overset{\displaystyle O}{\|\|}}{RCCl}} + R'NH_2 \longrightarrow \overset{\overset{\displaystyle O}{\|\|}}{RCNHR'}$$ $$\overset{\overset{\displaystyle O\ \ \ O}{\|\|\ \ \|\|}}{RCOCR}$$	Preparation of amides.
$$\underset{\text{or}}{\overset{\overset{\displaystyle O}{\|\|}}{RCOR'}} \xrightarrow[\text{2) } H_3O^{\oplus}]{\text{1) LiAlH}_4} RCH_2OH$$ $$\overset{\overset{\displaystyle O}{\|\|}}{RCOH}$$	Reduction of esters or acids to alcohols.
$$\overset{\overset{\displaystyle O}{\|\|}}{RCNHR'} \xrightarrow[\text{2) } H_2O]{\text{1) LiAlH}_4} RCH_2NHR'$$	Reduction of amides to amines.
$$RC\equiv N \xrightarrow[\text{2) } H_2O]{\text{1) LiAlH}_4} RCH_2NH_2$$	Reduction of nitriles to primary amines.
$$\overset{\overset{\displaystyle O}{\|\|}}{RCCl} \xrightarrow[-78°C]{\text{LiAlH}(Ot\text{-Bu})_3} \overset{\overset{\displaystyle O}{\|\|}}{RCH}$$	Reduction of acyl chlorides to aldehydes.
$$\overset{\overset{\displaystyle O}{\|\|}}{RCOR} \xrightarrow[\text{2) } H_3O^{\oplus}]{\text{1) DIBAH, } -78°C} \overset{\overset{\displaystyle O}{\|\|}}{RCH}$$	Reduction of esters to aldehydes.
$$\overset{\overset{\displaystyle O}{\|\|}}{RCOR'} \xrightarrow[\text{2) } H_3O^{\oplus}]{\text{1) 2 R''MgX}} R-\overset{\overset{\displaystyle OH}{\|}}{\underset{\underset{\displaystyle R''}{\|}}{C}}-R''$$	Preparation of alcohols from esters.
$$\overset{\overset{\displaystyle O}{\|\|}}{RCCl} \xrightarrow{R_2CuLi} \overset{\overset{\displaystyle O}{\|\|}}{RCR'}$$	Preparation of ketones from acyl chlorides.
$$RC\equiv N \xrightarrow[\text{2) } H_3O^{\oplus}]{\text{1) R'MgX}} \overset{\overset{\displaystyle O}{\|\|}}{RCR'}$$	Preparation of ketones from nitriles.

After completing this chapter you should be able to:

1. Show the products of any of the reactions discussed in this chapter.
2. Show the mechanism for these reactions.

3. Predict the effect of a change in structure on the rate and equilibrium of a reaction.

4. Use these reactions to interconvert any of the carboxylic acid derivatives and to prepare aldehydes, ketones, alcohols, and amines.

5. Use these reactions in combination with reactions from previous chapters to synthesize compounds.

End-of-Chapter Problems

15.28   Show the products of these reactions.

a) △—CCl (C=O) + CH_3CH_2OH $\longrightarrow$

b) (cyclohexanol with OH) + CH_3CO_2H $\xrightarrow{H_2SO_4}$

c) (toluene ring with CH_3 and CN) $\xrightarrow[\text{long time}]{\begin{array}{c}H_2O\\H_2SO_4\end{array}}$

d) (benzene ring with CH_2COCH_3) $\xrightarrow[\text{2) } H_3O^\oplus]{\text{1) LiAlH}_4}$

e) $CH_3CH_2\overset{O}{\overset{\|}{C}}OCH_3$ $\xrightarrow[\text{2) NH}_4\text{Cl, H}_2\text{O}]{\text{1) 2 CH}_3\text{CH}_2\text{CH}_2\text{CH}_2\text{MgBr}}$

15.29   Show the products of these reactions.

a) (toluene ring with CH_3 and NH_2) + $CH_2\overset{O}{\overset{\|}{C}}O\overset{O}{\overset{\|}{C}}CH_3$ $\longrightarrow$

b) (H_3C ring with $\overset{O}{\overset{\|}{C}}OCH_2CH_3$) $\xrightarrow[\text{H}_2\text{O}]{\text{NaOH}}$

c) $CH_3\overset{CH_3}{\overset{|}{C}}HCO_2H$ $\xrightarrow[\text{2) NH}_3\text{, H}_2\text{O}]{\text{1) SOCl}_2}$

d) (toluene ring with CH_3 and CN) $\xrightarrow[\text{2) H}_2\text{O}]{\text{1) LiAlH}_4}$

e) (cyclopentene ring with $\overset{O}{\overset{\|}{C}}$—Cl) + $(CH_3)_2CuLi$ $\longrightarrow$

15.30 Show the products of these reactions.

a) 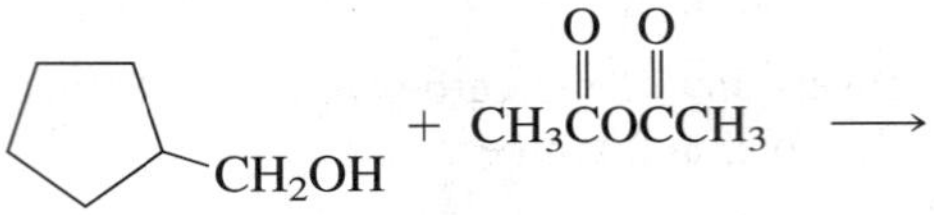 [cyclopentyl-CH₂OH] $+ CH_3\overset{O}{\overset{\|}{C}}O\overset{O}{\overset{\|}{C}}CH_3 \longrightarrow$

b) 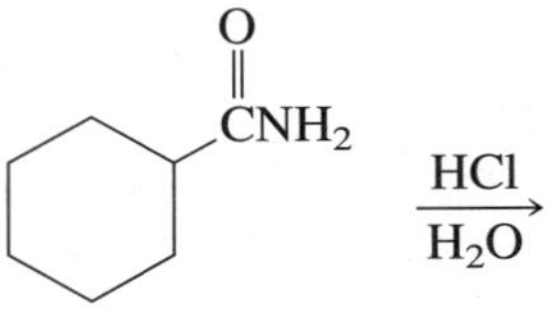 [cyclohexyl with $\overset{O}{\overset{\|}{C}}NH_2$] $\xrightarrow[H_2O]{HCl}$

c) $CH_3CH_2\overset{O}{\overset{\|}{C}}NH_2 \quad \xrightarrow[\text{2) } H_2O]{\text{1) LiAlH}_4}$

d) 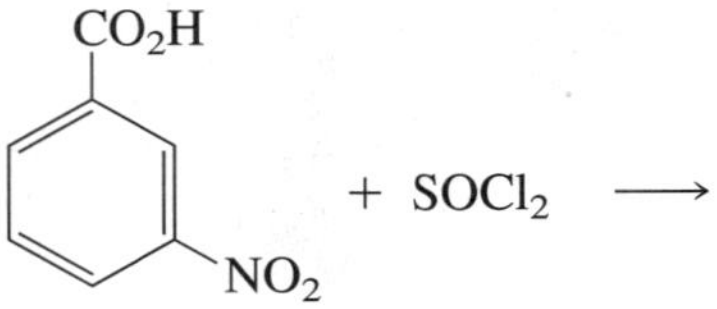 [benzene ring with CO_2H and NO_2] $+ SOCl_2 \longrightarrow$

e) $CH_3\overset{CH_3}{\overset{|}{C}}=CHCH_2CH_2\overset{O}{\overset{\|}{C}}Cl \quad \xrightarrow[-78°C]{\text{LiAlH(O}t\text{-Bu)}_3}$

f) 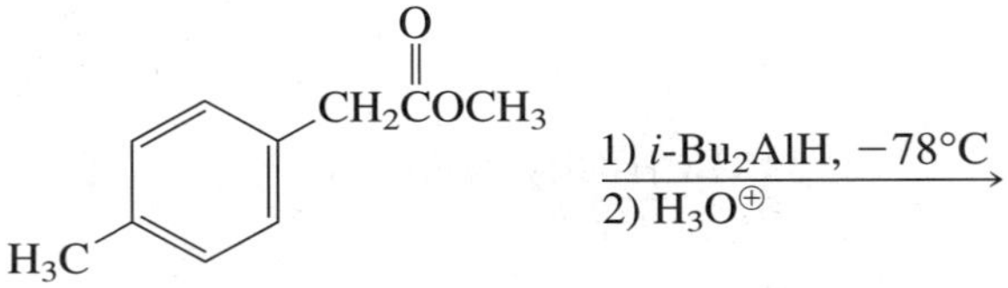 [benzene ring with H_3C and $CH_2\overset{O}{\overset{\|}{C}}OCH_3$] $\xrightarrow[\text{2) } H_3O^{\oplus}]{\text{1) } i\text{-Bu}_2\text{AlH, } -78°C}$

15.31 Arrange these compounds in order of increasing rate of saponification:

[benzene ring with $\overset{O}{\overset{\|}{C}}OCH_3$]   [benzene ring with O_2N and $\overset{O}{\overset{\|}{C}}OCH_3$]   [benzene ring with CH_3O and $\overset{O}{\overset{\|}{C}}OCH_3$]

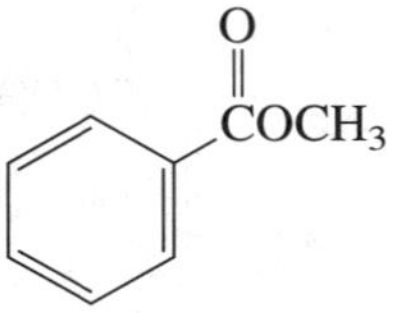

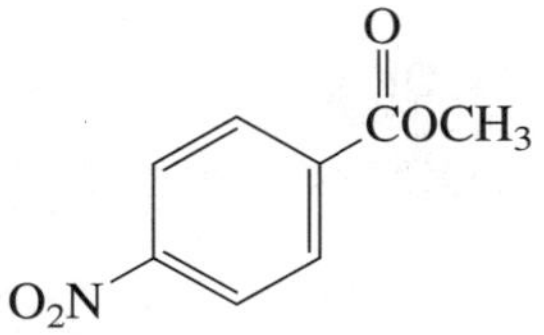

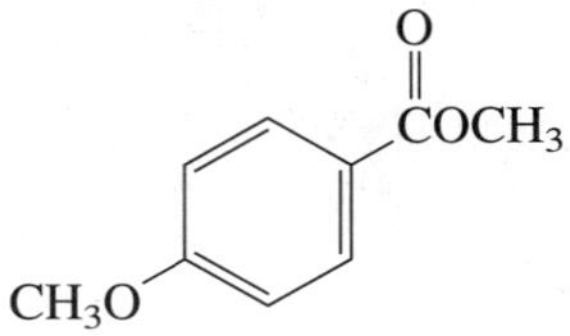

15.32 Arrange these compounds in order of increasing rate of saponification:

$CH_3-\overset{CH_3}{\underset{CH_3}{\overset{|}{\underset{|}{C}}}}-\overset{O}{\overset{\|}{C}}-OCH_2CH_3$ $CH_3-CH_2-\overset{O}{\overset{\|}{C}}-OCH_2CH_3$ $CH_3\overset{CH_3}{\overset{|}{CH}}-\overset{O}{\overset{\|}{C}}-OCH_2CH_3$

15.33 Suggest syntheses of these compounds starting from hexanoic acid.

a) 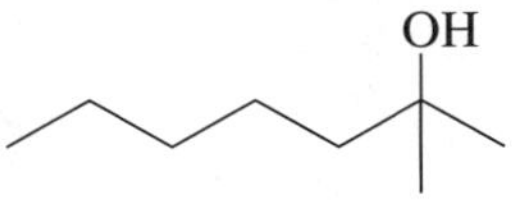 [structure with OH]   b) 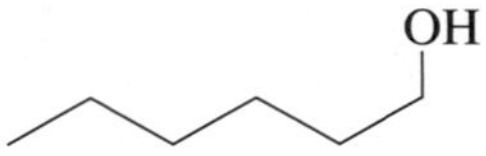 [structure with OH]

examples are shown in the following equations. The first example shows the alkylation of a β-ketoester. Close examination shows the similarity of the starting material to ethyl acetoacetate. Although sodium hydride is used as a base in this example, sodium ethoxide could also be employed.

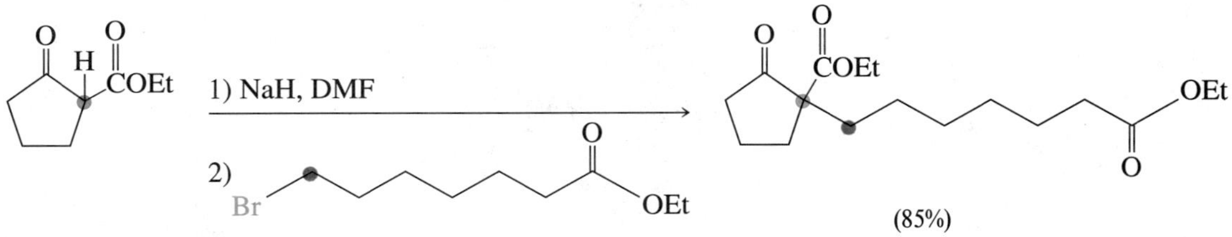

(85%)

This next example shows the alkylation of a β-diketone. Because this compound is more acidic than a β-ketoester or a β-diester, the weaker base potassium carbonate was used. However, sodium ethoxide would also be satisfactory as the base for this reaction:

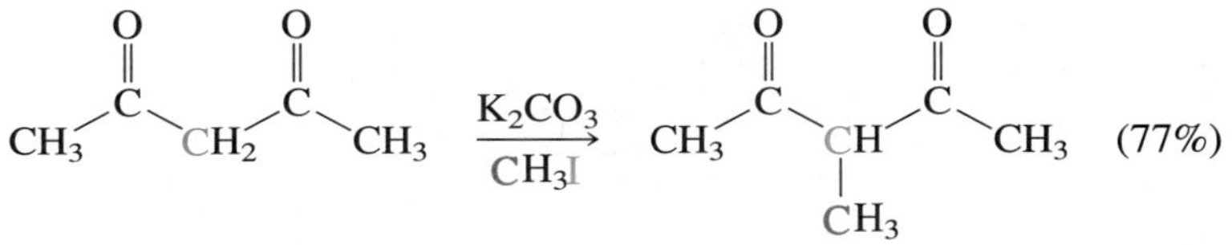

(77%)

This last example shows the addition of two alkyl groups to a dinitrile. Because the alkyl groups to be added are identical, they do not have to be added in sequence. Instead, the reaction is conducted by adding two equivalents of base and two equivalents of the alkylating agent, benzyl chloride, simultaneously:

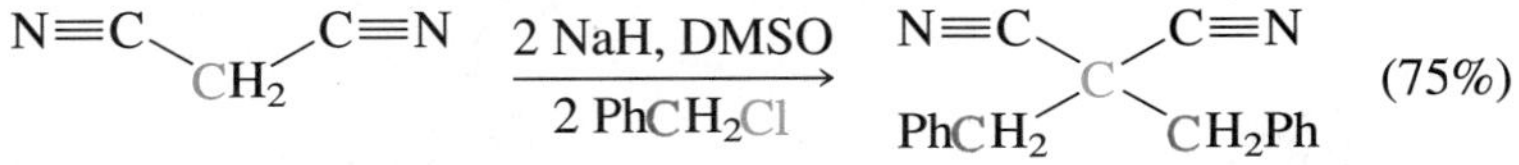

(75%)

Problems

Problem 16.5 Show the products of these reactions.

a) CH_3CCH_2COEt
  1) NaOEt, EtOH   1) NaOH, H_2O
  2) $PhCH_2Br$ 2) $H_3O^{\oplus}$, Δ

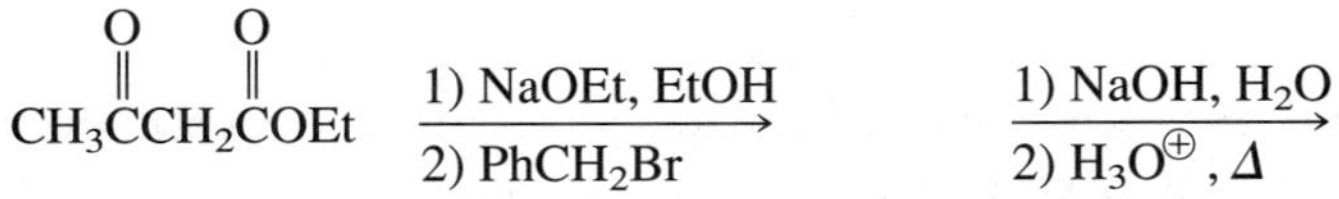

b) $EtOCCH_2COEt$
  1) NaOEt, EtOH   1) NaOH, H_2O
  2) $CH_3CH_2CH_2Br$ 2) $H_3O^{\oplus}$, Δ

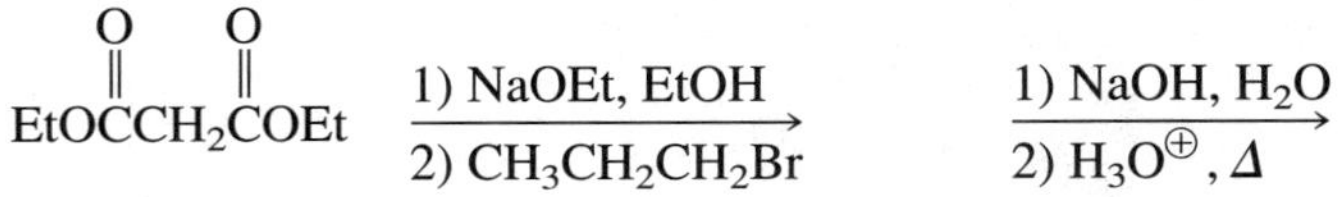

c)
  1) NaOEt, EtOH
  2) CH_3CH_2Br

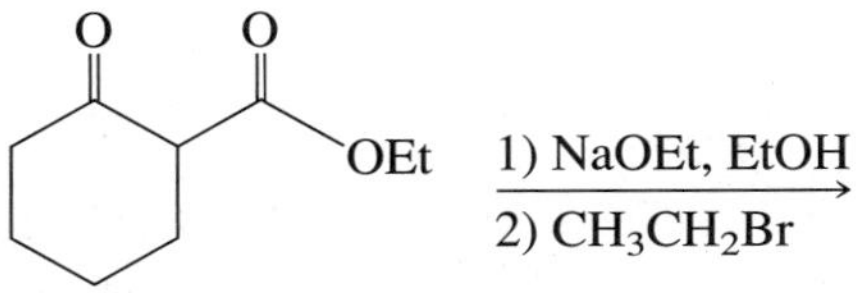

d) $CH_3\overset{\overset{\displaystyle O}{\|}}{C}CH_2\overset{\overset{\displaystyle O}{\|}}{C}OEt$
$\xrightarrow[\substack{\text{2) } CH_3CH_2CH_2CH_2Br \\ \text{3) NaOEt, EtOH} \\ \text{4) } CH_3I}]{\text{1) NaOEt, EtOH}}$
$\xrightarrow[\text{2) } H_3O^\oplus,\, \Delta]{\text{1) NaOH, } H_2O}$

e) $EtO\overset{\overset{\displaystyle O}{\|}}{C}CH_2\overset{\overset{\displaystyle O}{\|}}{C}OEt$
$\xrightarrow{\text{1) NaOEt, EtOH}}$
2)

f)
$\xrightarrow[\text{2) } CH_3I]{\text{1) NaOEt, EtOH}}$
$\xrightarrow[\text{2) } H_3O^\oplus,\, \Delta]{\text{1) NaOH, } H_2O}$

g) $N\equiv CCH_2\overset{\overset{\displaystyle O}{\|}}{C}CH_3$
$\xrightarrow[\text{2) } CH_3CH_2CH_2Cl]{\text{1) NaOEt, EtOH}}$

Problem 16.6   Show syntheses of these compounds using the acetoacetic or malonic ester syntheses.

a) $CH_3CH_2CH_2CH_2CH_2\overset{\overset{\displaystyle O}{\|}}{C}CH_3$

b) $CH_3\overset{\overset{\displaystyle CH_3}{|}}{C}HCH_2CO_2H$

c) $PhCH_2\overset{\overset{\displaystyle CH_3}{|}}{C}H\!-\!\overset{\overset{\displaystyle O}{\|}}{C}CH_3$

d) $CH_3CH_2\overset{\overset{\displaystyle CH_3CH_2}{|}}{C}H\!-\!CO_2H$

e) —$CH_2CH_2CO_2H$

f) —$\overset{\overset{\displaystyle O}{\|}}{C}CH_3$

Solution

a)   The acetoacetic ester synthesis is used to prepare methyl ketones such as this. (The malonic ester synthesis is used to prepare carboxylic acids.) The new carbon–carbon bond formed during this process is between the α-carbon and a group attached to it.

$$CH_3CH_2CH_2CH_2\!-\!\!-\!\!\!\overset{\overset{\displaystyle O}{\|}}{}\,CH_2CCH_3$$

This is the bond to be formed in this acetoacetic ester synthesis.

In this case a butyl group must be attached to the enolate nucleophile derived from ethyl acetoacetate. The alkylated product is then decarboxylated:

$$CH_3CCH_2COEt \xrightarrow[\text{2) } CH_3CH_2CH_2CH_2Br]{\text{1) NaOEt, EtOH}} CH_3CCHCOEt \xrightarrow[\text{2) } H_3O^{\oplus}, \Delta]{\text{1) NaOH, } H_2O} CH_3CCH_2CH_2CH_2CH_2CH_3$$

with the middle product having a $CH_2CH_2CH_2CH_3$ substituent.

Problem 16.7 Show how these compounds could be synthesized using alkylation reactions.

a) $CH_3CH_2CCHCCH_2CH_3$ with $CH_2CH_2CH_3$ substituent

b) cyclopentanone ring with CH_2Ph and $—CO_2Et$ substituents

c) $CH_3CHCH_2CH_2CHCN$ with CH_3 and CN substituents

d) $CH_3CHCH_2CH_2CCN$ with CH_3, CN and CH_3 substituents

16.5 The Aldol Condensation

You may have noticed that aldehydes were conspicuously absent from the examples of alkylation reactions presented in Sections 16.3 and 16.4. This is due to the high reactivity of the carbonyl carbon of an aldehyde as an electrophile. When an enolate anion nucleophile is generated from an aldehyde, under most circumstances it rapidly reacts with the electrophilic carbonyl carbon of an un-ionized aldehyde molecule. Although this reaction, known as the **aldol condensation,** interferes with the alkylation of aldehydes, it is a very useful synthetic reaction in its own right. The aldol condensation of ethanal is shown in the following equation:

$$2 \ CH_3CH \xrightarrow{\text{NaOH}} CH_3CH—CH_2CH \quad (75\%)$$

with the product having an OH group.

aldol

The product, 3-hydroxybutanal, is also known as aldol, and gives rise to the name for the whole class of reactions.

The mechanism for this reaction is shown in Figure 16.3. For this mechanism to occur, both the enolate anion derived from the aldehyde and the un-ionized aldehyde must be present. To ensure that this is the case, hydroxide ion is most commonly used as the base. Because hydroxide ion is a weaker base than the aldehyde enolate anion, only a small amount of the enolate anion is produced. Most of the aldehyde remains un-ionized and is available for reaction as the electrophile. Note that the strong bases described in Section 16.3, which would tend to convert most of the aldehyde molecules to enolate ions, are not used in aldol condensations. The addition of the enolate nucleophile to the aldehyde follows the same mechanism as the addition of other

In the first step, the base, hydroxide ion, removes an acidic hydrogen from the α-carbon of the aldehyde. The conjugate base of the aldehyde is a stronger base than hydroxide, so the equilibrium for this first step favors the reactants.

However, enough enolate ion nucleophile is present to react with the electrophilic carbonyl carbon of a second aldehyde molecule.

Figure 16.3

Mechanism for the Aldol Condensation.

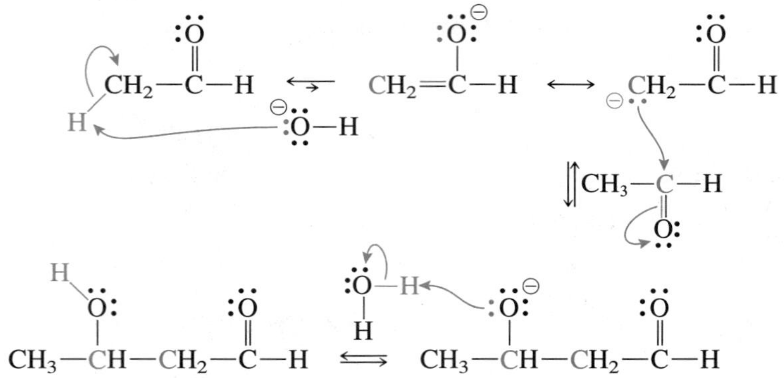

This part of the mechanism is just like the mechanism for the addition reactions of Chapter 14. After the nucleophile adds, the negative oxygen removes a proton from water.

nucleophiles that were described in Chapter 14. Remember that the α-carbon of one aldehyde molecule bonds to the carbonyl carbon of a second aldehyde molecule, as illustrated in the following example:

$$2\ CH_3CH_2CH_2\overset{\overset{\textstyle O}{\|}}{C}H \xrightarrow{\ KOH\ } CH_3CH_2CH_2\overset{\overset{\textstyle OH}{|}}{C}H-\overset{}{\underset{\underset{\textstyle CH_3CH_2}{|}}{C}}H\overset{\overset{\textstyle O}{\|}}{C}H \quad (75\%)$$

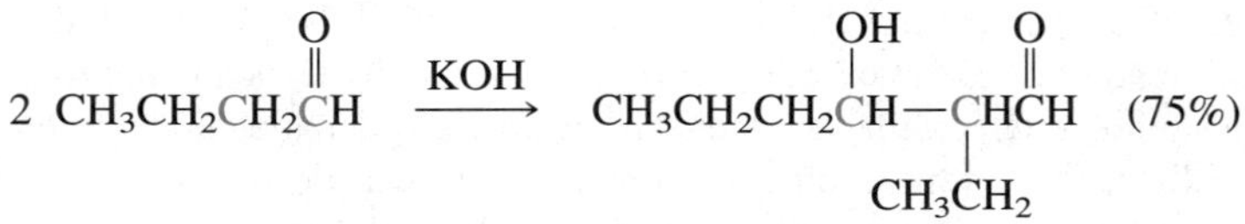

If the aldol condensation is conducted under more vigorous conditions (higher temperature, longer reaction time, and/or stronger base), elimination of water to form an α,β-unsaturated aldehyde usually occurs. This elimination is illustrated in the following example. Note that the α-carbon of one molecule is now doubly bonded to the carbonyl carbon of the other. (This text will use the symbol for heat, Δ, to indicate the vigorous conditions that cause eliminations to occur in these aldol condensations, even though other conditions might have been used.)

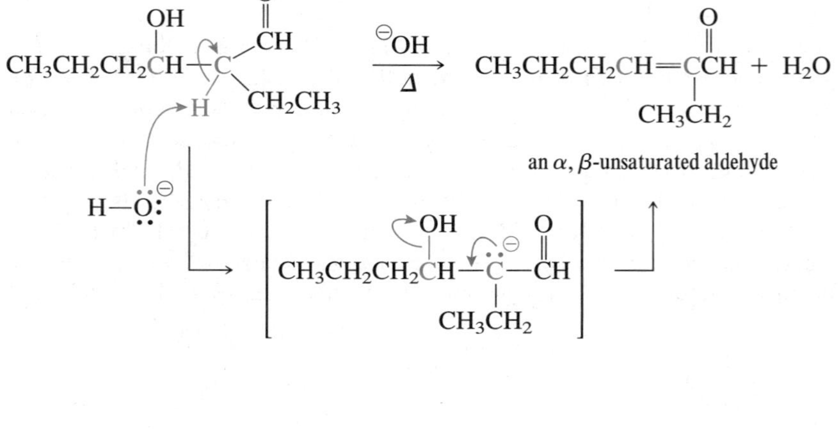

an α,β-unsaturated aldehyde

This elimination occurs by a somewhat different mechanism than those described in Chapter 8. Because the hydrogen on the α-carbon is relatively acidic, it is removed by the base in the first step to produce an enolate anion. Then hydroxide ion is lost from the enolate ion in the second step. Because this step is intramolecular and the product is stabilized because of the conjugation of its carbon–carbon double bond with the carbon–oxygen double bond of the carbonyl group, even a poor leaving group such as hydroxide ion can leave. (This is an example of the E1cb mechanism described in the elaboration on page 330 in Chapter 8.) Most aldol condensations are run under conditions that favor dehydration because the stability of the product helps to drive the equilibrium in the desired direction, resulting in a higher yield. For example, the reaction of butanal shown previously results in a 75% yield of the aldol product. If the reaction is conducted so that dehydration occurs, the yield of the conjugated product is 97%.

$$2\ CH_3CH_2CH_2\overset{\overset{\large O}{\|}}{C}H \xrightarrow[\Delta]{NaOH} CH_3CH_2CH_2CH{=}\underset{\underset{CH_3CH_2}{|}}{\overset{\overset{\large O}{\|}}{C}}CH + H_2O \quad (97\%)$$

Another example is shown in the following equation:

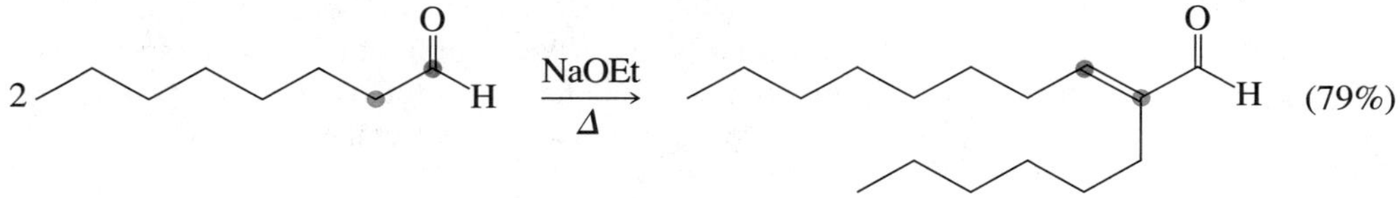

Problems

Problem 16.8 Show the products of these reactions.

a) $2\ CH_3CH_2\overset{\overset{\large O}{\|}}{C}H \xrightarrow{NaOH}$

b) 2 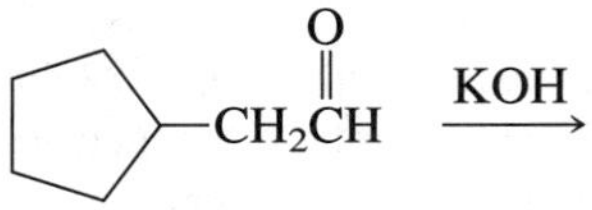

c) $2\ CH_3\overset{\overset{CH_3}{|}}{C}HCH_2CH_2\overset{\overset{\large O}{\|}}{C}H \xrightarrow[\Delta]{NaOH}$

d) $2\ PhCH_2\overset{\overset{\large O}{\|}}{C}H \xrightarrow[\Delta]{NaOH}$

Solution

a) The key to determining the products of an aldol condensation is to remember that the nucleophile is an enolate anion, which is formed at the α-carbon of the aldehyde, and the electrophile is the carbonyl carbon of another aldehyde molecule. Therefore the product has the α-carbon of one aldehyde molecule bonded to the carbonyl carbon of another aldehyde molecule. Under milder conditions an OH group remains on the carbonyl carbon of the electrophile, whereas under vigorous conditions the α-carbon and the carbonyl carbon are connected by a double bond.

$$CH_3CH_2\overset{\overset{\displaystyle O}{\|}}{C}H \quad CH_3\overset{\ominus}{\overset{..}{C}}HCH \longrightarrow \longrightarrow CH_3CH_2\overset{\overset{\displaystyle OH}{|}}{C}H\overset{\overset{\displaystyle O}{\|}}{C}HCH$$

$$\overset{\displaystyle CH_3}{|}$$

carbonyl carbon α-carbon
electrophile   nucleophile

Problem 16.9   Show all of the steps in the mechanism for this reaction:

$$2 \;\; CH_3CH_2CH_2\overset{\overset{\displaystyle O}{\|}}{C}H \;\; \xrightarrow{\text{NaOH}} \;\; CH_3CH_2CH_2\overset{\overset{\displaystyle OH}{|}}{C}H\overset{\overset{\displaystyle O}{\|}}{C}HCH$$

$$\overset{\displaystyle CH_3CH_2}{|}$$

Ketones are less reactive electrophiles than aldehydes. Therefore the aldol condensation of ketones is not often used because the equilibrium is unfavorable. However, the intramolecular condensation of diketones is useful if the size of the resulting ring is favorable (formation of five-, six-, and seven-membered rings).

(42%)

Often, it is desirable to conduct an aldol condensation in which the nucleophile and the electrophile are derived from different compounds. In general, such **mixed aldol condensations,** involving two different aldehydes, result in the formation of several products and for this reason are not useful. For example, the reaction of ethanal and propanal results in the formation of four products because there are two possible enolate nucleophiles and two carbonyl electrophiles:

$$\begin{array}{c} CH_3\overset{\overset{\displaystyle O}{\|}}{C}H \\ + \\ CH_3CH_2\overset{\overset{\displaystyle O}{\|}}{C}H \end{array} \xrightarrow[\Delta]{\text{NaOH}} \begin{array}{l} CH_3CH{=}CH\overset{\overset{\displaystyle O}{\|}}{C}H + CH_3CH_2CH{=}CH\overset{\overset{\displaystyle O}{\|}}{C}H \\[2mm] + CH_3CH{=}\overset{\overset{\displaystyle O}{\|}}{C}H + CH_3CH_2CH{=}\overset{\overset{\displaystyle O}{\|}}{C}H \\ \qquad\qquad \overset{\displaystyle CH_3}{|} \qquad\qquad\qquad \overset{\displaystyle CH_3}{|} \end{array}$$

Mixed aldol condensations can be employed if one of the aldehydes has no hydrogens on the α-carbon so it cannot form an enolate ion and can only act as the electrophilic partner in the reaction. Aromatic aldehydes are especially useful in this role because the dehydration product has additional stabilization from the conjugation of the newly formed carbon–carbon double bond with the aromatic ring. This stabilization makes the equilibrium for the formation of this product more favorable.

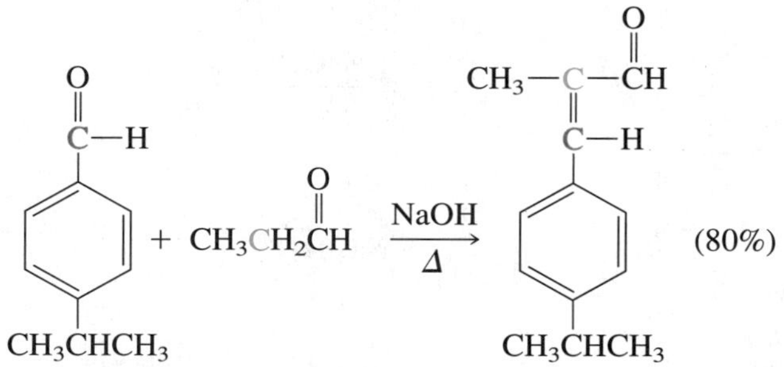

(80%)

With an aromatic aldehyde as the electrophilic partner, the nucleophilic enolate ion can also be derived from a ketone or a nitrile. As illustrated in the following examples, this enables the aldol condensation to be used to form a wide variety of compounds:

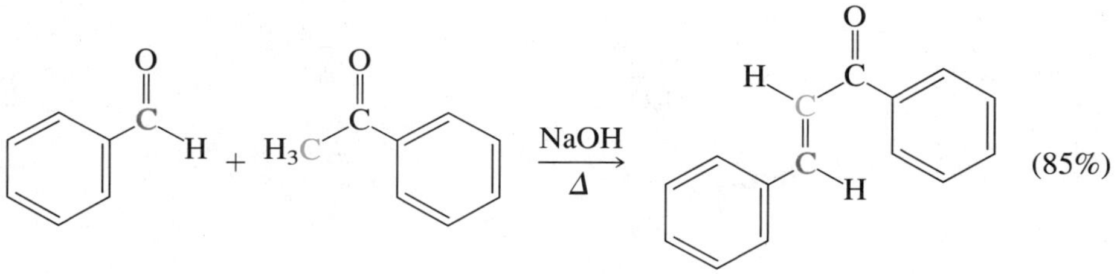

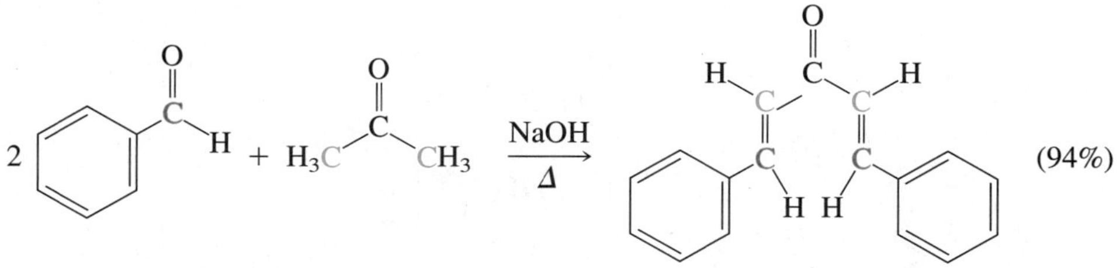

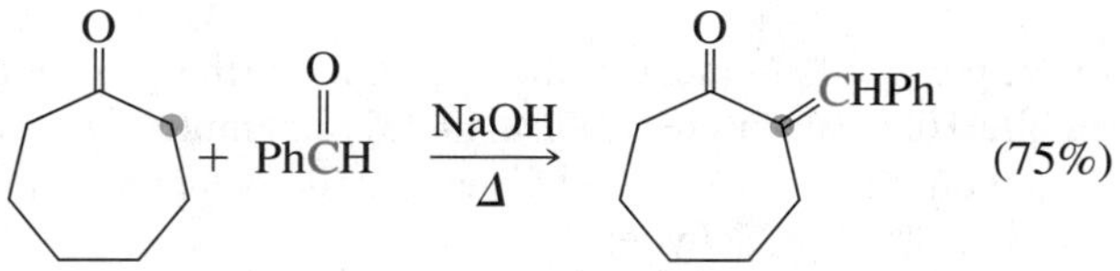

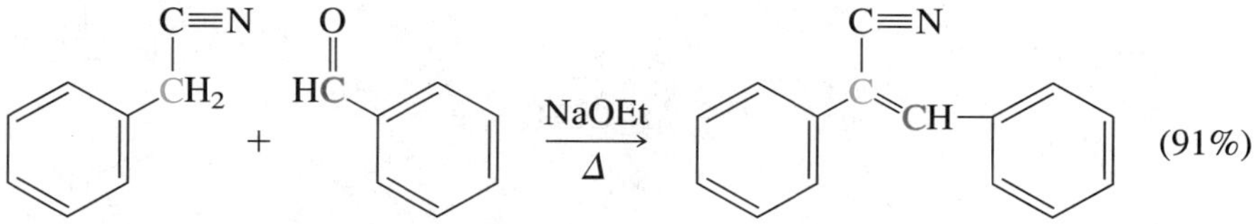

(85%)

(94%)

(75%)

(91%)

Problems

Problem 16.10   Show the products of these reactions.

a) (benzaldehyde) $+ CH_3CCH_3$ $\xrightarrow[\Delta]{NaOH}$

b) (furan-2-carbaldehyde) $+ CH_3CPh$ $\xrightarrow[\Delta]{NaOH}$

c) (4-methylbenzaldehyde) $+ CH_3CH_2$—(phenyl ketone) $\xrightarrow[\Delta]{NaOH}$

d) (2,5-hexanedione) $\xrightarrow[H_2O, \Delta]{NaOH}$

e) $NCCH_2COEt$ + (4-bromobenzaldehyde) $\xrightarrow[EtOH, \Delta]{KOH}$

f) (phenylacetonitrile, CH_2CN) + (benzaldehyde) $\xrightarrow[H_2O, \Delta]{KOH}$

Solution

a)   Again, the key is to identify the nucleophile (the enolate anion) and the electrophile (the carbonyl carbon). In this example the enolate anion can be derived only from acetone. The electrophile is the more reactive carbonyl carbon, that of benzaldehyde:

Problem 16.11   Show all of the steps in the mechanism for this reaction:

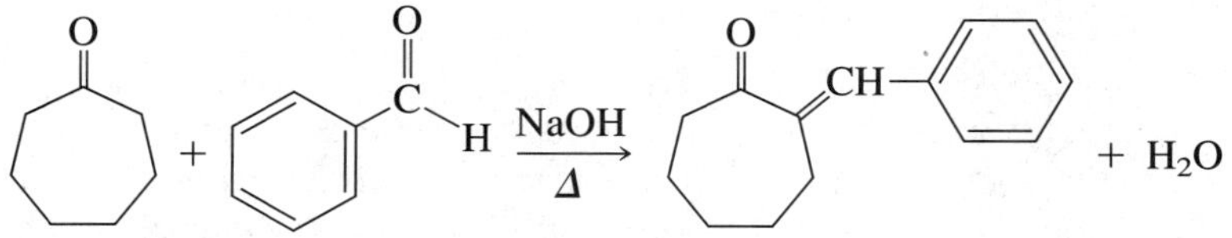

Problem 16.12   Show how the aldol condensation could be used to synthesize these compounds.

a) 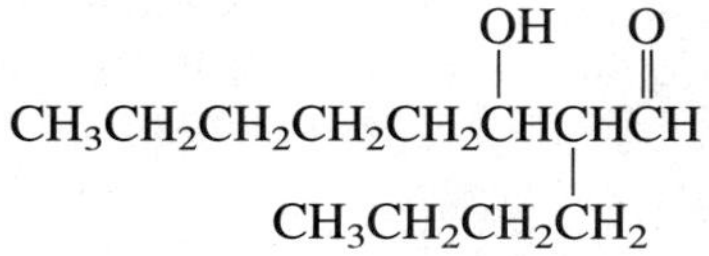 $CH_3CH_2CH_2CH_2CH_2\overset{OH}{\underset{\underset{CH_3CH_2CH_2CH_2}{|}}{CH}}\overset{O}{CHCH}$

b) 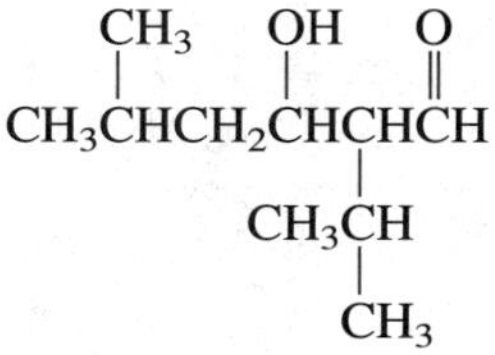 $CH_3\overset{CH_3}{\underset{|}{CH}}CH_2\overset{OH}{\underset{\underset{CH_3CH}{|}}{CH}}\overset{O}{CHCH}$
$\underset{\underset{CH_3}{|}}{}$

c)

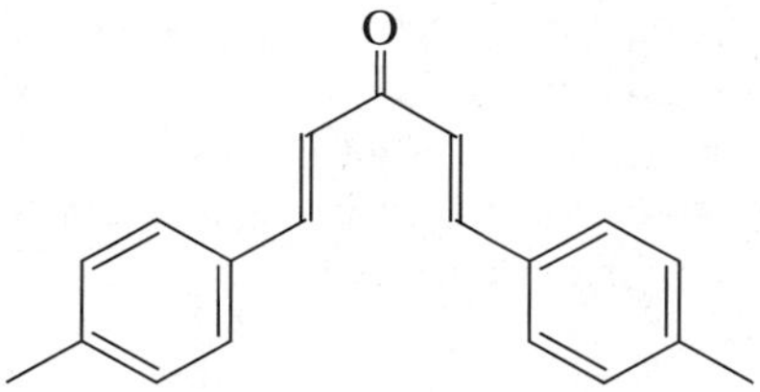

d)

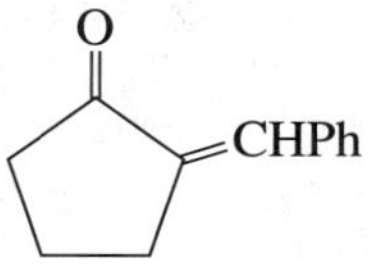

e)

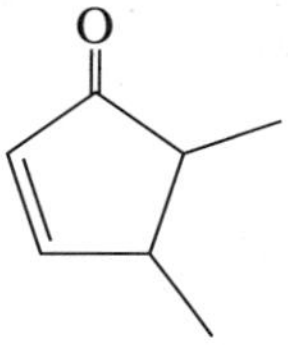

Solution

a) First identify the carbon–carbon bond that could be formed in an aldol condensation. This is the bond between the α-carbon of the carbonyl group of the product and the carbon that is either doubly bonded to it or has a hydroxy substituent. Disconnection of this bond gives the fragments needed for the aldol condensation.

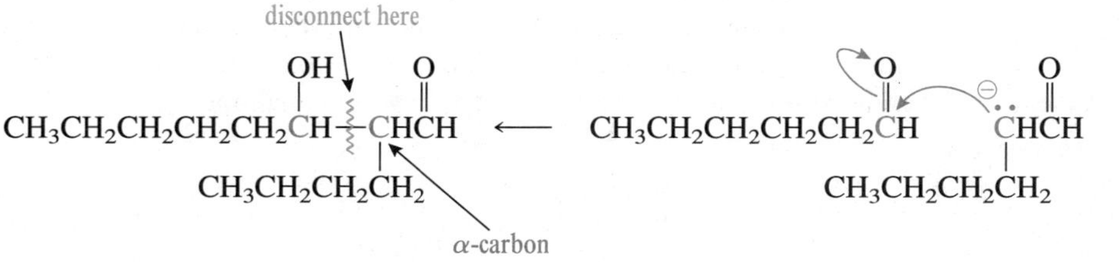

So the synthesis is

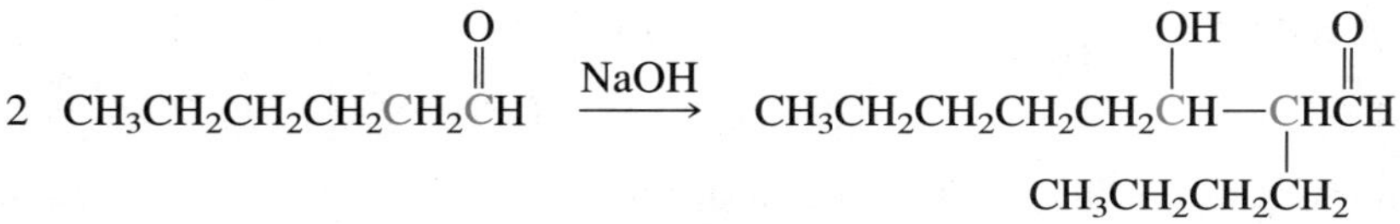

ELABORATION

The Reverse Aldol Reaction in Metabolism

The initial product of an aldol condensation has a hydroxy group on the β-carbon to a carbonyl group. Sugars also have hydroxy groups on the β-carbon to their carbonyl groups, and so they can be viewed as products of aldol condensations. In fact, a reverse aldol condensation is used in the metabolism of glucose (glycolysis) to cleave this six-carbon sugar into two three-carbon sugars.

To cleave a six-carbon sugar into two three-carbon fragments by a reverse aldol condensation, there must be a carbonyl group at C-2 and a hydroxy group at C-4. Therefore glucose is first isomerized to fructose during its metabolism. This is the role of the enzymatic reaction described in the elaboration on page 756. The substrate for the cleavage reaction is the diphosphate ester fructose-1,6-bisphosphate. In the first step, a proton is removed from the hydroxy group on C-4. The bond between C-3 and C-4 is then broken in a step that is the reverse of the aldol condensation, producing glyceraldehyde-3-phosphate (GAP) and the enolate ion of dihydroxyacetone phosphate (DHAP). This enolate ion is protonated in the final step. GAP and DHAP can be interconverted by the same process that interconverts glucose and fructose and thus provide a common intermediate for further metabolism.

16.6   Ester Condensations

So far, we have seen that an enolate anion is able to act as a nucleophile in an S_N2 reaction (Sections 16.3 and 16.4) and also in an addition reaction to the carbonyl group of an aldehyde in the aldol condensation (Section 16.5). It also can act as a nucleophile in a substitution reaction with the carbonyl group of an ester as the electrophile. When an ester is treated with a base such as sodium ethoxide, the enolate ion that is produced can react with another molecule of the same ester. The product has the α-carbon of one ester molecule bonded to the carbonyl carbon of a second ester molecule, replacing the alkoxy group. Examples of this reaction, called the **Claisen ester condensation,** are provided by the following equations:

$$2\ CH_3-\overset{O}{\overset{\|}{C}}-OEt \xrightarrow[\text{2) H}_3\text{O}^\oplus]{\text{1) NaOEt, EtOH}} CH_3-\overset{O}{\overset{\|}{C}}-CH_2-\overset{O}{\overset{\|}{C}}-OEt + EtOH\quad(80\%)$$

$$2\ CH_3CH_2CH_2\overset{O}{\overset{\|}{C}}OEt \xrightarrow[\text{2) H}_3\text{O}^\oplus]{\text{1) NaOEt, EtOH}} CH_3CH_2CH_2\overset{O}{\overset{\|}{C}}\underset{\underset{CH_3CH_2}{|}}{C}H\overset{O}{\overset{\|}{C}}OEt + EtOH\quad(76\%)$$

The mechanism for this reaction, shown in Figure 16.4, has similarities to those of both an aldol condensation (see Figure 16.3) and an ester saponification (see

c)
$$\text{PhCHO} + CH_3CH_2CCH_2CH_3 \xrightarrow[\Delta]{\begin{array}{c}NaOH\\H_2O\end{array}}$$

(benzaldehyde + 3-pentanone)

d)
(cycloheptanone) $+ \text{EtOCOEt} \xrightarrow{\begin{array}{c}1)\ NaOEt,\ EtOH\\2)\ H_3O^{\oplus}\end{array}}$

e) $PhCH_2COH \xrightarrow{\begin{array}{c}1)\ 2\ LDA\\2)\ PhCH_2Br\end{array}}$

f) $CH_3CCH_2COEt \xrightarrow{\begin{array}{c}1)\ NaH\\2)\ BuLi\\3)\ PhCH_2CH_2CH_2Cl\end{array}}$

16.30 Suggest a mechanism for this reaction:

$$\text{(phenol)} \xrightarrow[]{^\ominus OH} \text{(cyclohex-2-enone)}$$

16.31 Suggest a mechanism for this reaction. (The reaction does not occur by a carbocation rearrangement.)

$$\text{(cyclohex-3-enone)} \xrightarrow[]{H_3O^{\oplus}} \text{(cyclohex-2-enone)}$$

16.32 Show all of the steps in the mechanism for this reaction:

$$2\ CH_3CH_2CH_2COEt \xrightarrow{\begin{array}{c}1)\ NaOEt,\ EtOH\\2)\ H_3O^{\oplus}\end{array}} CH_3CH_2CH_2CCHCOEt$$
$$\underset{CH_3CH_2}{|}$$

16.33 Show all of the steps in the mechanism for this reaction:

$$2\ CH_3CH_2CH \xrightarrow{NaOH} CH_3CH_2CHCHCH$$
$$\underset{CH_3}{|}$$

with OH and O groups shown on product.

16.34 Show syntheses of these compounds from propanal.

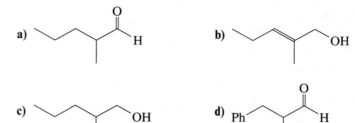

16.35 Show syntheses of these compounds from ethyl propanoate.

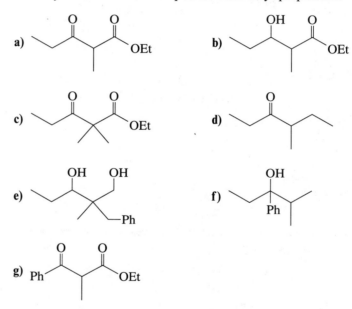

16.36 Show syntheses of these compounds from ethyl acetoacetate.

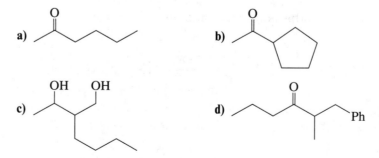

16.37 Optically active ketone **A** undergoes racemization in basic solution. Show a mechanism for this process. Explain whether ketone **B** would also racemize in basic solution.

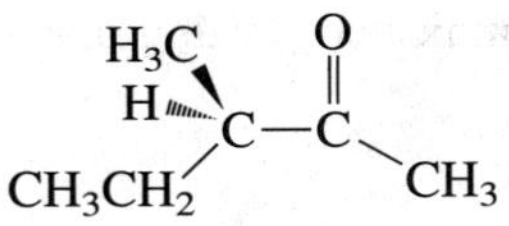

A

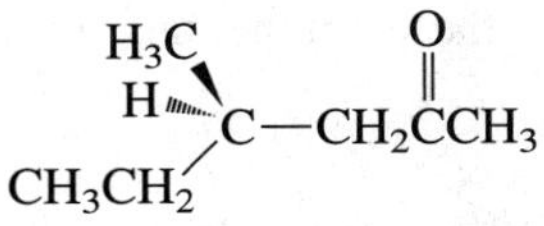

B

16.38 Show the products of these reactions.

a)

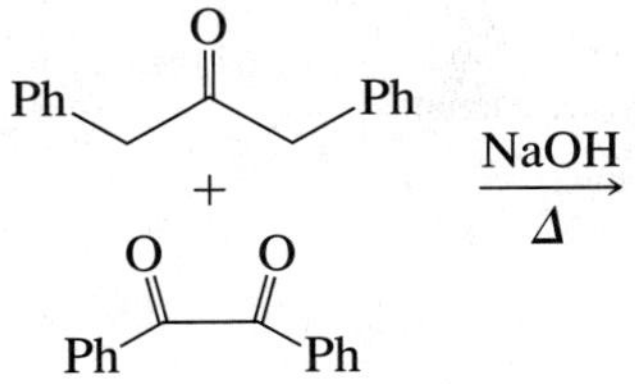

NaOH
$\xrightarrow{\Delta}$

b)

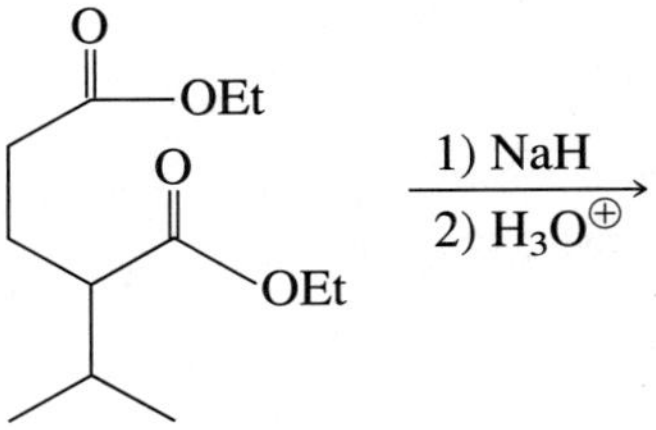

1) NaH
$\xrightarrow{\text{2) H}_3\text{O}^{\oplus}}$

c)

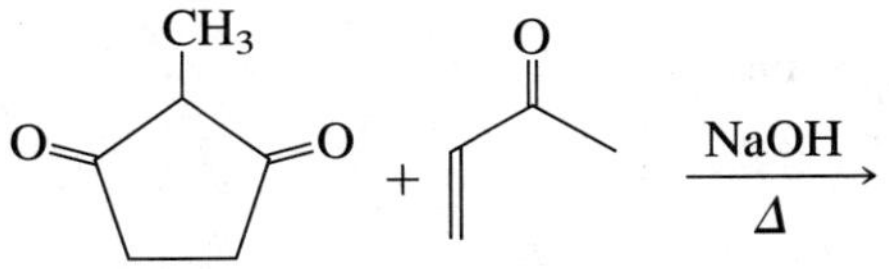

NaOH
$\xrightarrow{\Delta}$

d)

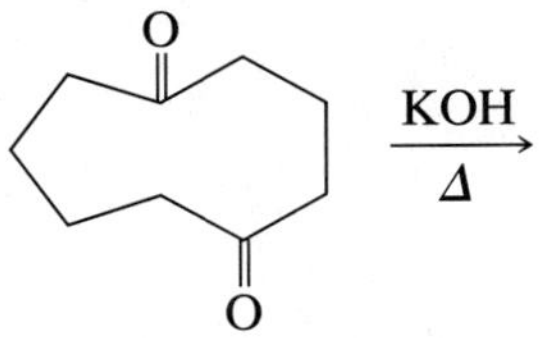

KOH
$\xrightarrow{\Delta}$

e)

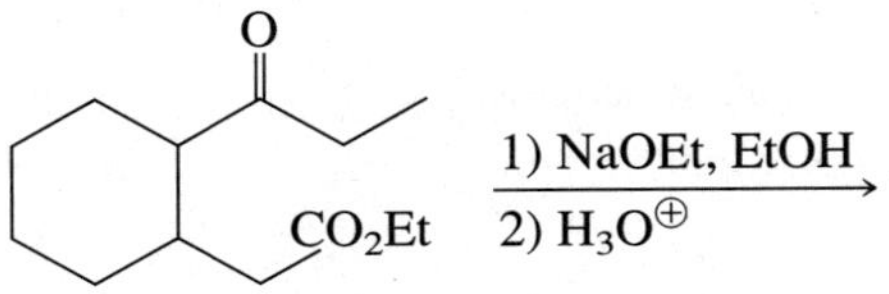

1) NaOEt, EtOH
$\xrightarrow{\text{2) H}_3\text{O}^{\oplus}}$

f) 2

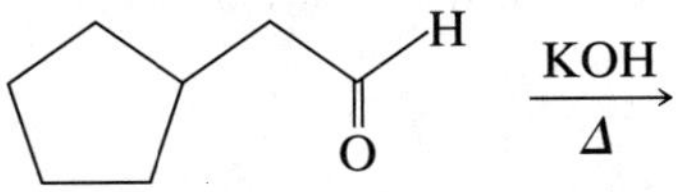

KOH
$\xrightarrow{\Delta}$

g)

$+ \ CH_3OCCH_2COCH_3$

NaOCH$_3$
$\xrightarrow{}$

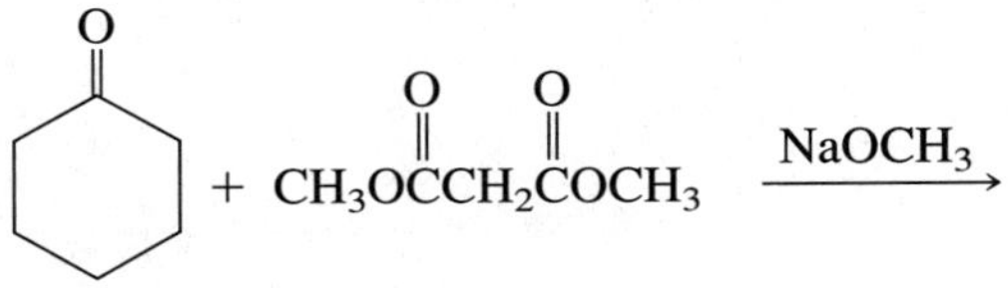

16.39 Show syntheses of these compounds from the indicated starting materials.

a) 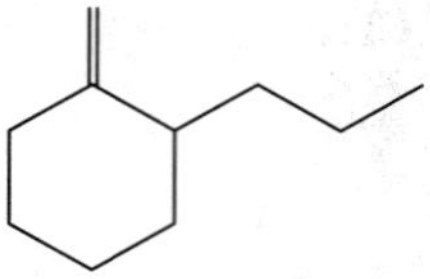 from cyclohexanone

b) 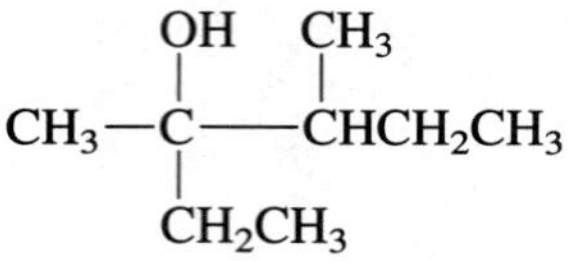 CH$_3$—C($\overset{OH}{}$)($\underset{CH_2CH_3}{}$)—CHCH$_2$CH$_3$ ($\overset{CH_3}{}$) from ethyl acetoacetate

c) 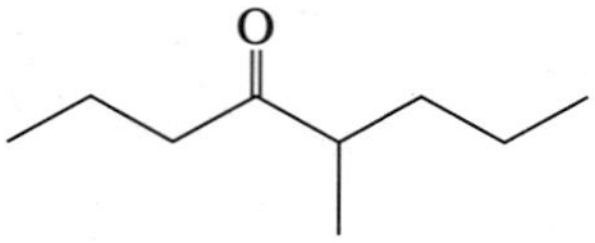 from butanoic acid

d) 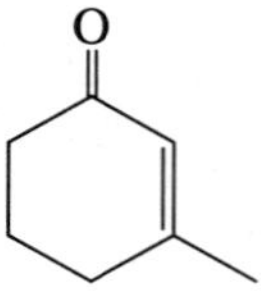 from compounds without a ring

e) 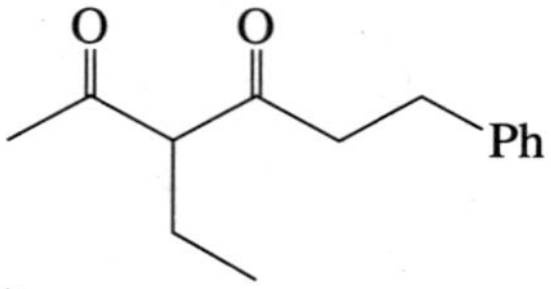 Ph from 2,4-pentanedione

f) 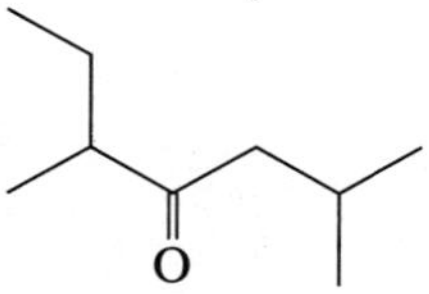 from propanoic acid

g) 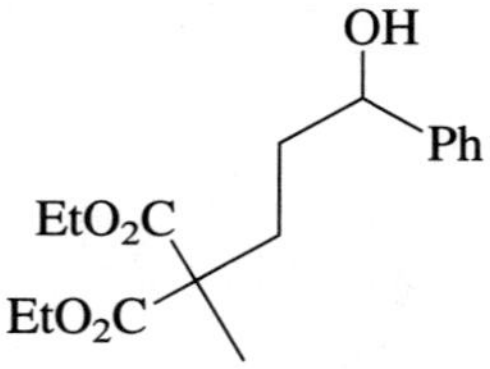 EtO$_2$C, EtO$_2$C ($\overset{OH}{}$)($\underset{Ph}{}$) from diethyl malonate

16.40 Show syntheses of these compounds from the indicated starting materials.

a) 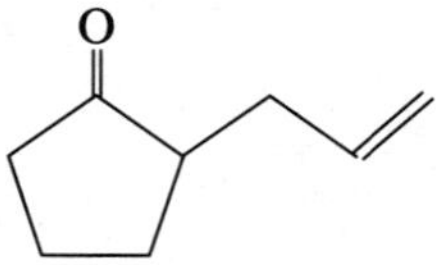 from EtOC(CH$_2$)$_4$COEt

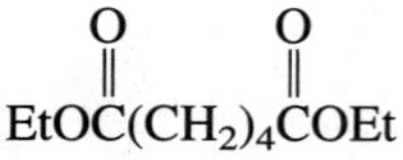

b) 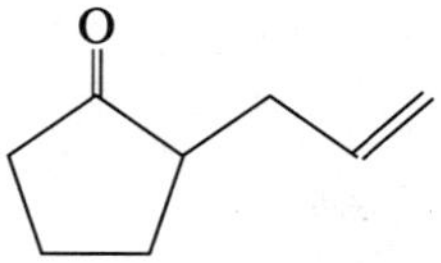 from

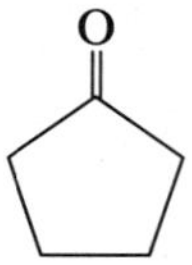

c) 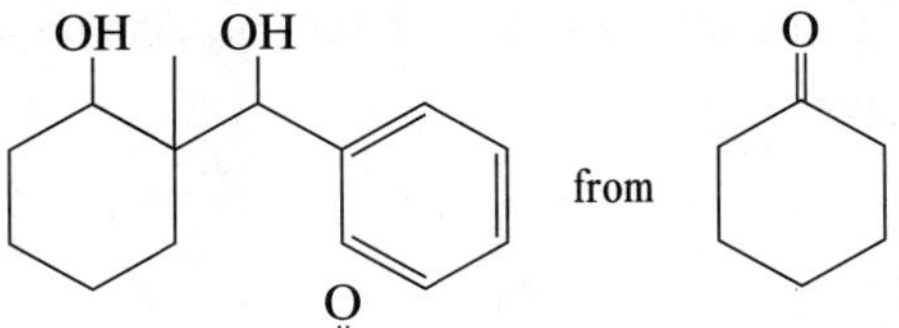 from

d) 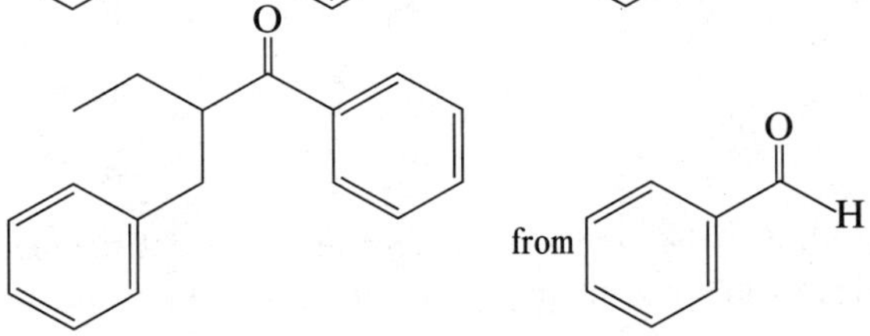 from and compounds with less than six carbons

e) $CH_3\overset{O}{\overset{\|}{C}}CH\overset{Ph}{\underset{|}{CH}}CH_2\overset{O}{\overset{\|}{C}}CH_3$ from $CH_3\overset{O}{\overset{\|}{C}}CH_3$
$\qquad\qquad\underset{|}{CO_2Et}$

16.41 Show all of the steps in the mechanism for this reaction:

$\xrightarrow[\text{2) } H_3O^{\oplus}, \Delta]{\text{1) NaOH, } H_2O}$ $+ CO_2 + EtOH$

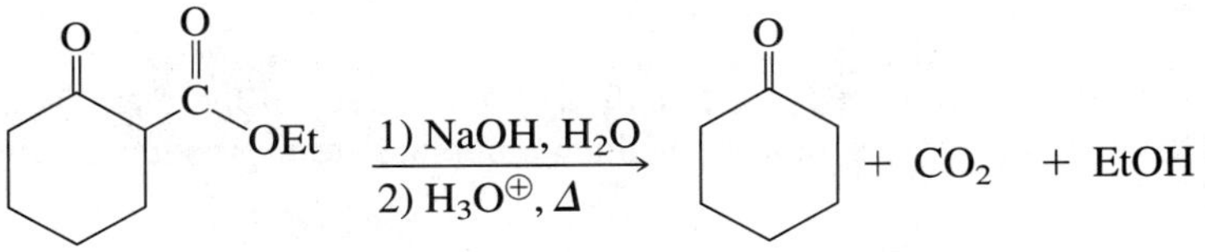

16.42 Show all of the steps in the mechanism for this reaction:

$CH_2{=}CH\overset{O}{\overset{\|}{C}}OEt + NCCH_2\overset{O}{\overset{\|}{C}}OEt \xrightarrow{\text{NaOEt}} Et\overset{O}{\overset{\|}{OC}}{-}\overset{CN}{\underset{|}{CH}}{-}CH_2CH_2\overset{O}{\overset{\|}{C}}OEt$

16.43 Show a mechanism for the interconversion of glyceraldehyde-3-phosphate (GAP) and dihydroxyacetone phosphate (DHAP) in basic solution:

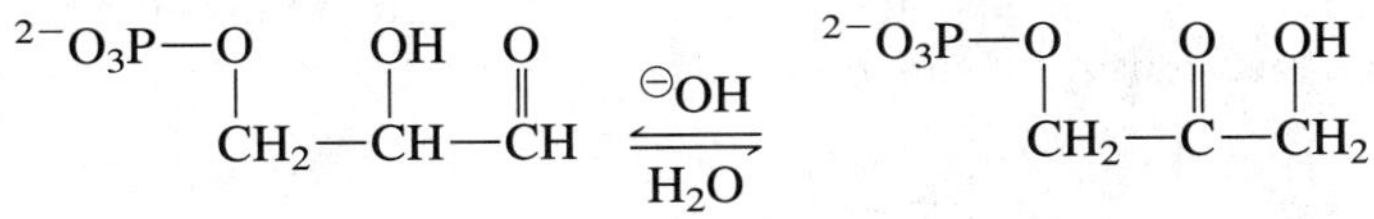

$^{2-}O_3P{-}O \quad \underset{|}{\overset{OH}{|}} \quad O$
$\qquad\underset{|}{}\qquad\underset{|}{}\qquad\overset{\|}{}$
$\qquad CH_2{-}CH{-}CH \underset{H_2O}{\overset{\ominus OH}{\rightleftharpoons}}$ $^{2-}O_3P{-}O \quad O \quad \overset{OH}{\underset{|}{|}}$
$\qquad\qquad\qquad\qquad\qquad CH_2{-}C{-}CH_2$

GAP   DHAP

16.44 Show the missing products, **A** and **B,** in this reaction scheme and explain the regiochemistry of the reactions.

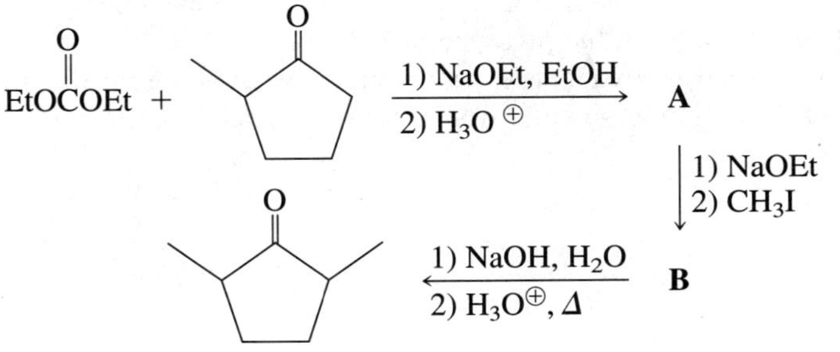

$EtO\overset{O}{\overset{\|}{C}}OEt +$ $\xrightarrow[\text{2) } H_3O^{\oplus}]{\text{1) NaOEt, EtOH}}$ **A**

$\qquad\qquad\qquad\qquad\qquad\qquad\qquad\Big\downarrow \begin{array}{l}\text{1) NaOEt}\\\text{2) } CH_3I\end{array}$

$\xleftarrow[\text{2) } H_3O^{\oplus}, \Delta]{\text{1) NaOH, } H_2O}$ **B**

16.45 Show syntheses of these compounds using the Robinson annulation reaction.

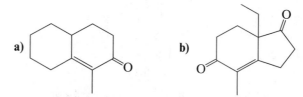

a)

b)

16.46 2-Ethyl-1,3-hexanediol is the active ingredient in the insect repellant "6-12." Suggest a synthesis of this compound from precursors with four or fewer carbons.

2-ethyl-1,3-hexanediol

16.47 2-Ethyl-1-hexanol is used industrially as a plasticizer in the manufacture of plastics. Suggest a synthesis of this compound from precursors with four or fewer carbons.

2-ethyl-1-hexanol

16.48 β-Ketoesters that have two substituents on the α-carbon undergo fragmentation when treated with ethoxide anion as shown in the following equation. Suggest a mechanism for this reaction.

$$CH_3\overset{O}{\overset{\|}{C}}-\underset{\underset{CH_3}{|}}{\overset{\overset{CH_3}{|}}{C}}-\overset{O}{\overset{\|}{C}}OEt + EtOH \xrightarrow{\text{NaOEt}} CH_3\overset{O}{\overset{\|}{C}}OEt + CH_3\underset{\underset{CH_3}{|}}{\overset{\overset{CH_3}{|}}{C}}H-\overset{O}{\overset{\|}{C}}OEt$$

16.49 Deuterium can be incorporated at the positions α to a carbonyl group by reaction with D_2O in the presence of acid. Show a mechanism for this process. If the reaction were continued, what is the maximum number of deuterium atoms that would be incorporated into a single molecule?

$$CH_3CH_2\overset{O}{\overset{\|}{C}}CH_2CH_3 + D_2O \xrightarrow{D_3O^{\oplus}} CH_3CH_2\overset{O}{\overset{\|}{C}}CHDCH_3$$

16.50 Suggest a mechanism for this reaction:

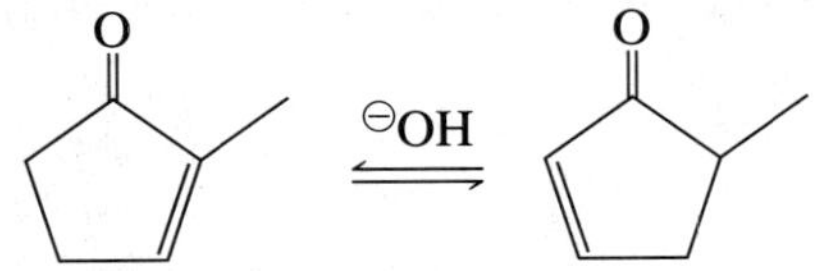

16.51 Intramolecular aldol condensations often present the possibility of the formation of several products. The reaction of 6-oxoheptanal gives the product shown in the following equation:

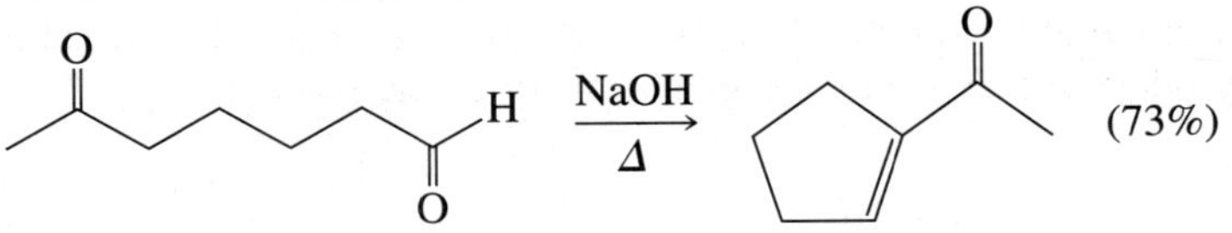

(73%)

a) Show structures for the other α,β-unsaturated products that could be formed in this reaction and explain why the observed product is formed preferentially.

b) Predict the preferred product in this aldol cyclization.

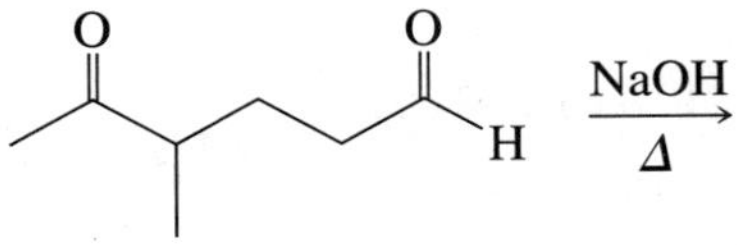

16.52 Show the missing products, **A** and **B**, in this reaction scheme.

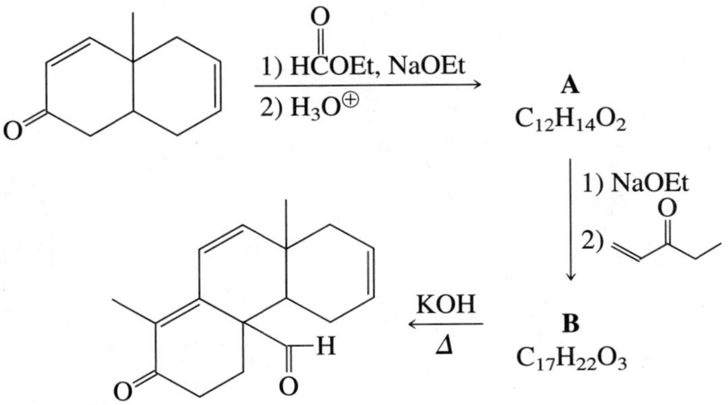

1) HCOEt, NaOEt
2) $H_3O^{\oplus}$

A
$C_{12}H_{14}O_2$

1) NaOEt
2)

B
$C_{17}H_{22}O_3$

$\xleftarrow{\text{KOH} \atop \Delta}$

Problems Involving Spectroscopy

16.53 2,5-Heptanedione forms two products upon reaction with NaOH. Both products show a strong absorption near 1715 cm^{-1} in their IR spectra, and neither shows any absorption bands in the region of 3600–3300 cm^{-1}. The major product has singlets at 1.90 and 1.65 δ in its ^{1}H-NMR spectrum. Show structures for these products.

16.54 The product from the reaction of 1-phenyl-2-butanone with LDA and methyl iodide shows a quartet (1 H), a quartet (2 H), a doublet (3 H), and a triplet (3 H) in the alkyl region of its ^{1}H-NMR spectrum. Show the structure of this product and explain the regiochemistry of the reaction.

16.55 The IR and ^{1}H-NMR spectra of the product of this reaction follow. The formula of the product is $C_9H_8O_2$. Show the structure of the product.

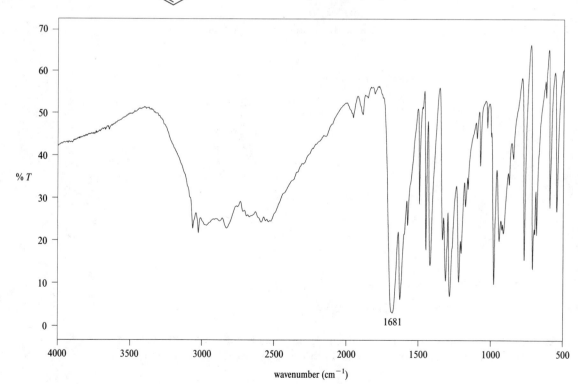

17.11 Annulenes

The general name **annulene** is sometimes given to rings that contain alternating single and double bonds in a single Lewis structure. Thus benzene can be called [6]annulene, and cyclooctatetraene can be called [8]annulene. A number of larger annulenes have been prepared to determine if they follow Hückel's rule and are aromatic when they have $4n + 2$ electrons in the cycle.

The larger members of this series would have considerable angle strain if they were planar and had only *cis* double bonds. The incorporation of *trans* double bonds provides a way to relieve this angle strain, although this often introduces steric strain resulting from atoms on opposite sides of the ring being forced into the same region of space. Consider, for example, [10]annulene. With two *trans* double bonds it has no angle strain, but the two hydrogens that point into the interior of the ring cause so much steric strain that attempts to prepare this compound have not yet been successful. However, the compound with a CH_2 bridge in place of the offending hydrogens has been prepared. Although the bridge causes the ring to be somewhat distorted from planarity, the compound does show the presence of a diamagnetic ring current typical of an aromatic compound. The hydrogens on the periphery of the ring appear at 6.9–7.3 δ, and the hydrogens on the bridge, which are held over the face of the ring, appear at the substantially upfield position of -0.5 δ. (The hydrogens of a typical CH_2 group attached to a carbon–carbon double bond appear near 2 δ.)

[10]annulene not yet prepared   a bridged [10]annulene

The hydrogens inside the ring of the [14]annulene with four *trans* double bonds appear at 0.0 δ and the hydrogens on the outside of the ring appear at 7.6 δ, indicating the presence of a diamagnetic ring current. However, the steric strain caused by the hydrogens inside the ring make this compound quite reactive. The bridged [14]annulene, where these steric interactions are absent, is quite stable and has many characteristics of an aromatic compound. The bond distances are all near 1.4 Å, it undergoes substitution reactions rather than addition reactions, the outer hydrogens appear at 8.1–8.7 δ, and the hydrogens of the methyl groups appear at -4.25 δ. As a final example, the hydrogens on the inside of the ring of [18]annulene appear at -3 δ and the outside hydrogens appear at 9 δ.

[14]annulene   a bridged [14]annulene

H ← 9 δ

H ← −3 δ

[18]annulene

Problem

Problem 17.12 The ^{1}H-NMR spectrum of this compound shows absorptions in the region of 9.5 δ and other absorptions in the region of −7 δ. Explain which hydrogens are responsible for each of these absorptions.

Exercise

Exercise 17.1 Build models of [10]annulene and the bridged [10]annulene discussed on the previous page and examine the strain and planarity of each.

17.12 Aromatic and Antiaromatic Ions

Rings containing an odd number of carbon atoms can be aromatic, or antiaromatic, if they are planar and have a conjugated p orbital on each ring atom. To have an even number of electrons in their odd number of p orbitals, these species must be ionic. They must be carbocations or carbanions.

The simplest example of such an ion is the cyclopropenyl carbocation:

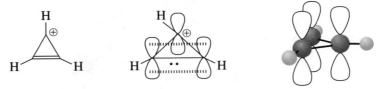

the cyclopropenyl carbocation

Since a carbocation is sp^2 hybridized, with trigonal planar geometry and an empty *p* orbital, this ion has a cycle of three *p* orbitals. (Remember that it is not the number of orbitals that determines whether a compound is aromatic or not, but rather the number of electrons in the pi MOs.) The cyclopropenyl carbocation has two electrons in its three pi MOs, so it fits Hückel's rule and should be aromatic. In fact, cyclopropenyl carbocations are significantly more stable than other carbocations, even though they have considerable angle strain. For example, most carbocations react rapidly with water, a weak nucleophile. In contrast, tri-*tert*-butylcyclopropenyl perchlorate, a carbocation salt, is stable enough to be recrystallized from water.

tri-*tert*-butylcyclopropenyl perchlorate

Another example is provided by the acidity of cyclopentadiene. This compound is approximately as strong an acid as water and is many orders of magnitude more acidic than other hydrocarbons because its conjugate base, with six pi electrons, is aromatic.

cyclopentadiene   the aromatic cyclopentadienyl anion $pK_a = 15$

The cycloheptatrienyl carbocation, also known as the tropylium cation, has six pi electrons. It is also aromatic and is quite stable. In fact, 7-bromo-1,3,5-cyclohepta-triene actually exists as an ionic compound.

7-bromo-1,3,5-cycloheptatriene   the aromatic tropylium cation

In contrast, the cyclopentadienyl carbocation, which has four pi electrons and is antiaromatic, is quite unstable. Thus 5-iodo-1,3-cyclopentadiene is unreactive under conditions in which iodocyclopentane reacts rapidly by an S_N1 mechanism.

5-iodo-1,3-
cyclopentadiene

the antiaromatic
cyclopentadienyl cation

Problems

Problem 17.13   Explain which of these compounds is a stronger acid.

Problem 17.14   Explain which of these compounds has the faster rate of substitution by the S_N1 mechanism.

17.13   Summary

The concept of aromaticity is very important in understanding the chemical behavior of cyclic, conjugated compounds. It is most important with benzene and its derivatives, but it also has applications to many other types of compounds. Whenever a reactant, product, or intermediate contains a planar cycle of p orbitals, the effect of aromaticity (or antiaromaticity) on the reaction must be considered.

After completing this chapter you should be able to:

1.   Show the MO energy levels for planar, cyclic, conjugated compounds.

2.   Apply Hückel's rule and recognize whether a particular compound is aromatic, antiaromatic, or neither.

3.   Understand how aromaticity and antiaromaticity affect the chemistry (and NMR spectra) of compounds.

End-of-Chapter Problems

17.15   Show the energy levels for the pi MOs of these species. Show the electrons occupying these MOs.

a) 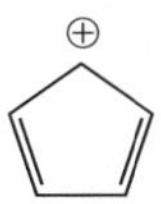   b) 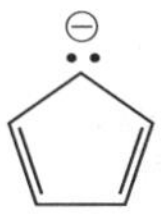   c)

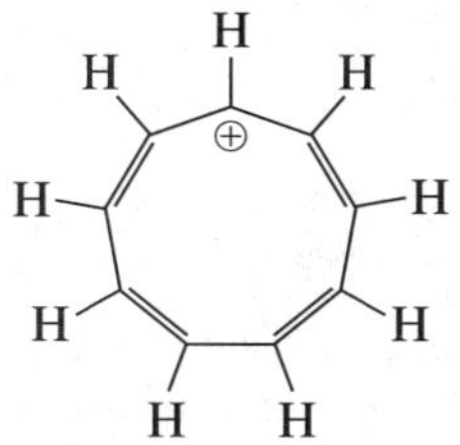

17.16   Explain whether each of these species is aromatic, antiaromatic, or neither.

a) 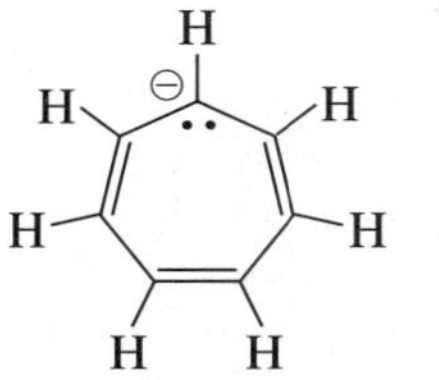   b)

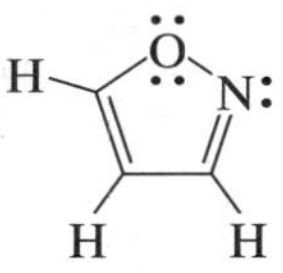

c) 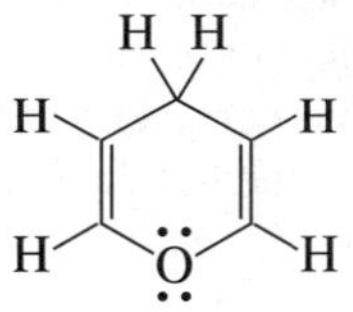   d)

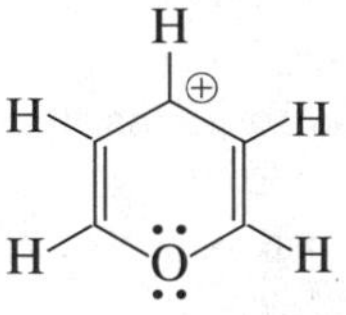

e) 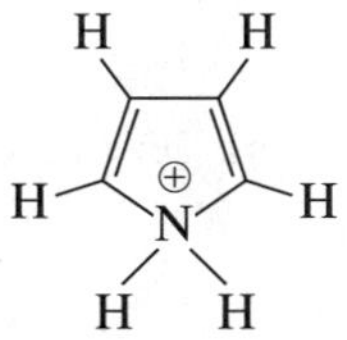   f)

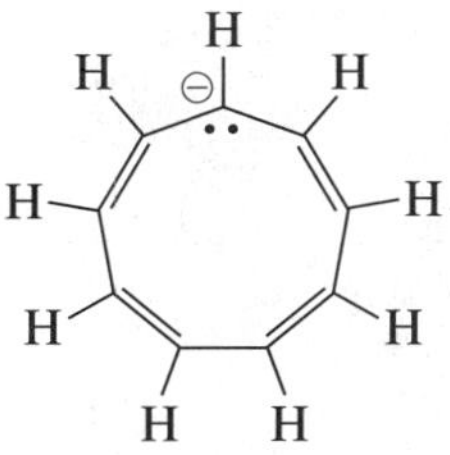

17.17   Cyclooctatetraene readily reacts with potassium metal to form a dianion. Discuss the electronic structure and geometry of this dianion and explain why it is formed so readily.

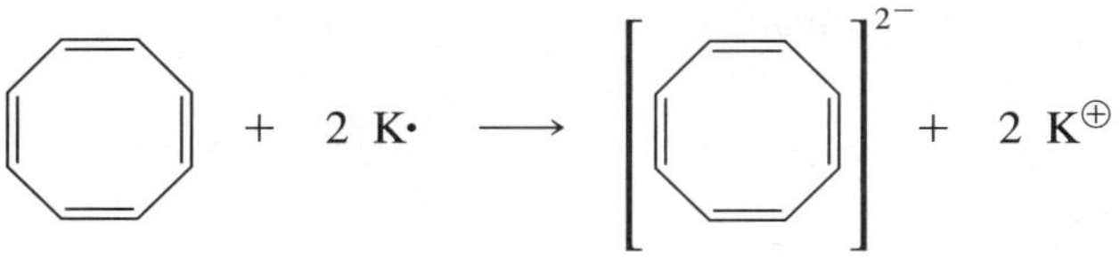

$$\text{(cyclooctatetraene)} \quad + \quad 2\ K\cdot \quad \longrightarrow \quad \left[\ \text{(cyclooctatetraene)}\ \right]^{2-} \quad + \quad 2\ K^{\oplus}$$

17.18   2,4-Cyclopentadienone is unstable and cannot be isolated, whereas 2,4,6-cycloheptatrienone is quite stable and is readily isolated. Use arguments based on resonance and aromaticity to explain these experimental observations. (*Hint*: Recall a resonance structure that is commonly written for the carbonyl group.)

2,4,6-cycloheptatrienone   2,4-cyclopentadienone
       stable   unstable

17.19 Cyclopropanone is a highly reactive ketone, presumably because of the extra angle strain introduced into the three-membered ring by the sp^2 hybridized carbonyl carbon. Cyclopropenone is much less reactive even though it has more angle strain. Offer an explanation for this experimental observation.

more reactive   less reactive

17.20 Hückel's rule applies only to compounds with a single ring, such as benzene and cyclobutadiene. However, it can be used with multiple-ring compounds if the resonance structure with all of the double bonds on the periphery of the ring is considered. For example, using such a structure for naphthalene shows ten electrons in the cycle, so it is predicted to be aromatic.

Use this to explain why two of the following compounds are very reactive whereas one is quite stable.

          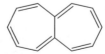

17.21 Coumarin is an important natural flavoring. Explain whether the oxygen-containing ring has any aromatic character.

coumarin

17.22 The conjugate acid of this ketone is quite stable. Explain.

17.23 Benzocyclobutadiene is a very reactive compound. Explain.

benzocyclobutadiene

17.24 Explain whether these compounds are aromatic or not.

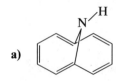

a)   b)

17.25 Predict the approximate positions for the absorptions in the ^{1}H-NMR spectrum of this compound.

17.26 The amino acid histidine contains an imidazole ring. In many enzymes this ring acts as a basic catalyst. Explain which nitrogen of the imidazole ring is more basic and show the structure of the conjugate acid of imidazole.

$$\text{N} \qquad -CH_2CHCOH \qquad \overset{\displaystyle O}{\overset{\|}{}}$$

$$\underset{H}{\text{N}} \qquad \underset{NH_2}{}$$

histidine

17.27 This compound is a stronger acid than nitric acid. Explain.

$$F_3C \qquad CF_3$$
$$F_3C \qquad CF_3$$
$$H \qquad CF_3$$

17.28 The heat of hydrogenation of pyrrole is 31.6 kcal/mol (132 kJ/mol). The heat of hydrogenation of cyclopentene is 26.6 kcal/mol (111 kJ/mol). Calculate the resonance stabilization of pyrrole.

17.29 The following compound reacts with butyllithium to form $C_8H_6^{2-}$. Suggest a structure for this dianion and explain its ready formation.

$$\xrightarrow{\text{2 BuLi}} C_8H_6^{2-}$$

17.30 Fused heterocyclic compounds are similar to polycyclic aromatic compounds except that one or more of the fused rings is a heterocycle. Explain whether or not the heterocyclic rings of these compounds are aromatic.

a) indole   b) benzimidazole   c) quinoline

17.31 Adenine is an important base that is found as a component of DNA. Explain whether adenine is aromatic or not.

adenine

17.32 Derivatives of this compound have been found to have large dipole moments. Use resonance and the theory of aromaticity to explain this observation.

Problems Involving Spectroscopy

17.33 When 7-fluoro-1,3,5-cycloheptatriene is dissolved in SbF_5/SO_2, a species is produced that shows a singlet near 9.2 δ in its ^{1}H-NMR spectrum. Show the structure of this species and explain why the absorption is so far downfield.

$$\xrightarrow[\text{SO}_2]{\text{SbF}_5}$$

17.34 When this dichloride is dissolved in SbF_5/SO_2, a species is produced that shows only two absorptions in its ^{13}C-NMR spectrum. One of these signals appears at 209 δ. Show the structure of this species and explain why the signal is so far downfield.

17.35 The fragmentation in the mass spectrometer of the molecular ion of aromatic compounds to produce benzylic carbocations was discussed in Section 13.7. Actually, the benzylic carbocation is thought to rearrange to an even more stable carbocation, also with m/z 91, that is the actual species that is detected. Suggest a structure for this carbocation and explain why it is so stable.

2,4,5-Trichlorophenol is no longer produced by heating 1,2,4,5-tetrachloroben-zene with sodium hydroxide because a small amount of tetrachlorodibenzo-*p*-dioxin (TCDD) is produced in this process, as shown in the following equation:

tetrachlorodibenzo-*p*-dioxin
(TCDD)

The mechanism for the formation of TCDD is not known with certainty but probably involves some variation of a nucleophilic aromatic substitution reaction. The amount of TCDD that is produced can be minimized by careful control of the reaction conditions, especially by not allowing the temperature to become too high.

TCDD, also known as dioxin, is extremely toxic to some animal species and is considered extremely dangerous although its exact health effect on humans is still somewhat controversial. The Agent Orange that was used in Vietnam con-tained about 10 ppm (parts per million) of dioxin, which may be responsible in part for the adverse health effects associated with exposure to this herbicide.

There have been two notable cases of TCDD contamination. One occurred in Seveso, Italy, in 1976 when a factory producing 2,4,5-trichlorophenol exploded. It seems that workers had left for the weekend without completely stopping the reac-tion. As the exothermic reaction proceeded, the temperature increased until an explosion occurred. The dioxin, which was produced in larger amounts at the ele-vated temperatures reached in the reaction, was spread over a wide area. Fortu-nately, no serious effects on human health were observed for many years, although there are now some indications of a possible increase in cancer among those who were exposed.

Another incident occurred in Missouri in the early 1970s. Waste oil that was sprayed on horse arenas and dirt roads to keep down dust was later found to be contaminated with dioxin. In March 1983 the Environmental Protection Agency bought out the whole town of Times Beach, Missouri, for $33 million.

Problem

Problem 18.28   Show the steps for the formation of 2,4,5-trichlorophenol from 1,2,4,5-tetrachlorobenzene using an addition–elimination mechanism. What argu-ments can be made against the involvement of an elimination–addition mechanism involving a benzyne intermediate?

The first of these reactions converts a nitro group to an amino group. This reduction can be accomplished using hydrogen and a catalyst or by using acid and a metal (Fe, Sn, or $SnCl_2$). Examples are provided in the following equations:

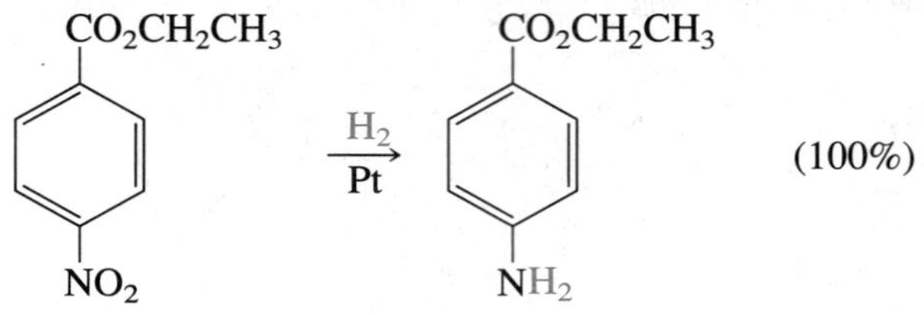

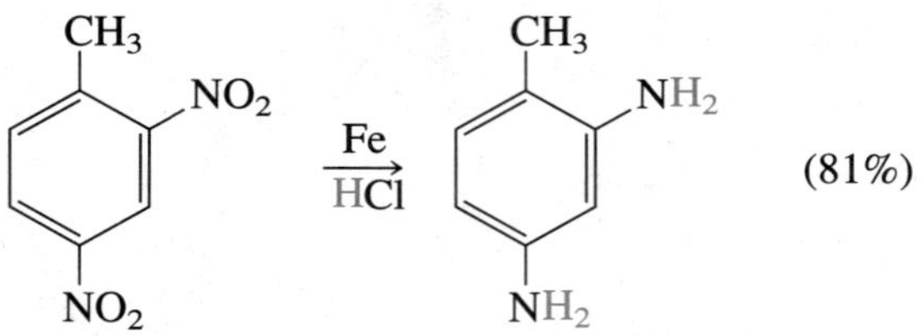

This reaction is important because it provides a method to place an amino substituent onto the benzene ring, a substitution that cannot be accomplished directly by electrophilic attack. And, as illustrated in the following example, this opens all of the substitution reactions that can be accomplished through diazonium ion reactions.

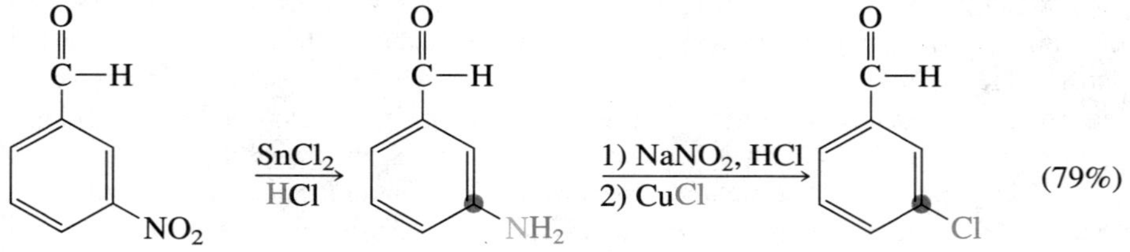

There are several procedures that allow the conversion of the carbonyl group of an aldehyde or ketone to a methylene group. One reaction, known as the Clemmensen reduction, employs amalgamated zinc (zinc plus mercury) and hydrochloric acid as the reducing agent. An example is provided by the following equation:

1-phenylbutanone   butylbenzene

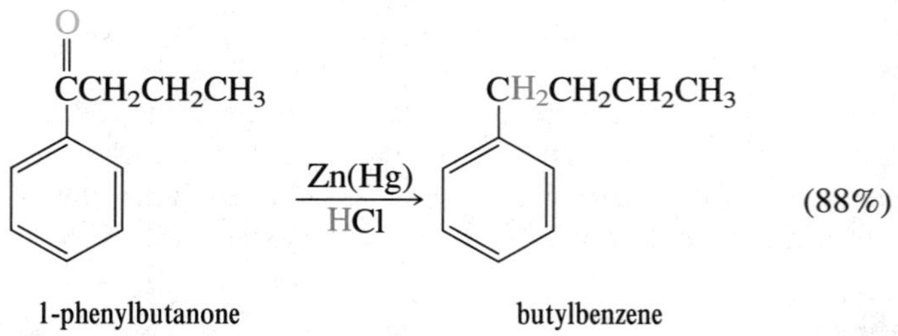

Another reaction that can be used to accomplish the same transformation is the Wolff-Kishner reduction. In this procedure the aldehyde or ketone is heated with hydrazine and potassium hydroxide in triethylene glycol as the solvent. An example is provided in the following equation. The Clemmensen reduction and the Wolff-Kishner reduction are complementary because one employs acidic conditions and the other employs basic conditions.

The reduction of the carbonyl group of an aromatic ketone to a methylene group can also be accomplished by catalytic hydrogenation. An example of this method is shown in the following equation. Note that the carbonyl group in this reaction must be attached directly to the aromatic ring. The Clemmensen and Wolff-Kishner reductions do not have this restriction.

These reactions are quite useful in the preparation of aromatic compounds substituted with primary alkyl groups. For example, suppose a synthesis of butylbenzene is required. We might first consider preparing this compound by a Friedel-Crafts alkylation reaction. However, using a primary alkyl halide in this reaction invariably results in carbocation rearrangement. The reaction of benzene with 1-chlorobutane produces a mixture of butylbenzene (34%) and *sec*-butylbenzene (66%) (see page 857). The low yield of the desired primary product and the difficulty in obtaining it pure from the product mixture make this an unacceptable synthetic route. A much better synthesis can be accomplished in two steps by first preparing 1-phenylbutanone by a Friedel-Crafts acylation reaction using benzene and butanoyl chloride, followed by conversion of the carbonyl group to a methylene group by one of these reduction reactions. As shown in the preceding equation, the Clemmensen reduction accomplishes this transformation in 88% yield.

The final reaction in this section provides a method to prepare aromatic rings bonded to a carboxylic acid group. Since we do not have a direct way for attaching this group, this procedure is very useful. The reaction is usually accomplished by oxidation of a methyl group to the carboxylic acid employing hot potassium permanganate in basic solution:

Although methyl groups are most commonly oxidized in these reactions, other alkyl groups can also be employed, as long as the carbon that is bonded to the aromatic

ring is not tertiary. Note that the use of aromatic compounds with larger alkyl groups still gives the same product as would be produced from the oxidation of the compound substituted with a methyl group. The extra carbons are lost as carbon dioxide:

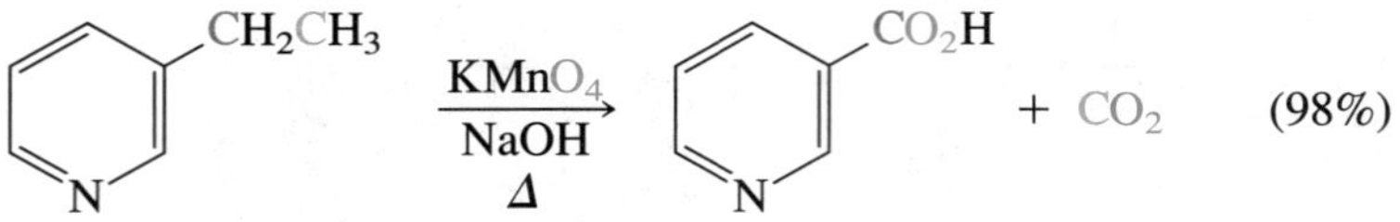

$$\xrightarrow[\substack{\text{NaOH} \\ \Delta}]{\text{KMnO}_4}$$

$+\ CO_2$ (98%)

Problems

Problem 18.29 Show the products of these reactions.

a)

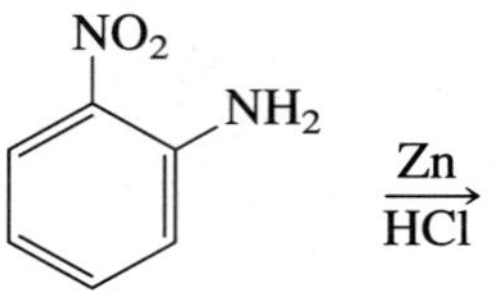

$\xrightarrow[\text{HCl}]{\text{Zn}}$

b)

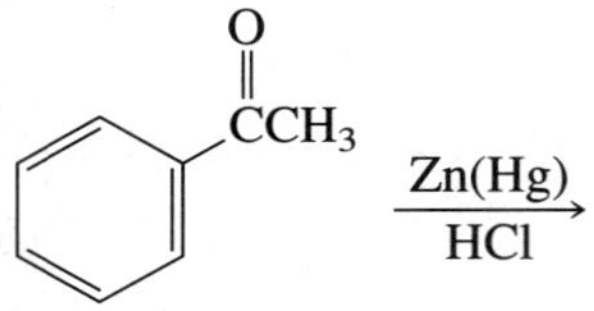

$\xrightarrow[\text{HCl}]{\text{Zn(Hg)}}$

c)

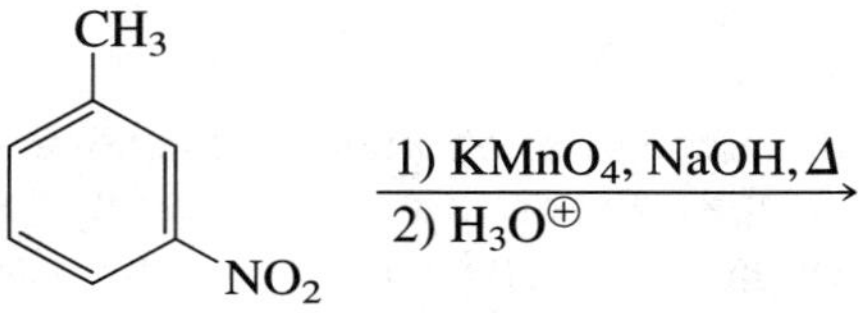

$$\xrightarrow[\text{2) } H_3O^{\oplus}]{\text{1) KMnO}_4,\ \text{NaOH},\ \Delta}$$

d)

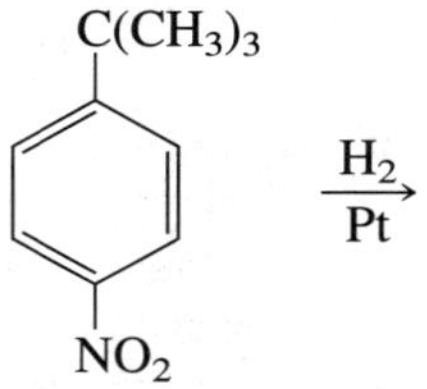

$\xrightarrow[\text{Pt}]{H_2}$

e)

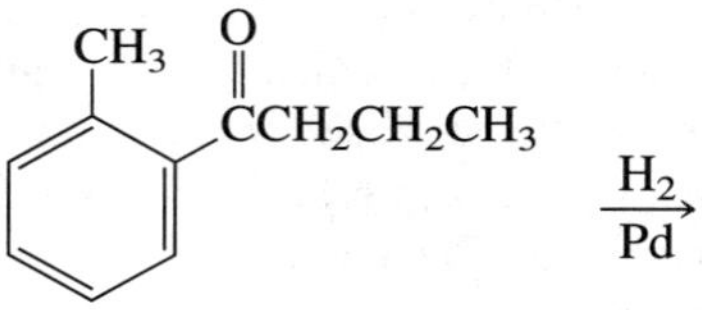

$\xrightarrow[\text{Pd}]{H_2}$

f)

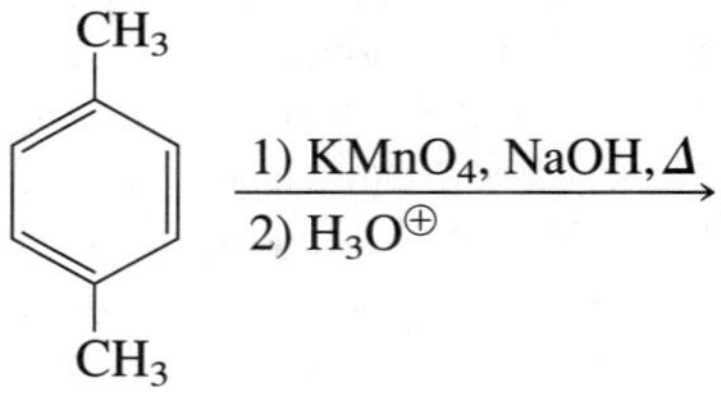

$$\xrightarrow[\text{2) } H_3O^{\oplus}]{\text{1) KMnO}_4,\ \text{NaOH},\ \Delta}$$

g)

$$\xrightarrow[\text{triethylene glycol}]{\substack{\text{NH}_2\text{NH}_2 \\ \text{KOH}}}$$

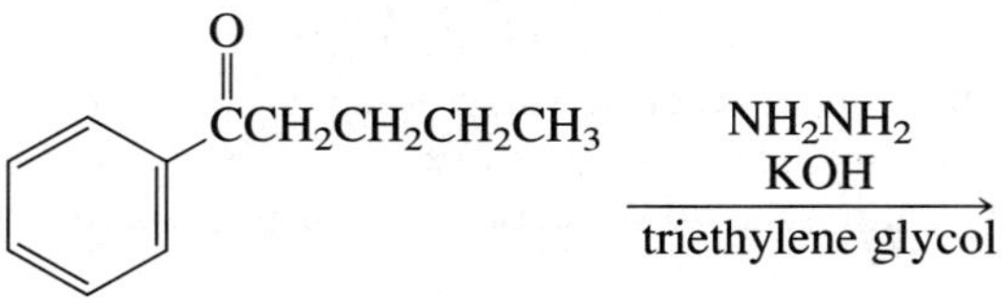

Problem 18.30 Show syntheses of these compounds from the indicated starting materials.

a)

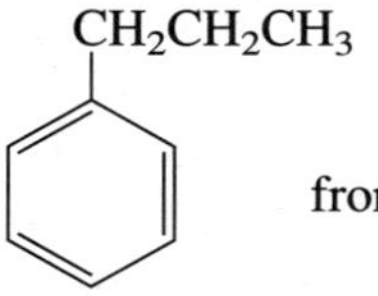

from benzene

b)

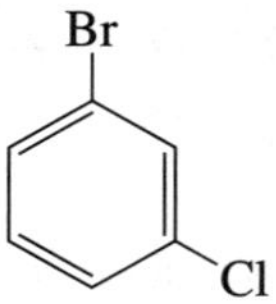

from *m*-chloronitrobenzene

18.15 Synthesis of Aromatic Compounds

The previous sections presented a powerful array of reactions that can be used to substitute almost any type of group onto an aromatic ring. Figure 18.7 summarizes the transformations that can be accomplished by using these reactions. Several

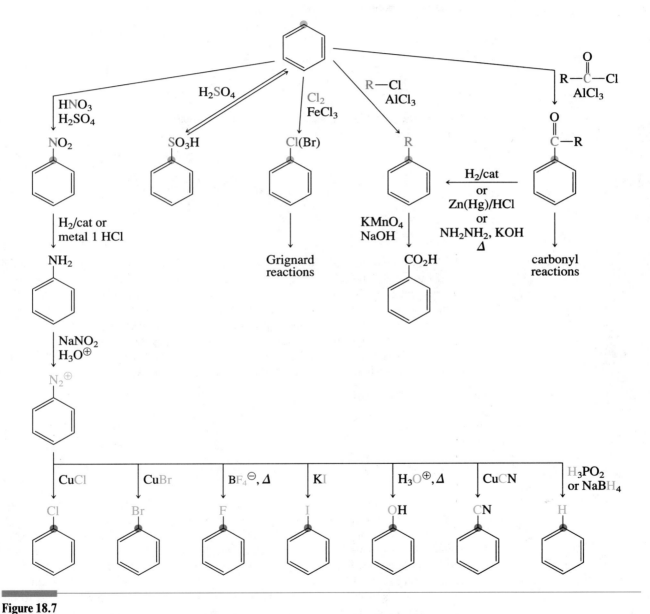

Figure 18.7

A Summary of Reactions Available for Use in the Synthesis of Aromatic Compounds.

syntheses of aromatic compounds using these reactions are presented in this section. This provides an excellent opportunity to develop and practice the strategy employed in synthesis of relatively complex compounds.

The first examples illustrate that the order of addition of the substituents is important in controlling their orientation. For example, suppose we needed to prepare *m*-chloronitrobenzene from benzene. Since the chlorine is an *ortho/para* director and the nitro group is a *meta* director, it is apparent that the nitro group must be added first if the *meta* product is desired:

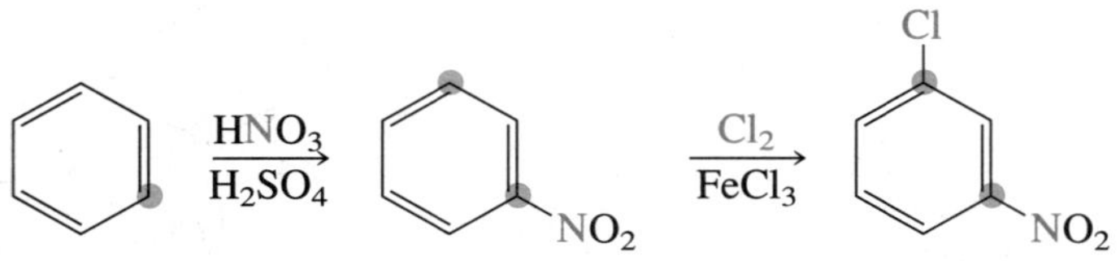

On the other hand, if the *para* isomer is desired, then the chlorine must be substituted onto the ring first:

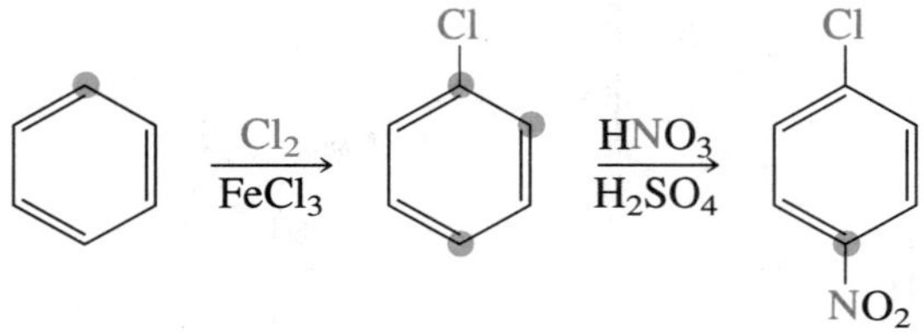

The *ortho* isomer of a disubstituted benzene is often difficult to prepare in good yield because of the steric effect that favors the *para* isomer. In such situations it is advantageous to place a sulfonic acid group at the *para* position, thus blocking this position from further reaction. After the desired group has been added to the *ortho* position, the sulfonic acid group can be removed by treatment with water and sulfuric acid. An example of the application of this strategy to the synthesis of *o*-bromophenol is shown in the following equation. In the first step of this synthesis the conditions are adjusted so as to introduce two sulfonic acid groups:

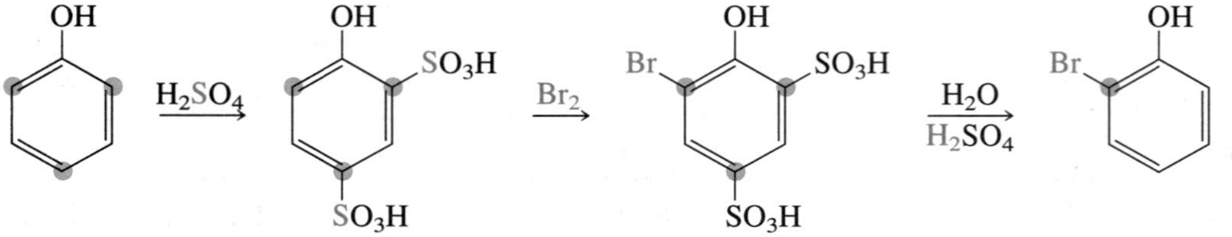

A nitro group or an amino group can be employed to direct an incoming group to the desired position. Then the nitro or amino group can be changed to a different group by converting it to the diazonium ion. This strategy is especially useful when neither of the groups in the final product will direct the other to the desired position. For example, suppose the synthetic target is *m*-bromochlorobenzene. Since both of the halogens are *ortho/para* directors, how can they be placed in a *meta* orientation? The solution to this problem is to use the nitro group to direct one of the halogens to the *meta* position and then change the nitro group into the other halogen. This synthesis is outlined in the following equation. A variation of this strategy can be employed to place two *meta* directing groups *para* to each other.

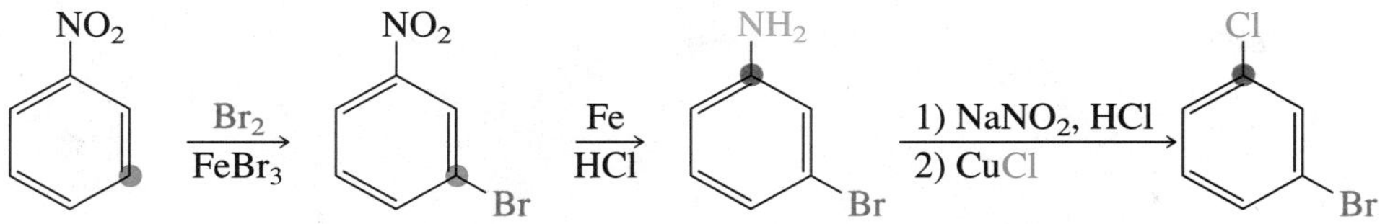

In another approach, an amino group can be used to obtain the desired regio-chemistry for the product. Then it can be removed via the diazonium ion. For example, suppose that it is desired to prepare *m*-bromotoluene starting from toluene. The difficulty is that the methyl group is an *ortho/para* director. The solution to this problem is to add an amino group *para* to the methyl group. This strong activating group can then be used to direct the bromine to the position *ortho* to itself and *meta* to the methyl group. The amino group is then removed. An application of this strategy is illustrated in the following example. Note that it is necessary to decrease the reactivity of the amine by converting it to the amide before the bromine is added:

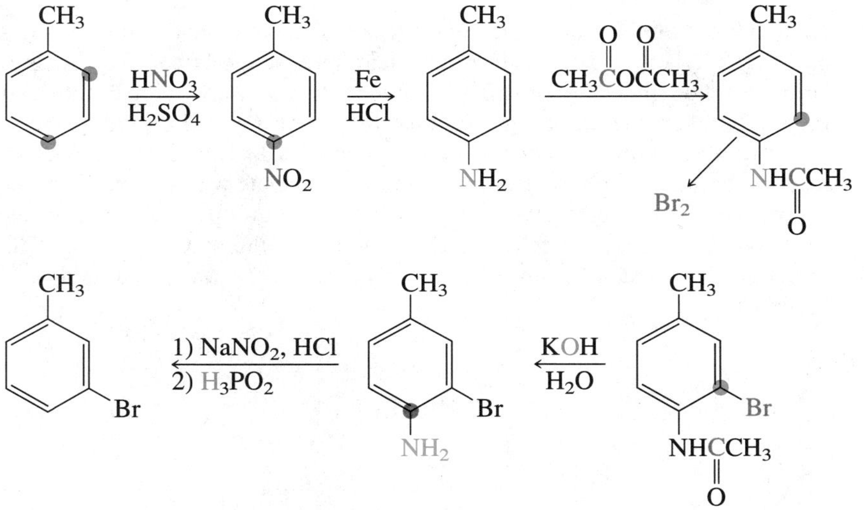

These examples show that a synthesis must be carefully planned. The regio-chemistry of each step must be considered as well as the compatibility of the substituents already on the ring with the reaction conditions. However, when completed, a cleverly crafted synthesis is a thing of beauty! Chemists often describe such a synthesis as elegant.

Problems

Problem 18.31 Show syntheses of these compounds from benzene.

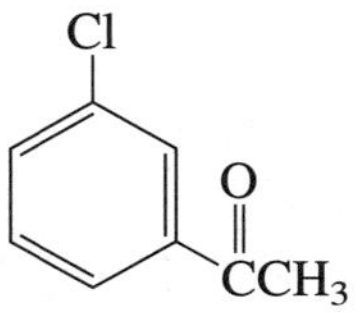

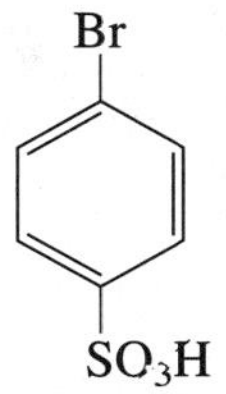

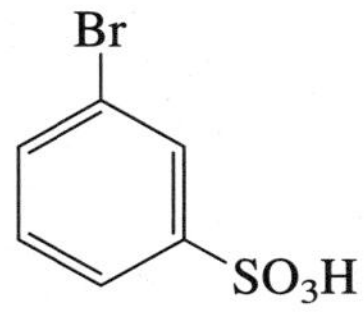

d) (aniline ring with NH$_2$ at top, Br at bottom)

e) (benzene ring with pentyl chain)

f) (ring with $\overset{O}{\overset{\|}{C}}CH_2CH_3$ at top, NO$_2$ at bottom)

g) (ring with CH$_2$CH$_2$CH$_2$CH$_3$ at top, SO$_3$H at bottom)

h) (ring with Cl and Cl, meta)

i) (ring with CN at top, NO$_2$ at bottom)

Solution

g)   Although the SO$_3$H group is a *meta* director, this compound cannot be prepared by alkylation of benzenesulfonic acid because the Friedel-Crafts alkylation reaction (and the acylation reaction) do not work with deactivated rings. Therefore the carbon group must be put on first. The primary butyl group cannot be directly added in good yield by a Friedel-Crafts alkylation reaction because of rearrangement of the intermediate primary butyl carbocation. The best way to attach the primary alkyl group is with a Friedel-Crafts acylation reaction followed by reduction of the carbonyl group. The SO$_3$H group is added before the reduction because the acyl group is a *meta* director.

Problem 18.32   Show syntheses of these compounds from benzene.

a) (ring with Br at top, CH$_2$CH$_2$CH$_2$CH$_2$CH$_3$ and NO$_2$ at bottom)

b) (ring with Cl at top, O$_2$N and NO$_2$)

c) (tetralone — bicyclic with ketone)

d) (ring with CO$_2$H at top, O$_2$N and NO$_2$, and NO$_2$ at bottom)

The mechanism then follows the same radical chain mechanism that is followed in other brominations with the exception that a radical initiator, such as AIBN or a peroxide, is usually employed to initiate the reaction. The initiation steps are shown in the following equations:

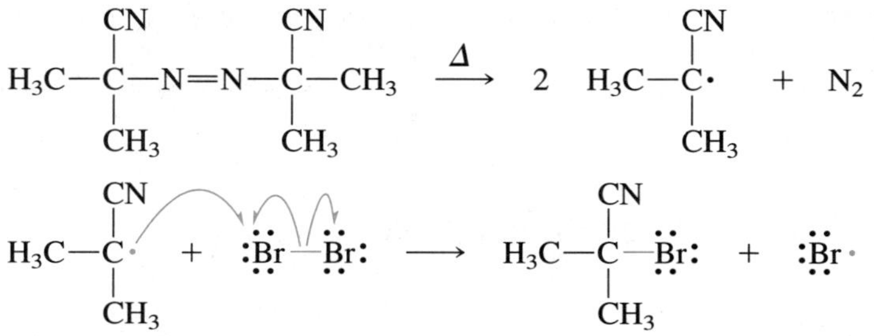

NBS has proved to be especially useful in brominations at allylic positions because competition from the addition of bromine to the double bond is not a problem. Apparently, the fact that only a low concentration of Br_2 is ever present in NBS brominations somehow inhibits the addition reaction. Examples are provided in the following equations. Note that the reaction is best if only a single type of allylic hydrogen is available to be abstracted. In addition, the resonance stabilized allylic radical provides two sites that can abstract a bromine atom. If these two sites are different, a mixture of products is formed, as shown in the second example:

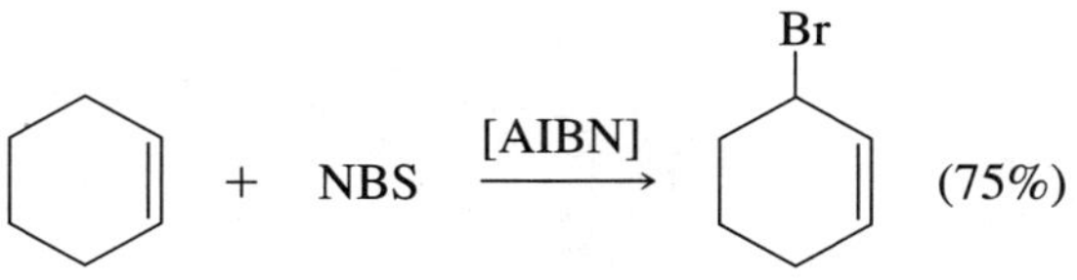

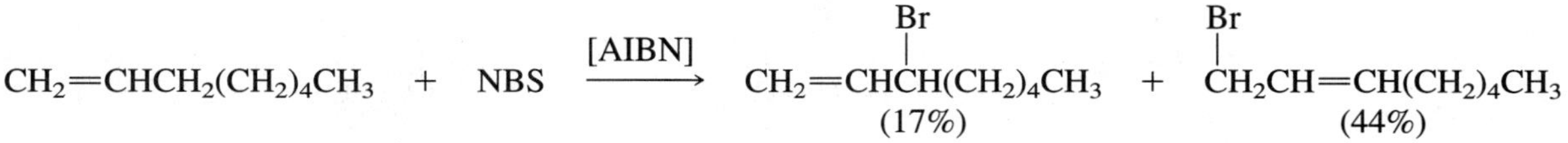

$$CH_2{=}CHCH_2(CH_2)_4CH_3 \ + \ NBS \ \xrightarrow{\text{[AIBN]}} \ CH_2{=}CHCH(CH_2)_4CH_3 \ + \ CH_2CH{=}CH(CH_2)_4CH_3$$

with Br substituents; (17%) and (44%)

Problems

Problem 19.11   Show the products of these reactions.

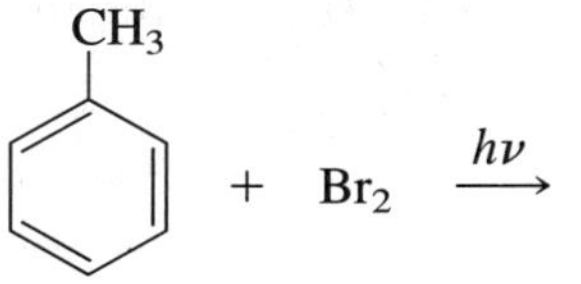

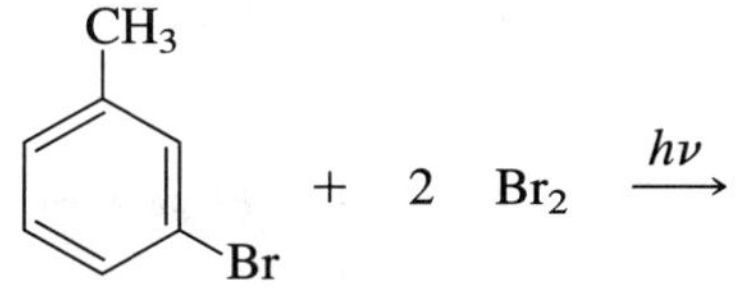

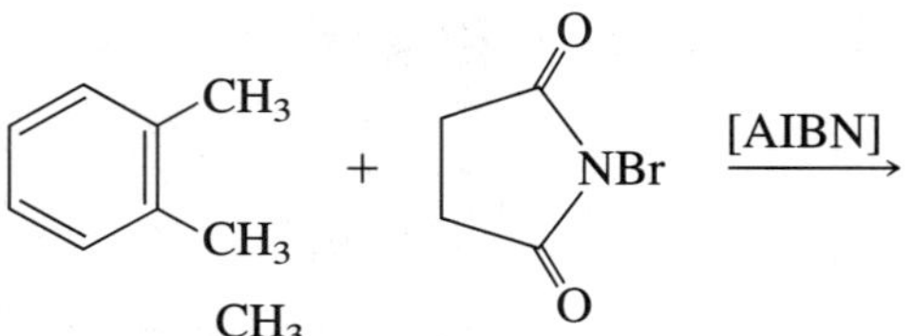

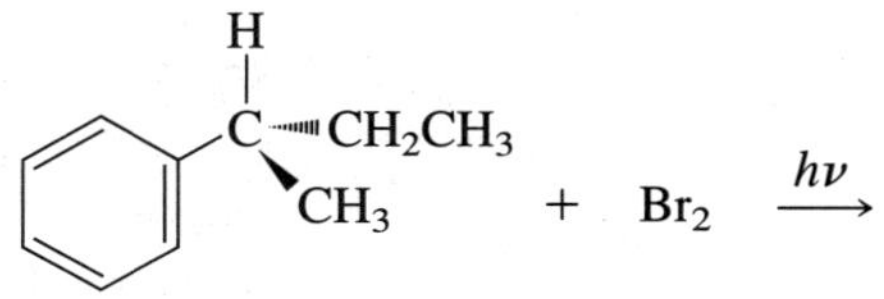

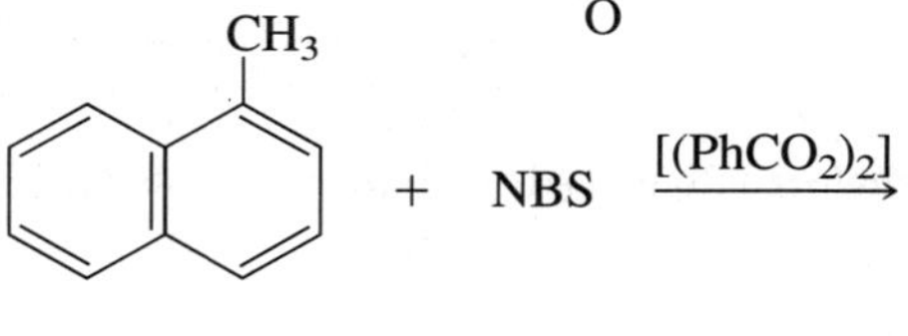

Problem 19.12 The following reaction produces 1-phenyl-3-bromopropene in good yield. What other product might be expected in this reaction? Explain why not much, if any, of this other product is formed.

$$PhCH\!=\!CHCH_3 \quad \xrightarrow[\text{[AIBN]}]{\text{NBS}} \quad PhCH\!=\!CHCH_2Br$$

19.9 Dehalogenation

The reaction of an alkyl halide with tributyltin hydride, using a radical initiator, results in the replacement of the halogen by hydrogen. The reaction follows a radical chain mechanism as outlined in Figure 19.3. Examples are provided in the following equations:

$$PhCH_2CH_2Br \;+\; Bu_3SnH \quad \xrightarrow{\text{[AIBN]}} \quad PhCH_2CH_3 \;+\; Bu_3SnBr \quad (85\%)$$

$$\underset{\displaystyle PhCCH_2Br}{\overset{\displaystyle O}{\parallel}} \;+\; Bu_3SnH \quad \xrightarrow{\text{[AIBN]}} \quad \underset{\displaystyle PhCCH_3}{\overset{\displaystyle O}{\parallel}} \;+\; Bu_3SnBr \quad (84\%)$$

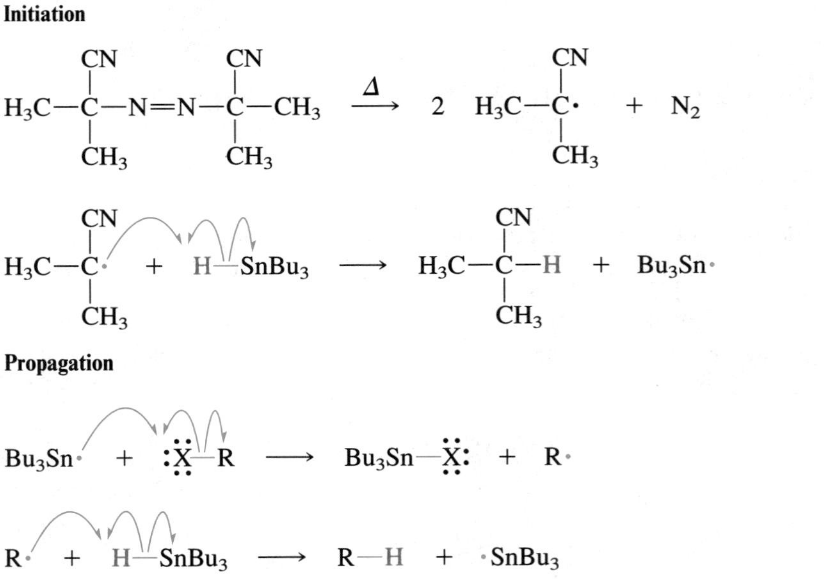

Figure 19.3

Mechanism for the Radical Chain Dehalogenation of Alkyl Halides by Tributyltin Hydride.

The various termination steps are not shown.

Problem

Problem 19.13   Show the products of these reactions.

a) [structure: cyclopentane with Br substituent] + Bu₃SnH $\xrightarrow{\text{[AIBN]}}$

b) [structure: benzene ring with CHCH₃ group bearing Cl] + Bu₃SnH $\xrightarrow{\text{[AIBN]}}$

19.10 Autoxidation

The slow oxidation of organic materials that are exposed to oxygen in the atmosphere is termed **autoxidation.** A simple example of this reaction is provided by the following equation:

[structure: tetralin + O₂ → tetralin with O—O—H group] $\xrightarrow{h\nu}$

This process follows a radical chain mechanism and is catalyzed by light. (This is one reason why compounds are often sold and stored in brown glass bottles.) In the first step of the mechanism an initiating radical (In·) is generated by light or some other means. This radical abstracts a hydrogen from the substrate to produce a carbon radical (R·):

$$R\!-\!H + \cdot In \longrightarrow R\cdot + H\!-\!In \qquad \text{initiation step}$$

In the propagation steps, the carbon radical first adds to the oxygen–oxygen double bond to produce a peroxide radical. This radical abstracts a hydrogen from another molecule of the substrate, generating a hydroperoxide and another carbon radical that can repeat the propagation cycle:

$$R\cdot + \ddot{:}\ddot{O}\!=\!\ddot{O}\ddot{:} \longrightarrow R\!-\!\ddot{O}\!-\!\ddot{O}\cdot \qquad \text{propagation steps}$$

$$R\!-\!\ddot{O}\!-\!\ddot{O}\cdot + H\!-\!R \longrightarrow R\!-\!\ddot{O}\!-\!\ddot{O}\!-\!H + R\cdot$$

$$R\!-\!H + \ddot{:}\ddot{O}\!=\!\ddot{O}\ddot{:} \longrightarrow R\!-\!\ddot{O}\!-\!\ddot{O}\!-\!H \qquad \text{overall reaction}$$

The autoxidation reaction is difficult to control, so it is not often used for synthetic purposes. However, it is a very important natural process. The slow deterioration of organic materials, such as rubber, paint, and oils, and that of many foods, such as butter and fats, is due to autoxidation. As one example, peroxides are formed in solvents such as diethyl ether or THF that are stored for long periods of time in contact with air:

$$CH_3CH_2OCH_2CH_3 \ + \ O_2 \ \longrightarrow \ \overset{\displaystyle OOH}{\overset{|}{CH_3CHOCH_2CH_3}}$$

Because peroxides are explosive, the use of such contaminated solvents leads to a very dangerous situation. Numerous explosions have resulted in the laboratory when peroxides have been concentrated as the solvent is removed by distillation during the workup of a reaction. Therefore it is important to be certain that ether and THF are free of peroxides before they are employed as solvents and solutions of ether, and THF should never be distilled to dryness.

The Industrial Preparation of Phenol

Phenol is used to make a variety of plastics and is an important industrial chemical. However, as was discussed in Chapter 18, a hydroxy group cannot be added to a benzene ring by electrophilic aromatic substitution. Therefore some indirect method must be used. The major industrial process that is used to prepare phenol is outlined in the following equations:

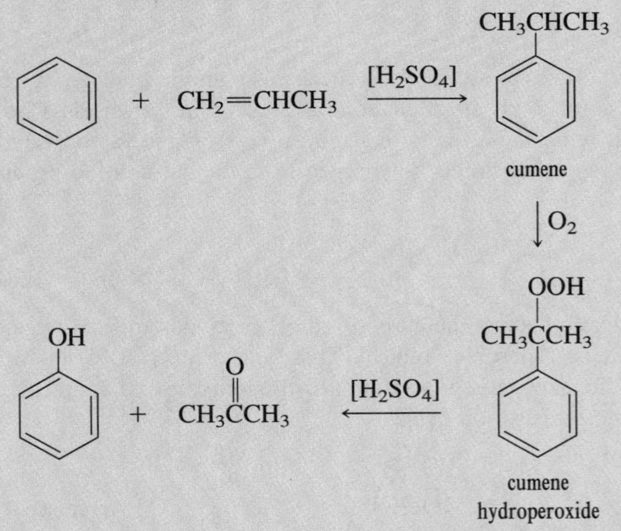

First, an isopropyl group is added to a benzene ring by a Friedel-Crafts alkylation to produce cumene. (Who would have thought that the best way to get a hydroxy group on the ring is to first add an isopropyl group?) Then autoxidation of cumene is employed to produce cumene hydroperoxide. The final step of the synthesis is the acid-catalyzed rearrangement of cumene hydroperoxide to phenol and acetone. This process is commercially feasible only because the acetone byproduct can also be sold.

Although we have not seen the mechanism for the rearrangement of cumene hydroperoxide, it has some similarities to a carbocation rearrangement. First, the oxygen of the hydroperoxide is protonated. Then water acts as a leaving group.

Since the formation of a positively charged oxygen with only six electrons in its valence shell is extremely unfavorable, rearrangement of the phenyl group to produce a resonance-stabilized carbocation occurs in a concerted manner as the water leaves. Attack of a water nucleophile, followed by the loss of a proton to some base in the reaction mixture, produces the hemiacetal of acetone and phenol. As discussed in Chapter 14, hemiacetals are unstable so hydrolysis to phenol and acetone occurs readily.

a hemiacetal

Problem

Problem 19.14 Show the mechanism for the oxidation of cumene to cumene hydroperoxide and explain why this reaction is more selective than most autoxidations.

Antioxidants, such as 2,6-di-*tert*-butyl-4-methylphenol (also known as butylated hydroxytoluene or BHT) and 2-*tert*-butyl-4-methoxyphenol (also known as butylated hydroxyanisole or BHA) are added to many organic materials to prevent autoxidation. They function by interfering with the autoxidation chain reaction. When a radical encounters an antioxidant molecule, such as BHA, it abstracts a hydrogen to produce a resonance-stabilized radical:

BHT   BHA

Because of this resonance stabilization and the steric hindrance provided by the bulky *tert*-butyl group, this radical is not very reactive. It acts as a chain terminator because it is not reactive enough to abstract a hydrogen or add to an oxygen–oxygen double bond and continue the autoxidation chain. The presence of a single molecule of BHA

Vitamin E and Lipid Autoxidation

Glycerophospholipids are major components of biological membranes. They resemble the fats described in the elaboration "Preparation of Soaps" on page 708 in that they are composed of esters of the triol glycerol with carboxylic acids (fatty acids) that have long hydrocarbon chains. In contrast to fats, however, glycerophospholipids have only two fatty acid groups attached to the glycerol. The third site is occupied by an ionic or highly polar group attached to the glycerol by an ester linkage. The hydrocarbon chains contribute hydrophobic character to the lipid, while the polar groups are hydrophilic. This combination of properties enables glycerophospholipids to form the bilayer membranes that are so important in biological systems.

The hydrocarbon chains of fatty acids often contain one or more *cis* double bonds. These unsaturated lipids are especially prone to autoxidation. An example is provided by the following equation, which shows an autoxidation reaction of the ester of α-linoleic acid. Note that this is a chain reaction, so one initiator can cause numerous lipid molecules to be oxidized.

Some radical initiates the chain by abstracting a hydrogen atom. The allylic hydrogens of unsaturated lipids are especially susceptible to this process.

The resulting radical adds to an oxygen molecule.

another unsaturated lipid

+

another lipid radical

The new lipid radical continues the chain process.

This peroxy radical abstracts a hydrogen from a neighboring lipid molecule.

or BHT can prevent the oxidation of thousands of other molecules by terminating a chain. Therefore only a small amount of an antioxidant need be added to a compound to provide protection against autoxidation. For example, the addition of a small amount of BHA to butter increases its storage lifetime from a few months to a few years.

The peroxidized tails of these lipids are more hydrophilic and try to migrate to the surface of the membrane. This disrupts the structure of the membrane and makes it "leaky." Similar peroxidation of unsaturated lipids in low-density lipoproteins is thought to contribute to arteriosclerosis and heart disease.

Nature has many defenses against unwanted oxidation reactions. Vitamin E, also known as α-tocopherol, helps to serve this role in human membranes. It is a hindered phenol, somewhat resembling BHT and BHA, and it is soluble in the membrane because of its nonpolar hydrocarbon tail. The radical that is produced when the hydrogen of its hydroxy group is abstracted is not very reactive, so vitamin E terminates autoxidation chains:

vitamin E

· OOR

+ ROOH

Radicals may have a role in a number of health problems, including those mentioned previously, along with cancer and the aging process. This has recently led to widespread use of antioxidant nutritional supplements. Unfortunately, little has yet been proven about the effectiveness of such supplements. A healthy lifestyle, including a good diet, will probably be more beneficial.

Problem

Problem 19.15 Explain why the radical produced from vitamin E is not very reactive and thus acts as a chain terminator.

Problem

Problem 19.16   Predict the major product from autoxidation of these compounds.

a)

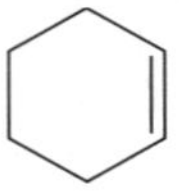

b)

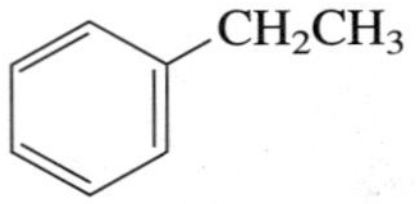

CH_2CH_3

19.11   Radical Additions to Alkenes

In Chapter 10 the electrophilic addition of hydrogen bromide to alkenes, which proceeds by an ionic mechanism, was discussed. Recall that this reaction follows Markovnikov's rule and the bromine adds to the more highly substituted carbon, as illustrated in the following example:

$$CH_3CH_2CH_2CH_2CH{=}CH_2 \ + \ HBr \ \longrightarrow \ CH_3CH_2CH_2CH_2\overset{\overset{\displaystyle Br}{|}}{C}HCH_3 \quad (88\%)$$

When a similar reaction occurs under conditions favoring the formation of radicals—that is, in the presence of light or a peroxide that can initiate the reaction—the addition still occurs, but with the opposite regiochemistry. The bromine adds to the less highly substituted carbon, and the addition is said to occur in an **anti-Markovnikov** manner. Examples are provided by the following equations:

$$CH_3CH_2CH{=}CH_2 \ + \ HBr \ \xrightarrow[\substack{or \\ h\nu}]{[ROOR]} \ CH_3CH_2CH_2\overset{\overset{\displaystyle Br}{|}}{C}H_2 \quad (95\%)$$

$$PhCH_2CH{=}CH_2 \ + \ H{-}Br \ \xrightarrow{[ROOR]} \ PhCH_2CH_2CH_2{-}Br \quad (80\%)$$

The change in regiochemistry is a result of a change in the mechanism of the reaction, from an ionic mechanism in the Markovnikov reaction to a radical chain mechanism in the anti-Markovnikov reaction. The radical chain mechanism for the addition of hydrogen bromide to 1-butene is outlined in the following equations:

Initiation:

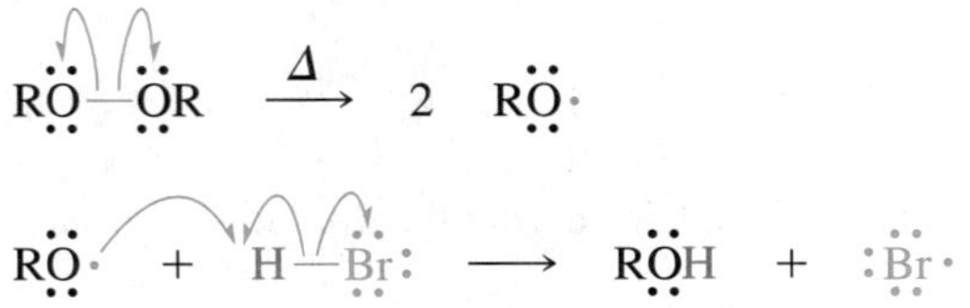

$$R\ddot{O}{-}\ddot{O}R \ \xrightarrow{\Delta} \ 2 \ R\ddot{O}\cdot$$

$$R\ddot{O}\cdot \ + \ H{-}\ddot{B}r\colon \ \longrightarrow \ R\ddot{O}H \ + \ \colon\!\ddot{B}r\cdot$$

arranged symmetrically about zero energy, so there are three bonding MOs and three antibonding MOs. The lowest-energy MO has zero nodes, the next has one node, and so on. There are six electrons in these pi MOs. The lowest-energy arrangement of electrons, called the **ground state,** has these six electrons in the three bonding pi MOs. (Any other arrangement of electrons is higher in energy and is termed an **excited state.**) For the ground state of 1,3,5-hexatriene the HOMO is π_3 and the LUMO is π_4^*.

As described in the discussion of UV-visible spectroscopy in Section 13.2, the absorption of a photon of light by a compound causes an electron to be excited from an occupied MO to an unoccupied MO. The lowest-energy excitation occurs when an electron in the HOMO is excited to the LUMO. In the case of 1,3,5-hexatriene an electron in π_3 is promoted to π_4^* to form the lowest-energy excited state (see Figure 20.3). In the excited state, the HOMO is π_4^* and the LUMO is π_5^*. It is this excited state that reacts in a photochemical reaction.

Conjugated carbocations, carbanions, and radicals that have an odd number of orbitals are also important. The same rules are used to construct the MOs for these odd orbital systems. As a simple example, consider the allyl radical, shown in Figure 20.4. The three p AOs result in the formation of three pi MOs. In order for these orbitals to be symmetrically placed about zero energy, one must occur at zero energy.

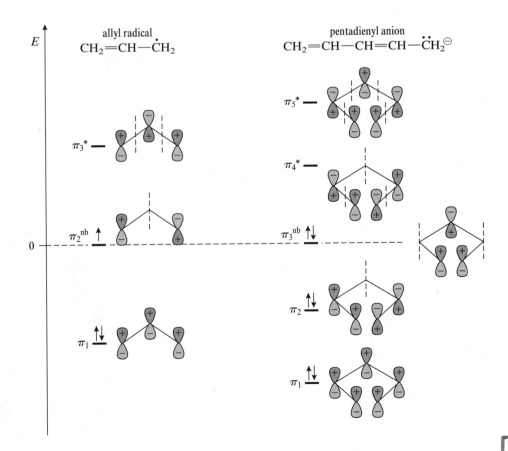

Figure 20.4

Energies and Nodal Properties for the Pi MOs of the Allyl Radical and the Pentadienyl Anion.

Therefore the allyl radical has a bonding MO, a nonbonding MO at zero energy, and an antibonding MO. The presence of a nonbonding MO is a characteristic of all odd orbital systems.

Another characteristic feature of odd orbital systems is nodes that pass through atoms. For example, π_2^{nb} of the allyl radical has one node. For this node to be symmetrically placed in the radical, it must pass through C-2, as shown in Figure 20.4. This means that the p orbital on C-2 does not contribute to the nonbonding MO at all. The p orbitals on C-1 and C-3 are too far apart to interact, so this MO has the same energy as an isolated p orbital and is nonbonding.

There are three electrons in the allyl radical, so two occupy π_1 and one occupies the nonbonding MO. The allyl cation and the allyl anion have exactly the same MOs. The cation, however, has only two electrons, which occupy π_1, while the anion has four electrons, which fill both the bonding MO and the nonbonding MO.

Figure 20.4 also shows the MOs for the pentadienyl anion. The five p orbitals of this carbanion combine to form five MOs: two bonding MOs, two antibonding MOs, and one nonbonding MO. The lowest-energy MO, π_1, has no nodes. The next MO, π_2, has one node that passes through the center of the carbanion at C-3. The nonbonding MO, π_3^{nb}, has two nodes, at C-2 and C-4. The next MO, π_4^*, has three nodes, one of which passes through the center of the molecule at C-3. Finally, π_5^* has four nodes. There are six electrons in this carbanion, so the nonbonding MO is occupied by a pair of electrons.

In general, these odd orbital systems all have a nonbonding MO that has nodes passing through the even-numbered carbons. A cation has no electrons in the nonbonding MO, the radical has one electron in this MO, and the anion has two electrons occupying this MO.

Problems

Problem 20.2 Show the energies of the pi MOs and the arrangement of electrons in them for these species.

a) ground state of $CH_2{=}CH{-}CH{=}CH{-}CH{=}CH{-}CH{=}CH_2$

b) ground state of $CH_2{=}CH{-}CH{=}CH{-}CH{=}CH{-}\overset{\oplus}{C}H_2$

c) lowest-energy excited state of $CH_2{=}CH{-}CH{=}CH_2$

d) ground state of $CH_2{=}CH{-}CH{=}CH{-}\overset{\bullet}{C}H_2$

e) lowest-energy excited state of $CH_2{=}CH{-}CH{=}CH{-}CH{=}CH{-}\overset{\overset{\displaystyle ..}{\ominus}}{C}H_2$

Problem 20.3 Show the nodal properties of these MOs.

a) LUMO of the ground state of $\oplus$

b) π_2 of

c) LUMO of lowest-energy excited state of

d) $\pi_4{}^{nb}$ of

e) HOMO of the ground state of

To determine whether a pericyclic reaction is favorable, we need to evaluate how the total energy of the electrons changes during the course of the reaction. As the reaction occurs, the MOs of the reacting molecule are converted into the MOs of the product. If this conversion is energetically favorable—that is, if the electrons in the occupied MOs do not increase in energy—then the reaction is likely to occur and is said to be *allowed*. If the electrons in the occupied MOs increase in energy, the reaction is unfavorable and is said to be *forbidden*. Therefore we need to evaluate what happens to the energies of the occupied MOs as the reaction proceeds.

The method developed by Woodward and Hoffmann uses the symmetry of the placement of the nodes in the MOs to determine how the reactant MOs are converted to the product MOs and how the total electron energy changes during the course of the reaction. Fukui's method concentrates on the so-called **frontier MOs**, the HOMO and the LUMO, because the energy changes of these are the key to whether the overall energy change is favorable. This method examines how the orbitals of the HOMO, or in some cases the orbitals of the HOMO of one component and the LUMO of the other, overlap to form the new bonds. If the new overlaps are favorable (bonding overlaps), then the reaction is allowed. And if the new overlaps are unfavorable (antibonding overlaps), then the reaction is forbidden. Fukui's method (the **frontier orbital method**) is a little simpler, so it is used in this book. The examples presented in the following sections illustrate how this method is applied.

20.4 Electrocyclic Reactions

Pericyclic reactions are commonly divided into three classes: electrocyclic reactions, cycloaddition reactions, and sigmatropic rearrangements. An **electrocyclic reaction** forms a sigma bond between the end atoms of a series of conjugated pi bonds within a molecule. The 1,3-butadiene to cyclobutene conversion is an example, as is the similar reaction of 1,3,5-hexatriene to form 1,3-cyclohexadiene:

As can be seen from these examples, the product has one more sigma bond and one less pi bond than the reactant.

Let's begin by considering the simplest electrocyclic reaction, the thermally induced interconversion of a diene and a cyclobutene. As illustrated in the following example, the reaction is remarkably stereospecific, occurring only by a conrotatory motion:

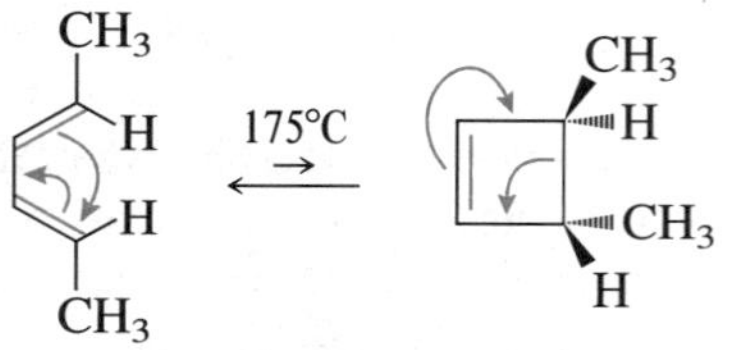

In electrocyclic reactions the end carbons of the conjugated system must rotate for the *p* orbitals on these carbons to begin to overlap to form the new carbon–carbon sigma bond. The preference for the stereochemistry of the rotation in these reactions can be understood by examination of the new orbital overlap in the HOMO as the rotation occurs. For the formation of the new sigma bond to be favorable, rotation must occur so that the overlap of the orbitals forming this bond is bonding in the HOMO.

For the ground-state reaction of a conjugated diene the HOMO is π_2. Conrotation of this MO causes the plus lobe of the *p* orbital on one end of the pi system to overlap with the plus lobe of the *p* orbital on the other end of the pi system:

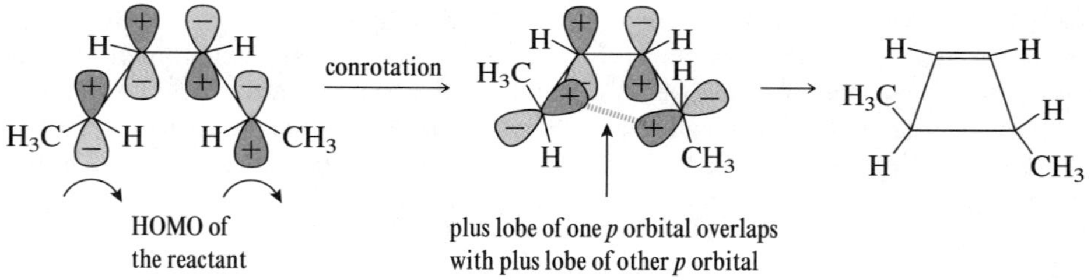

| HOMO of the reactant | plus lobe of one *p* orbital overlaps with plus lobe of other *p* orbital |

Conrotation of a diene is thermally allowed.

This bonding overlap in the HOMO when the rotation occurs makes the formation of the new sigma bond favorable. The conversion of a diene to a cyclobutene by a conrotatory motion is *thermally allowed*.

In contrast, disrotation causes the plus lobe of the *p* orbital on one end of the pi system of the diene to overlap with the minus lobe of the *p* orbital on the other end in the HOMO:

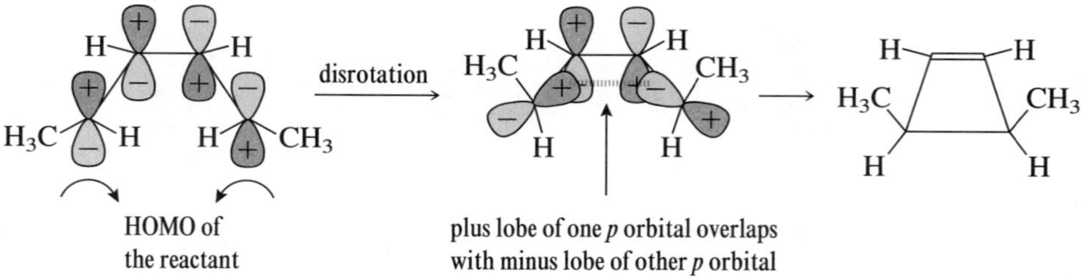

| HOMO of the reactant | plus lobe of one *p* orbital overlaps with minus lobe of other *p* orbital |

Disrotation of a diene is thermally forbidden.

The antibonding overlap in the HOMO when disrotation occurs makes the formation of the new sigma bond unfavorable. The disrotatory closure of a diene to a cyclobutene is *thermally forbidden*.

The requirement for a favorable bonding overlap of the orbitals forming the new sigma bond in the HOMO allows the preferred rotation to be predicted for any electrocyclic reaction. Let's consider the photochemical conversion of a butadiene to a cyclobutene. Since the reaction occurs through the excited state, the HOMO is π_3^*. As can be seen, disrotation is necessary here for the new overlap to be bonding:

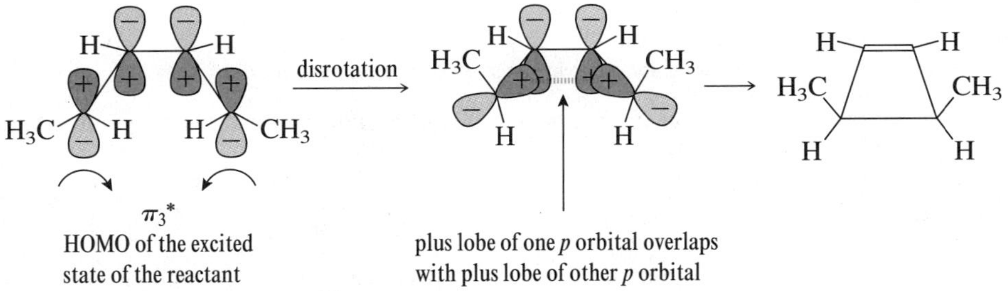

π_3^*
HOMO of the excited
state of the reactant

plus lobe of one *p* orbital overlaps
with plus lobe of other *p* orbital

Disrotation of a diene is photochemically allowed.

In accord with this analysis, numerous experiments have shown that the disrotatory closure of a diene to a cyclobutene is indeed the pathway that occurs when the compound is irradiated with UV light.

Problem

Problem 20.4   Use orbital drawings to show that conrotation of a diene is photochemically forbidden.

A similar analysis also correctly predicts the stereochemistry of the formation of a cyclohexadiene from a triene. For the thermal reaction the HOMO is π_3. Examination of this MO shows that disrotation is necessary for the overlap forming the new sigma bond to be bonding:

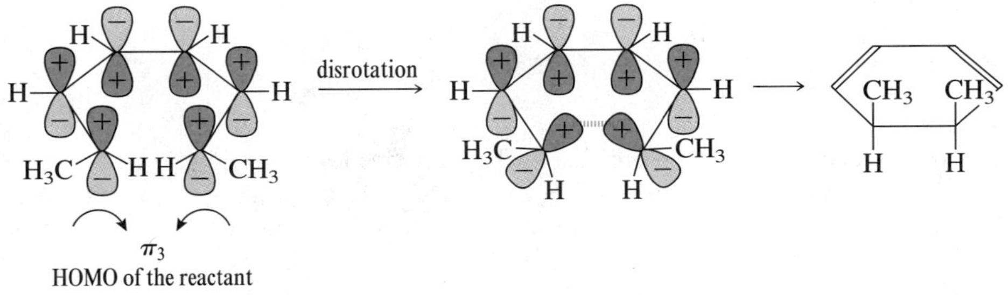

π_3
HOMO of the reactant

Disrotation of a triene is thermally allowed.

For the excited state reaction the HOMO is π_4^*. Here conrotation is the favored pathway:

Conrotation of a triene is photochemically allowed.

Both of these conclusions are in accord with the experimental results presented in the next section.

Problems

Problems 20.5   Use orbital drawings to show that conrotation of a triene is thermally forbidden and that disrotation is photochemically forbidden.

Problems 20.6   Use orbital drawings to determine whether these reactions are allowed or forbidden.

Solution

a)   This anion has five p orbitals, so there are five MOs (see Figure 20.4). There are six electrons in these MOs, so the HOMO for the thermal (ground state) reaction is π_3^{nb}. Conrotation is necessary to form the product with the methyl groups *trans*.

Because the overlap to form the new bond is antibonding, this reaction is thermally forbidden.

It is possible to generalize the preferences for conrotation or disrotation based on the number of electron pairs in the pi MOs of the reacting molecule. A diene, with two pi electron pairs (two pi bonds), has π_2 as its HOMO in the ground state. Since this MO has one node, conrotation is favored for the thermal reaction. A triene, with three pi electron pairs (three pi bonds), has π_3 as its HOMO. This MO has two nodes (one more than π_2 of a diene), so the opposite rotation, disrotation, is preferred. It is apparent that the favored rotation alternates with the number of pi electron pairs. Therefore a molecule with four pi bonds (π_4 is the HOMO) prefers conrotation. Furthermore, the preference for the excited state is just reversed from that of the ground state because the excited-state HOMO always has one more node than the ground-state HOMO. These preferences are summarized in the following chart:

Number of Electron Pairs	Disrotation	Conrotation
odd	thermally allowed	photochemically allowed
even	photochemically allowed	thermally allowed

(Note that an odd number of electron pairs equals $4n + 2$ electrons, whereas an even number of electron pairs equals $4n$ electrons, where n is any integer, including zero.)

Problems

Problem 20.7   Use the chart on this page to determine whether the reactions of problem 20.6 are allowed or forbidden.

Problem 20.8   Indicate the number of electron pairs and the type of rotation for these reactions and determine whether each is allowed or forbidden.

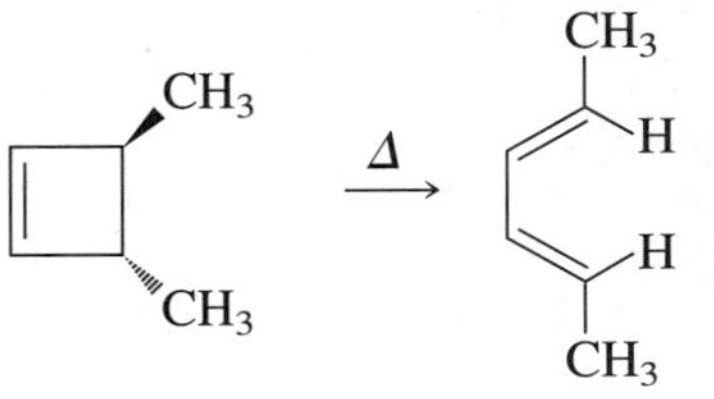

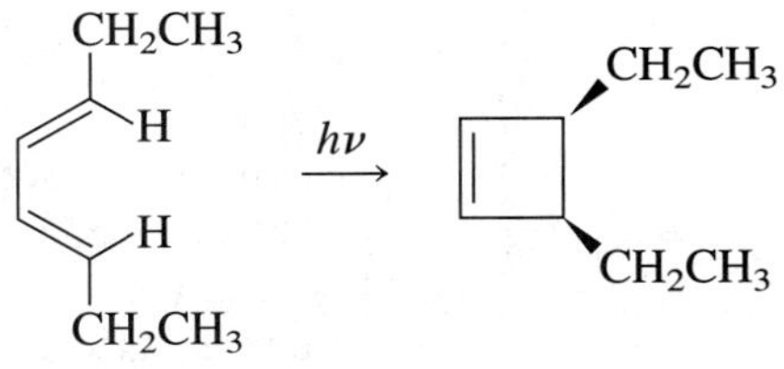

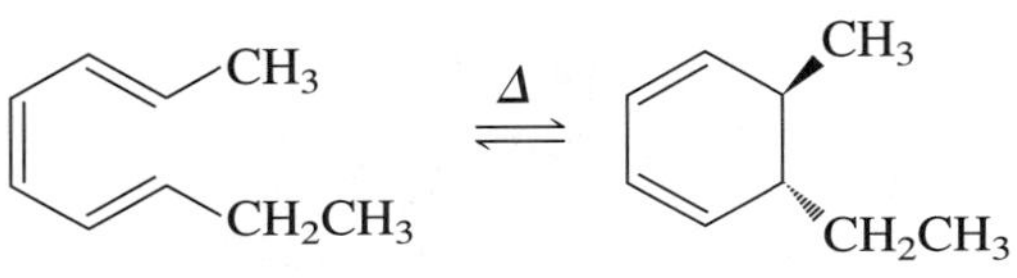

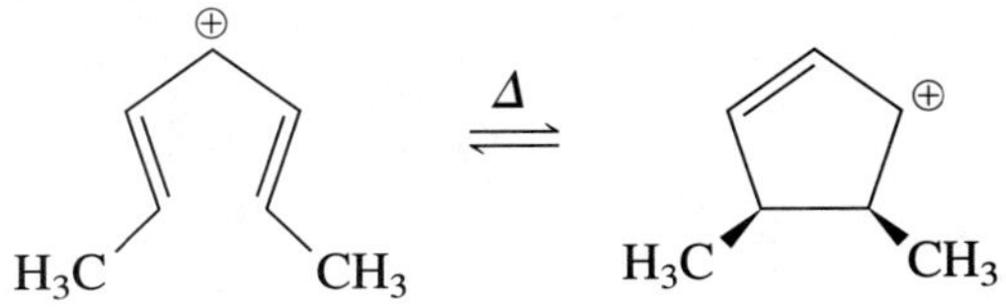

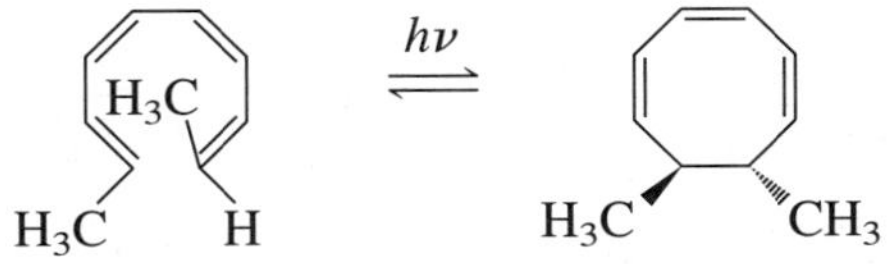

Solution

a)   We have usually analyzed this type of reaction from the other direction—
that is, the conversion of a diene to a cyclobutene. However, the same analysis works for either direction of a reaction. In this case an even number (two)
of electron pairs are involved; the sigma bond and the pi bond of the reactant are converted to the two pi bonds of the product. The reaction occurs by
a conrotation:

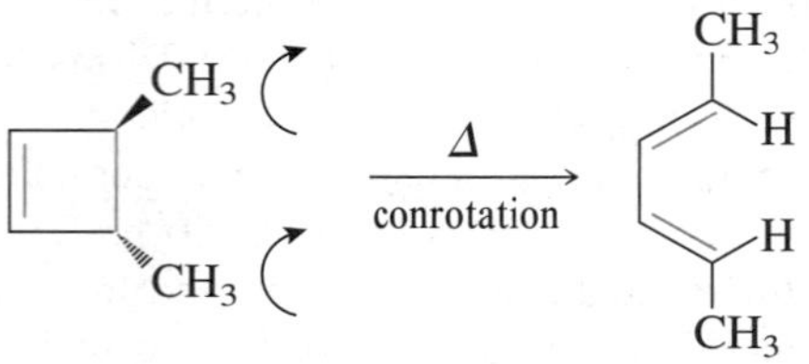

According to the chart, a conrotation involving an even number of electron
pairs is thermally allowed. Therefore the reaction shown is allowed. Note that
the same reaction is forbidden under photochemical conditions.

20.5   Examples of Electrocyclic Reactions

The analysis in Section 20.4 indicates that the thermal interconversion of a diene
with a cyclobutene should occur by conrotation. The reaction is allowed in both directions, as long as a conrotatory motion is followed. However, usually only the conversion of the cyclobutene to the diene is observed because the cyclobutene is destabilized by angle strain and is present only in trace amounts at equilibrium. An example
of the opening of a cyclobutene to form a diene is provided by the following equation:

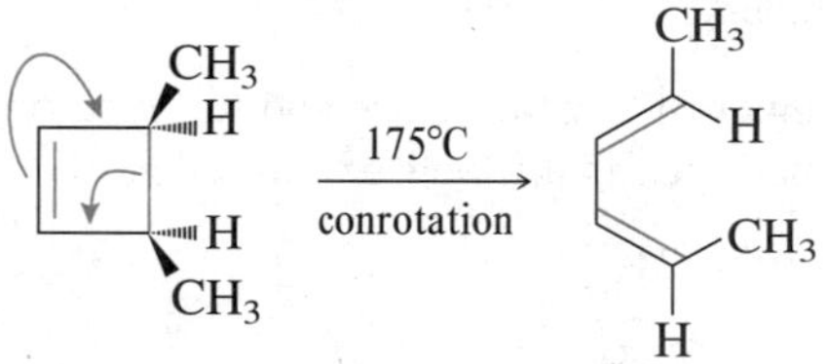

The reverse process, the conversion of a diene to a cyclobutene, can be accomplished photochemically. Although the cyclobutene is less stable, it is possible to selectively excite the diene because it absorbs longer-wavelength light (see Section 13.3).
An example is shown in the following equation:

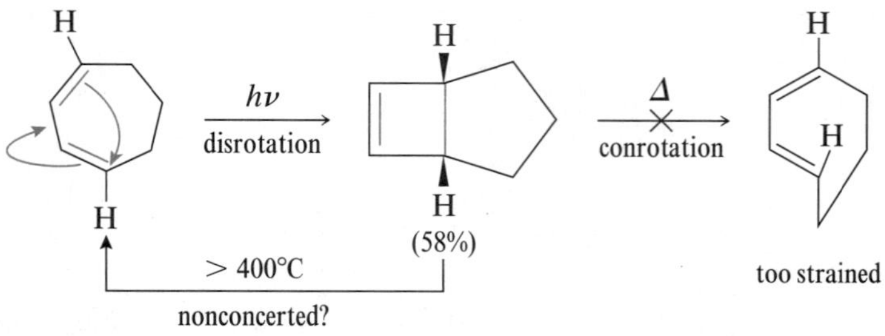

In this reaction, light of appropriate energy is used to selectively excite 1,3-cyclo-heptadiene. The diene closes to a cyclobutene by a disrotatory motion. Although the product, because of its strained cyclobutene ring, is much less stable than the reactant, it is unable to revert back to the diene by an allowed pathway. It does not absorb the light used in the reaction, so the photochemically allowed disrotatory pathway is not available. A conrotatory opening is thermally allowed but results in a cyclohepta-diene with a *trans*-double bond. Such a compound is much too strained to form. Therefore the product can be readily isolated. If it is heated to a high enough temperature, greater than 400°C, the strained ring does break to produce 1,3-cycloheptadiene. This reaction might be occurring by the forbidden disrotatory pathway, or, more likely, it may involve a nonconcerted mechanism.

Exercise

Exercise 20.2 Build a model of the following compound. Then replace the connector for the bond that is part of both rings with two separate connectors. Do a conrotation and a disrotation to see the strain that is incurred in each process.

Dewar Benzene

In the mid-1800s, chemists were struggling to determine the structure of benzene. One of the structures that was proposed at that time is known as Dewar benzene, after the chemist who suggested it:

Dewar benzene

As chemists learned more about the effects of structure on the stability of organic compounds, it became apparent that Dewar benzene is much less stable than benzene. Not only does it have a considerable amount of angle strain, but it also has none of the stabilization due to aromaticity that benzene has. Because of these factors, Dewar benzene is 71 kcal/mol (297 kJ/mol) less stable than benzene. Because the conversion of Dewar benzene to benzene is so exothermic and involves an apparently simple electron reorganization, many chemists believed that the isolation of this strained isomer would prove to be impossible. They thought that if it were prepared, it would rapidly convert to benzene. In support of this view, numerous attempts to synthesize Dewar benzene met with failure.

Once the theory of pericyclic reactions was developed, it was recognized that the conversion of Dewar benzene to benzene is an electrocyclic reaction. This conversion involves two pairs of electrons: one pair of pi electrons and one pair of sigma electrons of the Dewar benzene. (The third pair of electrons is located in exactly the same place in both the reactant and the product and so is not involved in the reaction.) An electrocyclic reaction involving two pairs of electrons must occur by a conrotatory motion if it is to be thermally allowed. However, the conrotatory opening of Dewar benzene is geometrically impossible, since it would result in a benzene with a *trans*-double bond, a compound with too much angle strain to exist.

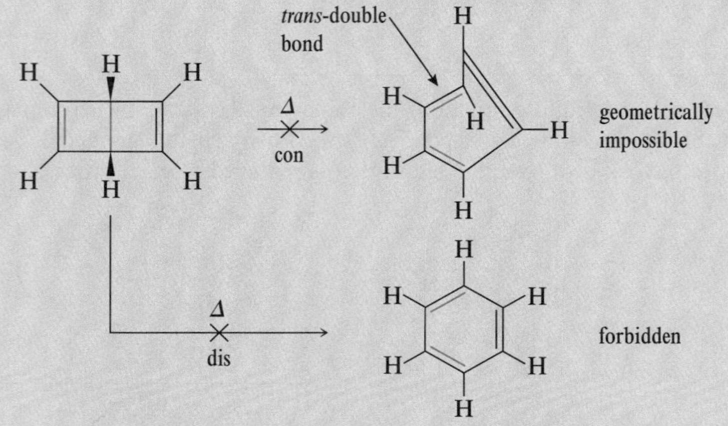

Disrotatory opening of Dewar benzene, which would produce benzene, is thermally forbidden. Therefore even though the conversion of Dewar benzene to benzene is quite exothermic, it might indeed prove possible to isolate Dewar benzene because there is no low-energy pathway for its conversion to benzene.

The first derivative of Dewar benzene was prepared in 1962 by irradiation of 1,2,4-tri-*t*-butylbenzene:

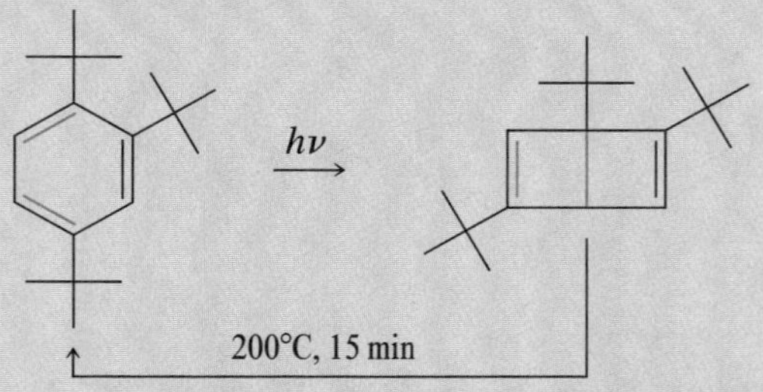

Disrotatory closure of the substituted benzene to produce a Dewar benzene is photochemically allowed, as is, of course, the reverse process. However, because benzene is conjugated, it absorbs UV light at longer wavelengths than the Dewar benzene isomer. Therefore it is possible to selectively excite the benzene chromophore and produce the less stable Dewar isomer. In this particular case the *tert*-butyl groups favor the reaction because they destabilize the benzene isomer somewhat, owing to steric hindrance. Since the two adjacent *tert*-butyl groups in the Dewar isomer do not lie in the same plane, this steric strain is decreased in the product. Because of this steric effect and the forbidden nature of the conversion back to benzene, the Dewar isomer is relatively stable. However, when it is heated to 200°C, it is rapidly converted to the benzene isomer, probably by a nonconcerted pathway.

Shortly after this, in 1963, the parent Dewar benzene was prepared by the following reaction sequence:

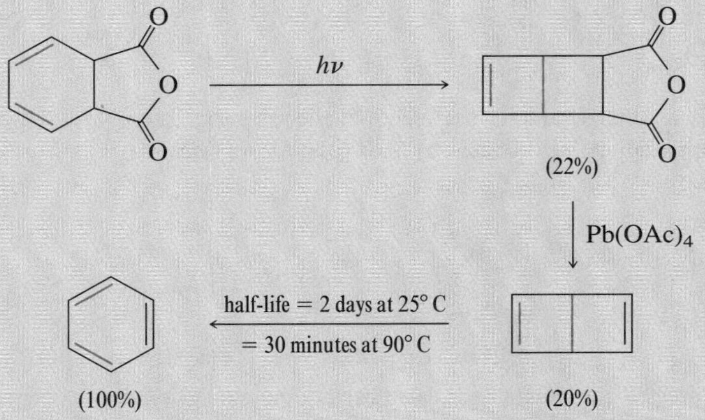

(22%)

Pb(OAc)₄

half-life = 2 days at 25° C
= 30 minutes at 90° C

(100%)   (20%)

Irradiation of the substituted cyclohexadiene resulted in the formation of the Dewar benzene skeleton by a disrotatory ring closure. Reaction with lead tetraacetate (a reaction that is not covered in this book) was used to remove the anhydride group and introduce the final double bond of Dewar benzene. Again, because of the forbidden nature of the conrotatory opening to benzene, Dewar benzene has an appreciable lifetime. At 25°C the half-life for its conversion to benzene is 2 days, and at 90°C its half-life is 30 minutes.

Exercise

Exercise 20.3   Build a model of Dewar benzene. Then replace the connector for the bond that is part of both rings with two separate connectors. Do a conrotation and a disrotation to see the strain that is incurred in each process. Also build a model of 1,2,4-tri-*t*-butylbenzene and the Dewar benzene formed from it and examine the steric strain that is present in each compound.

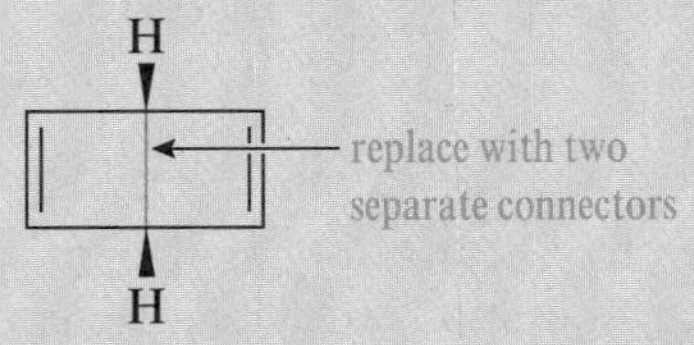

Explain why the following Dewar benzene is not produced in the photochemical reaction of 1,2,4-tri-*t*-butylbenzene:

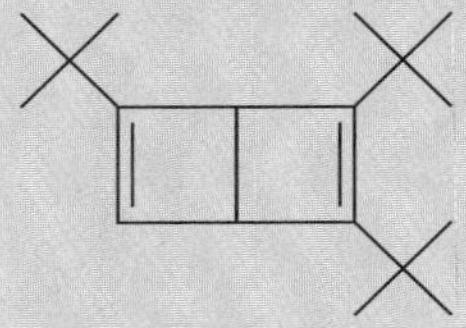

The thermally allowed cyclization of a triene to form a cyclohexadiene occurs by a disrotatory motion, as illustrated in the following equation:

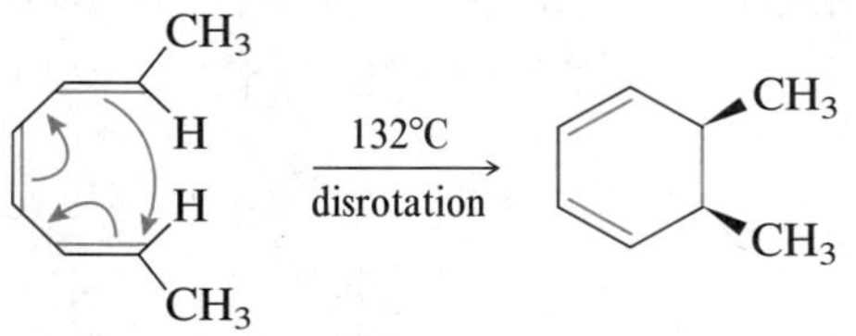

As another example, the electrocyclic interconversion of 1,3,5-cycloheptatriene and norcaradiene also occurs by a disrotatory motion:

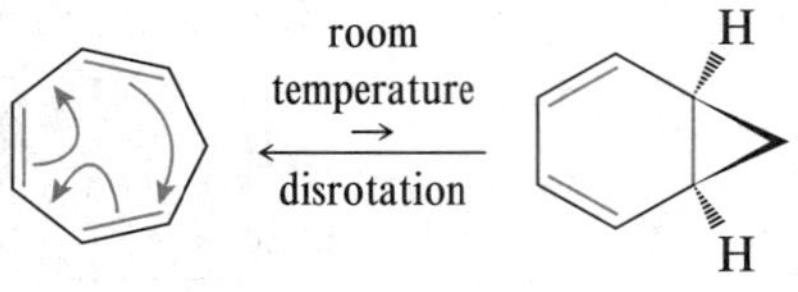

1,3,5-cycloheptatriene   norcaradiene

Norcaradiene cannot be isolated because its conversion to cycloheptatriene is fast at room temperature and its concentration in the equilibrium mixture is very low because of ring strain. In a related example, 1,3,5-cyclononatriene cyclizes by a disrotatory motion, as shown in the following equation. Again, the reaction is quite facile, as illustrated by its half-life of about 1 h at 50°C. In this case the product, with a five- and a six-membered ring, is more stable than the starting material with its nine-membered ring:

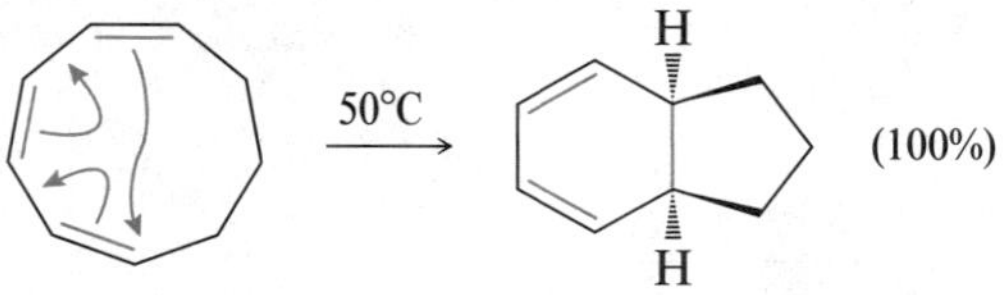

In the case of a tetraene, conrotation is again the thermally allowed pathway for cyclization. The following equation provides an example:

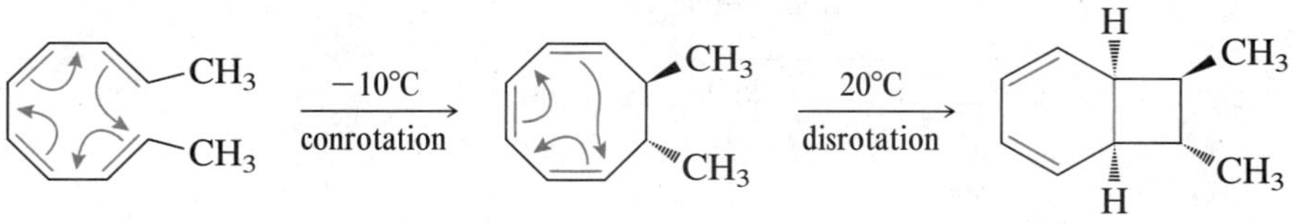

decatetraene   cyclooctatriene   (100%)

The decatetraene cyclizes at −10°C, by a conrotatory motion, to produce the cyclooctatriene with the two methyl groups in a *trans* orientation. When the cyclooctatriene is warmed to 20°C, it undergoes a thermally allowed disrotatory ring closure to give the final product.

Problem

Problem 20.9   Show the products of these reactions.

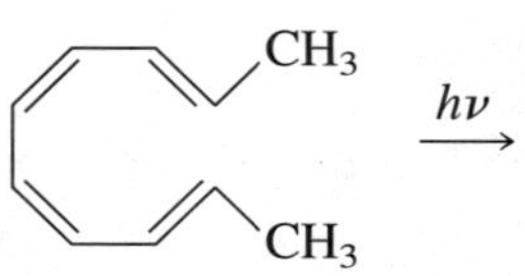

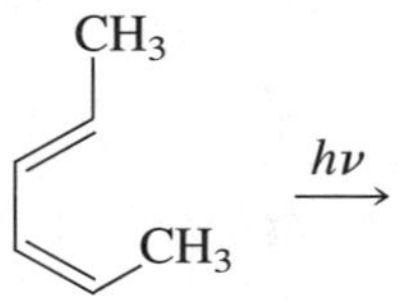

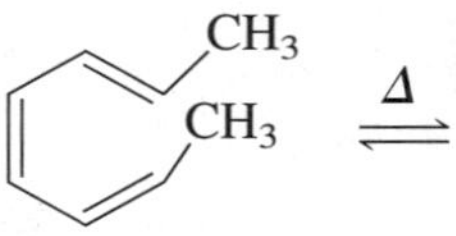

20.30 Show syntheses of these compounds from the indicated starting materials. Reactions from previous chapters may be needed in some cases.

a) [cyclobutene fused structure] from [cyclohexadiene]

b) [bicyclic CN structure] from [NC—CH=CH—CN]

c) [cyclohexene with methyl and side chain structure] from [CH₂=CH—C(=O)—OEt]

HO—C—CH₂CH₃
|
CH₂CH₃

d) [cyclohexadiene with CO₂Et groups] from compounds without rings

e) [bicyclic structure] from [cyclopentene]

f) [decalindione structure] from a compound with one ring

g) [bicyclic CH₂OH structure] from [diester structure] and compounds without rings

20.31 Suggest a method to convert (2E,4E)-hexadiene to (2E,4Z)-hexadiene in a manner so that the stereochemistry is controlled at each step of the process:

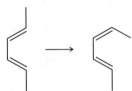

20.32 Explain which compound is more reactive as a diene in the Diels-Alder reaction.

a) [cyclohexadiene] or [butadiene]   **b)** [butadiene] or [substituted diene]   **c)** [octahydronaphthalene] or [octahydronaphthalene]

20.33 Suggest a method for the synthesis of *o*-eugenol from *o*-methoxyphenol:

OH
[aromatic ring with allyl and OCH₃ substituents]

o-eugenol

20.34 Explain why *trans*-3,4-dimethylcyclobutene produces (2*E*,4*E*)-hexadiene upon heating, whereas (2*Z*,4*Z*)-hexadiene is not formed, even though its formation is an allowed reaction.

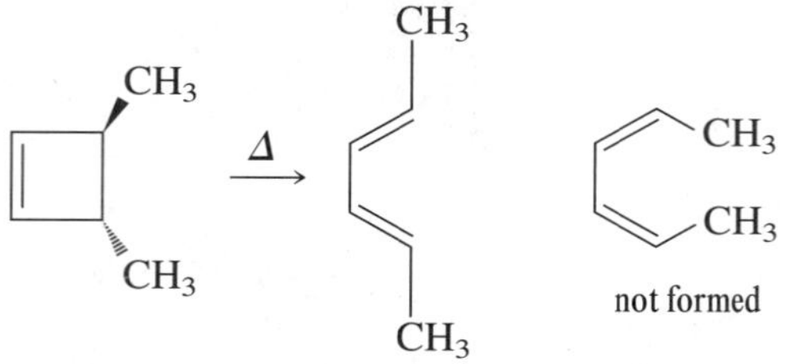

not formed

20.35 Show all of the steps in the mechanism for this reaction and explain why the aldehyde is formed rather than the ketone:

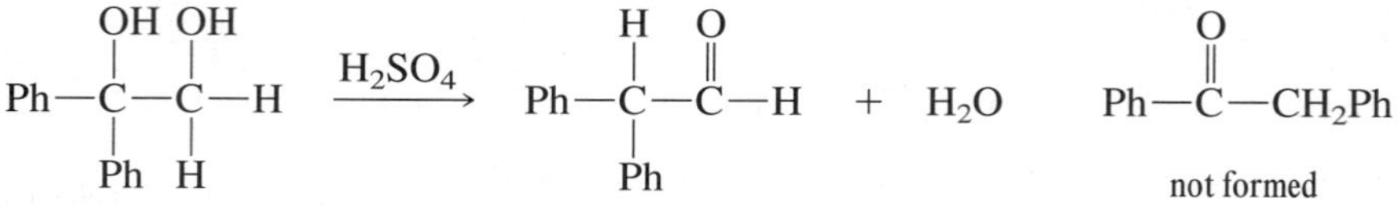

not formed

20.36 Show all of the steps in the mechanism for this reaction:

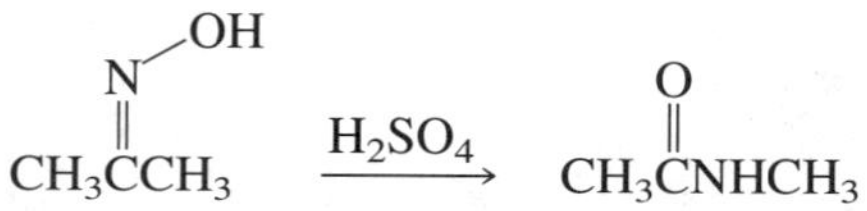

20.37 Show all of the steps in the mechanism for this reaction:

20.38 Attempts to prepare cyclobutadiene usually result in the isolation of this compound. Explain how the formation of this dimer from cyclobutadiene is allowed.

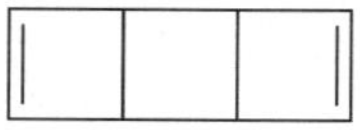

20.39 This reaction has been shown to occur in two thermally allowed steps. Show the structure of the intermediate in the reaction and explain why each step is allowed.

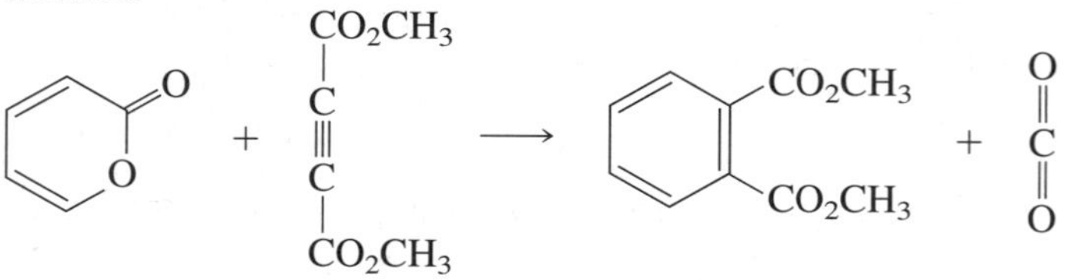

20.40 Suggest a mechanism for this reaction:

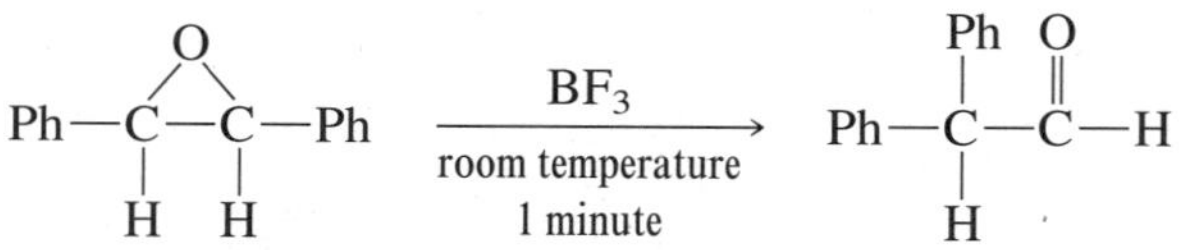

20.41 Suggest a mechanism for this reaction:

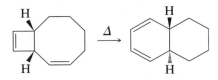

20.42 Explain the stereochemistry of this reaction:

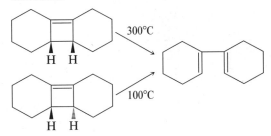

20.43 This reaction has been shown to occur in two thermally allowed steps. Show the structure of the intermediate in the reaction and explain why each step is allowed.

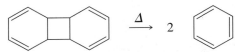

20.44 Suggest a mechanism for this reaction:

20.45 Explain why this reaction is highly exothermic. Then offer a reason why the reactant is quite stable at room temperature.

20.46 Explain the large difference in the temperatures required for these reactions.

20.47 Cyclopropyl carbocations react rapidly to form allyl carbocations. Explain how this process is allowed. Explain why the allyl carbocation is more stable than the cyclopropyl carbocation. Predict the stereochemistry of the allyl carbocation that is formed from the following cis-dimethylcyclopropyl carbocation:

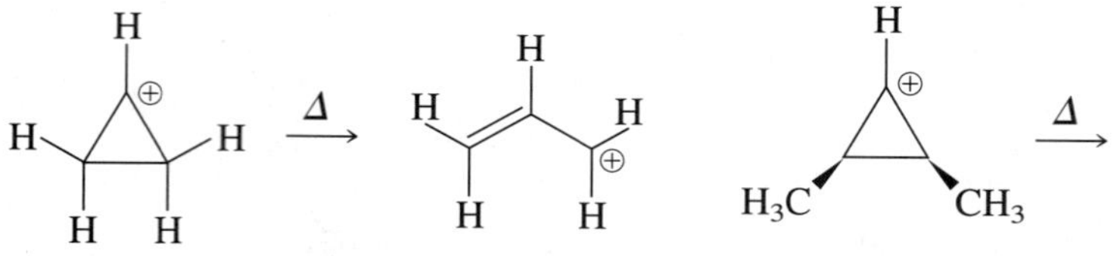

Problems Involving Spectroscopy

20.48 The reaction of anthracene with benzyne gives a product, $C_{20}H_{14}$, which shows only four peaks in its ^{13}C-NMR spectrum. Show the structure of this product and explain its formation.

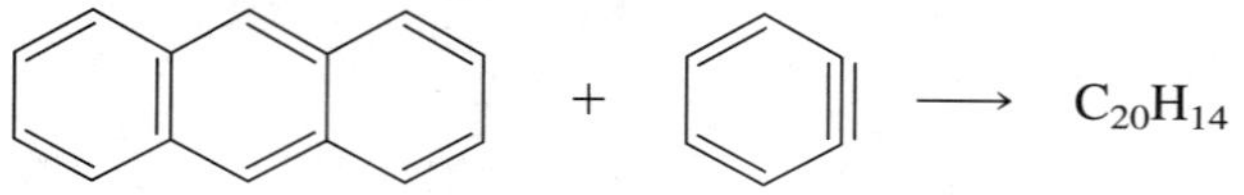

20.49 The mass spectrum of cyclohexene shows a peak at m/z 54. Show a structure for this fragment and suggest a mechanism for its formation.

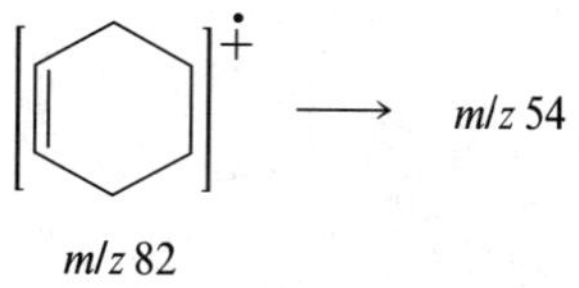

m/z 82

20.50 With unsymmetrical ketones the Baeyer-Villiger reaction can, in principle,
give two products. Usually, one product predominates. The ^{1}H-NMR spectrum of the product isolated from the Baeyer-Villiger reaction of 3-methyl-
2-butanone is shown below. Show the structure of this product.

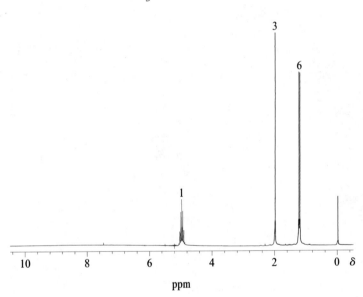

The Synthesis of Organic Compounds

21.1   Introduction

Each of the preceding chapters that presented new reactions contained a number of problems that required using these reactions to synthesize organic compounds. You have probably become at least somewhat comfortable in working this type of problem, even though they are among the more difficult ones in this text. In this chapter the multitude of reactions that have been presented in the previous chapters are applied to the synthesis of more complex compounds. The syntheses here are longer and require somewhat more thought than those done previously, although the general approach and strategies remain the same as those you have already learned.

Groups that can be used to protect alcohols, aldehydes and ketones, carboxylic acids, and amines during the course of a synthesis are presented first in this chapter. Then the strategy of synthesis is discussed in somewhat more detail. This is followed by the presentation of several syntheses. At the end of this chapter are tables that summarize all of the reactions that have been presented in the previous chapters and are of significant synthetic utility. Table 21.1 lists important carbon–carbon bond-forming reactions. Table 21.2 lists all of the reactions, including the carbon–carbon bond-forming reactions of Table 21.1, arranged according to the functional group that is produced in the reaction. Thus it is possible by consulting Table 21.2 to quickly identify the various reactions that can be used to prepare a particular functional group. Note that this table also constitutes a review of most of the reactions presented in earlier chapters.

21.2   Protective Groups for Alcohols

As discussed in Section 14.10, a protective group is used to protect one functional group in a complex compound from reacting while a reaction is occurring at another functional group in the molecule. For this process to be useful the functional

Table 21.2 Preparation of Functional Groups

	Reaction	Section	Comments
Alkanes	$R-L \xrightarrow[\text{or NaBH}_4]{\text{LiAlH}_4} R-H$	9.8	S_N2, 1° and 2° OK
	$RHC{=}CHR \xrightarrow[\Delta \text{ or } h\nu \text{ or Cu}^{2+}]{\text{CH}_2\text{N}_2}$ (cyclopropane: H H on C, RHC—CHR)	10.9	*syn* addition
	$RHC{=}CHR \xrightarrow{\frac{\text{CHCl}_3}{\text{NaOH}}}$ (cyclopropane: Cl Cl on C, RHC—CHR)	10.9	*syn* addition
	$RHC{=}CHR \xrightarrow{\frac{\text{CH}_2\text{I}_2}{\text{Zn(Cu)}}}$ (cyclopropane: H H on C, RHC—CHR)	10.9	Simmons-Smith reaction, *syn* addition
	$\begin{matrix}RHC{=}CHR \\ \text{or} \\ RC{\equiv}CR\end{matrix} \xrightarrow[\text{Ni, Pd, or Pt}]{\text{H}_2} RCH_2CH_2R$	10.13	*syn* addition
	$\underset{\text{R}}{\overset{\text{O}}{\|}}{-}\overset{\|}{C}{-}R' \xrightarrow[\substack{\text{NH}_2\text{NH}_2 \\ \text{KOH, }\Delta}]{\substack{\text{Zn(Hg), HCl} \\ \text{or}}} R-CH_2-R''$	18.14	Clemmensen or Wolff-Kishner reduction (H_2/Pd can be used when R = Ar)
	$RBr \xrightarrow[\text{[AIBN]}]{\text{Bu}_3\text{SnH}} RH$	19.9	radical chain mechanism
Alkenes	$\overset{\text{H L}}{-}\overset{\|\ \ \|}{C}{-}\overset{\|\ \ \|}{C}{-} \xrightarrow[\text{or }^{\ominus}\text{OR}]{^{\ominus}\text{OH}} \ \ C{=}C$	9.13	E2, Zaitsev's rule, usually *anti* elimination
	$\overset{\text{H OH}}{-}\overset{\|\ \ \|}{C}{-}\overset{\|\ \ \|}{C}{-} \xrightarrow[\text{or H}_3\text{PO}_4]{\text{H}_2\text{SO}_4} \ \ C{=}C$	9.15	usually E1, Zaitsev's rule
	$R-C{\equiv}C-R \xrightarrow[\substack{\text{Lindlar} \\ \text{catalyst}}]{1\ \text{H}_2}$ (cis-alkene: R,R up; H,H down)	10.13	*syn* addition

Table 21.2 *Continued*

	Reaction	Section	Comments
Alkenes		14.8	Wittig reaction
		19.12	Birch reduction
	R—C≡C—R →	19.12	*anti* addition
Alkynes	R—L $\xrightarrow{{}^{\ominus}\text{:C}\equiv\text{CR}'}$ R—C≡CR'	9.9	S_N2, 1° R only
	$\xrightarrow[\text{or }{}^{\ominus}\text{NH}_2]{{}^{\ominus}\text{OH}}$ —C≡C—	9.14	E2
	$\xrightarrow[\text{or }{}^{\ominus}\text{NH}_2]{{}^{\ominus}\text{OH}}$ —C≡C—	9.14	E2
Alcohols	R—X $\xrightarrow{{}^{\ominus}\text{OH}}$ R—OH	9.3	S_N2, best with 1° R
	R—X $\xrightarrow{\text{H}_2\text{O}}$ R—OH	9.3	S_N1, OK with 2° and 3° R
	R—X $\xrightarrow[\text{2) KOH, H}_2\text{O}]{\text{1) CH}_3\text{CO}_2{}^{\ominus}}$ R—OH	9.3	S_N2, good with 1° and 2° R, synthetic equivalent of hydroxide ion
	$\xrightarrow{\text{Nu}}$	9.11	epoxide opening
	$RCH\text{=}CH_2 \xrightarrow[\text{H}_2\text{O}]{\text{Cl}_2}$	10.5	halohydrin formation, *anti* addition, OH on more substituted carbon

Table 21.2 *Continued*

	Reaction	Section	Comments
Alcohols	$RCH{=}CH_2 \xrightarrow[H_2SO_4]{H_2O}$ $\overset{OH}{\underset{}{RCH}}{-}CH_3$	10.6	hydration, low yields, rearrangements may occur, Markovnikov's rule
	$RCH{=}CH_2 \xrightarrow[\text{2) NaBH}_4,\text{ NaOH}]{\text{1) Hg(OAc)}_2,\text{ H}_2\text{O}}$ $\overset{OH}{\underset{}{RCH}}{-}CH_3$	10.7	oxymercuration–reduction, better yields, Markovnikov's rule
	$RCH{=}CH_2 \xrightarrow[\text{2) H}_2\text{O}_2,\text{ NaOH}]{\text{1) BH}_3,\text{ THF}}$ $\overset{H}{\underset{}{RCH}}{-}\overset{OH}{\underset{}{CH_2}}$	10.8	hydroboration–oxidation, *syn* addition, anti-Markovnikov
	$RCH{=}CHR \xrightarrow[\substack{\text{or}\\ \text{KMnO}_4,\text{ NaOH}}]{\text{OsO}_4}$ $\overset{OH}{\underset{}{RCH}}{-}\overset{OH}{\underset{}{CHR}}$	10.11	*syn* addition
	$R{-}\overset{O}{\overset{\|}{C}}{-}R' \xrightarrow[\text{2) H}_3\text{O}^{\oplus}]{\text{1) LiAlH}_4\text{ or NaBH}_4}$ $R{-}\overset{OH}{\underset{H}{\overset{\|}{C}}}{-}R'$	14.3	reduction of aldehydes and ketones
	$R{-}\overset{O}{\overset{\|}{C}}{-}R' \xrightarrow[\text{2) H}_3\text{O}^{\oplus}]{\text{1) R''MgX or R''Li}}$ $R{-}\overset{OH}{\underset{R''}{\overset{\|}{C}}}{-}R'$	14.7	addition of Grignard and organolithium reagents to aldehydes and ketones
	$R{-}\overset{O}{\overset{\|}{C}}{-}OR'$ or $R{-}\overset{O}{\overset{\|}{C}}{-}OH \xrightarrow[\text{2) H}_3\text{O}^{\oplus}]{\text{1) LiAlH}_4}$ $R{-}\overset{OH}{\underset{H}{\overset{\|}{C}}}{-}H$	15.8	reduction of carboxylic acids and esters, LiAlH$_4$ required
	$R{-}\overset{O}{\overset{\|}{C}}{-}OR' \xrightarrow[\text{2) H}_3\text{O}^{\oplus}]{\text{1) 2 R''MgX or R''Li}}$ $R{-}\overset{OH}{\underset{R''}{\overset{\|}{C}}}{-}R''$	15.10	addition of Grignard and organolithium reagents to esters, adds two identical R'' groups
Ethers	$R{-}L \xrightarrow{\ominus OR'} R{-}OR'$	9.4	Williamson ether synthesis, S$_N$2, best with 1° R
	$R{-}L \xrightarrow{HOR'} R{-}OR'$	9.4	S$_N$1, OK with 2° and 3° R

Table 21.2 *Continued*

	Reaction	Section	Comments
Epoxides	$-\overset{\text{OH}}{\underset{}{\text{C}}}-\overset{}{\underset{\text{X}}{\text{C}}}- \xrightarrow{\text{NaOH}} -\text{C}\overset{\text{O}}{\diagup \diagdown}\text{C}-$	10.5	intramolecular S_N2
	$\text{C}=\text{C} \xrightarrow{\text{RCO}_3\text{H}} -\text{C}\overset{\text{O}}{\diagup \diagdown}\text{C}-$	10.10	*syn* addition
Amines	$\text{R}-\text{L} \xrightarrow{\text{NH}_3} \text{R}-\overset{\oplus}{\text{NH}}_3$	9.7	S_N2, multiple alkylation problems
	$\text{R}-\text{L} \xrightarrow[\text{2) KOH, H}_2\text{O}]{1)} \text{R}-\text{NH}_2$	9.7	Gabriel synthesis, S_N2, 1° and 2° R OK
	$\overset{\text{O}}{\underset{}{\text{R}-\text{C}-\text{R}'}} \xrightarrow[\substack{\text{NaBH}_3\text{CN} \\ \text{or H}_2/\text{Ni}}]{\text{R''NH}_2} \text{R}-\overset{\text{NHR''}}{\underset{}{\text{CH}}}-\text{R}'$	14.9	reductive amination
	$\xrightarrow{\text{RNH}_2}$	14.11	conjugate addition
	$\overset{\text{O}}{\underset{}{\text{RCNHR}'}} \xrightarrow[\text{2) H}_2\text{O}]{\text{1) LiAlH}_4} \text{RCH}_2\text{NHR}'$	15.8	reduction of amides
	$\text{R}-\text{C}\equiv\text{N} \xrightarrow[\text{2) H}_2\text{O}]{\text{1) LiAlH}_4} \text{RCH}_2\text{NH}_2$	15.8	reduction of nitriles
	$\text{C}_6\text{H}_5\text{NO}_2 \xrightarrow[\substack{\text{or} \\ \text{H}_2/\text{cat}}]{\text{Fe or Sn, HCl}} \text{C}_6\text{H}_5\text{NH}_2$	18.14	reduction of aromatic nitro compounds
Halides	$\text{R}-\text{OH} \xrightarrow{\text{HX}} \text{R}-\text{X}$	9.6	chlorides require ZnCl_2 for 1° and 2° R

Table 21.2 *Continued*

	Reaction	Section	Comments
Halides	$R-OTs \xrightarrow{X^{\ominus}} R-X$	9.6	S_N2 conditions
	$R-OH \xrightarrow{SOCl_2} R-Cl$	9.6	thionyl chloride
	$R-OH \xrightarrow{PBr_3} R-Br$	9.6	phosphorus tribromide
	$R-OH \xrightarrow{PI_3} R-I$	9.6	phosphorus triiodide
	$R-CH=CH_2 \xrightarrow{HX} R-\overset{X}{\underset{}{C}}H\overset{H}{\underset{}{C}}H_2$	10.3	Markovnikov's rule
	$RCH=CHR' \xrightarrow[or\ Br_2]{Cl_2} RCH-CHR'$ with X substituents	10.4	*anti* addition
	$RCH=CHR' \xrightarrow[H_2O]{X_2} RCH-CHR'$ with X and OH	10.5	halohydrin formation, *anti* addition, Markovnikov's rule
	ring $\xrightarrow[FeX_3\ or\ AlX_3]{Cl_2\ or\ Br_2}$ ring-X	18.6	electrophilic aromatic substitution
	ring-$NH_2 \xrightarrow[H^{\oplus}]{NaNO_2}$ ring-$\overset{\oplus}{N_2} \xrightarrow[\ominus BF_4,\ \Delta]{CuCl,\ CuBr,\ KI,\ or}$ ring-X	18.11	substitutions via diazonium ion
	ring-$CH_3 \xrightarrow[NBS,\ initiator]{Br_2,\ h\nu\ or}$ ring-CH_2Br	19.8	used for allylic bromides also, radical chain mechanism
	$RCH=CH_2 \xrightarrow[initiator]{HBr} RCH-CH_2$ with H and Br	19.11	anti-Markovnikov regiochemistry, HBr only, radical chain mechanism

Table 21.2 *Continued*

	Reaction	Section	Comments
Aldehydes		9.16	oxidation using chromium trioxide–pyridine complex
		9.16	oxidation using pyridinium chlorochromate (PCC)
		10.8	hydroboration of terminal alkynes
		15.9	selective reduction of acyl chlorides
		15.9	selective reduction of esters with diisobutyl aluminum hydride
		16.5	aldol condensation
Ketones		9.16	oxidation
		10.7	hydration of alkynes, Markovnikov regiochemistry, alkyne should be terminal or symmetrical
		10.8	hydroboration of alkynes, alkyne should be symmetrical

Table 21.2 *Continued*

	Reaction	Section	Comments
Ketones	$R-\overset{\overset{\text{O}}{\|}}{C}-Cl \xrightarrow{R'_2CuLi} R-\overset{\overset{\text{O}}{\|}}{C}-R'$	15.11	addition of organocuprate reagents to acyl chlorides
	$R-C\equiv N \xrightarrow[\text{2) H}_3\text{O}^\oplus]{\text{1) R'MgBr}} R-\overset{\overset{\text{O}}{\|}}{C}-R'$	15.11	addition of Grignard reagents to nitriles
	$R'-\overset{\overset{\text{O}}{\|}}{C}-CH_2R \xrightarrow[\text{2) R''X}]{\text{1) LDA}} R'-\overset{\overset{\text{O}}{\|}}{C}-\overset{\overset{\text{R''}}{\|}}{C}HR$	16.3	alkylation of ketones, S_N2, R'' must be 1°, regiochemistry must be controlled somehow
	$H_3C-\overset{\overset{\text{O}}{\|}}{C}-CH_2-\overset{\overset{\text{O}}{\|}}{C}-OEt \xrightarrow[\substack{\text{2) RL} \\ \text{3) KOH, H}_2\text{O} \\ \text{4) H}_3\text{O}^\oplus, \Delta}]{\text{1) NaOEt}} H_3C-\overset{\overset{\text{O}}{\|}}{C}-CH_2R$	16.4	acetoacetic ester synthesis, S_N2, R must be 1° or 2°
	$R\overset{\overset{\text{O}}{\|}}{C}H + R'CH_2\overset{\overset{\text{O}}{\|}}{C}-R'' \xrightarrow{\text{base}} R'\overset{\underset{\|}{C}C}{\|}-R'' \quad \overset{\overset{\text{O}}{\|}}{\underset{RCH}{}}$	16.5	aldol condensation
	$RCH_2\overset{\overset{\text{O}}{\|}}{C}R' \xrightarrow[\substack{\text{2) R''X} \\ \text{3) H}_3\text{O}^\oplus}]{\substack{\text{1)} \overset{\overset{\bigcirc}{N}}{\underset{H}{}} \text{HA}}} R\overset{\overset{\text{O}}{\|}}{C}H\overset{\overset{\text{O}}{\|}}{C}R' \;\underset{R''}{}$	16.7	enamine reaction, R''X must be very reactive
	$R\overset{\overset{\text{O}}{\|}}{C}H \xrightarrow[\substack{\text{2) BuLi} \\ \text{3) R'X} \\ \text{4) Hg}^{2+}, \text{H}_2\text{O}}]{\substack{\text{1)} \overset{\frown}{SH\;SH} \; BF_3}} RC\overset{\overset{\text{O}}{\|}}{}-R'$	16.8	dithiane alkylation, acyl anion equivalent, S_N2, R' must be 1° or 2°
	$\bigcirc \xrightarrow[\text{AlCl}_3]{\overset{\overset{\text{O}}{\|}}{RCCl}} \overset{\overset{\text{O}}{\|}}{\underset{\bigcirc}{C-R}}$	18.9	Friedel-Crafts acylation, does not work when ring has strong deactivating substituents

Table 21.2 *Continued*

	Reaction	Section	Comments
Carboxylic Acids	$R\text{--}\overset{\displaystyle O}{\overset{\|}{C}}\text{--}H \xrightarrow[\text{or } KMnO_4]{Ag_2O, NaOH} R\text{--}\overset{\displaystyle O}{\overset{\|}{C}}\text{--}OH$	9.16	oxidation of aldehydes
	$RMgX \text{ or } RLi \xrightarrow[\text{2) } H_3O^{\oplus}]{\text{1) } CO_2} R\text{--}CO_2H$	14.7	reaction of Grignard reagents with carbon dioxide
	$R\text{--}\overset{\displaystyle O}{\overset{\|}{C}}\text{--}L \xrightarrow[H_3O^{\oplus} \text{ or } {}^{\ominus}OH]{H_2O} R\text{--}\overset{\displaystyle O}{\overset{\|}{C}}\text{--}OH$	15.6	any carboxylic acid derivative can be hydrolyzed to a carboxylic acid
	$R\text{--}C{\equiv}N \xrightarrow[H_3O^{\oplus} \text{ or } {}^{\ominus}OH]{H_2O} R\text{--}\overset{\displaystyle O}{\overset{\|}{C}}\text{--}OH$	15.6	nitriles can also be hydrolyzed to carboxylic acids
	$EtO\text{--}\overset{\displaystyle O}{\overset{\|}{C}}\text{--}CH_2\text{--}\overset{\displaystyle O}{\overset{\|}{C}}\text{--}OEt \xrightarrow[\substack{\text{2) RL} \\ \text{3) } KOH, H_2O \\ \text{4) } H_3O^{\oplus}, \Delta}]{\text{1) NaOEt}} HO\text{--}\overset{\displaystyle O}{\overset{\|}{C}}\text{--}CH_2R$	16.4	malonic ester synthesis, S_N2, R must be 1° or 2°
	$\text{C}_6\text{H}_5\text{--}CH_3 \xrightarrow[\text{2) } H_3O^{\oplus}]{\text{1) } KMnO_4, NaOH} \text{C}_6\text{H}_5\text{--}CO_2H$	18.14	any alkyl group, except tertiary ones, can be oxidized to the benzoic acid derivative
Acyl Chlorides	$R\text{--}\overset{\displaystyle O}{\overset{\|}{C}}\text{--}OH \xrightarrow[\text{or } PCl_5]{SOCl_2} R\text{--}\overset{\displaystyle O}{\overset{\|}{C}}\text{--}Cl$	15.3	
Anhydrides	$R\text{--}\overset{\displaystyle O}{\overset{\|}{C}}\text{--}Cl + HO\text{--}\overset{\displaystyle O}{\overset{\|}{C}}\text{--}R \longrightarrow R\text{--}\overset{\displaystyle O}{\overset{\|}{C}}\text{--}O\text{--}\overset{\displaystyle O}{\overset{\|}{C}}\text{--}R$	15.4	only symmetrical anhydrides are usually prepared
	$2\,R\text{--}\overset{\displaystyle O}{\overset{\|}{C}}\text{--}OH \xrightarrow{CH_3\text{--}\overset{\displaystyle O}{\overset{\|}{C}}\text{--}O\text{--}\overset{\displaystyle O}{\overset{\|}{C}}\text{--}CH_3} R\text{--}\overset{\displaystyle O}{\overset{\|}{C}}\text{--}O\text{--}\overset{\displaystyle O}{\overset{\|}{C}}\text{--}R$	15.4	anhydride exchange

Table 21.2 *Continued*

	Reaction	Section	Comments
Amides	$R-\overset{O}{\overset{\|}{C}}-Cl \xrightarrow{NH_2R'} R-\overset{O}{\overset{\|}{C}}-NHR'$	15.7	
	$R-\overset{O}{\overset{\|}{C}}-O-\overset{O}{\overset{\|}{C}}-R \xrightarrow{NH_2R'} R-\overset{O}{\overset{\|}{C}}-NHR'$	15.7	
Esters	$R-L \xrightarrow{\overset{\ominus}{O}-\overset{O}{\overset{\|}{C}}-R'} R-O-\overset{O}{\overset{\|}{C}}-R'$	9.5	S_N2, OK for 1° and 2°
	$R-\overset{O}{\overset{\|}{C}}-Cl + R'OH \longrightarrow R-\overset{O}{\overset{\|}{C}}-OR'$	15.5	
	$R-\overset{O}{\overset{\|}{C}}-O-\overset{O}{\overset{\|}{C}}-R + R'OH \longrightarrow R-\overset{O}{\overset{\|}{C}}-OR'$	15.5	
	$R-\overset{O}{\overset{\|}{C}}-O-H + R'OH \xrightarrow{HA} R-\overset{O}{\overset{\|}{C}}-OR'$	15.5	Fischer esterification
	$R'O-\overset{O}{\overset{\|}{C}}-CH_2R \xrightarrow[2)\ R''X]{1)\ LDA} R'O-\overset{O}{\overset{\|}{C}}-\overset{R''}{\overset{\|}{C}}HR$	16.3	S_N2, R" must be 1°
	$RCOEt + R'CH_2\overset{O}{\overset{\|}{C}}-OEt \xrightarrow[2)\ H_3O^{\oplus}]{1)\ NaOEt} \underset{R-C=O}{R'CH\overset{O}{\overset{\|}{C}}-OEt}$	16.6	ester condensation
Nitriles	$R-L + {}^{\ominus}CN \longrightarrow R-CN$	9.9	S_N2, OK with 1° and 2°
	$R-\overset{O}{\overset{\|}{C}}-R' \xrightarrow[{[CN^{\ominus}]}]{HCN} R-\overset{OH}{\underset{CN}{\overset{\|}{C}}}-R'$	14.5	cyanohydrin formation
		14.11	conjugate addition

Table 21.2 *Continued*

	Reaction	Section	Comments
Nitriles	$RCH_2-CN \xrightarrow[\text{2) R'X}]{\text{1) LDA}} RCH(R')-CN$	16.3	S_N2, R' must be 1°
	(aniline) $NH_2 \xrightarrow[\text{2) CuCN}]{\text{1) NaNO}_2, \text{HA}}$ (benzonitrile) CN	18.11	via diazonium ion
Aromatic Compounds	benzene $\xrightarrow[\text{H}_2\text{SO}_4]{\text{HNO}_3}$ nitrobenzene NO_2	18.5	electrophilic aromatic substitution
	benzene $\xrightarrow[\text{FeX}_3 \text{ or AlX}_3]{\text{Cl}_2 \text{ or Br}_2}$ (halobenzene) X	18.6	electrophilic aromatic substitution
	benzene $\xrightarrow{\text{H}_2\text{SO}_4}$ benzenesulfonic acid SO_3H	18.7	electrophilic aromatic substitution, reversible in H_2O/H_2SO_4
	benzene $\xrightarrow[\text{AlCl}_3]{\text{RX}}$ (alkylbenzene) R	18.8	Friedel-Crafts alkylation, rearrangements, multiple alkylation, fails with deactivated rings
	benzene $\xrightarrow[\text{AlCl}_3]{\text{RCCl (O)}}$ aryl ketone C=O, —R	18.9	Friedel-Crafts acylation, fails with strongly deactivated rings
	(aniline) $NH_2 \xrightarrow[\text{H}^{\oplus}]{\text{NaNO}_2}$ ($\oplus N_2$) $\xrightarrow[\text{KI, or }^{\ominus}\text{BF}_4, \Delta]{\text{CuCl, CuBr,}}$ (halobenzene) X	18.11	via diazonium ion

Table 21.2 *Continued*

Reaction		Section	Comments
Aromatic Compounds	⬡—NH₂ $\xrightarrow[\text{2) CuCN}]{\text{1) NaNO}_2,\ \text{HA}}$ ⬡—CN	18.11	via diazonium ion
	⬡—NH₂ $\xrightarrow[\substack{\text{2) NaBH}_4 \\ \text{or} \\ \text{H}_3\text{PO}_2}]{\text{1) NaNO}_2,\ \text{HA}}$ ⬡—H	18.11	via diazonium ion
	O=C—R ⬡ $\xrightarrow[\substack{\text{or} \\ \text{Zn(Hg), HCl} \\ \text{or} \\ \text{NH}_2\text{NH}_2,\ \text{KOH},\ \Delta}]{\text{H}_2/\text{Pt or Pd}}$ CH₂—R ⬡	18.14	Clemmensen reduction, hydrogenolysis, or Wolff-Kishner reduction
	⬡—NO₂ $\xrightarrow[\substack{\text{or} \\ \text{Sn, HCl}}]{\text{Fe, HCl}}$ ⬡—NH₂	18.14	reduction of nitro groups
	⬡—CH₃ $\xrightarrow[\text{2) H}_3\text{O}^{\oplus}]{\substack{\text{1) KMnO}_4, \\ \text{NaOH}}}$ ⬡—CO₂H	18.14	oxidation of any alkyl group except tertiary

Problems

Problem 20.11 Suggest syntheses of these compounds. Each synthesis should involve the formation of at least two carbon–carbon bonds. Make sure that your synthesis produces the correct stereochemistry where it is shown in the target compound.

a)
$$\underset{\text{CH}_3\text{CH}_2}{\overset{H}{}}\!\!C=C\!\!\underset{\text{CH}_2\text{CH}_3}{\overset{H}{}}$$

b) PhCH=CHCPh (with O above the last C)

c) PhCH₂CCH₃ (with OH above and CH₃ below the central C)

d) CH₃CH₂C—C—COCH₂CH₃ (with O on first C, CH₃ above central C, O on third C, CH₂CH₃ below central C)

e)

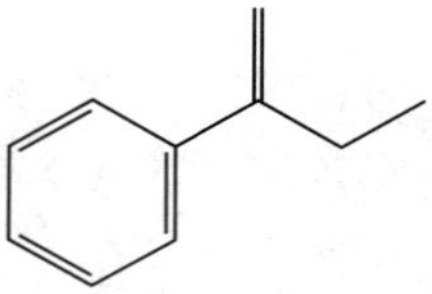

f) 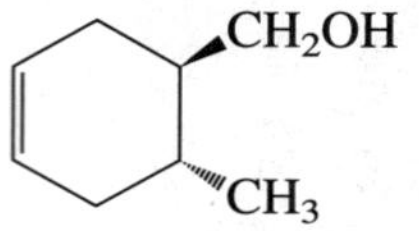
CH₂OH
CH₃

g) CH₃CH₂CHCCH₃
| ||
CH₃   O

h)

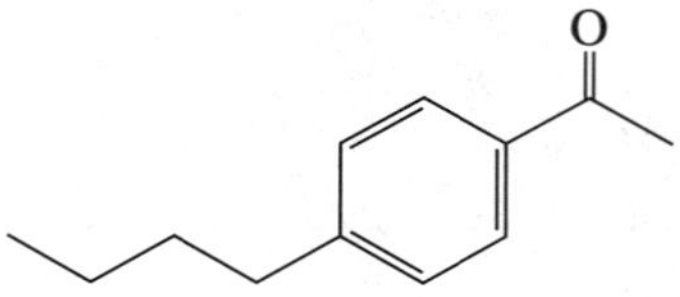

Problem 20.12   Show syntheses of these compounds using methanol, ethanol, and benzene as the only sources for the carbons in the compounds. You may use any solvents or inorganic reagents that you need. Once you have made a compound, you may use it in the syntheses of any subsequent compound.

a) $CH_3OCH_2CH_3$

b) $CH_3\overset{O}{\overset{||}{C}}OCH_3$

c)

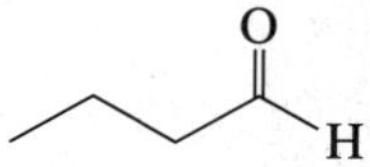

d)

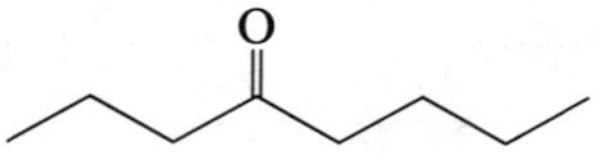

e)

f)

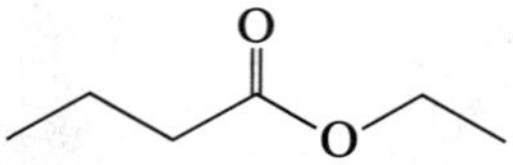

g)

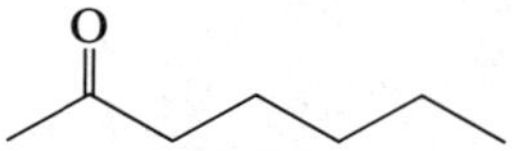

h)

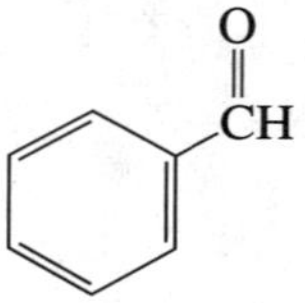

i)

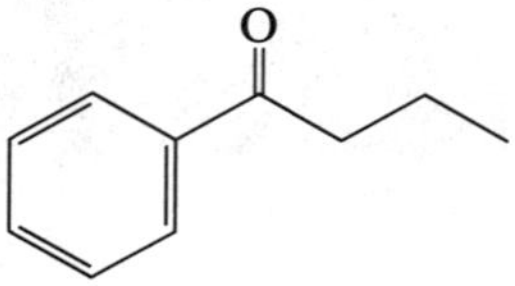

j) 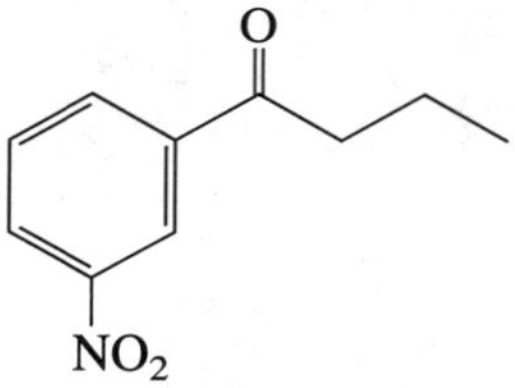
NO₂

k)

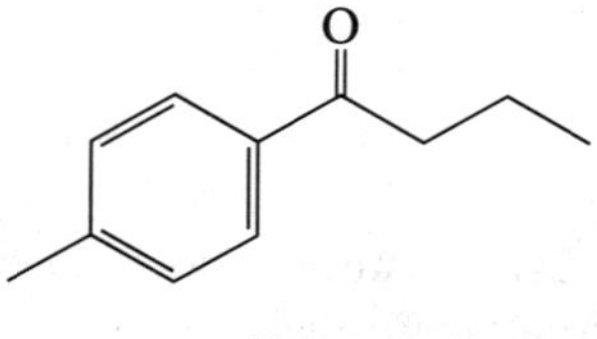

l) 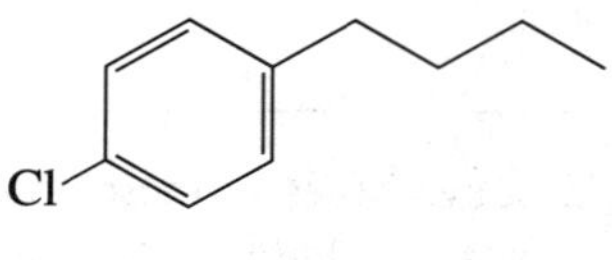
Cl

m) 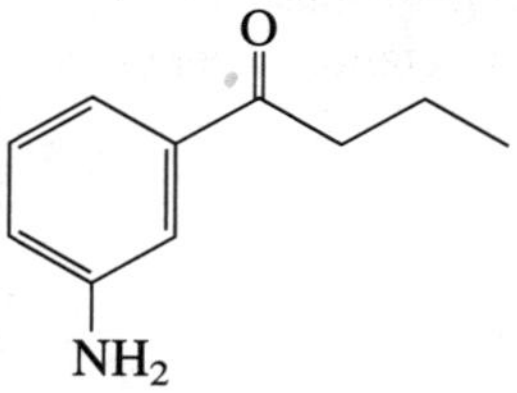
NH₂

n) 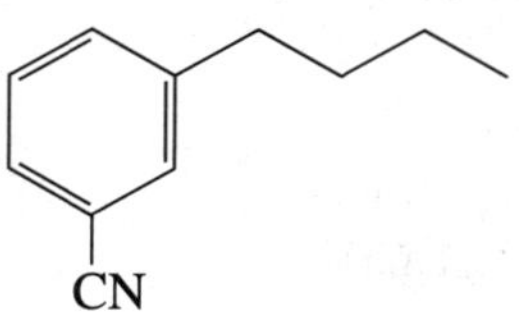
CN

o)

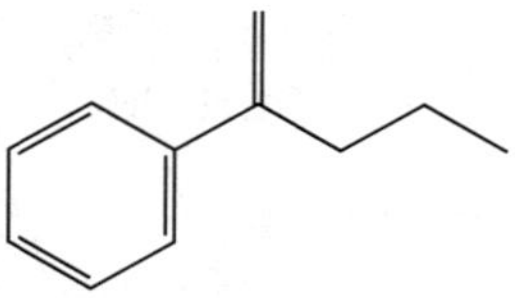

p) 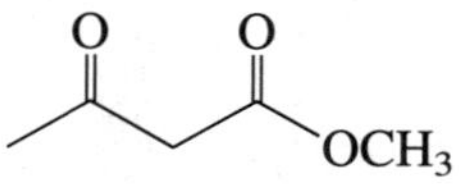
OCH₃

q)

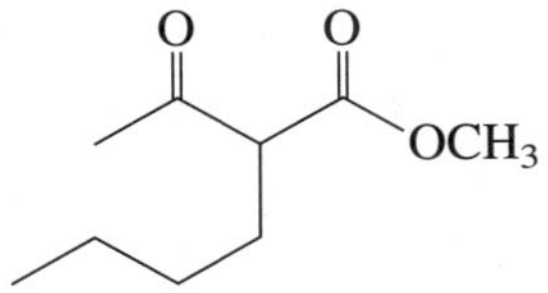

r)

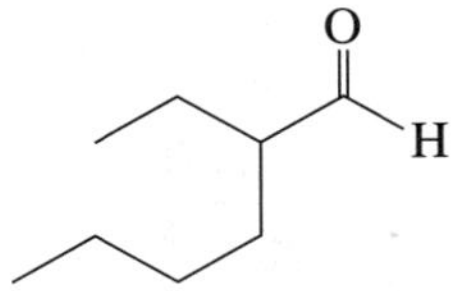

s)

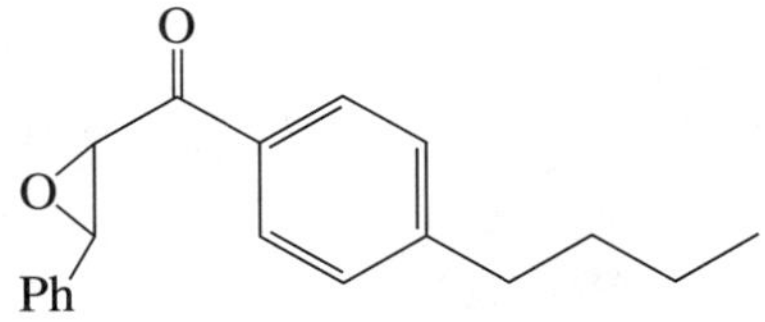

t)

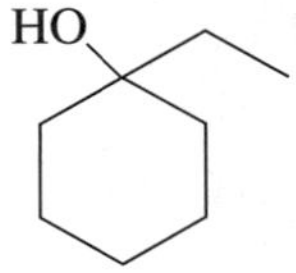

21.10   Summary

This chapter discussed the use of groups to protect alcohols, aldehydes, ketones, carboxylic acids, and amines during the course of a synthesis. Methods to add and to remove the protective groups were presented. Retrosynthetic analysis was discussed in more detail than previously, and several example syntheses were presented. The purpose of this was to make you more comfortable with, and competent in, designing syntheses of organic compounds.

After completing this chapter you should be able to:

1. Recognize when a protective group is needed and how to use it in the synthesis of an organic compound.

2. Use retrosynthetic analysis to design syntheses of more complex organic compounds.

End-of-Chapter Problems

21.13   Show the products of these reactions.

a) 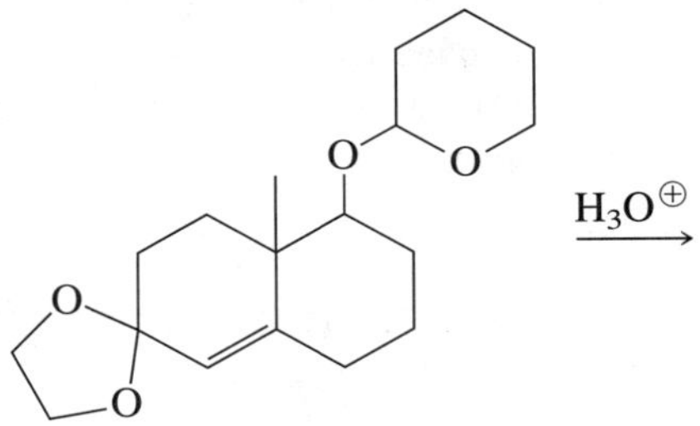 $\xrightarrow{H_3O^{\oplus}}$

b) 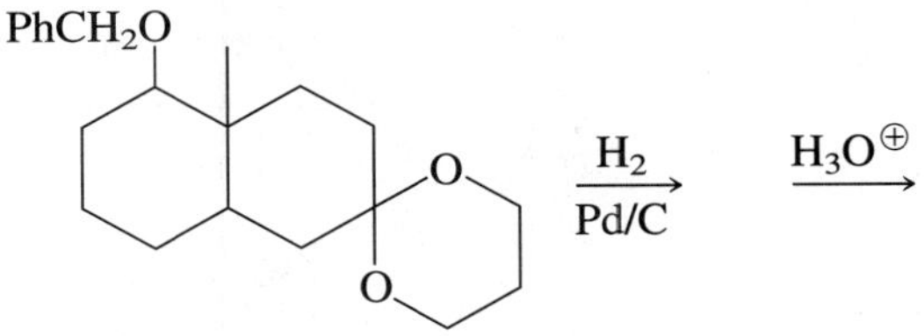 $\xrightarrow[\text{Pd/C}]{H_2}$ $\xrightarrow{H_3O^{\oplus}}$

c) $\xrightarrow[\text{TsOH}]{\text{OH OH}}$

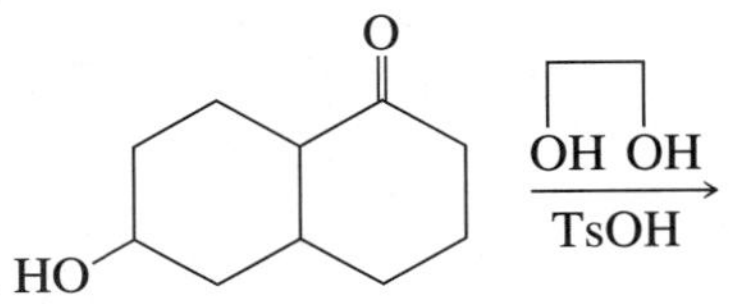

d) $CH_3OCH_2OCH_2CH_2CH_2\overset{\overset{\displaystyle O}{\|}}{C}OCH_2Ph$ $\xrightarrow[\text{Pd/C}]{H_2}$ $\xrightarrow{H_3O^{\oplus}}$

e)
$$\text{CH}_3\overset{\displaystyle O}{\overset{\|}{C}}\text{---}\overset{\displaystyle O}{\overset{\|}{C}}\text{OC(CH}_3)_3 \quad \xrightarrow[\text{H}_2\text{O}]{\text{HCl}}$$

f) (cyclohexanol) + (dihydropyran) $\xrightarrow{\text{TsOH}}$

g)
cyclopentane ring with CH₃ and C(=O)CH₃, and HO substituent $\xrightarrow[\text{2) ClCH}_2\text{OCH}_3]{\text{1) NaH}}$

h)
cyclopentane ring with CH₃ and C(=O)CH₃, and HO substituent $\xrightarrow[\text{Et}_3\text{N}]{\text{(CH}_3)_3\text{SiCl}}$

i)
cyclohexane-NH₂ + $(\text{CH}_3)_3\text{CO}\overset{\displaystyle O}{\overset{\|}{C}}\text{O}\overset{\displaystyle O}{\overset{\|}{C}}\text{OC(CH}_3)_3$ $\xrightarrow[\text{DMF}]{\text{Et}_3\text{N}}$

j)
decalin system with NHCOC(CH₃)₃ group and H₂N---C(=O) group $\xrightarrow[\text{H}_2\text{O}]{\text{HCl}}$

k)
cyclopentane-CH₂CH₂-NH₂ + PhCH₂OCCl (with C=O) $\xrightarrow{\text{Na}_2\text{CO}_3}$

l)
cyclohexane with NHCOCH₂Ph group and CH₃OC(=O) group $\xrightarrow[\text{Pd/C}]{\text{H}_2}$

21.14   Use disconnections of carbon–carbon bonds to generate electrophile and nucleophile synthon fragments that could be used to prepare these compounds. Then show an actual reaction suggested by these synthons that would provide a method for the synthesis of each compound. Design at least three different routes to each compound.

a) $\text{CH}_3\text{CH}_2\text{CH}_2\text{CH}_2\overset{\displaystyle O}{\overset{\|}{C}}\text{---}\overset{\displaystyle \text{CH}_2\text{CH}_3}{\overset{|}{\text{CH}}}\text{CH}_2\text{CH}_3$

b) $\text{CH}_3\text{CH}_2\text{C}\equiv\text{CCH}_2\text{CH}_2\overset{\displaystyle O}{\overset{\|}{C}}\text{CH}_3$

c)
cyclohexane with $\overset{\displaystyle \text{OH}}{\overset{|}{\text{CH}}}\text{CH}_2\text{CH}_3$ group

d)
H_3C-benzene with $\text{CH}=\overset{\displaystyle \text{Ph}}{\overset{|}{\text{C}}}\text{---}\overset{\displaystyle O}{\overset{\|}{\text{C}}}\text{CH}_2\text{Ph}$ group

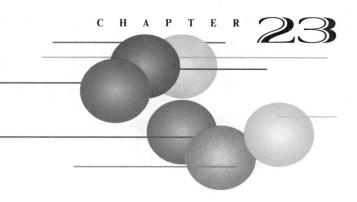

Synthetic Polymers

23.1 Introduction

Polymers are very large molecules made up of repeating units. A majority of the industrial chemicals that were described in Chapter 22 are ultimately used to prepare polymers. These human-made or synthetic polymers are the plastics (polyethylene, polystyrene), the adhesives (epoxy glue), the paints (acrylics), and the fibers (polyester, nylon) that we encounter many times each day. It is difficult to picture our lives without these materials. In addition to these synthetic polymers, natural polymers such as wood, rubber, cotton, and wool are all around us. And, of course, life itself depends on polymers such as carbohydrates, proteins, and DNA. This chapter discusses synthetic polymers. Naturally occurring polymers are presented in Chapters 24, 25, and 26.

First, a common method of forming polymers by a radical reaction is discussed. After the structures of the addition polymers made by this method are examined, several other procedures that can be used to prepare these or similar polymers are presented. Next, the effect of the structure of a polymer on its physical properties is discussed. This provides a basis for understanding the properties and uses of a number of other addition polymers. Rubbers (elastomers) are then discussed, followed by condensation polymers and thermosetting polymers. The chapter concludes with a brief examination of the chemical properties of polymers.

23.2 Radical Chain Polymerization

Polymers can be considered to be formed from the bonding of a large number of individual units, called monomers, together. This can be represented by the following general equation, in which some large number (n) of monomers (A) are combined to form a polymer:

$$n\,A \longrightarrow \cdots A-A-A-A-A-A\cdots \quad \text{or} \quad \left(\!A\!\right)_n$$

monomers   polymer

Because the number of monomers that combines to form an individual polymer molecule is usually very large, hundreds or even thousands, polymers are extremely large and are termed **macromolecules.**

Polymers are often classified as **addition polymers** or **condensation polymers** according to the general mechanism by which they are prepared. Most addition polymers are prepared by the reaction of an alkene monomer as illustrated in the following equation:

$$
\text{Initiator} \quad CH_2{=}\overset{\overset{\displaystyle R}{|}}{CH} \quad CH_2{=}\overset{\overset{\displaystyle R}{|}}{CH} \quad CH_2{=}\overset{\overset{\displaystyle R}{|}}{CH} \quad CH_2{=}\overset{\overset{\displaystyle R}{|}}{CH} \quad \text{etc.}
$$

$$\downarrow$$

$$
\text{Initiator}{-}CH_2{-}\overset{\overset{\displaystyle R}{|}}{CH}{-}CH_2{-}\overset{\overset{\displaystyle R}{|}}{CH}{-}CH_2{-}\overset{\overset{\displaystyle R}{|}}{CH}{-}CH_2{-}\overset{\overset{\displaystyle R}{|}}{CH}{-}\text{etc.}
$$

In this process, some initiator molecule adds to one carbon of the carbon–carbon double bond of the monomer to generate a reactive site, such as a radical or a carbocation, at the other carbon. This reactive carbon species then adds to another monomer to produce another reactive carbon species, and the process continues until a large number of monomers have been connected. Another way to represent this reaction shows the repeating unit that is formed when the monomers react to give the polymer:

$$
n \quad CH_2{=}\overset{\overset{\displaystyle R}{|}}{CH} \quad \xrightarrow{\text{Initiator}} \quad \left(CH_2{-}\overset{\overset{\displaystyle R}{|}}{CH}\right)_n
$$

monomer   repeat unit
                            of polymer

These addition polymers are often called **vinyl polymers** because of the vinyl group that is present in the monomers.

Although there are several variations, depending on the reactive intermediate that is present, all addition polymers are formed by chain mechanisms, in which one initiator molecule causes a large number of monomers to react to form one polymer molecule. For this reason these polymers are also known as **chain-growth polymers.** To better understand how this process occurs, let's examine a specific case, the formation of polyethylene by a radical chain mechanism, as shown in the following equation:

$$
n \quad CH_2{=}CH_2 \quad \xrightarrow{\overset{\displaystyle \overset{O}{\|}\;\;\overset{O}{\|}}{PhCOOCPh}} \quad \left(CH_2{-}CH_2\right)_n
$$

polyethylene

As is the case with other reactions that proceed by a radical chain mechanism (see Chapter 19), this reaction involves three kinds of steps: initiation, propagation, and termination.

Initiation

Recall that the initiation step generates the radicals. In this case the weak oxygen–oxygen bond of the initiator, dibenzoyl peroxide, cleaves to produce two benzoyloxy radicals. A benzoyloxy radical then adds to the carbon–carbon double bond of an ethylene molecule:

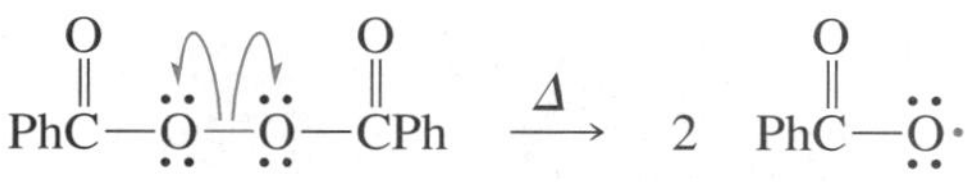

dibenzoyl peroxide　　　benzoyloxy radical

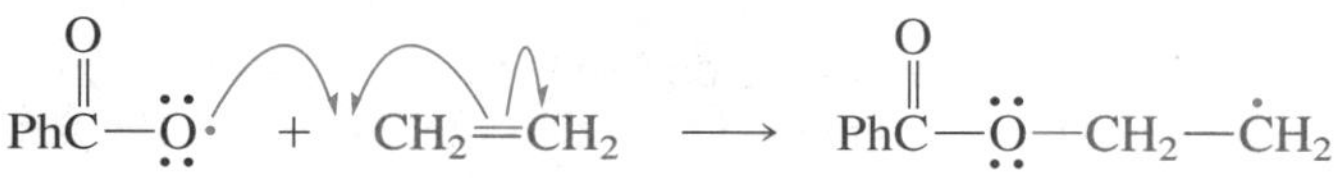

Propagation

In the propagation steps, a carbon radical adds to the double bond of a monomer to produce a larger carbon radical. This radical adds to another ethylene to produce an even larger radical, and the process continues until the radical is somehow destroyed.

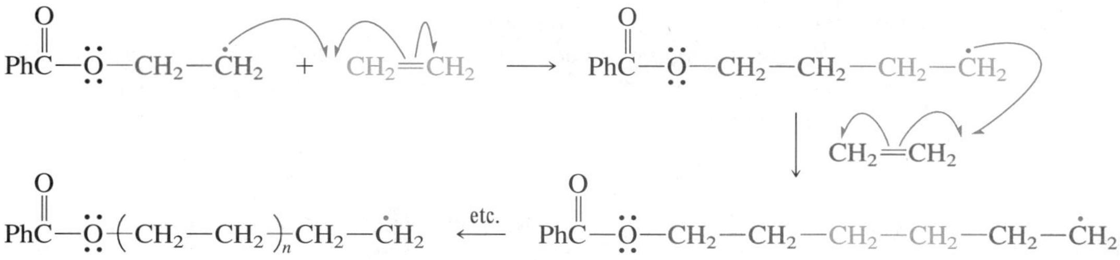

Termination

Various types of termination reactions can occur. Two radicals can couple to form nonradical products:

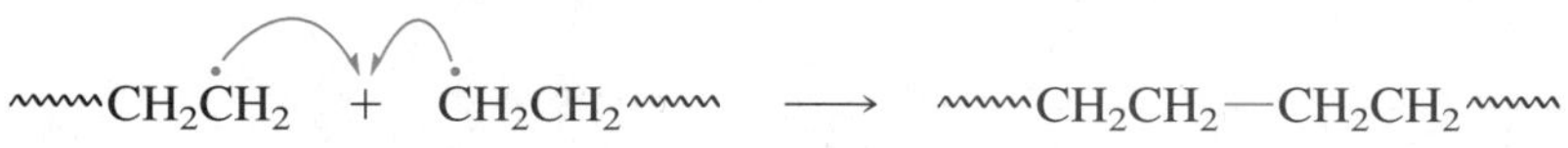

Or one radical can abstract a hydrogen from the carbon adjacent to the radical center of another radical in a disproportionation process:

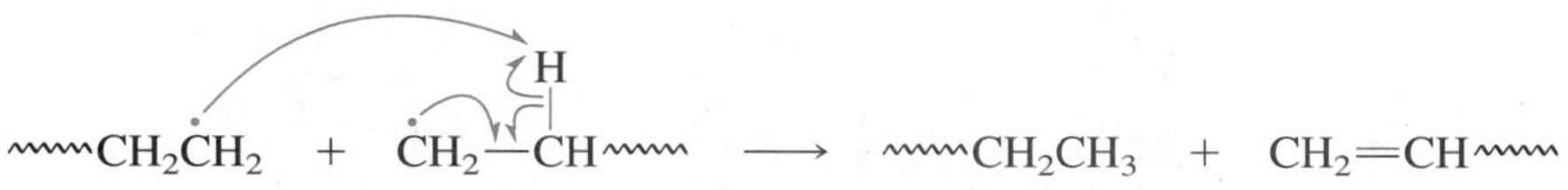

The average length of the polymer depends on the average number of propagation cycles that occur before a termination occurs. As is the case with many radical chain reactions, terminations are relatively rare because the concentration of radicals is extremely low. Therefore the probability of one radical encountering another is also quite low. This means that a typical polymer molecule is composed of thousands of monomers. For a hydrocarbon polymer such as polyethylene, useful mechanical properties are not present until the polymer contains more than approximately 100 monomer units.

Another polymer that can be produced by radical chain polymerization is polystyrene:

$$n \quad CH_2{=}\overset{\underset{\displaystyle |}{Ph}}{\underset{\alpha}{CH}} \quad \xrightarrow{\text{[ROOR]}} \quad \left(CH_2{-}\overset{\underset{\displaystyle |}{Ph}}{CH}\right)_n$$

styrene   polystyrene

In this case the two carbons of the double bond of the monomer are not identical. The polymer is found to have a regular structure in which the α-carbon of one monomer is always bonded to the β-carbon of the next, as shown in the following partial structure for polystyrene. The polymer is said to be formed by head-to-tail bonding of the monomer units.

$$\text{\textasciitilde\textasciitilde\textasciitilde}CH_2{-}\overset{\underset{\displaystyle |}{Ph}}{CH}{-}CH_2{-}\overset{\underset{\displaystyle |}{Ph}}{CH}{-}CH_2{-}\overset{\underset{\displaystyle |}{Ph}}{CH}{-}CH_2{-}\overset{\underset{\displaystyle |}{Ph}}{CH}\text{\textasciitilde\textasciitilde\textasciitilde}$$

Examination of the propagation steps in the mechanism for the formation of polystyrene makes the cause of this regularity readily apparent. Each time a new monomer is added to the growing polymer chain, the new bond is formed to the β-carbon. The resulting radical, with the odd electron located on the α-carbon, is stabilized by resonance involving the phenyl group and the odd electron:

$$\text{\textasciitilde\textasciitilde\textasciitilde}CH_2{-}\overset{\underset{\displaystyle |}{Ph}}{\overset{}{CH}} \ + \ CH_2{=}\overset{\underset{\displaystyle |}{Ph}}{CH} \ \longrightarrow \ \text{\textasciitilde\textasciitilde\textasciitilde}CH_2{-}\overset{\underset{\displaystyle |}{Ph}}{CH}{-}CH_2{-}\overset{\underset{\displaystyle |}{Ph}}{\overset{}{CH}}$$

As was the case with the radical addition reactions discussed in Chapter 19, the addition always occurs so as to produce the more stable radical. Since most other substituents also stabilize a radical on the carbon to which they are attached, vinyl polymers typically result from head-to-tail coupling of their monomer units.

Problems

Problem 23.1   Show all of the steps in the mechanism for the radical polymerization of propylene and explain why the polymer is formed by head-to-tail bonding of the monomer units.

$$n \quad CH_2{=}\overset{\underset{\displaystyle |}{CH_3}}{CH} \quad \xrightarrow{\left[\overset{\displaystyle O \quad O}{\overset{\displaystyle \| \quad \|}{PhCOOCPh}}\right]} \quad \left(CH_2{-}\overset{\underset{\displaystyle |}{CH_3}}{CH}\right)_n$$

Problem 23.2   Show the repeat unit of the polymers that would result from addition polymerization of these monomers.

a)
$$\underset{\text{vinyl chloride}}{CH_2{=}\overset{\overset{\displaystyle Cl}{|}}{CH}}$$

b)
$$\underset{\text{acrylonitrile}}{CH_2{=}\overset{\overset{\displaystyle C{\equiv}N}{|}}{CH}}$$

c)
$$\underset{\text{vinylidene chloride}}{CH_2{=}\overset{\overset{\displaystyle Cl}{|}}{\underset{\underset{\displaystyle Cl}{|}}{C}}}$$

Problem 23.3   Show the monomers that could be used to prepare these polymers.

a)
$$\underset{\text{poly(vinyl acetate)}}{\left(\!\!CH_2{-}\overset{\overset{\displaystyle O}{\overset{\displaystyle \|}{\overset{\displaystyle OCCH_3}{|}}}}{CH}\!\!\right)_{\!n}}$$

b)
$$\underset{\text{polypropylene}}{\left(\!\!CH_2{-}\overset{\overset{\displaystyle CH_3}{|}}{CH}\!\!\right)_{\!n}}$$

23.3   Structures of Polymers

Let's consider the structure of polyethylene in more detail. First, it is important to note that the number of monomers varies widely from one macromolecule to another. The termination steps, which stop the growth of individual polymer chains, occur at random times during the polymerization of those chains. Thus one reacting chain may terminate early, resulting in a polymer molecule that contains relatively few monomer units, while another may terminate much later in the chain process, resulting in a molecule that contains many more monomer units. While it is possible to exercise some experimental control over the average molecular mass of the polymer—that is, the average number of monomer units in each macromolecule—all synthetic polymers are composed of a variety of individual macromolecules of differing molecular masses.

The structure at one end of a polymer chain depends on the initiator that started that particular chain, and the structure at the other end depends on how that chain terminated. However, most polymers are so large that the ends do not have much effect on their properties. Therefore the structure of a polymer is usually represented by its repeat unit, and the ends are not specified.

The radical chain mechanism shown in Section 23.2 implies that polyethylene and polystyrene are composed of linear macromolecules—that is, that the carbons of the vinyl groups of the monomers are connected in a straight chain. In fact, there are two processes that can cause individual macromolecules to have branched structures. The first of these occurs when a growing radical chain abstracts a hydrogen from a random position in the interior of another macromolecule. This process, called chain transfer, occurs when the radical does not find another monomer unit to which to add nor another radical center so that termination can occur. Chain transfer results in the termination of the original chain but forms a new radical in the interior of the other macromolecule. Polymerization can occur at this new radical site, resulting in the formation of a long branch on the macromolecule. The formation of such a long branch in polyethylene is outlined in Figure 23.1.

A second process results in the formation of shorter branches that contain only four carbons. These result when the radical end of a growing polymer chain reaches back and abstracts a hydrogen from itself. Since the cyclic transition state for this abstraction is most favorable when it contains six atoms, four-carbon butyl group branches are formed. The mechanism for the formation of these butyl branches is outlined in Figure 23.2.

These two branching processes decrease the regularity of the polyethylene macromolecules. Individual polymer chains may have long branches or butyl branches that occur at random positions. As we will see shortly, this irregularity in the structure dramatically affects the physical properties of the polymer.

Problem

Problem 23.4   In addition to the four-carbon branches shown in Figure 23.2, polyethylene has a smaller number of branches containing three carbons. Show the steps in the mechanism for the formation of these three-carbon branches.

Another type of irregularity results if the vinyl monomer that is used to make an addition polymer has two different substituents on one end of the double bond.

The radical end of a growing polymer chain may occasionally abstract a hydrogen from the interior of another chain. This process is favorable because a primary radical in the reactant is converted to a secondary radical in the product. This terminates the original polymer and forms a reactive radical in the interior of the second chain that can serve as a site to initiate polymerization.

The new radical can add to an ethylene monomer, resulting in a polymer chain growing from the interior of this macromolecule. Although a secondary radical is converted to a primary radical, this step is still exothermic because a weak pi bond is broken and a stronger sigma bond is formed.

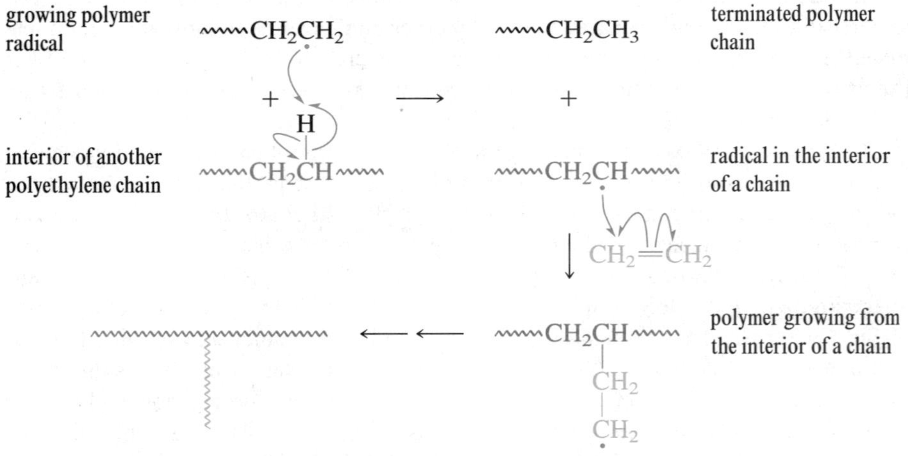

Figure 23.1

Mechanism for the Formation of a Long Branch in the Radical Polymerization of Ethylene.

This process results in the formation of a long branch on the original polymer chain. This is represented schematically by the structure at the left.

Styrene, with a hydrogen and a phenyl group on one of the vinyl carbons, provides an example. When such a monomer polymerizes, a new stereocenter (an asymmetric carbon) is created each time a new monomer is added:

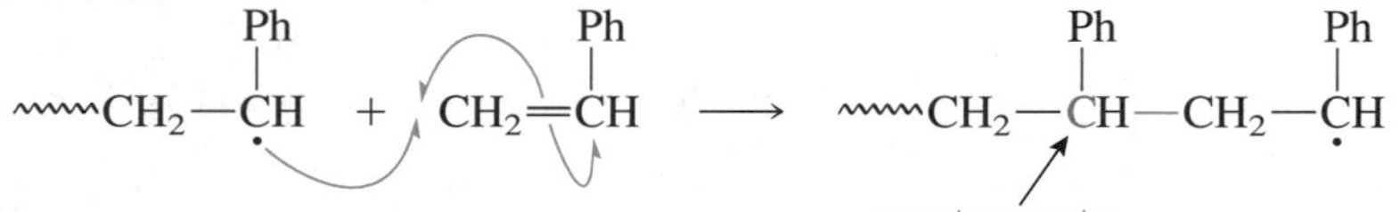

new stereocenter

Since there is no preference for one absolute configuration over the other at this new stereocenter, the resulting macromolecule has random configuration at its many stereocenters and is termed **atactic**. A partial structure for atactic polystyrene is shown in the following diagram:

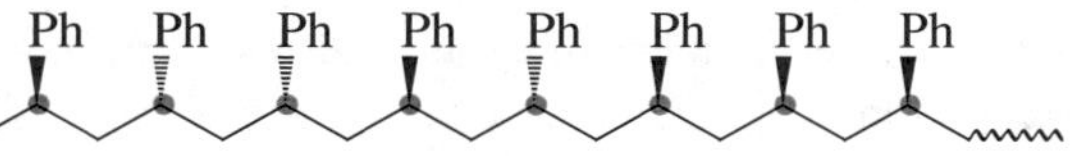

atactic polystyrene
(random configuration at the carbons bonded to the phenyl groups)

Problem

Problem 23.5 Which of these monomers could form an atactic polymer?

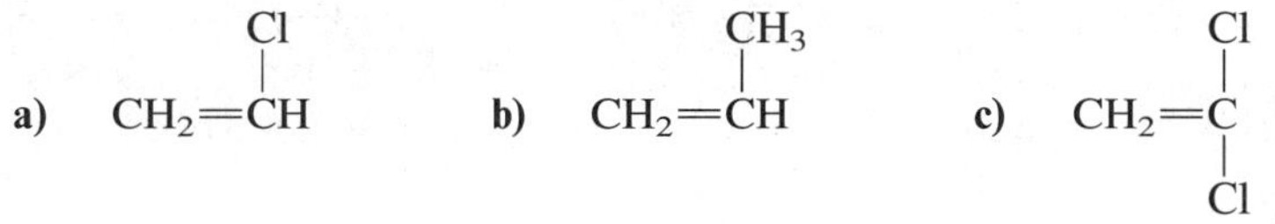

The radical carbon at the end of a growing polymer chain bends back and abstracts a hydrogen from the interior of the molecule. The most favorable size for the cyclic transition state is six atoms. This process is favorable because the original primary radical is converted to a more stable secondary radical.

Polymerization can continue at the site of the new secondary radical.

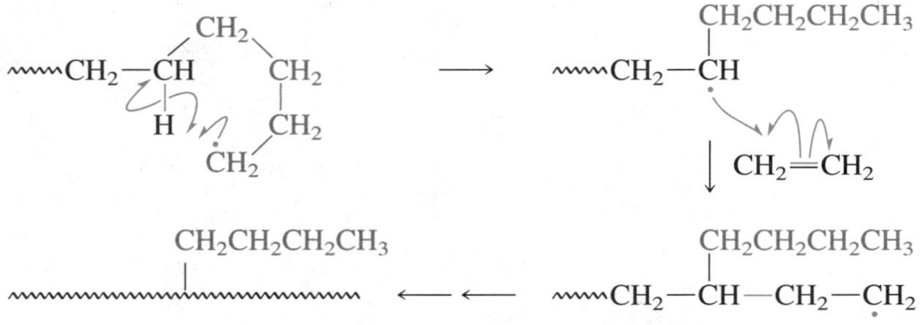

This results in a four-carbon branch on the ultimate macromolecule.

Figure 23.2

Mechanism for the Formation of a Butyl Branch During the Polymerization of Ethylene.

23.4 Ionic Polymerization

As well as mechanisms involving radical intermediates, addition polymers can also be made by mechanisms involving cationic or anionic intermediates. For example, polyisobutylene is made by treating isobutylene with a small amount of boron trifluoride and water:

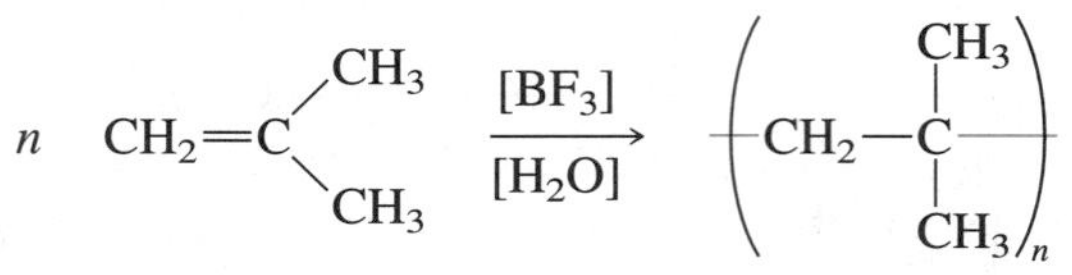

$$n \quad CH_2{=}C \overset{CH_3}{\underset{CH_3}{\diagdown}} \quad \xrightarrow[{[H_2O]}]{[BF_3]} \quad \left(CH_2{-}\overset{CH_3}{\underset{CH_3}{\overset{|}{\underset{|}{C}}}} \right)_n$$

isobutylene   polyisobutylene

The Lewis acid–base complex of water and boron trifluoride donates a proton to an isobutylene monomer to produce a carbocation. This carbocation adds to another isobutylene monomer to produce a larger carbocation, and the process continues, producing the polymer:

Super Glue

Super glue is a polymer of methyl cyanoacrylate. Because both the cyano and carbonyl groups of the monomer help to stabilize carbanions, this compound is very sensitive to polymerization by the anionic mechanism. The tube of glue contains very pure monomer, which does not polymerize until it contacts an initiator. However, contact with any nucleophile causes rapid polymerization. Therefore when the tube is opened, polymerization is initiated by water in the air, by SiOH groups on a glass surface, by FeOH groups on an iron surface, or by various nucleophiles that are part of the proteins in skin. The adhesion between the polymer and the surface to which it is applied is very strong because the polymer chains are covalently linked to the nucleophiles that are part of the surface!

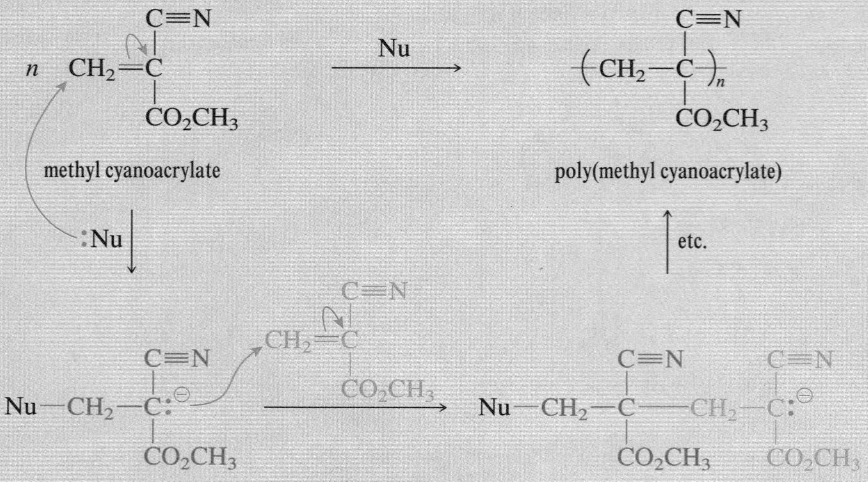

$$n \quad CH_2{=}\overset{C{\equiv}N}{\underset{CO_2CH_3}{\overset{|}{\underset{|}{C}}}} \quad \xrightarrow{\quad Nu \quad} \quad \left(CH_2{-}\overset{C{\equiv}N}{\underset{CO_2CH_3}{\overset{|}{\underset{|}{C}}}} \right)_n$$

methyl cyanoacrylate   poly(methyl cyanoacrylate)

:Nu $\downarrow$ $\uparrow$ etc.

$$Nu{-}CH_2{-}\overset{C{\equiv}N}{\underset{CO_2CH_3}{\overset{|}{\underset{|}{C}}}}\overset{\ominus}{:} \quad \xrightarrow{\quad\quad} \quad Nu{-}CH_2{-}\overset{C{\equiv}N}{\underset{CO_2CH_3}{\overset{|}{\underset{|}{C}}}}{-}CH_2{-}\overset{C{\equiv}N}{\underset{CO_2CH_3}{\overset{|}{\underset{|}{C}}}}\overset{\ominus}{:}$$

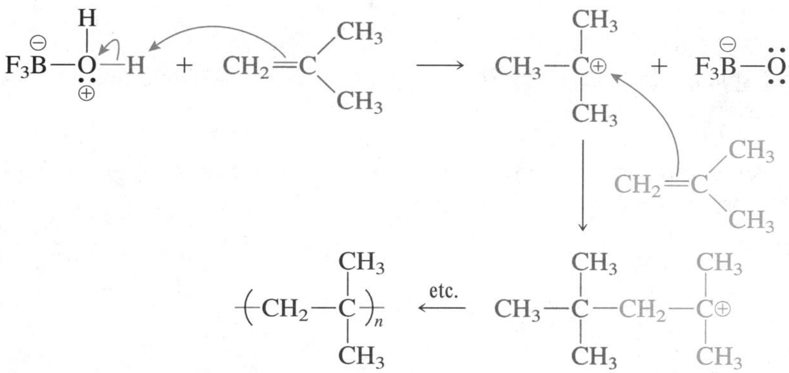

Since each addition occurs so as to form the more stable tertiary carbocation, the monomers are connected in a regular head-to-tail fashion. Cationic polymerization can be used only when one vinyl carbon of the monomer is substituted with groups that can stabilize the intermediate carbocation, as is the case with the two methyl groups of isobutylene.

Addition polymerization can also occur by a mechanism involving anionic intermediates. For example, styrene can be polymerized by the addition of a small amount of sodium amide. In this case the amide anion adds to the double bond to produce a carbanion. This carbanion then adds to another styrene molecule to form a larger carbanion, and the process continues to form polystyrene:

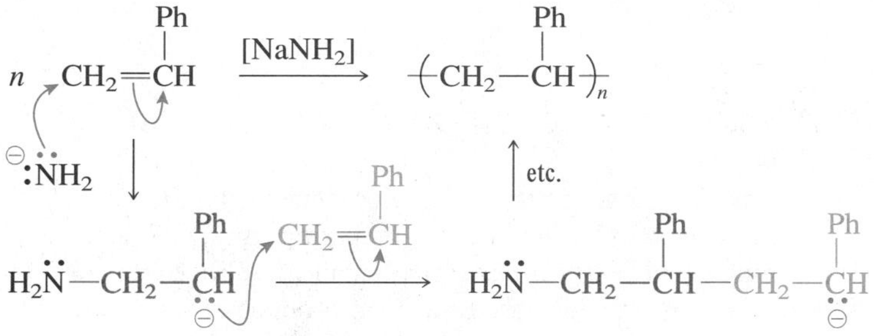

Most addition polymers are prepared from vinyl monomers. However, another type of addition polymer can be formed by ring-opening reactions. For example, the polymerization of ethylene oxide can be accomplished by treatment with a small amount of a nucleophile, such as methoxide ion. The product, a polyether, is formed by a mechanism involving anionic intermediates:

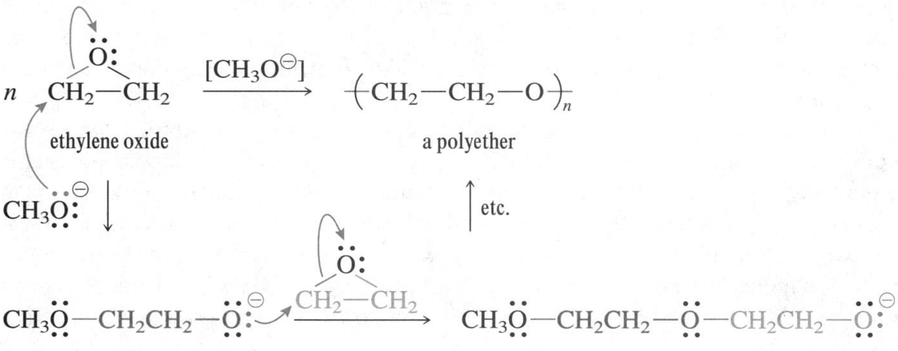

Problems

Problem 23.6   Explain why polyethylene cannot be prepared by cationic polymerization whereas polystyrene can.

Problem 23.7   Anionic polymerization is a good method for the preparation of polyacrylonitrile but not polyisobutylene. Explain.

$$
\begin{array}{cc}
& C\equiv N \\
& | \\
-\!\!\left(CH_2-CH\right)_{\!n} &
\end{array}
\qquad
\begin{array}{cc}
& CH_3 \\
& | \\
-\!\!\left(CH_2-C\right)_{\!n} \\
& | \\
& CH_3
\end{array}
$$

polyacrylonitrile   polyisobutylene

Problem 23.8   In contrast to the formation of a polyether from the reaction of ethylene oxide with a small amount of methoxide ion, a similar reaction using THF does not result in the formation of a polymer. Explain.

$$
\underset{\text{O}}{\bigcirc} \xrightarrow[\quad\times\quad]{[CH_3O^{\ominus}]} -\!\!\left(CH_2CH_2CH_2CH_2O\right)_{\!n}
$$

23.5   Coordination Polymerization

Perhaps the most important development in the area of addition polymerization was the discovery of a method for preparing vinyl polymers using metal catalysts. This breakthrough earned Karl Ziegler and Giulio Natta the 1963 Nobel Prize in chemistry and is probably used today to produce a larger amount of vinyl polymers than all other methods combined. While various catalysts have been employed, a typical one uses an organometallic compound, such as triethylaluminum, and a transition metal halide, such as titanium tetrachloride. Although the mechanism is complex, a simplified version that has the correct general features is shown in Figure 23.3 for the polymerization of propylene. Basically, the mechanism proceeds by coordination of a monomer to a titanium that has an alkyl group sigma bonded to it. The alkyl group then migrates to one carbon of the double bond of the coordinated monomer, while the other carbon forms a sigma bond to the titanium. This step regenerates the original catalyst, but with a larger alkyl group bonded to it. Additional monomer units are added in a similar fashion.

What makes the method using Ziegler-Natta catalysts so important? The resulting polymers are much more regular than those produced by other methods. For example, polyethylene produced by using a coordination catalyst is linear. It does not have the short or long branches that characterize polyethylene that is produced by a radical initiator. Furthermore, coordination polymerization can be used to prepare stereoregular polymers. For example, polypropylene can be prepared with identical configuration at all of the stereocenters. The resulting **isotactic** polypropylene has very

The active catalytic species results from the transfer of an ethyl group from an aluminum to form a sigma bond to a titanium species. Propylene forms a "pi complex" with the titanium by interaction of its pi MOs with a vacant coordination site on the metal.

The ethyl group migrates to one carbon of the double bond of the coordinated propylene, and the other carbon forms a sigma bond to the titanium. This creates a vacant coordination site on the metal, so the process can occur again.

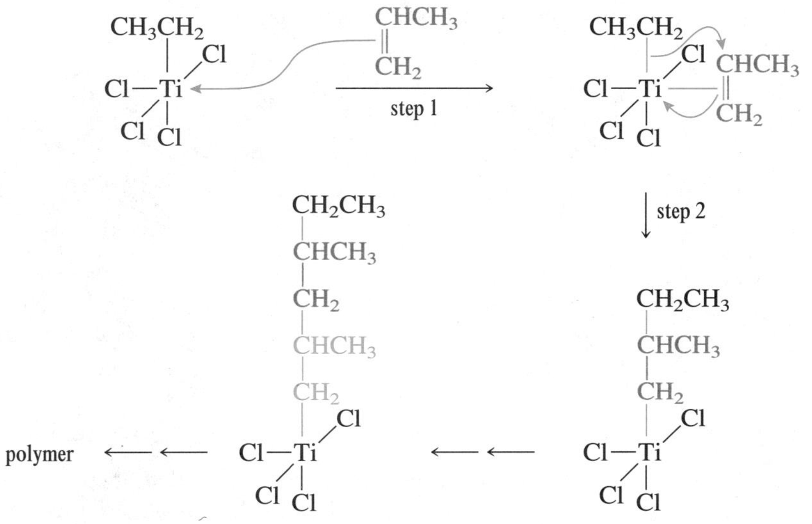

Continuation of this process produces polypropylene.

The process of steps 1 and 2 is repeated as another propylene coordinates to the titanium, followed by migration of the new, larger alkyl group.

Figure 23.3

A Simplified Version of the Mechanism for Polymerization Involving a Metal Coordination Catalyst.

different, and much more useful, properties than the atactic polypropylene that is produced by radical polymerization. Let's see how the structure and regularity of these polymers affect their physical properties.

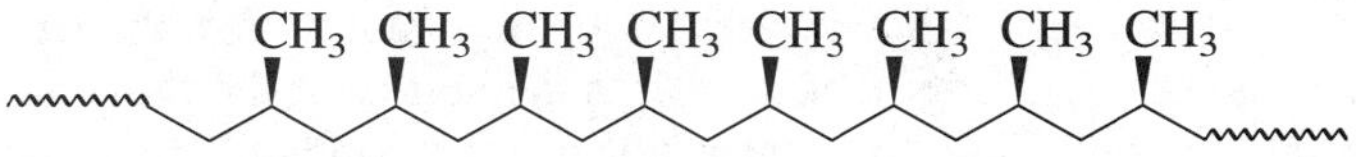

isotactic polypropylene
(identical configuration at all of the stereocenters)

23.6 Physical Properties of Polymers

The molecules of most nonpolymeric compounds are arranged with a very high degree of order in the solid state. Such compounds are said to form **crystalline solids.** Some compounds, however, have no order in the solid state. These compounds form glassy solids with a random arrangement of molecules and are said to be **amorphous.**

The enormous molecules of a polymer such as polyethylene are too long to form a completely crystalline solid. However, many polymers are semicrystalline; that is, they have both crystalline and amorphous regions. For example, solid polyethylene has crystalline regions, called **crystallites,** where the chains are arranged in a very ordered manner, along with amorphous regions that are completely disordered. The part of a polyethylene molecule that is in a crystallite has an *anti* conformation about each of its carbon–carbon bonds. The resulting zigzag chain can pack well with the zigzag chains of other molecules. A two-dimensional picture of a crystallite can be represented schematically as shown in the following diagram:

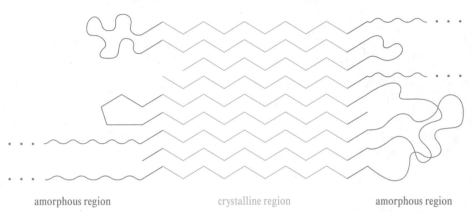

amorphous region   crystalline region   amorphous region

These crystallites vary in size and shape and are much smaller than the crystals of a normal organic compound. An individual crystallite has dimensions on the order of 10^3 Å. Recall that the length of most covalent bonds is slightly greater than 1 Å. Therefore the ordered part of a crystallite extends over a region containing tens to hundreds of bonds. When a chain reaches the boundary of a crystallite, it may bend back and become part of the crystallite again. Other chains wander off into the amorphous region. Some of these remain in the amorphous region; others return to the same crystallite, and some even become part of another crystallite. In general, polymers that have more and larger crystalline regions are stiffer and stronger and are more useful for many applications.

Regularity in the structure of a polymer favors crystallinity because the chains can pack closer together. The head-to-tail bonding of the individual monomer units is one type of regularity that is present in most addition polymers. The presence of branches is an irregularity that decreases the ability of the polymer chain to pack into crystalline regions. Therefore linear polyethylene prepared by polymerization using a Ziegler-Natta catalyst is more crystalline than the highly branched polyethylene produced by a radical mechanism. In addition, polymers must have a regular stereochemistry if they are to be crystalline. Atactic polymers are completely amorphous. Again, the ability to use coordination polymerization for the preparation of stereoregular polymers is extremely valuable.

Crystallinity is also favored by strong forces between polymer chains. In a nonpolar polymer such as polyethylene, the only forces holding the chains in place are van der Waals forces (see Section 2.6). Although these attractions are relatively weak, there are many of them, so their total force can be quite large. However, polymers with polar groups have stronger intermolecular forces and are more crystalline than nonpolar polymers, other factors being equal.

Most addition polymers are **thermoplastics;** that is, they are hard at room temperature but soften and eventually melt as they are heated. At low temperatures there is very little motion of the molecules, and the polymer is glasslike and brittle. As the temperature of the polymer is raised, it passes through its glass transition temperature (T_g). Above T_g, more motion of the chains is possible, and the polymer is a rubbery solid. Eventually, the polymer passes through its crystalline melting point (T_m), whereupon it melts to form a viscous liquid. Many semicrystalline polymers are most useful at temperatures between T_g and T_m. Both T_g and T_m increase as the crystallinity of the polymer increases and as the strength of the intermolecular forces between the polymer chains increases. The total intermolecular force increases as the length of the polymer chains increases.

In general, polymers have no useful mechanical properties until the chains reach a certain average length. Above this minimum length, the strength increases as the polymer gets longer, but it also becomes more difficult to process the polymer. Therefore the average molecular mass is usually a compromise, large enough that the polymer has useful mechanical properties but not so large that it cannot be molded, extruded, or drawn into fibers.

Problem

Problem 23.9 Explain which of these monomers produces the more crystalline polymer.

a) $CH_2{=}CH$ (Cl) or $CH_2{=}C$ (Cl, Cl) using radical polymerization

b) $CH_2{=}CH$ (CH_3) or $CH_2{=}CH$ (CO_2CH_3) using radical polymerization

c) $CH_2{=}CH$ (Ph) using radical polymerization or coordination polymerization

d) $CH_2{=}CH$ (CH_3) or $CH_2{=}C$ (CH_3, CH_3) using radical polymerization

23.7 Major Thermoplastic Addition Polymers

Four thermoplastic addition polymers—polyethylene, poly(vinyl chloride), polypropylene, and polystyrene—comprise the majority of the total amount of polymers manufactured in the United States. In 1994 a total of 49.5 billion pounds of these plastics were produced, distributed as shown in Table 23.1.

Two types of polyethylene are manufactured: high-density polyethylene (HDPE) and low-density polyethylene (LDPE). HDPE is produced by coordination polymerization

Table 23.1   Output of Thermoplastic Polymers.

Polymer	Structure	Amount (billion lb)
polyethylene	$\left(\!CH_2\!-\!CH_2\!\right)_n$	23.5
poly(vinyl chloride)	$\left(\!CH_2\!-\!\overset{\displaystyle Cl}{\underset{\displaystyle \vert}{CH}}\!\right)_n$	10.9
polypropylene	$\left(\!CH_2\!-\!\overset{\displaystyle CH_3}{\underset{\displaystyle \vert}{CH}}\!\right)_n$	9.5
polystyrene	$\left(\!CH_2\!-\!\overset{\displaystyle Ph}{\underset{\displaystyle \vert}{CH}}\!\right)_n$	5.6

using a Ziegler-Natta type catalyst. Its regularity and the absence of branches make it more crystalline, and the resulting closer packing of its chains results in a higher density. It is strong and rigid, with a higher T_m. It is used to make a variety of containers and plastic items, such as bottles, mixing bowls and other kitchen items, and toys. LDPE is produced by radical polymerization and has both long and short branches. As a result, it is more amorphous and less dense because its chains are not packed as closely. It is weaker and less rigid than HDPE. A majority of it is used as film in packaging prod-

ELABORATION

Teflon

Teflon is formed by radical polymerization of tetrafluoroethylene. The story of its discovery in a research laboratory at DuPont provides an interesting tale of chemical serendipity. Tetrafluoroethylene was synthesized, but attempts to polymerize it were not successful, so the gaseous compound was stored in a small cylinder. Some time later, when a sample was needed for another experiment, no gas came forth when the cylinder was opened. However, rather than presuming that the gas had leaked from the container and discarding it, the chemist weighed the cylinder. The weight was identical to that of the cylinder plus the gas that had originally been added to it. Obviously, the gaseous tetrafluoroethylene had decomposed to form nongaseous products. Out of curiosity, the chemist sawed open the cylinder. The polymer that he found had interesting properties and spurred additional efforts that soon resulted in a method for preparing the polymer. Because it is linear and has no stereochemical complications, Teflon has a high melting point of 327°C. It is insoluble and very chemically inert. It is used to make valves and chemically resistant coatings. And because of its nonstick characteristics and high temperature stability, it has found wide application as a coating for cooking utensils.

ucts, such as garbage bags, household plastic wrap, and the transparent film used to cover trays of meat in the supermarket. Its major advantage is its low cost.

Poly(vinyl chloride), also known as PVC, is prepared by radical polymerization to produce material composed of an average of 10,000–24,000 monomer units. It is atactic and therefore amorphous, but it has a relatively high T_g because of the large size of its molecules and its polar carbon–chlorine bonds. It is a rigid material and is used to make pipe, panels, and molded objects. About 68% of PVC is used in the building and construction industry. A more flexible form of PVC is produced by adding a plasticizer such as dioctyl phthalate (see the Elaboration "Dioctyl Phthalate Plasticizer" on page 1060). This is used to prepare electric wire coatings, film, and simulated leather or "vinyl."

Polypropylene owes its current market success to the development of coordination polymerization. Before 1957 it was not produced commercially because radical polymerization gives an atactic polymer that is amorphous and has poor mechanical properties. Using a coordination catalyst, however, enables the production of an isotactic polymer that is semicrystalline. This material is stiff and hard and has a high tensile strength. Among its many useful products are rope, molded objects, and furniture.

Polystyrene is made by radical polymerization and therefore is atactic and amorphous. Incorporation of small air bubbles produces a foam (styrofoam) that finds a major use in packaging materials and insulation.

Many other addition polymers are manufactured commercially, although in much smaller amounts than those described above. For example, poly(methyl methacrylate) is prepared by radical polymerization of the methyl ester of methacrylic acid:

$$n \quad CH_2{=}C{\overset{CH_3}{\underset{CO_2CH_3}{}}} \quad \xrightarrow{\text{initiator}} \quad {-}(CH_2{-}\overset{CH_3}{\underset{CO_2CH_3}{C}}){-}_n$$

methyl methacrylate   poly(methyl methacrylate)

Although it is atactic and amorphous, its polar groups cause it to have a relatively high T_g (110°C), and it is rigid and glasslike at normal temperatures. It is used to make products such as Plexiglas and Lucite.

$$n \quad {\overset{F}{\underset{F}{}}}C{=}C{\overset{F}{\underset{F}{}}} \quad \xrightarrow[\text{initiator}]{\text{radical}} \quad {-}(\overset{F}{\underset{F}{C}}{-}\overset{F}{\underset{F}{C}}){-}_n$$

tetrafluoroethylene   poly(tetrafluoroethylene)
                                       Teflon

Problem

Problem 23.10 Explain why Teflon is linear and has no stereochemical complications even though it is prepared by radical polymerization.

Figure 24.1 shows all of the possible D-aldoses with three to six carbons. Of these, only D-glucose is commonly found as a monosaccharide in nature. Several others, including D-glyceraldehyde, D-mannose, D-galactose, and D-ribose, are found as part of polysaccharides or in other biological molecules. In addition, the ketoses D-fructose, D-ribulose and D-xylulose (2-ketopentoses with the same configuration at the other carbons as ribose and xylose), and dihydroxyacetone (1,3-dihydroxy-2-propanone) are also common.

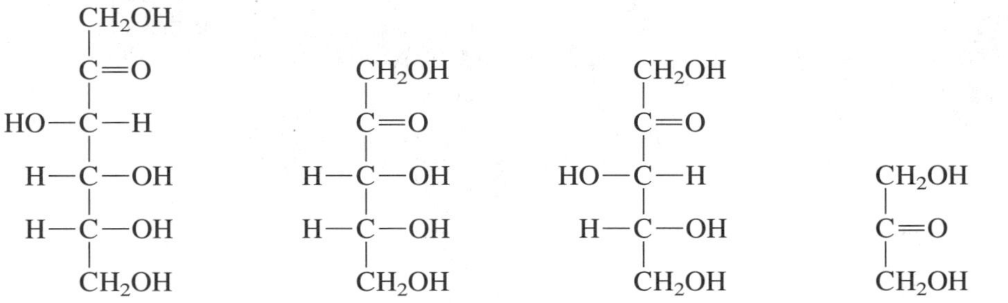

D-fructose   D-ribulose   D-xylulose   dihydroxyacetone

The use of D and L to designate the absolute stereochemistry of naturally occurring compounds is firmly entrenched and is actually simpler in many ways than the R, S method. However, it is informative to see how molecules such as the aldohexoses would be named by using the R, S system. Since they differ only in the configuration at their stereocenters, all 16 aldohexoses would be named as 2,3,4,5,6-pentahydroxyhexanals. To distinguish among them it is necessary only to designate the configuration at the four stereocenters.

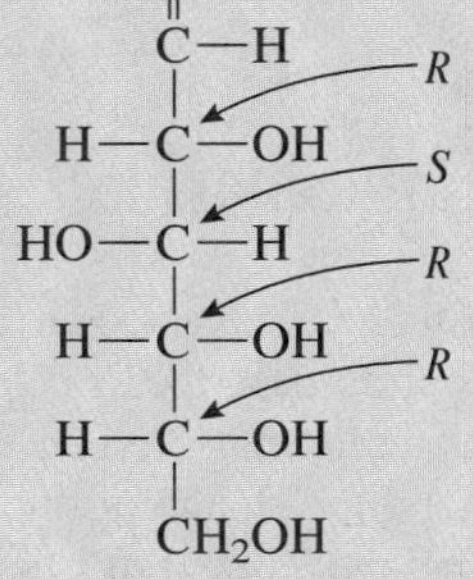

R and S Nomenclature Applied to Sugars

D-glucose

Thus, D-glucose is $(2R),(3S),(4R),(5R)$,6-pentahydroxyhexanal. D-Mannose, which has configuration opposite to D-glucose at carbon 2, is $(2S),(3S),(4R),(5R)$,6-pentahydroxyhexanal. Note that all of the D-aldohexoses have the R configuration at carbon 5. While it is possible, using the R, S system, to write a name that correctly designates the stereochemistry of any aldose, or other carbohydrate for that matter, by simply examining its structure, the name D-glucose conveys information more rapidly than does the name $(2R),(3S),(4R),(5R)$,6-pentahydroxyhexanal. Most chemists, even experienced ones, would need to carefully write out the structure of the compound with the latter name before they would be able to determine that it is D-glucose rather than one of its stereoisomers.

Problems

Problem 24.3   Draw Fischer projections for these monosaccharides.

    a) L-glyceraldehyde   **b)** L-mannose

Problem 24.4   Determine the identity of each of these carbohydrates.

a)

$$
\begin{array}{c}
\overset{\displaystyle O}{\underset{\displaystyle \|}{}} \\
CH \\
| \\
HO-C-H \\
| \\
HO-C-H \\
| \\
CH_2OH
\end{array}
$$

b)

$$
\begin{array}{c}
\overset{\displaystyle O}{\underset{\displaystyle \|}{}} \\
CH \\
| \\
HO-C-H \\
| \\
HO-C-H \\
| \\
H-C-OH \\
| \\
HO-C-H \\
| \\
CH_2OH
\end{array}
$$

c)

$$
\begin{array}{c}
CH_2OH \\
| \\
HO-C-H \\
| \\
HO-C-H \\
| \\
HO-C-H \\
| \\
H-C-OH \\
| \\
CH \\
\| \\
O
\end{array}
$$

(Hint: This must be rotated first.)

24.4   Cyclization of Monosaccharides

Up to this point, the structure of glucose has been shown as an aldehyde with hydroxy groups on the other carbons. However, as described in Section 14.10, aldehydes and ketones react with alcohols to form hemiacetals. When this reaction is intermolecular—that is, when the aldehyde group and the alcohol group are in different molecules—then the equilibrium is unfavorable, and the amount of hemiacetal that is present is very small. However, when the aldehyde group and the alcohol group are contained in the same molecule, as is the case in the second equation below, the intramolecular reaction is much more favorable (because of entropy effects; see Sections 7.14 and 14.10), and the hemiacetal is the predominant species present at equilibrium.

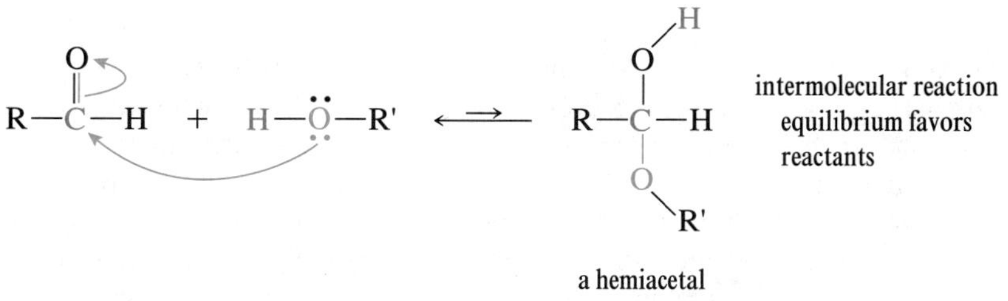

a hemiacetal

intermolecular reaction
equilibrium favors
reactants

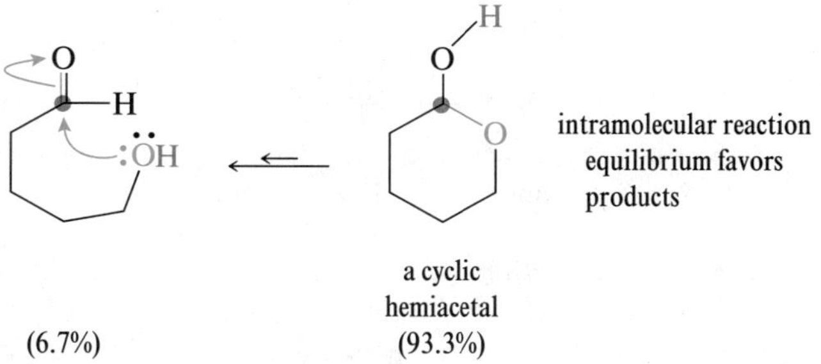

a cyclic
hemiacetal
(93.3%)

(6.7%)

intramolecular reaction
equilibrium favors
products

Because glucose and the other monosaccharides contain both a carbonyl group and hydroxy groups, they exist predominantly in the form of cyclic hemiacetals.

Let's consider the cyclization of glucose in more detail. There are five different hydroxy groups that might react with the aldehyde group. However, because five- and six-membered rings are much more stable than others, these are the only ring sizes that need be considered. These two possibilities are illustrated for glucose in the following equation:

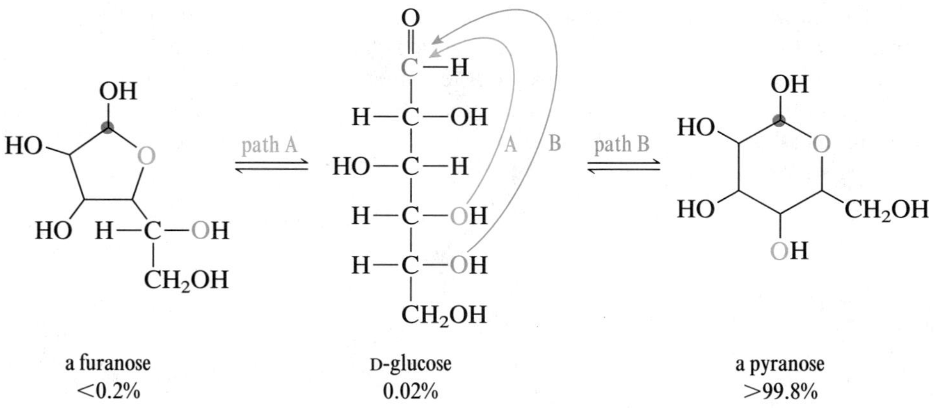

a furanose
<0.2%

D-glucose
0.02%

a pyranose
>99.8%

The cyclic acetal that has a five-membered ring is called a **furanose.** This name is derived from that of the five-membered, oxygen-containing heterocyclic compound **furan.** The hemiacetal with a six-membered ring is known as a **pyranose** after the heterocycle **pyran.**

furan   pyran

As we saw in Chapter 6, six-membered rings are generally more stable than five-membered rings, primarily because of increased torsional strain in the latter. Therefore it is not surprising that the pyranose form of a monosaccharide is usually more stable than the furanose form. At equilibrium, glucose exists primarily as the pyranose (>99.8%), with little, if any, furanose (<0.2%) present. There is also a trace amount (0.02%) of the uncyclized aldehyde present. Of course, this equilibrium depends on the structure of the monosaccharide, and some other sugars have larger amounts of the furanose form.

Problems

Problem 24.5   Show the steps in the mechanism for the cyclization of the open form of D-glucose to the pyranose form. (Use the acid-catalyzed mechanism of Chapter 14.)

Problem 24.6   At equilibrium, D-fructose exists 67.5% in the pyranose form, 31.5% in the furanose form, and 1% in the open, uncyclized form. Draw the pyranose and furanose forms. Explain why D-fructose has more of the uncyclized form present at equilibrium than does D-glucose.

Problem 24.7   How much pyranose form, furanose form, and uncyclized aldehyde would be present at equilibrium for L-glucose?

Problem 24.8   Show the structure of the hemiacetal formed from D-erythrose.

Let's now address the stereochemistry of the cyclization of D-glucose to a pyranose. Note that carbon 1, the hemiacetal carbon, becomes a new stereocenter when the cyclization occurs. Therefore two diastereomers of the pyranose, with different configurations at the new stereocenter, are formed when D-glucose cyclizes. Such diastereomers are called **anomers**. The two anomers for the pyranose form of D-glucose are shown in the following equation:

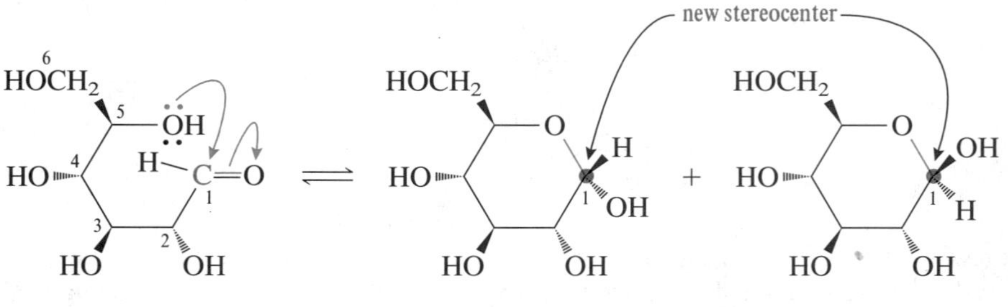

D-glucose   α-D-glucopyranose   β-D-glucopyranose
                          36.4%   63.6%

By convention, when the sugar is drawn as shown in the preceding equation and the hydroxy group at the new stereocenter projects down, the compound is designated as the α-stereoisomer. When the hydroxy group at the hemiacetal carbon projects up, it is the β-stereoisomer. The full names for these two anomers of glucose are α-D-glucopyranose and β-D-glucopyranose.

Problem

Problem 24.9 D-Mannose differs from D-glucose only in its configuration at C-2. Show the formation of α-D-mannopyranose and β-D-mannopyranose from the uncyclized form of D-mannose in the same manner as done for D-glucose.

Carbohydrate chemists often represent these compounds using **Haworth projections.** In this method the ring is drawn flat and viewed partly from the edge. The bonds to the substituents are shown as coming straight up or straight down from the carbons. While Haworth projections have the geometry at the carbons distorted, they are easy to draw and make the stereochemical relationships among the substituents readily apparent.

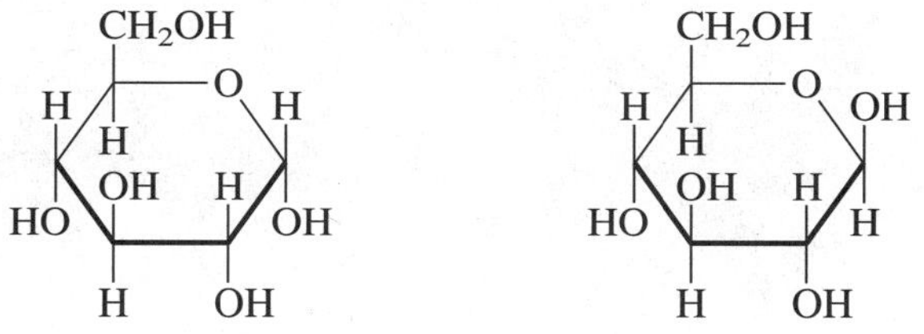

α-D-glucopyranose β-D-glucopyranose

Haworth projections

Problem

Problem 24.10 Show Haworth projections for α-D-mannopyranose and its β-anomer. (Remember that D-mannose differs from D-glucose only in its configuration at C-2.)

Haworth projections show the absolute stereochemistries at the various stereocenters of glucose quite well, but the six-membered rings of pyranoses are, of course, not planar. The presence of an oxygen atom in these rings causes only a slight perturbation, so most of the discussion about cyclohexane rings presented in Chapter 6 also applies to pyranose rings. The cyclization of D-glucose to form chair conformations of α- and β-D-glucopyranose is shown in the following diagrams:

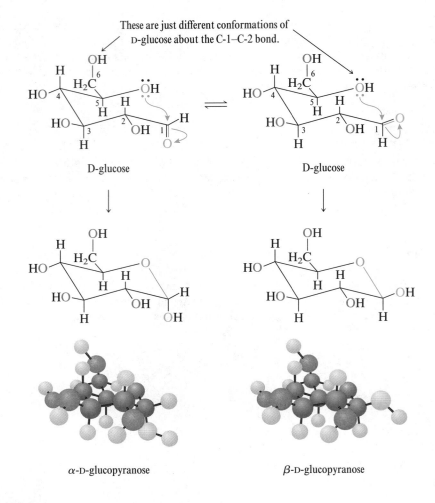

α-D-glucopyranose   β-D-glucopyranose

Problem

Problem 24.11   Show the chair conformations for α-D-mannopyranose and its β-anomer.

As was the case for cyclohexane derivatives, the chair conformer that has the larger groups equatorial is usually more stable. Therefore the preceding chair conformers, which have most or all of the larger substituents equatorial, are more stable than the conformers obtained by ring flips. The α- and β-anomers differ only in the stereochemistry of the groups at the hemiacetal carbon. In the α-anomer the hydroxy group on this carbon is axial, and in the β-anomer it is equatorial.

Both α-D-glucopyranose and β-D-glucopyranose can be isolated in pure form. Because they are diastereomers, they have different physical properties. For example, the α-stereoisomer has a specific rotation of +112.2°, while that of the β-isomer is +18.7°. However, if either of these pure stereoisomers is dissolved in water, the specific rotation slowly changes, over a period of several hours, to a value of +52.7°.

This process, termed **mutarotation,** results from the formation of an equilibrium mixture that consists of 36.4% of the α-isomer and 63.6% of the β-isomer. (Of course, the same equilibrium mixture results starting from either of the anomers.) In fact, it is the specific rotation at equilibrium that is used to calculate the equilibrium concentration of the two stereoisomers.

Problems

Problem 24.12   Given that the rotation of α-D-glucopyranose is $+112.2°$, the rotation of β-D-glucopyranose is $+18.7°$, and the rotation of an equilibrium mixture of the two anomers is $+52.7°$, calculate the percentages of each anomer present at equilibrium.

Solution
Let x equal the decimal fraction of the α-isomer that is present at equilibrium. Then $1 - x$ equals the decimal fraction of the β-isomer present. The rotations of each must total to $+52.7°$:

$$x(+112.2°) \quad + \quad (1-x)(+18.7°) \quad = \quad +52.7°$$

$$112.2x \quad + \quad 18.7 \quad - \quad 18.7x \quad = \quad 52.7$$

$$93.5x \quad = \quad 34.0$$

$$x \quad = \quad 0.364$$

Therefore 36.4% of the α-isomer and 63.6% of the β-isomer are present at equilibrium.

Problem 24.13   D-Mannose exists entirely in pyranose forms. The specific rotation for the α-anomer is $+29°$, and that for the β-isomer is $-17°$. The rotation of the equilibrium mixture is $+14°$. Calculate the percentages of each anomer in the equilibrium mixture.

24.5 Reactions of Monosaccharides

Although monosaccharides exist predominantly as hemiacetals, enough aldehyde or ketone is present at equilibrium that the sugars give most of the reactions of these functional groups. In addition, monosaccharides exhibit the reactions of alcohols. Of course, the presence of both functional groups may perturb the reactions of either of them.

Alcohol and aldehyde groups are readily oxidized. Reaction with nitric acid oxidizes both the aldehyde group and the primary alcohol group of an aldose to produce a dicarboxylic acid. As an example, D-galactose is oxidized to the dicarboxylic acid known as galactaric acid:

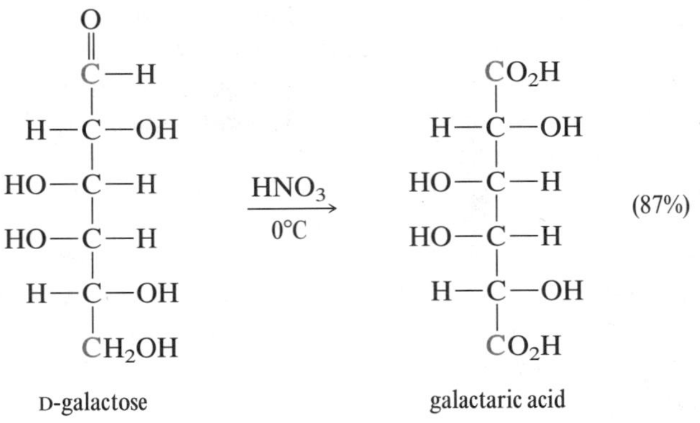

D-galactose   galactaric acid

Problems

Problem 24.14 Although D-galactose rotates plane-polarized light, its oxidation product, galactaric acid, does not. Explain.

Problem 24.15 Identify all of the D-aldopentoses from Figure 24.1 that, on oxidation with nitric acid, give diacids that do not rotate plane-polarized light.

Determination of Anomer Configuration

The determination of the configuration at the anomeric carbon of a cyclic sugar such as D-glucopyranose is a difficult problem. Usually, both stereoisomers must be isolated. Then comparison of their properties to those of compounds of known stereochemistry often enables the correct configuration to be determined.

Nuclear magnetic resonance spectroscopy can be used to answer this question in some cases. It has been demonstrated that the coupling constant between two hydrogens depends on their dihedral angle. This relationship, known as the Karplus correlation, is shown on the next page. Note that the coupling constant is maximum at dihedral angles near 180° and is minimum at angles near 90°. In the case of hydrogens on a six-membered ring, this correlation is especially useful. The dihedral angle for axial hydrogens on adjacent carbons is 180°, resulting in a relatively large coupling constant of 8–10 Hz. In contrast, if one hydrogen is axial and one is equatorial, or if both are equatorial, then their dihedral angle is 60°. This results in a much smaller coupling constant, usually about 2–3 Hz.

Consider the two anomers α- and β-D-glucopyranose. The peak for the hydrogen on the anomeric carbon, carbon 1, is separated from those for the other hydrogens because of the downfield shift caused by the two electronegative oxygens attached to the carbon. The hydroxy group on carbon 1 of the α-stereoisomer is axial, so the hydrogen on this carbon is equatorial. The hydroxy group on carbon 2 is equatorial, so the hydrogen on this carbon is axial. As can be seen from the Newman projections, the dihedral angle between the hydrogens is 60°. The coupling constant is 3 Hz in this case. The β-isomer has the opposite configuration at carbon 1, the anomeric carbon. Therefore the hydrogen on this carbon is axial. The dihedral angle between this hydrogen and the hydrogen on carbon 2, which is also axial, is 180°. This results in a larger coupling constant of 8 Hz for the β-isomer. This method works for other stereoisomers of glucose also, as long as the hydrogen on carbon 2 is axial.

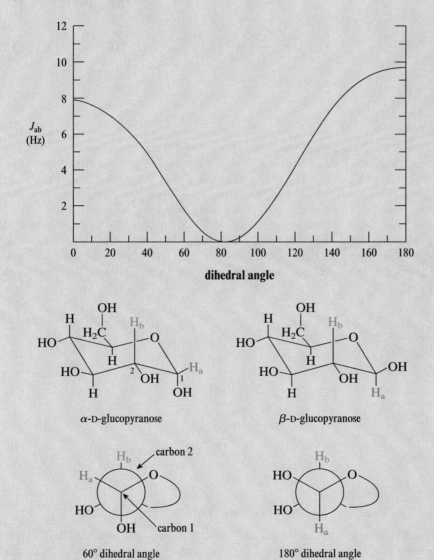

α-D-glucopyranose

β-D-glucopyranose

60° dihedral angle
$J_{ab} = 3$ Hz

180° dihedral angle
$J_{ab} = 8$ Hz

It is also possible to selectively oxidize only the aldehyde group with a milder oxidizing agent such as bromine. The oxidation of D-glucose with bromine produces D-gluconic acid in high yield. The reaction of the carboxylic acid group with one of the hydroxy groups to form an intramolecular ester (a lactone) occurs readily. Both five-membered and six-membered lactone rings can be formed.

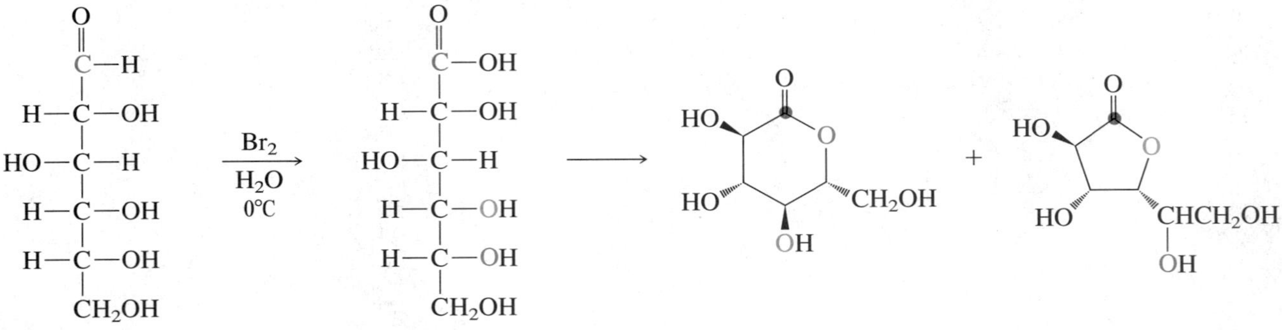

D-glucose D-gluconic acid (96%)

Oxidation reactions are often used as a test for the presence of an aldehyde group in a carbohydrate. In Tollins' test, the compound is treated with a solution of $Ag^{\oplus}$ ion in aqueous ammonia. The aldehyde group is oxidized, and the $Ag^{\oplus}$ ion is reduced to metallic silver. The formation of a silver mirror constitutes a positive test. Benedict's test and Fehling's test both employ a solution of Cu^{2+} ion in aqueous base. When the carbohydrate is oxidized, the blue Cu^{2+} ion is reduced to Cu_2O, which forms a brick-red precipitate. (Benedict's reagent has been used in a self-test kit for sugar in the urine caused by diabetes.) Carbohydrates that give a positive test in these reactions are called **reducing sugars** because they reduce the metal ion.

$$RCHO + Ag^{\oplus} \xrightarrow[\ominus OH,\ H_2O]{NH_3} RCO_2H + \text{other oxidation products} + \underset{\text{silver mirror}}{Ag\ (s)}$$

Tollins' test

$$RCHO + Cu^{2+} \xrightarrow[H_2O]{\ominus OH} RCO_2H + \text{other oxidation products} + \underset{\text{brick-red}}{Cu_2O\ (s)}$$

Benedict's test and Fehling's test

The aldehyde group can be readily reduced by catalytic hydrogenation or with reagents such as sodium borohydride. The reduction of xylose produces xylitol, which is used as a sweetener in "sugarless" gum:

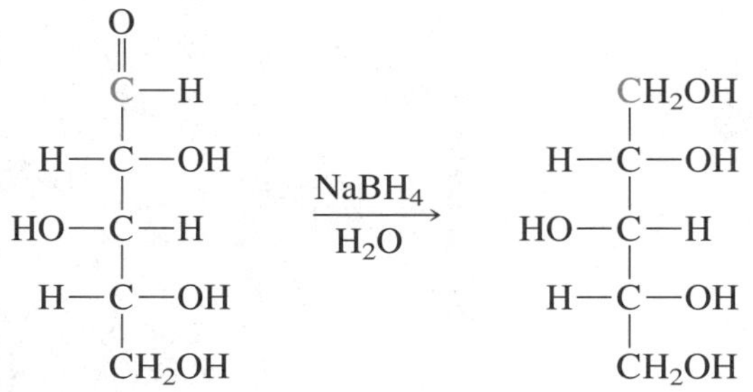

NaBH$_4$ / H$_2$O

D-xylose → xylitol

Problems

Problem 24.16   Why is xylitol not called D-xylitol?

Problem 24.17   Identify all of the D-aldohexoses from Figure 24.1 that, on reduction with NaBH$_4$, give products that do not rotate plane-polarized light.

Artificial Sweeteners

ELABORATION

Roses are red, violets are blue, sugar is sweet, and so are—many other organic compounds. For example, many compounds that have hydroxy groups on adjacent carbons have a sweet taste. The simplest of these is ethylene glycol or 1,2-ethane-diol, the main component of antifreeze. Although this diol is sweet, it is also toxic. Every year, numerous cases of poisonings occur when animals, especially dogs, ingest antifreeze that was carelessly discarded.

Xylitol, another polyhydroxy compound, is used as a sweetener in "sugarless" gum. It has approximately the same number of calories per gram as does sucrose and is not a low-calorie sweetener. However, because it does not have a carbonyl group, it is not fermented by bacteria in the mouth and does not promote tooth decay.

In recent years there has been a great search for a safe, calorie-free sweetener to help diabetics who need to control their sugar intake and, more recently, to meet the demands of health- and diet-conscious consumers. One such sweetener that was marketed in the late 1960s was calcium cyclamate. However, cyclamates were banned in the United States in 1970 because of a suspected link with cancer.

Saccharin is 300 times as sweet as sucrose on a weight basis. Like cyclamate, it is a low-calorie sweetener, not because it has fewer calories than sucrose on a weight basis, but because it is so much sweeter that only a small amount need be used. Saccharin is another artificial sweetener that is suspected of causing cancer in laboratory animals at very high doses. Although the U.S. Food and Drug Administration moved to prohibit the use of saccharin in the late 1970s, the ban was

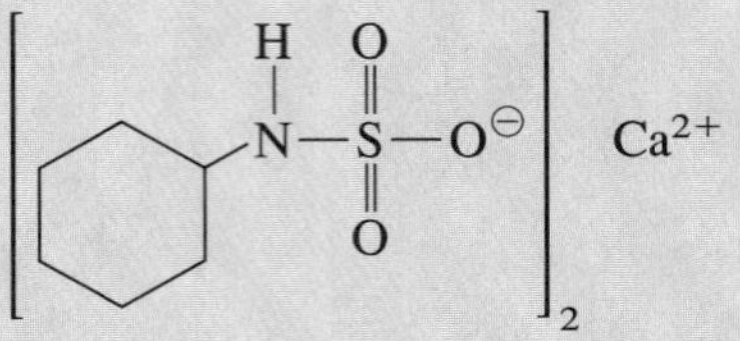

calcium cyclamate

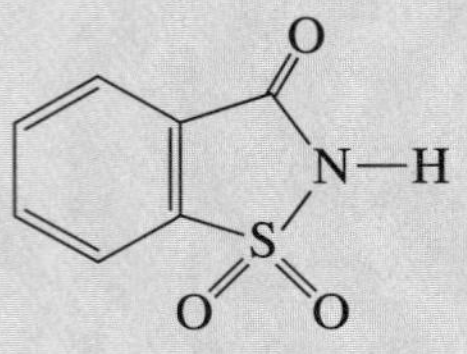

saccharin
(300×)

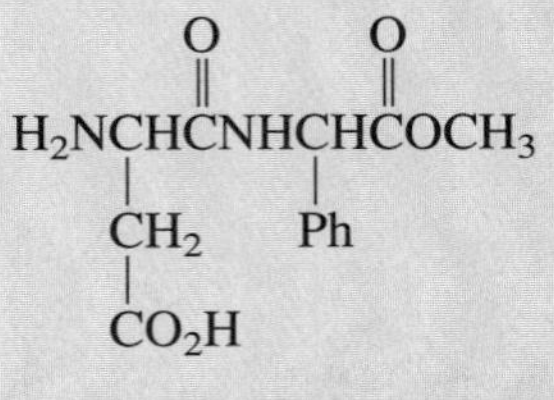

aspartame
(180×)

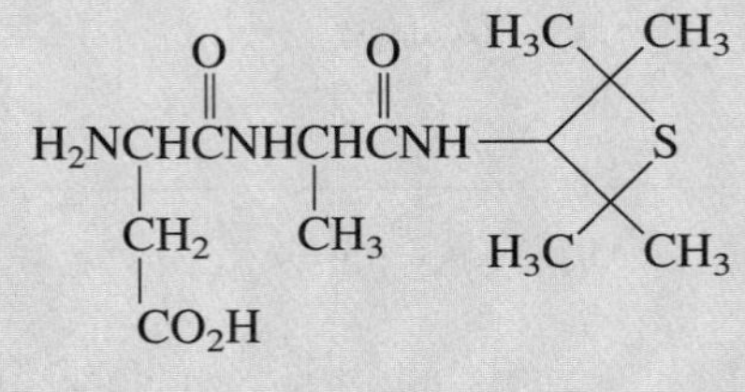

alitame
(2000×)

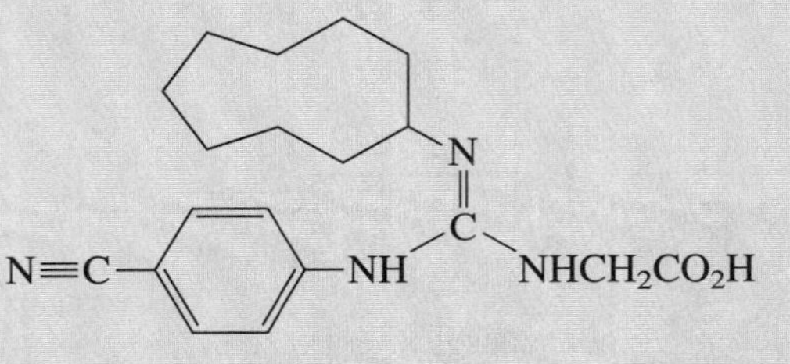

Some Artificial Sweeteners.

Numbers in parentheses indicate
how much sweeter than sucrose
these compounds are.

sucrononic acid
(200,000×)

blocked by Congress because saccharin was the only artificial sweetener available at that time.

The sweetener aspartame was discovered in 1965 and approved by the FDA in 1981. It is the methyl ester of a dipeptide formed from the amino acids aspartic acid and phenylalanine. Since both of these amino acids occur naturally and are part of nearly every protein, there is much less reason to be concerned about the health effects of this compound. Nevertheless, it has been extensively tested. Aspartame is about 180 times sweeter than sucrose, so the amount that is needed to sweeten a can of a soft drink, for example, is so small that it contributes only negligible calories to the diet. In addition, the taste profile of aspartame is much closer to sugar than is that of saccharin. Aspartame, sold under the brand name NutraSweet, has been an enormous financial success, with sales of $949 million in 1991.

As might be expected, the search for an even better artificial sweetener continues. Alitame is a dipeptide formed from aspartic acid and alanine, with an unusual amide at the carboxylate end of the alanine. It is 2000 times as sweet as sucrose—one pound of alitame has the sweetening power of one ton of sucrose!

In addition, because an amide bond is more stable than an ester bond, alitame is more stable to hydrolysis than is aspartame. Therefore alitame keeps its sweetness in aqueous solution better than aspartame does. Even sweeter compounds have been discovered. For example, sucrononic acid is 200,000 times as sweet as sucrose. Neither alitame nor sucrononic acid has yet been approved for use.

Obtaining approval for the use of a new artificial sweetener is a very expensive proposition and requires considerable expenditure of time and scientific effort. This is absolutely necessary, though, because the number of people that would be exposed to products containing the sweetener is so large. (Of course, the fact that most products would contain only a very small amount of an artificial sweetener somewhat decreases the potential for problems.) In addition, certain groups, such as diabetics, might have consumption patterns that are much higher than other parts of the population. Although obtaining FDA approval is extremely expensive, we may indeed see the approval of another artificial sweetener in the future because the financial rewards are potentially so large.

Examination of the structures of these sweet compounds does not reveal any simple pattern for their structural features. Chemists are still attempting to determine what it is that makes a compound sweet. They are also trying to model the taste receptor that is responsible for detecting sweetness. Although some progress has been made in this area, there is still a long way to go.

Because of this difficulty in predicting what features are necessary for a compound to taste sweet, most of the discoveries of artificial sweeteners have been serendipitous. In fact, many of the early discoveries resulted from dangerous laboratory practices that we would not condone today. For example, the sweetness of saccharin was discovered in 1879 by a chemist who spilled some of the compound on his hand. Later, while eating lunch in the laboratory, he noticed the extremely sweet taste. The sweetness of cyclamate was discovered in 1937 by a chemist who tasted it on a cigarette that he had set on the lab bench. And aspartame was found to be sweet by a chemist who got some on his hand and later licked his finger before picking up a piece of paper. This resulted in a billion-dollar-per-year product!

The unsafe nature of these practices needs to be emphasized. Precautions should always be taken to minimize exposure to all laboratory compounds. Hands should be washed whenever a compound is spilled on them. Eating, drinking, and smoking in the laboratory are all extremely unsafe practices and should be forbidden. Even licking a finger to pick up a piece of paper should be avoided.

As is the case with other carbonyl compounds, the carbonyl group of a sugar causes any hydrogens on adjacent carbons to be weakly acidic. This provides a mechanism for the isomerization of sugars in basic solution. Thus D-glucose is isomerized to D-mannose and D-fructose under basic conditions.

Problems

Problem 25.2   Explain why the carboxylic acid group of the cationic form of glycine ($pK_a = 2.3$) is a stronger acid than the carboxylic acid group of acetic acid ($pK_a = 4.8$).

Problem 25.3   Explain the order of these pK_a values:

$$\overset{\oplus}{H_3N}CH_2\overset{\overset{O}{\|}}{C}OCH_3 \qquad \overset{\oplus}{H_3N}CH_2\overset{\overset{O}{\|}}{C}O^{\ominus}$$

$$pK_a = 7.8 \qquad\qquad pK_a = 9.8$$

There are two important points to be learned from this discussion. First, glycine (and other amino acids) is never present in aqueous solution in a neutral form with uncharged carboxylic acid and amino groups. It is present as a cationic form, a dipolar ion, or an anionic form, depending on the pH. Second, because the pH of most physiological solutions is near 7, which is close to the isoelectric point of glycine, it is commonly present as a dipolar ion in biological fluids.

Table 25.1 lists the pK_as of the amino acids along with their isoelectric points (pI). Both those with nonpolar side chains and those with polar side chains have pK_a values near those of glycine. Therefore these all have isoelectric points near 6 and exist predominantly as dipolar ions at neutral pH.

However, some amino acids have an acidic or basic functional group in their side chains. In these cases there is another acidity constant to consider. Let's examine the case of aspartic acid, which has two carboxylic acid groups and an amino group. At low pH, aspartic acid is present in a cationic form:

$$\overset{\oplus}{H_3N}\overset{\overset{O}{\|}}{C}HCOH \rightleftharpoons \overset{\oplus}{H_3N}\overset{\overset{O}{\|}}{C}H\overset{O}{C}O^{\ominus} \rightleftharpoons \overset{\oplus}{H_3N}\overset{\overset{O}{\|}}{C}HCO^{\ominus} \rightleftharpoons H_2N\overset{\overset{O}{\|}}{C}HCO^{\ominus}$$

CH_2	CH_2	CH_2	CH_2
COH	COH	$CO^{\ominus}$	$CO^{\ominus}$
O	O	O	O

$pK_{a3} = 9.9$ $pK_{a1} = 2.0$ $pK_{a2} = 3.9$

cationic form	dipolar ion	anionic form	dianionic form

As the pH is increased, the proton is removed first from the carboxylic acid group closest to the ammonium group to generate the dipolar ion, then from the other carboxylic acid group, to generate an anionic form, and finally from the ammonium group to generate a dianionic form. The pI is the average of pK_{a1} and pK_{a2} and equals 3.0. The concentration of the anionic form is a maximum at a pH of 6.9, the average of pK_{a2} and pK_{a3} . Therefore at neutral pH, aspartic acid is present predominantly in its anionic form. The situation is quite similar for glutamic acid.

The amino acids with basic side chains—lysine, arginine, and histidine—all have pI values greater than 7. In strongly acidic solution they exist in dicationic forms, and all have significant amounts of a cationic form and a dipolar ion present at neutral pH.

Problems

Problem 25.4   Explain why the carboxylic acid group in the side chain of aspartic acid ($pK_a = 3.9$) is a weaker acid than the main carboxylic acid group of the amino acid ($pK_a = 2.0$).

Problem 25.5   Explain why the carboxylic acid group in the side chain of glutamic acid is a weaker acid than the carboxylic acid group in the side chain of aspartic acid.

Problem 25.6   Show the structures of these species.

a)   the dipolar ion of proline

b)   the dianion form of tyrosine

c)   the anion form of cysteine (careful)

d)   the cation form of arginine

Solution

b)   The pK_as of the various groups of tyrosine are listed in Table 25.1. Starting from the cationic form that is present in strong acid, first a proton is removed from the carboxylic acid group ($pK_a = 2.2$) to form the dipolar ion. The next proton is removed from the α-$NH_3^{\oplus}$ ($pK_a = 9.2$) to give the anionic form. The final proton is removed from the OH group of the side chain ($pK_a = 10.5$) to give the dianionic form:

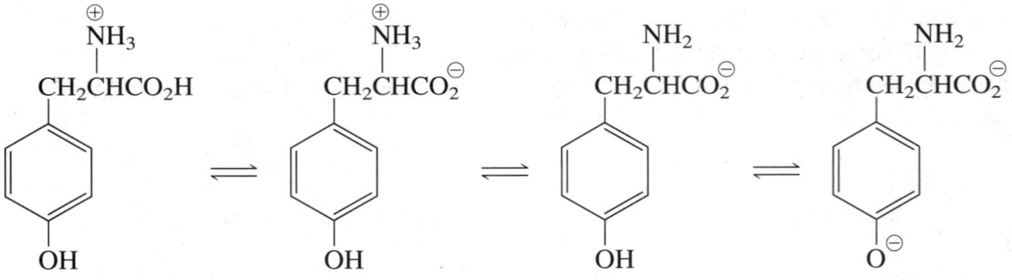

cationic form   dipolar ion   anionic form   dianionic form

Problem 25.7

a)   Show the four differently charged forms of lysine.

b)   Construct a diagram like Figure 25.1 for lysine.

25.4   Chemical Reactions of Amino Acids

Amino acids exhibit chemical reactions that are typical of both amines and carboxylic acids. For example, the acid can be converted to an ester by the Fischer method. This reaction requires the use of an excess of acid because one equivalent is needed to react with the amino group of the product. As another example, the amine can be converted to an amide by reaction with acetic anhydride. Additional examples are provided by the reactions that are used in the preparation of peptides from amino acids described in Section 25.8:

$$
\underset{\substack{| \\ \text{PhCH}_2\text{CH}}}{\overset{\text{NH}_2}{}}\overset{\text{O}}{\overset{\|}{\text{C}}}\text{—OH} + \text{CH}_3\text{OH} \xrightarrow[\text{HCl}]{\text{excess}} \underset{\substack{| \\ \text{PhCH}_2\text{CH}}}{\overset{\text{Cl}^{\ominus} \ {}^{\oplus}\text{NH}_3}{}}\overset{\text{O}}{\overset{\|}{\text{C}}}\text{—OCH}_3 + \text{H}_2\text{O}
$$

(89%)

$$
\text{NH}_2\text{CH}_2\overset{\text{O}}{\overset{\|}{\text{C}}}\text{—OH} + \text{CH}_3\overset{\text{O}}{\overset{\|}{\text{C}}}\text{—O}\text{—}\overset{\text{O}}{\overset{\|}{\text{C}}}\text{CH}_3 \longrightarrow \text{CH}_3\overset{\text{O}}{\overset{\|}{\text{C}}}\text{—NHCH}_2\overset{\text{O}}{\overset{\|}{\text{C}}}\text{—OH} + \text{CH}_3\overset{\text{O}}{\overset{\|}{\text{C}}}\text{—OH} \quad (92\%)
$$

Problem

Problem 25.8 Show the products of these reactions.

a) $\underset{\substack{| \\ \text{NH}_2\text{CH}}}{\overset{\text{CH}_3}{}}\overset{\text{O}}{\overset{\|}{\text{C}}}\text{OH} + \text{PhCH}_2\text{OH} \xrightarrow[\text{HCl}]{\text{excess}}$

b) $\underset{\substack{| \\ | \\ \text{NH}_2\text{CH}}}{\overset{\text{CH}_3\text{CHCH}_3}{\overset{\text{CH}_2}{}}}\overset{\text{O}}{\overset{\|}{\text{C}}}\text{OH} + \text{Ph}\overset{\text{O}}{\overset{\|}{\text{C}}}\text{Cl} \longrightarrow$

c) $\underset{\substack{| \\ \text{NH}_2\text{CH}}}{\overset{\text{CH}_3}{}}\overset{\text{O}}{\overset{\|}{\text{C}}}\text{OH} + \text{CH}_3\overset{\text{O}}{\overset{\|}{\text{C}}}\text{O}\overset{\text{O}}{\overset{\|}{\text{C}}}\text{CH}_3 \longrightarrow$

25.5   Laboratory Synthesis of Amino Acids

Because of their importance, a number of laboratory methods for the synthesis of amino acids have been developed. In the **Strecker synthesis** an aldehyde is treated with NaCN and NH_4Cl to form an aminonitrile, which is then hydrolyzed to the amino acid:

$$
\text{CH}_3\overset{\text{O}}{\overset{\|}{\text{C}}}\text{H} \xrightarrow[\text{NH}_4\text{Cl, H}_2\text{O}]{\text{NaCN}} \underset{\substack{| \\ \text{CH}_3\text{CH}}}{\overset{\text{NH}_2}{}}\text{—C}{\equiv}\text{N} \xrightarrow[\text{H}_2\text{O}]{\text{HCl}} \underset{\substack{| \\ \text{CH}_3\text{CH}}}{\overset{\text{NH}_2}{}}\text{—CO}_2\text{H} \quad (70\%)
$$

aminonitrile   alanine

$$
\text{NH}_4\text{Cl} \Big\downarrow \qquad \left[\underset{\text{CH}_3\text{CH}}{\overset{\text{NH}}{\overset{\|}{}}} \right] \xrightarrow[\text{H}_2\text{O}]{\text{NaCN}}
$$

This reaction proceeds by initial reaction of ammonium chloride with the aldehyde to form an imine (see Section 14.9). Then cyanide adds to the imine in a reaction that is exactly analogous to the addition of cyanide to an aldehyde to form a cyanohydrin (see Section 14.5). The final step in the Strecker synthesis is the hydrolysis of the nitrile to a carboxylic acid (see Section 15.6).

In another method the amino group is introduced onto the α-carbon of a carboxylic acid. To accomplish this, an H on the α-carbon is first replaced with a Br, which can then act as a leaving group. The Br is replaced with an amino group by an S_N2 reaction with NH_3 as the nucleophile:

The carboxylic acid reacts with PBr_3 to form an acyl bromide. The mechanism for this part of the reaction is very similar to that described for the formation of an acyl chloride in Section 15.3.

Step 2 follows the mechanism for acid-catalyzed enolization described in Section 16.2. The acyl bromide enolizes more readily than the carboxylic acid.

As can be seen by the resonance structure, the double bond of the enol is electron-rich, so bromine adds rapidly by an electrophilic mechanism. Loss of a proton produces the α-brominated acyl bromide.

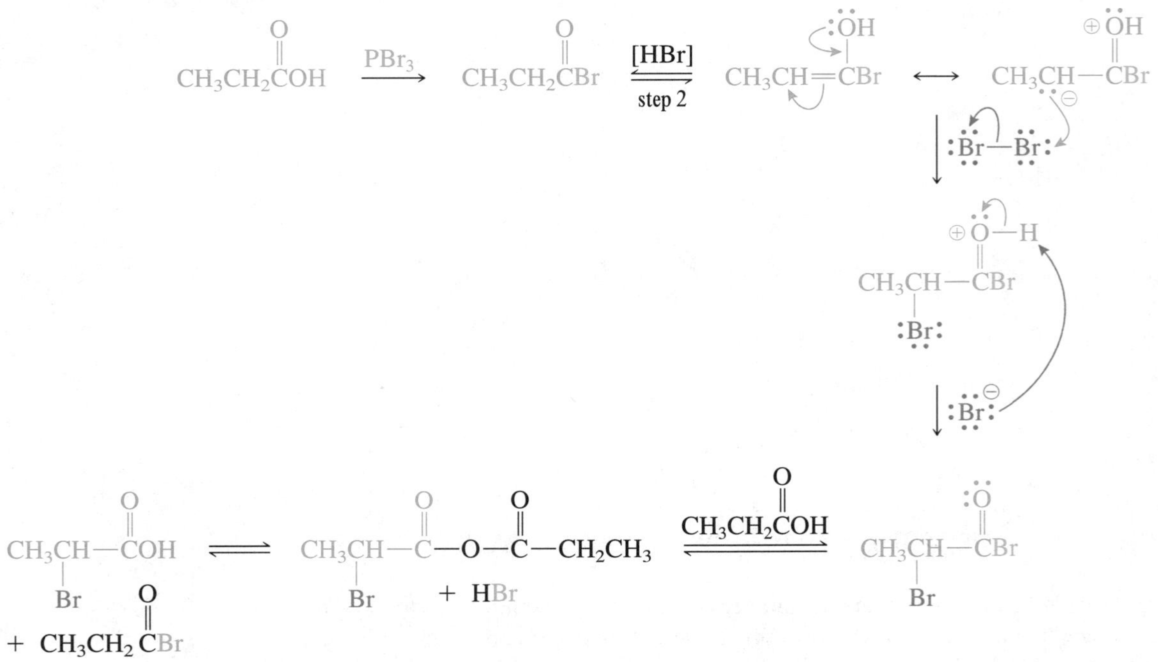

The acyl bromide can reenter the mechanism at step 2 and undergo enolization and bromination. Therefore, only a catalytic amount of PBr_3 is necessary.

The next part of the mechanism results in the exchange of the Br of the acyl bromide with the OH of another molecule of the carboxylic acid to produce the α-brominated acid and an acyl bromide. This exchange reaction proceeds through an anhydride intermediate.

Figure 25.2

Partial Mechanism for the Hell-Volhard-Zelinsky Reaction.

Not all of the steps in the mechanism are shown.

$$CH_3CH_2\overset{\overset{\displaystyle O}{\|}}{C}OH \xrightarrow[\text{[PBr}_3]]{Br_2} CH_3\overset{\overset{\displaystyle Br}{|}}{C}H\overset{\overset{\displaystyle O}{\|}}{C}OH \xrightarrow[\text{NH}_3]{\text{excess}} CH_3\overset{\overset{\displaystyle NH_2}{|}}{C}H\overset{\overset{\displaystyle O}{\|}}{C}OH$$

(70%)

The bromine is introduced onto the α-carbon by treating the carboxylic acid with Br_2 and a catalytic amount of PBr_3 in a process known as the **Hell-Volhard-Zelinsky reaction.** This reaction proceeds through an enol intermediate. Because carboxylic acids form enols only with difficulty, a catalytic amount of PBr_3 is added to form a small amount of the acyl bromide, which enolizes more readily than the acid. Addition of bromine to the enol produces an α-bromoacyl bromide. This reacts with a molecule of the carboxylic acid in a process that exchanges the Br and OH groups to form the product and another molecule of acyl bromide, which can go through the same cycle. Part of the mechanism for this process is outlined in Figure 25.2.

Problem

Problem 25.9 Show all of the steps in the mechanism for the acid-catalyzed enolization of the acyl bromide in the Hell-Volhard-Zelinsky reaction.

$$CH_3CH_2\overset{\overset{\displaystyle O}{\|}}{C}Br \underset{}{\overset{\text{[HBr]}}{\rightleftharpoons}} CH_3CH=\overset{\overset{\displaystyle OH}{|}}{C}Br$$

A third amino acid synthesis begins with diethyl α-bromomalonate. First the Br is replaced by a protected amino group using the Gabriel synthesis (see Section 9.7). Then the side chain of the amino acid is added by an alkylation reaction that resembles the malonic ester synthesis (see Section 16.4). Hydrolysis of the ester and amide bonds followed by decarboxylation of the diacid produces the amino acid. An example that shows the use of this method to prepare aspartic acid is shown in the following sequence:

(71%)

1) NaOEt
2) $ClCH_2CO_2Et$

(99%)

$$HO_2CCH_2\overset{\overset{\displaystyle NH_2}{|}}{C}HCO_2H \xleftarrow[\substack{H_2O \\ \Delta}]{HCl}$$

aspartic acid
(43%)

A drawback of all of these methods is that they produce racemic amino acids. If the product is to be used in place of a natural amino acid, it must first be resolved. This can be accomplished by the traditional method of preparing and separating diastereomeric salts. Alternatively, nature's help can be enlisted through the use of enzymes. In one method the racemic amino acid is converted to its amide by reaction with acetic anhydride. The racemic amide is then treated with a deacylase enzyme. This enzyme catalyzes the hydrolysis of the amide back to the amino acid. However, the enzyme reacts only with the amide of the naturally occurring L-amino acid. The L-amino acid is easily separated from the unhydrolyzed D-amide. The following equation illustrates the use of this process to resolve methionine:

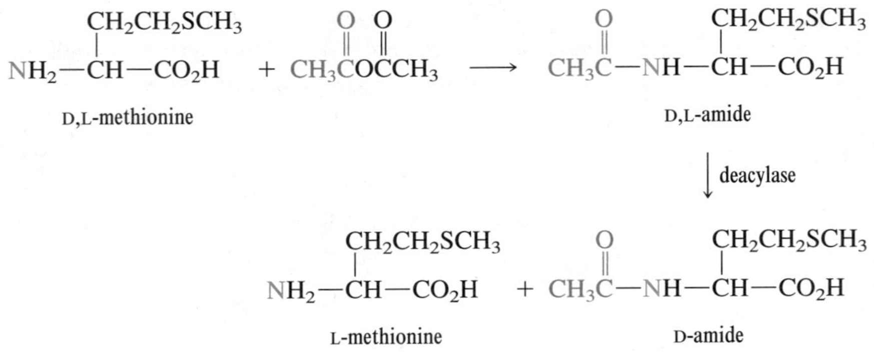

$$\underset{\text{D,L-methionine}}{\overset{\text{CH}_2\text{CH}_2\text{SCH}_3}{\underset{|}{\text{NH}_2-\text{CH}-\text{CO}_2\text{H}}}} + \underset{}{\overset{\text{O   O}}{\overset{||\ ||}{\text{CH}_3\text{COCCH}_3}}} \longrightarrow \underset{\text{D,L-amide}}{\overset{\text{O}\qquad\qquad\text{CH}_2\text{CH}_2\text{SCH}_3}{\overset{||}{\text{CH}_3\text{C}}-\text{NH}-\underset{|}{\text{CH}}-\text{CO}_2\text{H}}}$$

deacylase ↓

$$\underset{\text{L-methionine}}{\overset{\text{CH}_2\text{CH}_2\text{SCH}_3}{\underset{|}{\text{NH}_2-\text{CH}-\text{CO}_2\text{H}}}} + \underset{\text{D-amide}}{\overset{\text{O}\qquad\qquad\text{CH}_2\text{CH}_2\text{SCH}_3}{\overset{||}{\text{CH}_3\text{C}}-\text{NH}-\underset{|}{\text{CH}}-\text{CO}_2\text{H}}}$$

Problems

Problem 25.10   Show the products of these reactions.

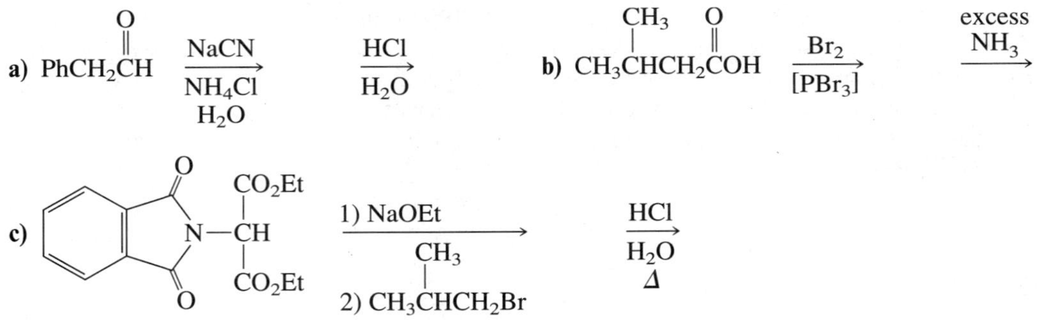

a) $\underset{}{\text{PhCH}_2\overset{\overset{\text{O}}{||}}{\text{CH}}}$ $\xrightarrow[\substack{\text{NH}_4\text{Cl} \\ \text{H}_2\text{O}}]{\text{NaCN}}$ $\xrightarrow[\text{H}_2\text{O}]{\text{HCl}}$

b) $\underset{}{\text{CH}_3\overset{\overset{\text{CH}_3}{|}}{\text{CH}}\text{CH}_2\overset{\overset{\text{O}}{||}}{\text{C}}\text{OH}}$ $\xrightarrow[\text{[PBr}_3\text{]}]{\text{Br}_2}$ $\xrightarrow{\substack{\text{excess} \\ \text{NH}_3}}$

c) (phthalimide structure) $\text{N}-\overset{\overset{\text{CO}_2\text{Et}}{|}}{\underset{\underset{\text{CO}_2\text{Et}}{|}}{\text{CH}}}$ $\xrightarrow[\substack{\text{2) CH}_3\text{CHCH}_2\text{Br} \\ \underset{\text{CH}_3}{|}}]{\text{1) NaOEt}}$ $\xrightarrow[\substack{\text{H}_2\text{O} \\ \Delta}]{\text{HCl}}$

Problem 25.11   Show syntheses of these amino acids.

a)   leucine by the Strecker synthesis

b)   phenylalanine by the Hell-Volhard-Zelinsky reaction

c)   tryptophan starting from diethyl α-bromomalonate

Problem 25.12   Show representative steps in the mechanism for this reaction:

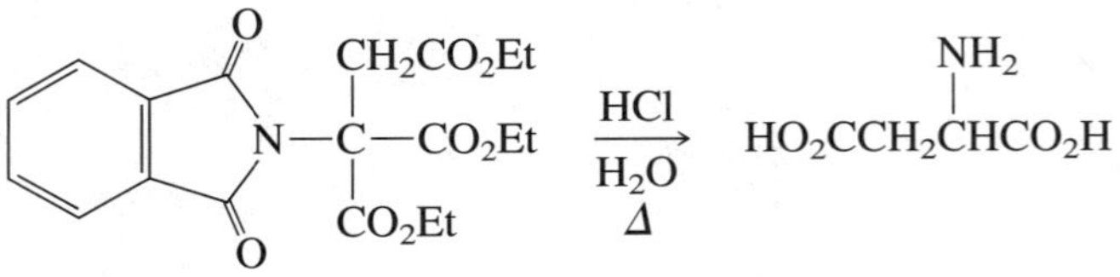

(phthalimide structure) $\text{N}-\overset{\overset{\text{CH}_2\text{CO}_2\text{Et}}{|}}{\underset{\underset{\text{CO}_2\text{Et}}{|}}{\text{C}}}-\text{CO}_2\text{Et}$ $\xrightarrow[\substack{\text{H}_2\text{O} \\ \Delta}]{\text{HCl}}$ $\underset{}{\text{HO}_2\text{CCH}_2\overset{\overset{\text{NH}_2}{|}}{\text{CH}}\text{CO}_2\text{H}}$

Solution

First all three ester bonds and both amide bonds are hydrolyzed to carboxylic acid groups by the aqueous acid. The mechanisms for these reactions are discussed in Section 15.6. The ester hydrolyses follow the exact reverse of the Fischer esterification mechanism shown in Figure 15.3, and the amide hydrolysis occurs by a very similar mechanism. The product of these hydrolysis steps has three carboxylic acid groups and one amino group. Two of these acid groups are attached to the same carbon so that one can be eliminated as carbon dioxide by the cyclic mechanism described in Section 16.4 for the malonic ester synthesis:

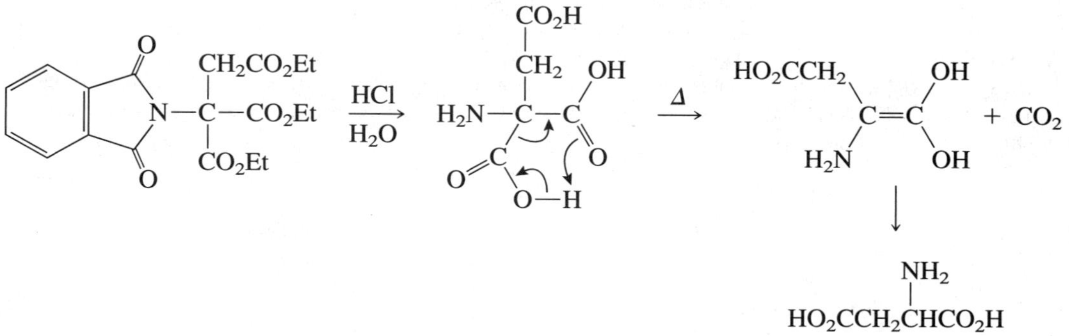

Asymmetric Synthesis of Amino Acids

Even if the resolution of an amino acid is relatively easy, the synthesis of a racemic mixture when only one enantiomer is desired is wasteful, since half of the product cannot be used. Recently, considerable effort has been devoted to the development of methods that produce only the desired enantiomer by so-called asymmetric synthesis. As was discussed in Chapter 6, one enantiomer of a chiral product can be produced only in the presence of one enantiomer of another chiral compound. In some asymmetric syntheses a chiral reagent is employed. In others a compound called a chiral auxiliary is attached to the achiral starting material and used to induce the desired stereochemistry into the product. The chiral auxiliary is then removed and recycled.

An example of the use of a chiral reagent to accomplish an asymmetric synthesis of an amino acid is provided in the following equation:

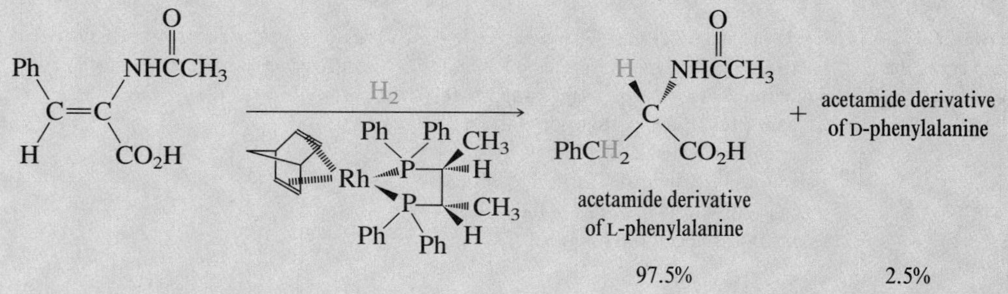

The stereocenter at the α-carbon is introduced by catalytic hydrogenation. To selectively produce one enantiomer of the product, the acetamide of phenylalanine, a chiral catalyst is employed. Rather than using a metal surface as a catalyst, as is common for hydrogenations, a metal complex that is soluble in the reaction solvent is

The starting material for the synthesis of L-alanine is 2-oxopropanoic acid, also known as pyruvic acid. The carbonyl group at the α-carbon will be replaced with a H and a NH_2 so that only one enantiomer is formed at the new chiral center. In the first step, an amide is formed by the reaction of the acid with the methyl ester of L-proline, using DCC as the coupling agent (see Section 25.8).

The product is then reacted with ammonia. The ammonia nucleophile attacks the carbonyl carbon of the ester group, resulting in the formation of an amide group. This product is not isolated but spontaneously proceeds to the next step.

An intramolecular nucleophilic attack by the newly introduced nitrogen at the α-carbonyl carbon produces a six-membered ring.

L-alanine   L-proline

Acid-catalyzed hydrolysis of the two amide bonds produces L-alanine and regenerates the L-proline so that it can be used again.

It is at this stage that the new stereocenter is introduced by catalytic hydrogenation of the double bond. The catalyst prefers to approach from the less hindered bottom side of the molecule, so predominantly, a single stereoisomer is produced. The yield of this step is quantitative, with over 90% of the product having the stereochemistry shown.

Trifluoroacetic acid causes the elimination of water to produce a carbon–carbon double bond.

employed. In this particular case the catalyst is rhodium complexed to a bicyclic diene and a chiral phosphorus-containing ligand. Similar to the reaction on the surface of a metal, the reaction occurs by initial coordination of the alkene and hydrogen to the metal atom in place of the bicyclic diene. The chiral phosphorus ligand causes the hydrogen to be transferred to the alkene so as to produce a single enantiomer of the product. A variety of chiral catalysts have been developed, so one can often be found that will accomplish a particular asymmetric hydrogenation. In addition, the use of a chiral catalyst has the advantage that a full equivalent of the expensive chiral reagent is not needed and the catalyst can often be recovered.

The use of proline methyl ester as a chiral auxiliary in the asymmetric synthesis of alanine is shown on the previous page. The idea is to start with 2-oxopropanoic acid, which has the correct carbon skeleton, and replace the oxygen on carbon 2 with an amino group and a hydrogen. This must be done in such a manner as to produce only the S-enantiomer of the amino acid, that is, L-alanine. This is accomplished by first attaching a chiral auxiliary, the methyl ester of L-proline, to the acid. In the critical step of the process, the catalytic hydrogenation, the chirality of the L-proline is used to induce the proper stereochemistry at the new stereocenter. To put this in terms used in Chapter 6, the α-carbon of proline has S stereochemistry. The new stereocenter generated in hydrogenation, which is the α-carbon of the alanine, could have either R or S stereochemistry. The potential products, with stereochemistries of S,S or S,R, are diastereomers. The hydrogenation occurs preferentially (greater than 90%) at the less hindered bottom side of the molecule, so the hydrogen at the α-carbon of the alanine is added *cis* to the hydrogen at the α-carbon of the proline. Hydrolysis produces the S-enantiomer of alanine, L-alanine, and regenerates the chiral auxiliary, L-proline, so that it can be used again.

Problem

Problem 25.13   What starting material would be used for the synthesis of L-phenylalanine by the method using proline methyl ester as a chiral auxiliary?

25.6   Peptides and Proteins

Because they contain two functional groups, amino acids can react to produce condensation polymers by forming amide bonds. These polymers are called peptides, polypeptides, or proteins. Although there is no universally accepted distinction, the term *protein* is usually reserved for naturally occurring polymers that contain a relatively large number of amino acid units and have molecular masses in the range of a few thousand or larger. The term *peptide* is used for smaller polymers.

As a simple example, the dipeptide formed by the reaction of two glycines has the following structure:

Biosynthesis of Amino Acids from α-Ketoacids

One of the methods that nature uses to make α-amino acids is reductive amination of α-ketoacids. In the process, a second amino acid is converted to an α-ketoacid so an exchange reaction, a transamination, actually occurs. This process is outlined in the following equation:

$$\underset{\text{amino acid-1}}{R-\underset{\underset{NH_2}{|}}{C}H-\underset{\underset{O}{\|}}{C}-OH} \; + \; \underset{\text{ketoacid-2}}{R'-\underset{\underset{O}{\|}}{C}-\underset{\underset{O}{\|}}{C}-OH} \;\rightleftharpoons\; \underset{\text{ketoacid-1}}{R-\underset{\underset{O}{\|}}{C}-\underset{\underset{O}{\|}}{C}-OH} \; + \; \underset{\text{amino acid-2}}{R'-\underset{\underset{NH_2}{|}}{C}H-\underset{\underset{O}{\|}}{C}-OH}$$

The reaction is, of course, catalyzed by an enzyme. The enzyme employs a cofactor, called a coenzyme, which is a small organic molecule that is intimately involved in the actual chemical transformation. The coenzyme in this case is pyridoxal phosphate, a phosphate ester of pyridoxine, also known as vitamin B_6:

The amino group of an amino acid reacts with the aldehyde carbonyl group of pyridoxal phosphate to form an imine. The formation of imines is discussed in Section 14.9.

The hydrogen on the α-carbon of the amino acid is more acidic, so it is readily removed by a base attached to the enzyme.

Several resonance structures for the conjugate base are shown. Reprotonation of the base at the carbon that was originally the carbonyl carbon of pyridoxal produces an isomeric imine, a proton tautomer of the original imine. In this imine the double bond is now between the nitrogen and the α-carbon of the amino acid.

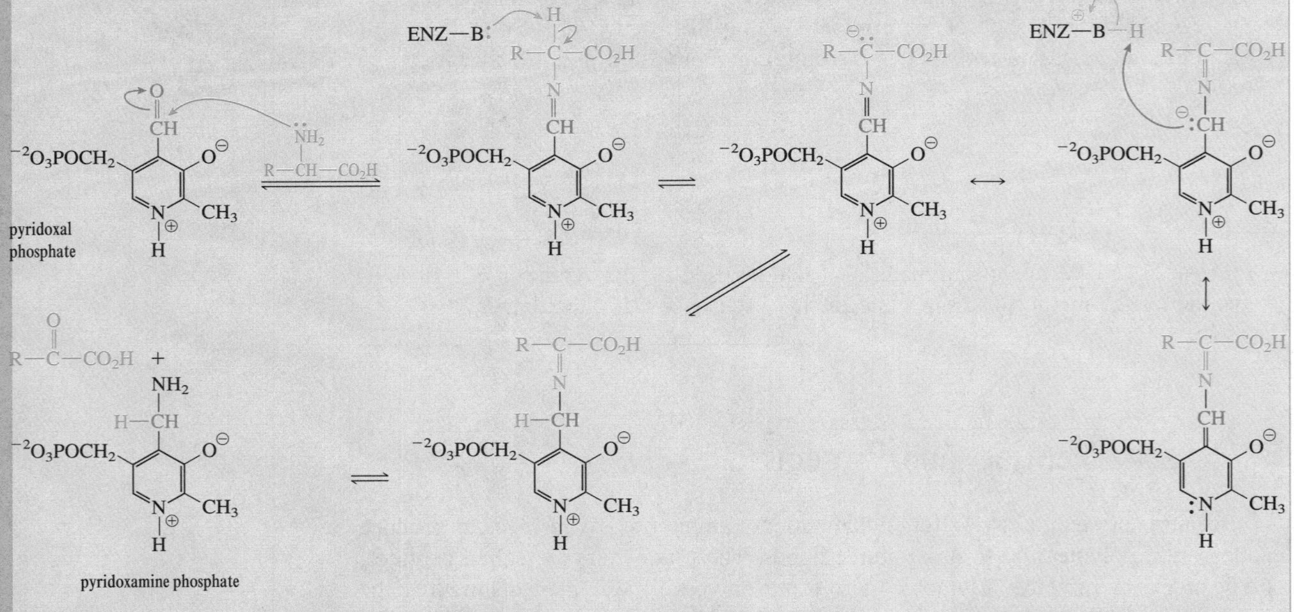

At this point, the exchange reaction is half completed. Next, a different α-ketoacid reacts with pyridoxamine phosphate, and the reverse of this sequence occurs, ultimately producing pyridoxal and the amino acid derived from the new α-ketoacid.

Hydrolysis of the imine produces pyridoxamine phosphate and the α-ketoacid derived from the amino acid.

This resonance structure illustrates why the protonated pyridine ring is such an excellent resonance electron-withdrawing group.

$$O$$
$$CH$$
$$HOCH_2 \quad\quad OH$$
$$CH_3$$
$$N$$

pyridoxine (vitamin B_6)

The mechanism by which pyridoxal phosphate catalyzes the transamination reaction is shown on the previous page. First the aldehyde group of the coenzyme reacts with the amino group of the amino acid to form an imine. The role of the protonated pyridine ring is to act as an excellent resonance electron-withdrawing group, thus making the hydrogen on the α-carbon of the bound amino acid more acidic than it was before the imine was formed. A base in the enzyme removes this proton, and the resulting conjugate base is reprotonated at a different site. This is just a proton tautomerization reaction with the proton removal step facilitated by the electron-withdrawing pyridine ring. Hydrolysis of the carbon–nitrogen double bond of the new imine results in the formation of pyridoxamine phosphate and the α-ketoacid derived from the original amino acid. A transamination between pyridoxal and the original amino acid has occurred. Next the process is reversed. Pyridoxamine phosphate and a different α-ketoacid react in a transamination to produce pyridoxal phosphate and an amino acid. Then the cycle can be repeated.

The imine produced from an amino acid and pyridoxal phosphate has the ability to form a relatively stable anion at the α-carbon of the amino acid. Pyridoxal phosphate also acts as a coenzyme in other processes in which stabilization of a negative charge on the α-carbon is needed. For example, the biosynthesis of histamine, which is involved in allergic responses, is accomplished by the decarboxylation of histidine by the enzyme histidine decarboxylase. The enzyme uses pyridoxal phosphate as a coenzyme to stabilize the anion that is produced during the decarboxylation. This reaction is outlined in the following equation:

histidine $\xrightarrow{\text{histidine decarboxylase}}$ histamine $+ CO_2$

histidine

histamine

pyridoxal phosphate

This anion is too unstable to produce directly.

$-^2O_3POCH_2$

$-^2O_3POCH_2$

This anion is much more stable, owing to a large amount of resonance stabilization.

$$NH_2CH_2\overset{\displaystyle O}{\overset{\|}{C}}—OH \; + \; NH_2CH_2\overset{\displaystyle O}{\overset{\|}{C}}—OH \; \longrightarrow \; NH_2CH_2\overset{\displaystyle O}{\overset{\|}{C}}—NHCH_2\overset{\displaystyle O}{\overset{\|}{C}}—OH \; + \; H_2O$$

amide bond
or
peptide bond

Biochemists say that the two amino acids are connected by a peptide bond, but, of course, the peptide bond is just an amide bond. A slightly more complex example, the phagocytosis-stimulating tetrapeptide known as tuftsin, derived from four amino acids, can be employed to illustrate some of the conventions that are used in writing the structures of polypeptides and proteins:

tuftsin

By convention, peptides are written so that the end with the free amino group, called the N-terminus, is on the left and the end with the free carboxyl group, the C-terminus, is on the right. Because it takes considerable space to show the structure of even a small polypeptide like this one, it is common to represent the structures of peptides and proteins by using the three-letter abbreviation for each amino acid (see Table 25.1). Thus tuftsin, with a threonine N-terminal amino acid, followed by lysine, proline, and, finally, arginine as the C-terminal amino acid, is represented as

Thr-Lys-Pro-Arg

Note that the N-terminal amino acid is on the left and the C-terminal amino acid is on the right in this abbreviated representation also. (For very large polypeptides the one-letter codes for the amino acids are used to save even more space.)

Problems

Problem 25.14   Draw the complete structure for Phe-Val-Asp.

Problem 25.15   Identify the amino acids in this polypeptide and show its structure using the three-letter abbreviations for the amino acids:

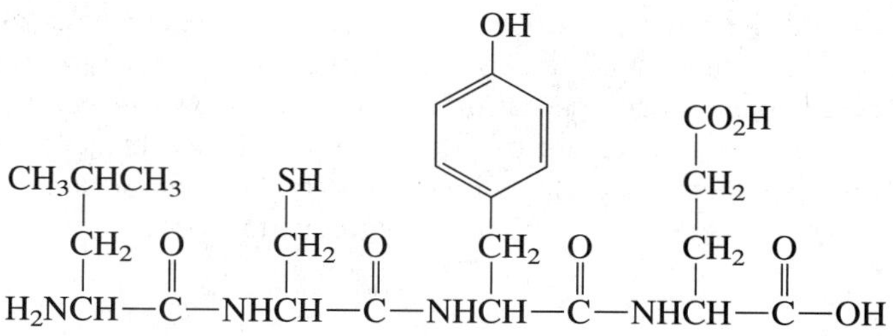

$$OH$$

$$CH_3CHCH_3 \quad\quad SH \quad\quad\quad\quad\quad\quad\quad\quad CO_2H$$

$$CH_2 \quad O \quad\quad CH_2 \quad O \quad\quad CH_2 \quad O \quad\quad CH_2 \quad O$$

$$H_2NCH-C-NHCH-C-NHCH-C-NHCH-C-OH$$

    Let's compare proteins to the polymers that were discussed in Chapter 23. One difference is that all the molecules of a particular protein are identical; that is, they have the same molecular mass and contain the same number of amino acids connected in the same sequence. Recall that a typical condensation polymer consists of molecules containing many different numbers of monomers. More important, proteins are enormously more complex than simple condensation polymers because they are formed from a combination of 20 different monomer units. And these monomers are not randomly distributed in the protein. Rather, each molecule of a particular protein has an identical sequence of amino acid units. The exact sequence is of critical importance because it is the order of the side chains that determines the shape and function of that particular protein.

    Because there are 20 different amino acids that can occupy each position in a polypeptide or protein, the number of possible structures is enormous. Consider, for example, a dipeptide. There are 20 possibilities for the N-terminal amino acid and 20 possibilities for the C-terminal amino acid. Therefore there are $(20)(20) = 20^2 = 400$ different dipeptides. The number of possibilities increases rapidly as the number of amino acids in the polymer increases. For a tripeptide there are $20^3 = 8000$ possibilities. And for a polypeptide that contains 100 amino acids (many proteins are considerably larger than this) there are $20^{100} = 1.27 \times 10^{130}$ possibilities. Such large numbers have little meaning for most of us, so let's try to put this number in perspective. It has been estimated that the total number of atoms in the universe is about 10^{80}. The number of possible polypeptides containing only 100 amino acids vastly exceeds the total number of atoms in the entire universe!

    The geometry of the amide bond helps to determine the overall shape of a peptide or protein. The nitrogen of an amide is sp^2 hybridized, so the electron pair on the nitrogen is in a p orbital that can overlap with the pi bond of the carbonyl group. The nitrogen is planar, and there is considerable double-bond character to the bond connecting it to the carbonyl carbon. In other words, the structure on the right makes an important contribution to the resonance hybrid for an amide:

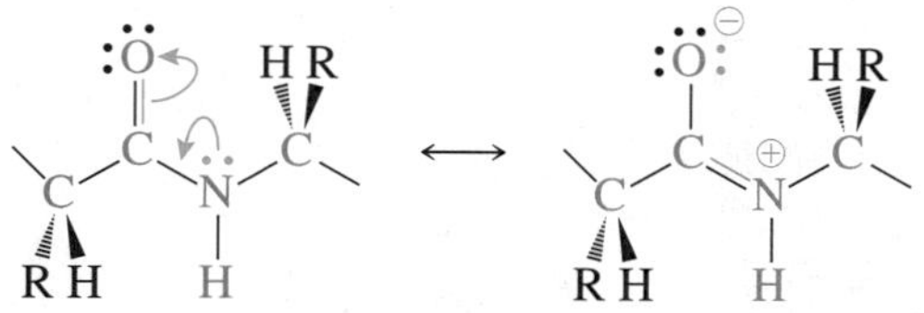

This requires that the carbonyl carbon, the nitrogen, and the two atoms attached to each of them (the α-carbon and the oxygen bonded to the carbonyl carbon and the hydrogen and the other α-carbon bonded to the nitrogen) must all lie in the same plane. The most stable conformation has the bulky α-carbons in a "trans" relationship about the carbon–nitrogen partial double bond, as shown in the preceding structure.

Another important feature of proteins is due to the presence of cysteine amino acids in the polymer chain. The SH groups of two cysteine residues can react to form a disulfide bond as shown in the following equation:

NMR Spectra of Amides

Because the bond between the nitrogen and the carbonyl carbon of an amide has considerable double-bond character, rotation about this bond is relatively slow. This slow rotation is evident in the NMR spectra of many amides. For example, the two methyl groups on the nitrogen of N,N-dimethylacetamide appear at different chemical shifts. If rotation about the carbon–nitrogen bond were fast, the methyl groups would be equivalent and would appear at the same chemical shift.

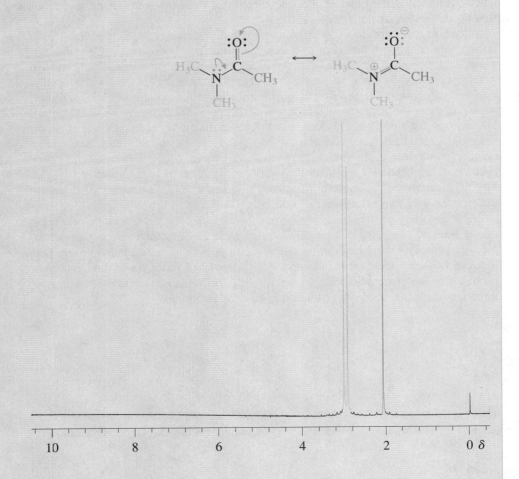

If the cysteines are part of different polypeptide chains, the resulting disulfide bond acts as a cross-link between the chains. If the cysteines are part of the same polypeptide chain, the large ring that is formed helps to determine the overall shape of the peptide.

25.7   Sequencing Peptides

The sequence of amino acids in a protein is of critical importance in determining the function of that protein. Therefore considerable effort has been devoted to the development of methods to determine amino acid sequences. The process usually begins with the determination of the relative numbers of each amino acid that are present in the protein. To accomplish this, a sample is completely hydrolyzed to its individual amino acid components by treatment with 6 **M** HCl at 100–120°C for 10–100 hours. These rather vigorous conditions are needed to completely hydrolyze the protein because amide bonds are rather unreactive. The amino acids are then separated by some type of chromatography, and the relative number of each is determined.

Problem

Problem 25.16   Explain why asparagine and glutamine are never found when a peptide is completely hydrolyzed by using HCl, H_2O, and elevated temperatures. What amino acids are found in place of these?

The separation and detection process has been automated. In the original amino acid analyzer, developed by W.H. Stein and S. Moore, who were awarded the 1972 Nobel Prize in chemistry for determining the structure of the enzyme ribonuclease, the amino acids are separated by ion-exchange chromatography. They are then reacted with ninhydrin, and the resulting purple derivatives are detected by visible spectroscopy. In a more modern version the amino acids are reacted with dansyl chloride, and the resulting derivatives are separated by high-performance liquid chromatography. The dansyl group is highly fluorescent, so very small amounts of the dansylated amino acids can be detected. With a modern amino acid analyzer the complete analysis of a hydrolyzed protein can be done in less than 1 hour. The method is sensitive enough to detect as little as 10^{-12} mol of an amino acid, so only a very small amount of the protein need be hydrolyzed.

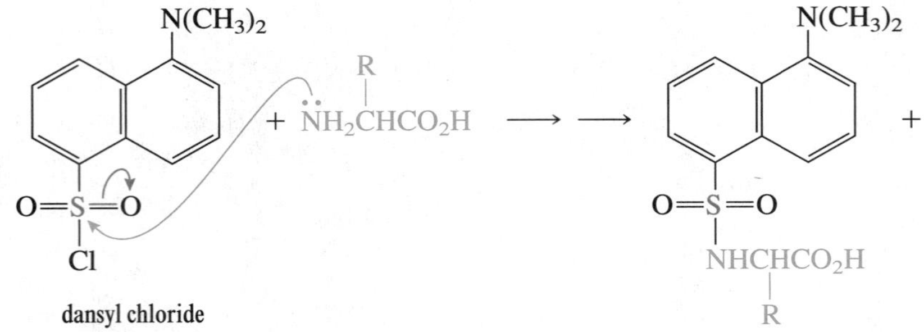

dansyl chloride

Of course, the determination of the number of each kind of amino acid that is present is only a small part of the solution to the structure of a protein. The sequence of the amino acids must also be determined. This is accomplished by taking advantage of the fact that only the N-terminal amino acid has a free NH_2 group that is nucleophilic. All of the other nitrogens are part of amide groups and are not nucleophilic (unless the side chain contains an amino group, as is the case with lysine).

In one method the polypeptide is reacted with Sanger's reagent, 2,4-dinitrofluorobenzene (DNFB). The nucleophilic nitrogen of the N-terminal amino acid displaces the fluorine in a nucleophilic aromatic substitution reaction. (This reaction follows an addition–elimination mechanism; see Section 18.12.) The polypeptide is then hydrolyzed to its individual amino acid components. Because the bond between the nitrogen and the dinitrophenyl group is resistant to hydrolysis, the N-terminal amino acid is labeled and can easily be identified in the hydrolysis mixture:

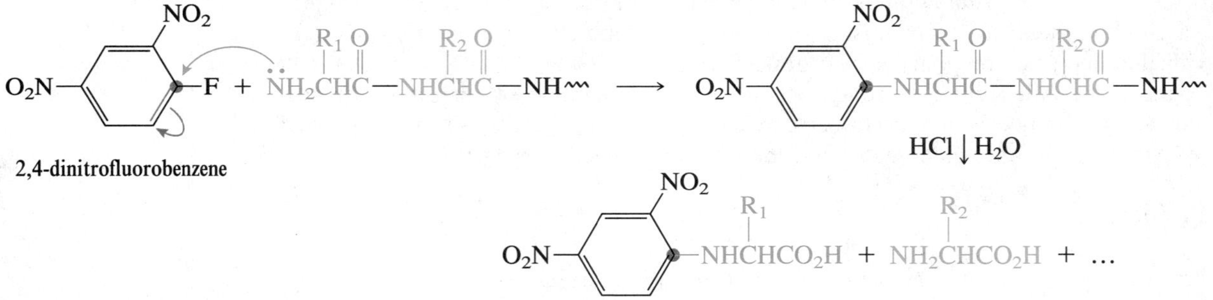

2,4-dinitrofluorobenzene

In another method the nitrogen of the N-terminal amino acid is reacted with dansyl chloride. The resulting sulfonamide bond is quite resistant to hydrolysis, so the N-terminal amino acid, labeled with the dansyl group, is readily determined after the peptide bonds have been hydrolyzed.

Problem

Problem 25.17   Show all of the steps in the mechanism for the reaction of an amino acid with Sanger's reagent and explain why the nitro groups are necessary for the reaction to occur.

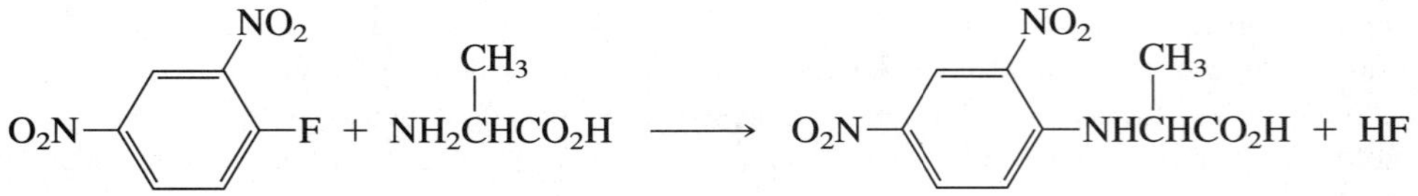

The most useful method of N-terminal analysis is called the **Edman degrada-tion.** This method allows the N-terminal amino acid to be removed and its identity to be determined without hydrolyzing the other peptide bonds. The reaction initially pro-duces a thiazolinone, which is rearranged by aqueous acid to a phenylthiohydantoin for identification by high-performance liquid chromatography:

$$\underset{\text{H}_2\text{N}-\overset{R_1}{\overset{|}{\text{CH}}}-\overset{O}{\overset{||}{\text{C}}}-\text{NH}-\overset{R_2}{\overset{|}{\text{CH}}}-\overset{O}{\overset{||}{\text{C}}}-\text{NH}\text{\textasciitilde}}{} \quad \xrightarrow[\text{2) HF}]{\text{1) Ph}-\text{N}=\text{C}=\text{S}} \quad \text{a thiazolinone} \quad + \quad \underset{\text{new polypeptide}}{\text{NH}_2\text{CHCNH}\text{\textasciitilde}}$$

a thiazolinone

new polypeptide

$$\text{HCl} \downarrow \\ \text{H}_2\text{O}$$

a phenylthiohydantoin

The feature that makes the Edman degradation so useful is that the new polypep-tide, with one fewer amino acid, can be isolated and submitted to the process again, allowing identification of its N-terminal amino acid. It is possible to continue removing and identifying the N-terminal amino acid for 40 or more cycles before impurities due to incomplete reactions and side reactions build up to the extent that the identification of the last removed amino acid is uncertain. The Edman degra-dation procedure has also been automated, so it is now possible to sequence a polypeptide from the N-terminal end at the rate of about one amino acid residue per hour.

The critical feature of the Edman degradation is that it allows the N-terminal amino acid to be removed without cleaving any of the other peptide bonds. Let's see how this occurs. The mechanism for the reaction is shown in Figure 25.3. First the nucleophilic nitrogen of the N-terminal amino acid attacks the electrophilic carbon of phenyl isothiocyanate. When anhydrous HF is added in the next step, the sulfur of the thiourea acts as an intramolecular nucleophile and attacks the carbonyl carbon of the closest peptide bond. It is the intramolecular nature of this step and the formation of a five-membered ring that result in the selective cleavage of only the N-terminal amino acid. The mechanism for this part of the reaction is very similar to that for acid-cat-alyzed hydrolysis of an amide (see Section 15.6). However, since no water is present, only the sulfur is available to act as a nucleophile. The sulfur is ideally positioned for intramolecular attack at the carbonyl carbon of the N-terminal amino acid, so only this amide bond is broken.

Problem

Problem 25.18   Show the products of these reactions.

a)

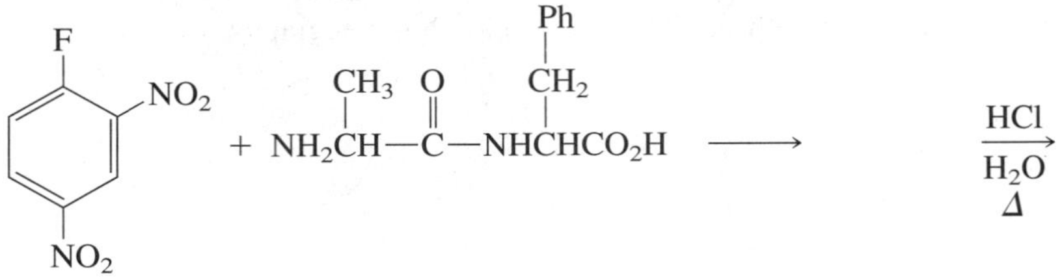

$$\text{(2,4-dinitrofluorobenzene)} + NH_2CH-\overset{O}{\underset{|}{C}}-NHCHCO_2H \quad (CH_3; Ph-CH_2 \text{ substituents}) \xrightarrow{\begin{array}{c} HCl \\ \hline H_2O \\ \Delta \end{array}}$$

b)

$$NH_2CH-\overset{O}{\underset{||}{C}}-NHCH_2-\overset{O}{\underset{||}{C}}-NHCH-\overset{O}{\underset{||}{C}}-OH \xrightarrow[\text{2) HCl, H}_2O, \Delta]{\text{1) (5-dimethylaminonaphthalene-1-sulfonyl chloride)}}$$

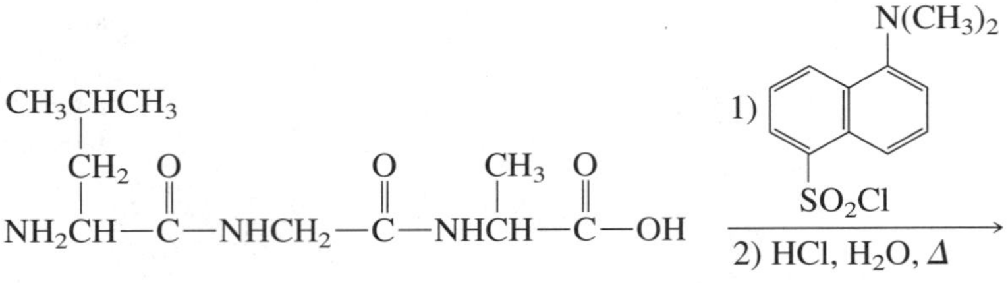

The nucleophilic nitrogen of the N-terminus of the peptide attacks the electrophilic carbon of phenyl isothiocyanate. This carbon resembles a carbonyl carbon and is quite electrophilic. After proton transfers, a thiourea is formed.

The thiourea is cleaved by treatment with anhydrous HF. First the oxygen of the carbonyl group is protonated.

Then the sulfur of the thiourea acts as an intramolecular nucleophile and attacks the carbonyl carbon. This process is especially favorable because a five-membered ring is formed.

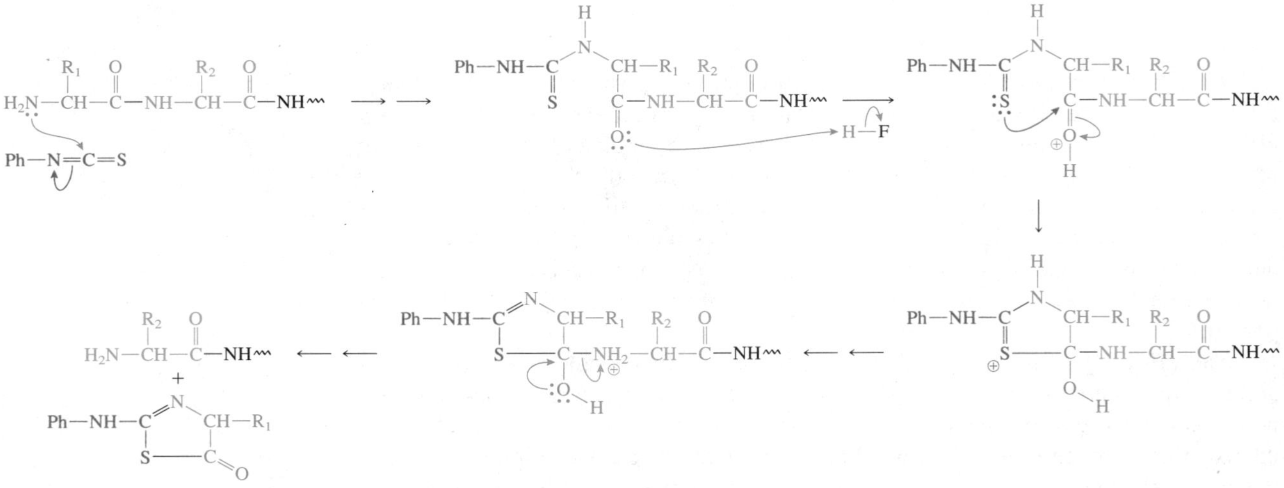

After several proton transfers, the nitrogen of the amide leaves. Overall, this mechanism is similar to acid-catalyzed amide hydrolysis (see Chapter 15), but the nucleophile is sulfur rather than water. No water is present, and the sulfur nucleophile can reach only the closest carbonyl group, so only the N-terminal amino acid is cleaved.

Figure 25.3

Partial Mechanism for the Edman Degradation.

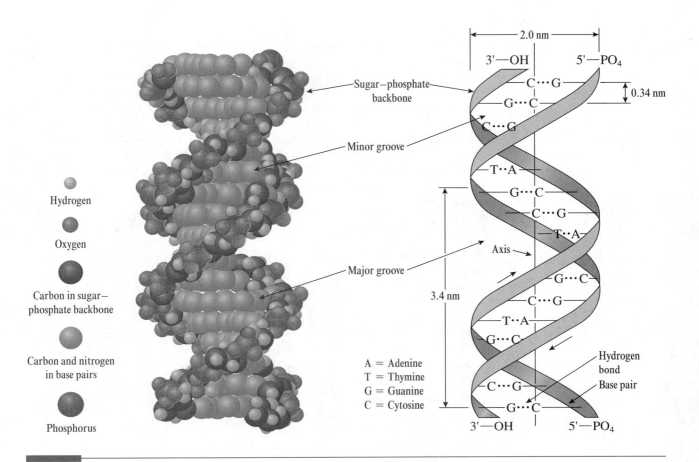

Figure 26.3

Model of the DNA Double Helix.

RNA has the same general structure as DNA with several exceptions. It employs the sugar ribose rather than deoxyribose. It has the base uracil in place of thymine. Note that thymine and uracil are both pyrimidine bases, and both hydrogen bond to adenine in the same manner. Unlike DNA, RNA usually occurs as a single-stranded molecule. Because RNA is used to transfer and translate the information that is stored in DNA the complementary copy is not necessary and is not synthesized by the cell. Finally, RNA is much smaller than DNA.

Problems

Problem 26.6   Use drawings to show the hydrogen bonding that occurs in a base pair formed from adenine and uracil.

Problem 26.7   Use drawings to explain why thymine and guanine do not form a base pair that has hydrogen bonds as strong as those between cytosine and guanine.

26.4   Replication, Transcription, and Translation

Genetic information is stored in DNA molecules. The sequence of bases in a particular piece of DNA specifies the sequence of amino acids in a particular protein. Exactly how the DNA is replicated when a cell divides and how the information in the DNA sequence is converted to an amino acid sequence of a protein are of critical importance to the functioning of a living organism and comprise a major part of any modern biochemistry textbook. We only have space here to briefly outline these processes.

The information for the amino acid sequence of a protein is stored in the base

Base-Catalyzed Hydrolysis of RNA

The presence of a 2'-hydroxy group makes RNA much more susceptible to base-catalyzed hydrolysis than DNA. As shown in the following equations, the base removes a proton from the 2'-hydroxy group. The oxygen anion then attacks the phosphorus in a favorable intramolecular nucleophilic displacement reaction. The resulting cyclic phosphodiester is hydrolyzed by water to a mixture of the 2'- and 3'-phosphate. The intramolecular displacement step of this mechanism, which makes the overall reaction much faster, requires the presence of the 2'-hydroxy group. Thus, base-catalyzed hydrolysis is much faster for RNA than for DNA, in which the necessary hydroxy group is absent. This is thought to be why DNA, rather than RNA, has evolved to be the storehouse for genetic information in most organisms.

sequence of DNA. However, 20 amino acids are found in proteins, while only four bases occur in DNA. This means that a single base cannot code for an individual amino acid. Likewise, a two-base code, which provides $4 \times 4 = 16$ combinations, still is not large enough to specify 20 different amino acids. The genetic code is actually based on a series of three bases, called a **codon**, which provides $4^3 = 64$ different possibilities. A codon for a particular amino acid is designated by listing the first letters of the three bases that compose it. Thus one codon for serine is UCA, which designates a base sequence of uracil, cytosine, and adenine. The code is degenerate; that is, most amino acids are specified by two or more codons. For example, the codons CCC, CCU, CCA, and CCG all specify the amino acid proline. In addition, the codons UAA, UAG, and UGA all specify a stop signal; that is, they indicate the end of the protein chain.

The two complementary strands of DNA provide a stable reservoir for genetic information, which must be preserved over the entire lifetime of the organism. Either of the strands, by itself, has all of the genetic information. Therefore if one strand is damaged, perhaps by random hydrolysis, the information to repair the damage is still present in the complementary strand. For example, suppose a G and a T were lost from the piece of DNA represented by the following schematic structure. When a repair enzyme encounters this damage, the C and A of the complementary strand ensure that the correct bases are inserted in the repair process.

C═══G T═══A C═══G A═══T G═══C G═══C A═══T	C═══G T═══A C A G═══C G═══C A═══T	C═══G T═══A C═══G A═══T G═══C G═══C A═══T	C═══G T═══A C═══G A═══T G═══C G═══C A═══T
double strand of DNA	damaged DNA	repair enzyme →	repaired DNA

The double-stranded nature of DNA also provides a method for **replication**, the process whereby DNA is duplicated so that two identical copies are available when a cell divides. In this process, the DNA unwinds, and each strand serves as a template for the synthesis of its complementary strand. When replication is completed, two identical versions of the original DNA helix are present. This process is represented schematically in the following diagram:

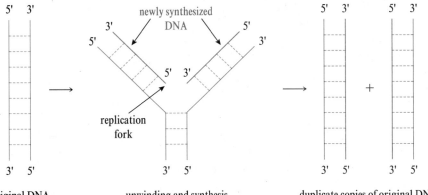

original DNA   unwinding and synthesis   duplicate copies of original DNA

Like all biological processes, replication is controlled by enzymes. There are enzymes that unwind the DNA double helix so that the individual chains are accessible. Then an enzyme called DNA polymerase attaches a new nucleotide to a deoxyribose in the growing chain of the new DNA strand.

In replication the nucleotide that is to be attached next is held in place by hydrogen bonding to its complementary base in the original DNA strand. As shown in Figure 26.4, the new nucleotide has a triphosphate group attached to its 5'-hydroxy group. The 3'-hydroxy group on the end of the growing DNA chain reacts with the triphosphate to form a phosphate ester in a reaction that is quite similar to the reaction of an alcohol with a carboxylic acid anhydride to form the ester of a carboxylic acid. Of course, these reactions are under enzymatic control.

Problem

Problem 26.8   The mechanism for the attachment of a new nucleotide to a growing DNA chain shown in Figure 26.4 is very similar to the mechanism for the formation of a carboxylic ester from an alcohol and a carboxylic anhydride. Suggest the steps in the mechanism for this reaction:

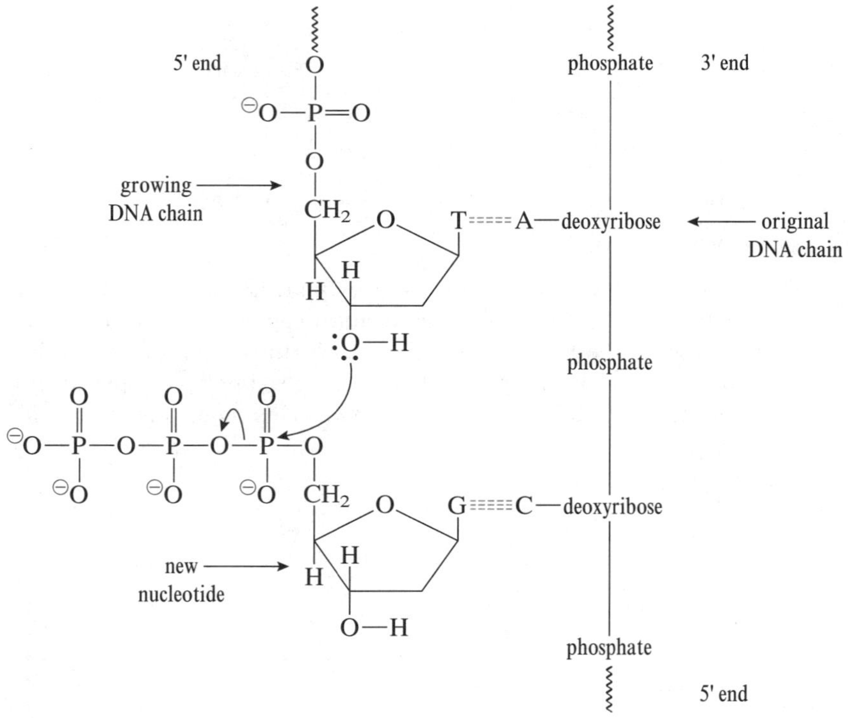

Figure 26.4

Attachment of a New Nucleotide to a Growing DNA Chain.

$$R-O-\overset{\overset{\displaystyle O}{\|}}{\underset{\underset{\displaystyle O_\ominus}{|}}{P}}-O-\overset{\overset{\displaystyle O}{\|}}{\underset{\underset{\displaystyle O_\ominus}{|}}{P}}-O-\overset{\overset{\displaystyle O}{\|}}{\underset{\underset{\displaystyle O_\ominus}{|}}{P}}-O^\ominus \;+\; R'OH \;\longrightarrow\; R-O-\overset{\overset{\displaystyle O}{\|}}{\underset{\underset{\displaystyle O_\ominus}{|}}{P}}-O-R' \;+\; H-O-\overset{\overset{\displaystyle O}{\|}}{\underset{\underset{\displaystyle O_\ominus}{|}}{P}}-O-\overset{\overset{\displaystyle O}{\|}}{\underset{\underset{\displaystyle O_\ominus}{|}}{P}}-O^\ominus$$

The transformation of the information in a DNA strand into a protein sequence involves several steps. First an RNA polymer, called **messenger RNA** (mRNA) that is complementary to the DNA is synthesized in much the same manner as the synthesis of new DNA described above. This process is called **transcription**. Messenger RNA then serves as a template for protein synthesis in a process called **translation**. Individual amino acids are attached to relatively small RNA molecules called **transfer RNA** (tRNA). Each amino acid has its own type of tRNA that has a three-base

The Treatment of AIDS with AZT

Azidothymidine, also known as AZT or zidovudine, was initially prepared in the 1960s with the hope that it might be useful in the treatment of cancer. It is very similar to the nucleoside deoxythymidine, except that it has an azido group in place of the 3'-hydroxy group. The reasoning was that this "fraudulent nucleoside" might be incorporated into new DNA chains that were being synthesized by the cancer cells. Since AZT has no 3'-hydroxy group, it would act to terminate the DNA chain and would therefore prevent cell division. Unfortunately, cancer cells recognize AZT as a fake and do not incorporate it.

azidothymidine
(AZT)

2',3'-dideoxycytidine
(DDC)

The AIDS virus is a retrovirus; that is, it stores its genetic information in the form of RNA. On infection it injects its RNA into the target cell and uses an enzyme called reverse transcriptase to synthesize DNA that is complementary to this RNA template. AZT is accepted by reverse transcriptase as a building block for this synthesis and slows or prevents the conversion of the viral RNA information into DNA. By disrupting this process, AZT slows the replication of the virus in the cell. In a similar strategy the pseudonucleoside DDC is used when AZT is ineffective. Although both of these drugs have some positive results, unfortunately neither is able to halt the infection.

ELABORATION

region, known as the **anticodon**, that is complementary to the codon for that amino acid. The tRNA with the correct amino acid forms three base pairs with the mRNA and brings the amino acid into position for attachment to the growing protein chain. The process for attaching a proline, followed by a phenylalanine, to a growing polypeptide chain is represented schematically in the following diagram:

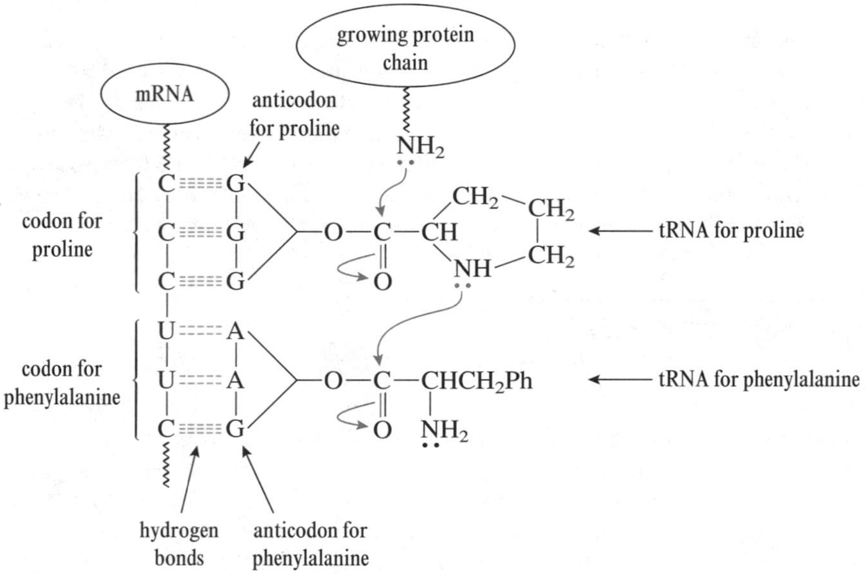

The tRNA for proline, which has the anticodon GGG, hydrogen bonds to the CCC codon for proline in the mRNA and brings its attached proline amino acid into position for attachment to the growing protein chain. Then the tRNA for phenylalanine hydrogen bonds to the codon for phenylalanine in the mRNA and brings its attached phenylalanine into position for attachment to the proline. This process continues until a stop signal is reached.

This has been a necessarily brief overview of the processes of replication, transcription, and translation. If you are interested in learning more, please consult a biochemistry textbook.

26.5 Sequencing DNA

Determination of the sequence of bases in DNA is accomplished in a manner similar to the determination of the sequence of amino acids in a protein. The problem is somewhat simpler in that there are only four possible bases. However, DNA is much larger than a protein, so there is considerably more information to obtain. First it is necessary to break the enormous DNA molecule into specific fragments of more manageable size. Then the sequence of bases in these fragments must be determined. Several different cleavage processes must be employed to produce the fragments so that one set overlaps the cleavage points of another and can be used to determine how the fragments were connected.

27.7   Synthesis of Steroids

Steroids are very powerful hormones and therefore are present in animals in only extremely minute concentrations. They are very difficult to obtain from natural sources. For example, four tons of sow ovaries (from 80,000 pigs) were required to isolate the 12 mg of estrogen that was used for the determination of its structure. Syntheses of most of these hormones have been accomplished from simple starting materials in the laboratory. Although these routes are too long and complex to provide a practical source of the hormones for medicinal uses, many are truly elegant. R.B. Woodward was awarded the 1965 Nobel Prize in chemistry for his syntheses of steroids and numerous other natural products.

The most viable method to obtain larger amounts of steroid hormones is to start with some readily available natural product with a structure that is similar to a steroid and convert it to the desired compound. Russell Marker, a professor at Pennsylvania State University, developed such a method to prepare progesterone from diosgenin, a material that is readily available from Mexican yams. His synthesis is outlined in Figure 27.9. However, he could not interest a major pharmaceutical company in his process, so in 1944 he founded his own company, Syntex, in Mexico City to develop it.

Problems

Problem 27.18   Show a mechanism for the elimination reaction that occurs in the first step of the process shown in Figure 27.9.

Problem 27.19   One way to accomplish the isomerization of a double bond into conjugation with a carbonyl group, one of the reactions needed for the last part of the

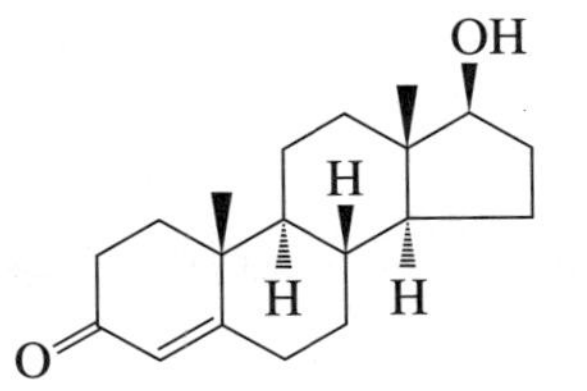

testosterone
(an androgen)

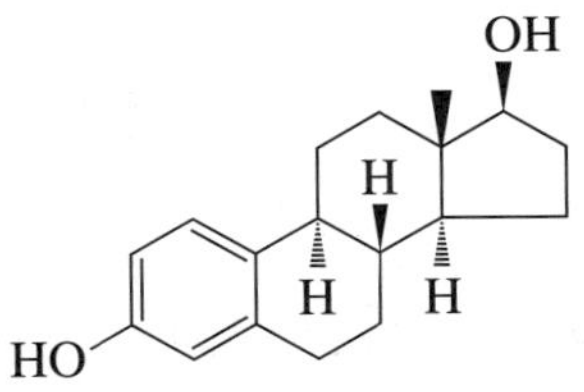

estradiol
(an estrogen)

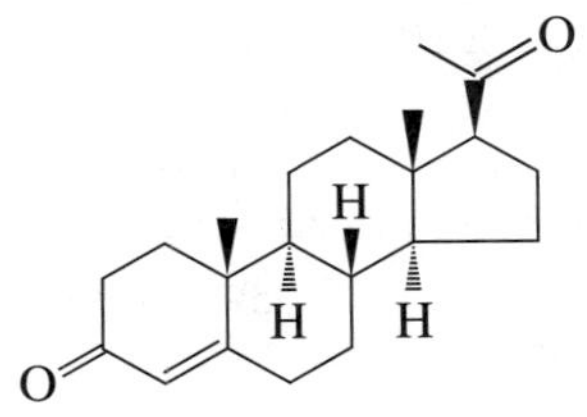

progesterone
(a progestin)

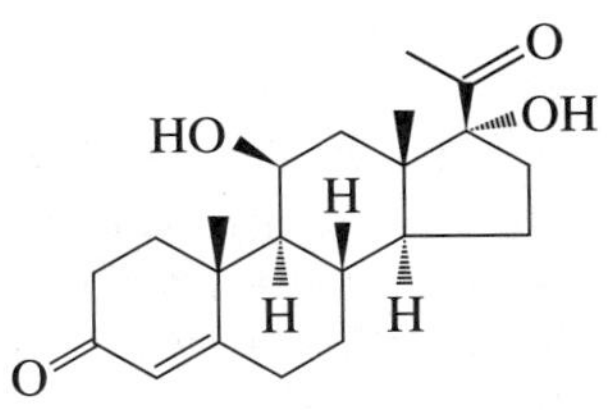

cortisol (hydrocortisone)
(an adrenocorticol hormone)

Figure 27.8
Some Steroidal Hormones.

This is an elimination reaction. The oxygen is acetylated to make it a better leaving group. Then an E1 elimination occurs.

The CrO_3 cleaves the double bond. Although this reaction has not been covered in this book, it is quite similar to the ozonolysis reaction (Section 10.12).

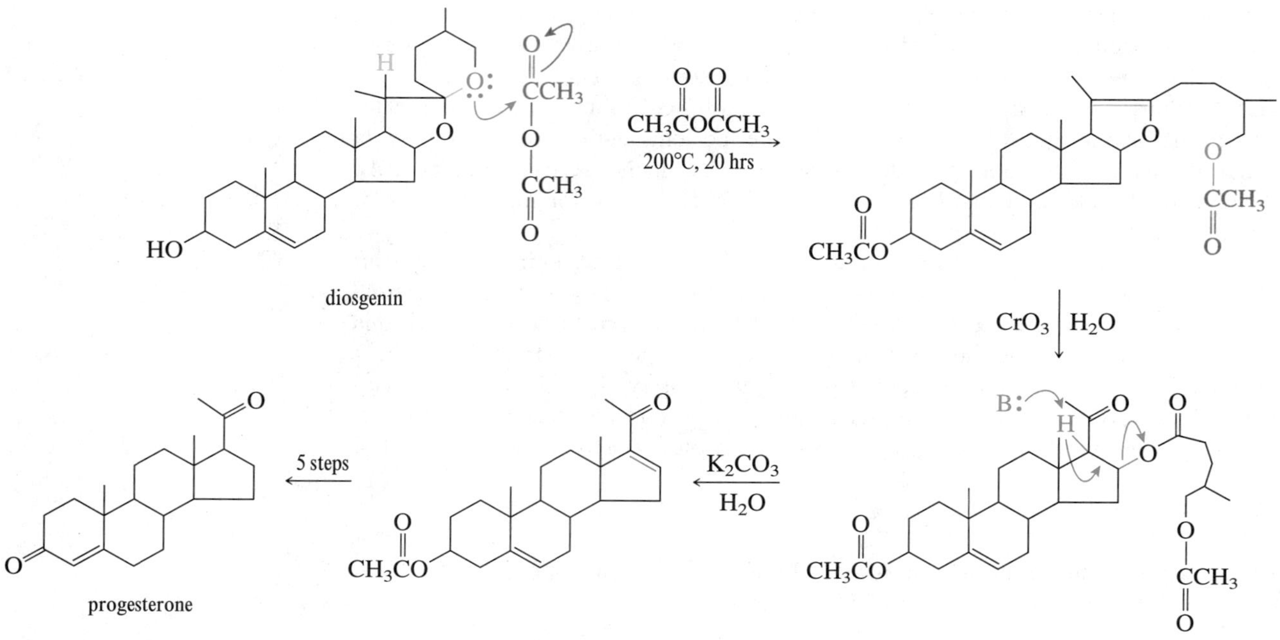

diosgenin

progesterone

To complete the synthesis of progesterone, it is necessary to selectively reduce one of the double bonds, hydrolyze the acetate group to a hydroxy group, oxidize the hydroxy group to a carbonyl group, and isomerize the remaining double bond into conjugation with the carbonyl group. This is accomplished in five steps.

The ester group is lost by an elimination reaction. The adjacent carbonyl group makes the hydrogen more acidic, so it is more readily removed by the base.

Figure 27.9

The Preparation of Progesterone from Diosgenin.

synthesis of progesterone from diosgenin, is to treat the compound with base as shown in the following equation. Show a mechanism for this reaction.

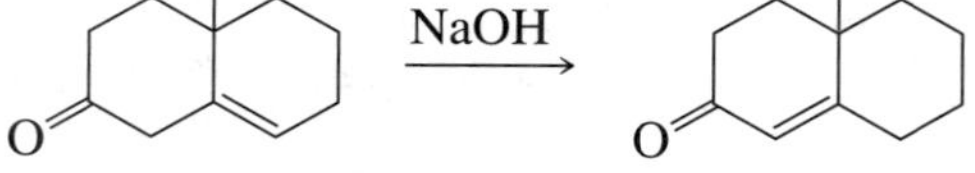

In the late 1940s, cortisone was the most sought-after steroid because its anti-inflammatory properties appeared to have amazing effects in the treatment of rheumatoid arthritis. The major difficulty in converting a readily available steroid, such as progesterone, to cortisone was the introduction of an oxygen functionality at carbon 11. As we have seen innumerable times, reactions occur at a functional group or sometimes at the carbon adjacent to it. Causing a reaction to occur specifically at an unactivated carbon presents quite a challenge. The problem was solved by a roundabout method that introduced a double bond between carbons 9 and 11. Several syntheses of cortisone were reported in 1951.

However, some chemists at Upjohn found a better method. They were able to introduce the troublesome oxygen at carbon 11 by the microbiological fermentation of progesterone to produce 11-hydroxyprogesterone. The microbe, with its highly specific enzymes, was able to accomplish in a single step what required numerous steps for the synthetic chemist. The completion of the synthesis of cortisone required only nine additional steps.

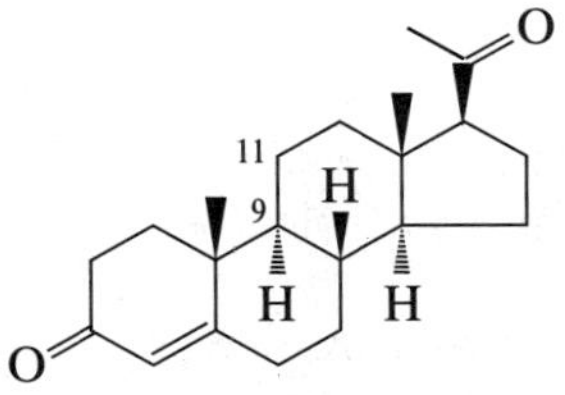

progesterone 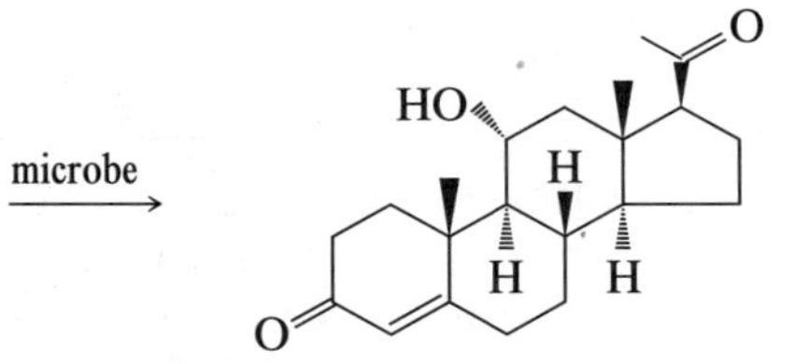   →   11-hydroxyprogesterone
                      microbe

9 steps

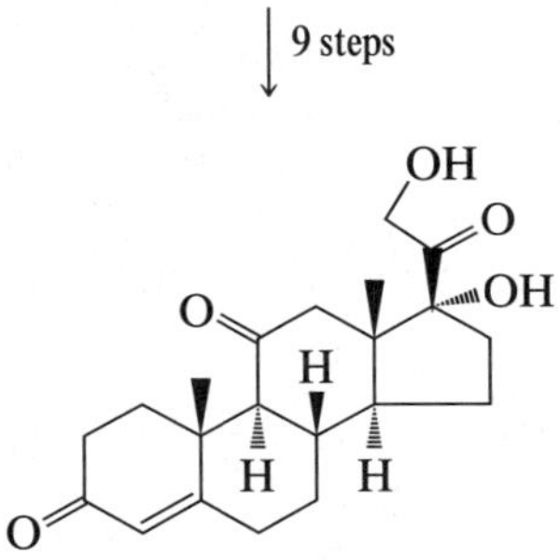

cortisone

The Birth Control Pill

Perhaps no single development has influenced today's society more than the birth control pill. This simple, effective, and inexpensive method to limit pregnancy helped to bring about both the sexual revolution and women's liberation.

By 1937 it was known that large doses of progesterone inhibited ovulation, and the possibility of its use in birth control was recognized. A major problem, however, was that progesterone displays only weak activity when administered orally. The idea of an injectable contraceptive did not win wide acceptance. Somewhat later, it was discovered that the removal of the methyl group from carbon 10 of progesterone makes it more active. Other researchers discovered that the incorporation of an ethynyl group at position 17 increased the oral activity of these drugs. Carl Djerassi at Syntex decided to combine both of these effects and prepared norethindrone. This compound was found to be an effective oral contraceptive and was quickly patented in 1951.

ELABORATION

norethindrone

Although norethindrone was the first oral contraceptive to be developed, it was not the first to be marketed. This was due mainly to the lack of both biological testing laboratories and marketing expertise at Syntex.

In 1953, G.D. Searle and Co. patented a related compound, norethynodrel. This steroid is also an active contraceptive when taken orally. This fact is not at all surprising to an experienced organic chemist because it is well known that a carbon–carbon double bond such as the one in norethynodrel is readily isomerized into conjugation with a carbonyl group. This acid-catalyzed isomerization occurs via an enol intermediate, and the equilibrium favors the more stable, conjugated compound. The acidic conditions in the human gastric system are quite sufficient to accomplish the transformation of norethynodrel to norethindrone. It seems that chemical reactions do not care much about patents!

norethynodrel   [HA]

norethindrone

Syntex licensed norethindrone to Parke-Davis for development. Eventually, Parke-Davis decided not to market it because of fears about a possible boycott of their other products by groups who were opposed to birth control. Searle had no such concerns and brought its pill to the market in 1960 under the name Enovid. In the meantime, Syntex was forced to find another partner and finally settled on the Ortho division of Johnson & Johnson. The norethindrone pill was first marketed in 1962 under the name Ortho-Novum. By 1965 "the pill" was the most popular form of birth control.

Progestin-based contraceptives work by inducing a state of pseudopregnancy. Therefore they are administered in a twenty-day cycle, followed by a break to allow normal menstruation. Today, most birth control pills use norethindrone or norgestrel, which is similar to norethindrone but has an ethyl group at position 13 rather than a methyl group. In addition, they contain a small amount of estrogen to reduce breakthrough bleeding. Interestingly, the usefulness of estrogen was discovered by accident when one batch of pills was synthesized with a small amount of estrogen as a contaminant.

There has been considerable concern about the health effects of the pill, and many studies have been done. These are extremely powerful compounds and are taken by a large number of healthy women over an extended period of time, not to cure disease, but to prevent pregnancy. Although there was some evidence for heightened risk of cardiovascular disease in early studies, this risk decreased as the amount of estrogen in the pill was decreased. Today, the amount of estrogen has been reduced from 150 μg per pill for Enovid to 30–35 μg per pill. The progestin component has also been reduced, and the pill is a relatively safe method of birth control.

27.8 Alkaloids

Alkaloids are a group of nitrogen-containing natural products that occur primarily in higher plants, although they are also found in some fungi, such as mushrooms. The name *alkaloid* (meaning "alkalilike") is applied because they are amines and thus basic. Their basic character allows them to be readily isolated from their plant source. The plant material is extracted with aqueous acid. This converts the alkaloid to an ammonium cation, which is water soluble. Neutralization of the acidic extract with base causes the alkaloid to precipitate.

Like terpenes, alkaloids have been important in the development of organic chemistry. Some, such as nicotine, have relatively simple structures. Others, such as morphine and strychnine, have very complex structures containing multiple rings. Most are quite physiologically active.

nicotine
(the addictive substance
in tobacco)

morphine
(a highly addictive painkiller
obtained from the opium poppy)

strychnine
(an extremely poisonous central nervous system stimulant)

The complex structures of alkaloids, along with their biological activity and relative ease of isolation, have kept the interest of organic chemists over the years. They have provided enormously challenging structure elucidation problems. For example, although strychnine was first isolated in 1818, its complete structure was not determined until 1946. (R. Robinson was awarded the 1947 Nobel Prize in chemistry for the determination of the structure of strychnine and other alkaloids as well as synthetic work in this area.) Of course, soon after this, synthetic chemists accepted the

Table 27.1 Some Important Classes of Alkaloids.

Amino Acid	Alkaloid Partial Structure	Example	
ornithine	pyrrolidine	cocaine	Cocaine is a central nervous system stimulant; it is obtained from the coca plant.
lysine	piperidine	coniine	Coniine is a major component of poisonous hemlock that was used to kill Socrates.
tyrosine	isoquinoline	emetine	Emetine is used to treat amebic dysentery.
tryptophane	indole	lysergic acid	Lysergic acid is an ergot alkaloid that is obtained from a fungus that grows on cereal. The diethyl amide is the hallucinogen LSD.

challenge of preparing this complicated compound in the laboratory. The first synthesis, reported by Woodward in 1953, required 28 steps (not very many when the complexity of the molecule is considered). More recently, strychnine has been prepared, enantiomerically pure, in 20 steps with an overall yield of 3%!

Just as terpenes could be viewed as being formed from isoprene units, alkaloids can be viewed as being derived from amino acids. There are four amino acids that give rise to important classes of alkaloids. As shown in Table 27.1, the pyrrolidine alkaloids are derived from the amino acid ornithine (not one of the 20 "standard" amino acids), the piperidine alkaloids from lysine, the isoquinoline alkaloids from tyrosine, and the indole alkaloids from tryptophan.

As was the case with terpenes, the function of alkaloids in plants is not known. It has been proposed that they are merely nitrogen-containing waste products of plants, like urea in animals. However, most plants reutilize nitrogen, rather than wasting it. Furthermore, it is difficult to imagine why such complex structures would be needed to store waste nitrogen. Like terpenes, alkaloids have been proposed to serve as protection from herbivores and insects. However, only a few examples of such protection can be demonstrated. Whatever the role of alkaloids is, some 70–80% of plants manage to do quite nicely without them.

27.9   Fats and Related Compounds

As discussed briefly in the Elaboration "The Preparation of Soap" in Chapter 15, fats are triesters formed from glycerol and long, linear-chain carboxylic acids, known as fatty acids. The resulting triesters are called **triacylglycerols.** Stearic, palmitic, and oleic acids are among the many different fatty acids that occur in nature:

a triacylglycerol

Fatty acids usually have an even number of carbons because they are biosynthesized from acetate ion, which has two carbons. Those with 14, 16, 18, and 20 carbons are most common. Their biosynthesis is outlined in Figure 27.10. They may be saturated, like stearic acid and palmitic acid, or they may have one or more carbon–carbon double bonds, like oleic acid. The double bonds invariably have *cis*-geometry, and the kink in the long chain caused by the *cis*-double bond prevents the chain from packing as well into a crystal lattice and decreases the melting point of the lipid. Thus oils tend to have a larger percentage of unsaturated fatty acids than fats.

Problems

Problem 27.20   Show a mechanism for step 1 of Figure 27.10. Ignore the participation of the enzyme.

Problem 27.21   Show a mechanism for step 3 of Figure 27.10. Assume that the reaction is catalyzed by base.

The first step is an ester condensation (see Section 16.6). The R group represents either an enzyme or coenzyme A.

The carbonyl group is reduced to a hydroxy group in step 2.

Water is eliminated to form an alkene in step 3. This reaction is very similar to the elimination of water from the product of an aldol condensation (see Section 16.5).

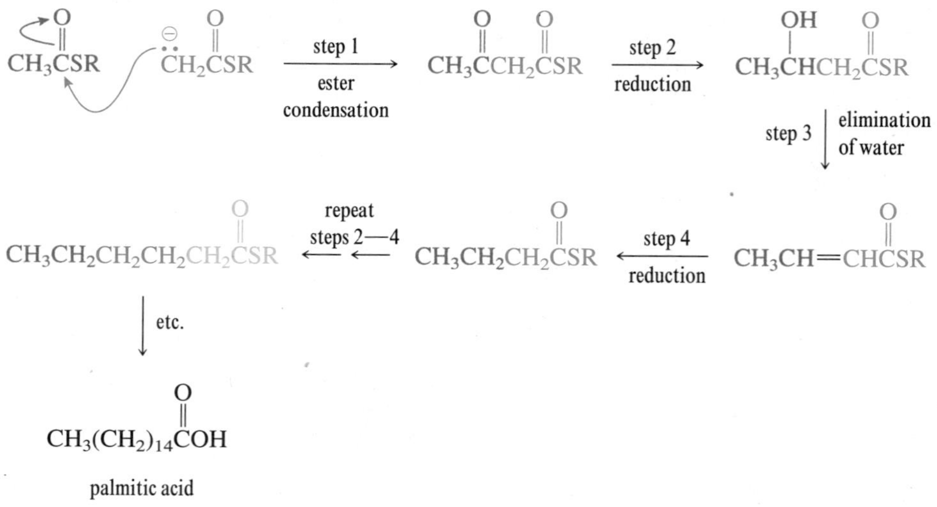

Additional cycles of this mechanism, adding two more carbons at a time, produce fatty acids with an even number of carbons, such as palmitic acid with 16 carbons.

The four-carbon acid group acts as the electrophilic component in step 1. After steps 2—4 are repeated, a six-carbon acid group is produced.

The double bond is saturated in step 4. The result of steps 1—4 is the conversion of two acetate groups to a four-carbon acid group in the form of its thiol ester.

Figure 27.10

The Biosynthesis of Fatty Acids.

A particular species of plant or animal contains a number of different fatty acid residues in its triacylglycerols. These are randomly distributed, so the individual molecules of a triacylglycerol are not all identical. For example, one molecule may have the structure shown previously, having been formed from one molecule each of stearic, palmitic, and oleic acid. Another may contain two molecules of stearic acid and one of oleic acid. Still another may contain entirely different fatty acids. The average composition of the fatty acid part of triacylglycerols varies with the species of the source. Triacylglycerols from plants usually contain more unsaturated fatty acid residues and

Partially Hydrogenated Vegetable Oil

Margarines are prepared from vegetable oils. However, most people do not like to spread liquid oil on their toast. The presence of *cis*-double bonds in the triacylglycerols causes kinks in the hydrocarbon tails of the fatty acid residues. These kinks prevent the triacylglycerol molecules from packing closely and lower the melting point. To raise the melting point of the oil so that it is a solid at room temperature, some of the double bonds are reduced by catalytic hydrogenation:

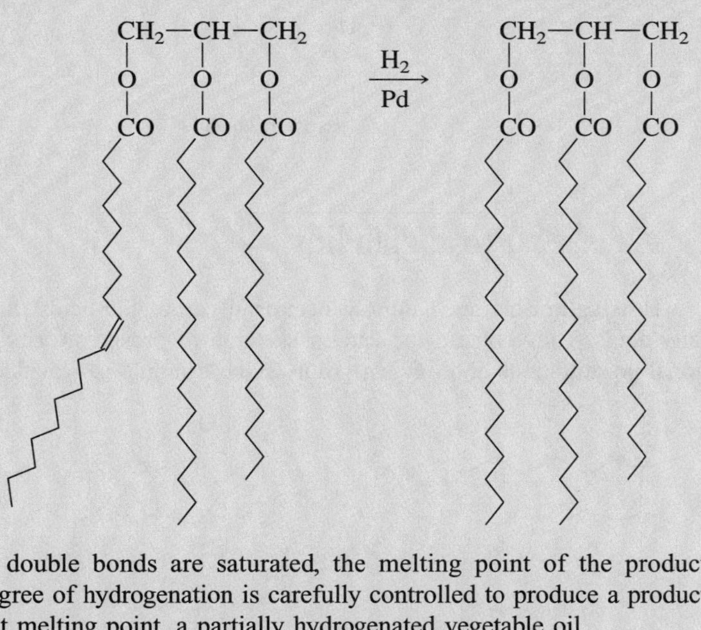

As more of the double bonds are saturated, the melting point of the product increases. The degree of hydrogenation is carefully controlled to produce a product with just the right melting point, a partially hydrogenated vegetable oil.

A product with a higher melting point is necessary for consumer acceptance. In addition, triacylglycerides that are more unsaturated tend to spoil more rapidly. This spoilage is due to oxidation caused by radical reactions. (This is an example of the autoxidation process described in Section 19.10 and the elaboration "Unsaturated Fats, Autoxidation, and Vitamin E" on page 924. The hydrogens on the allylic carbons of unsaturated fatty acid residues are more readily abstracted because the resulting radicals are stabilized by resonance, so these compounds oxidize and spoil faster.) However, there is a tradeoff, since it has been demonstrated that saturated fats have more deleterious health consequences than unsaturated fats do.

ELABORATION

have lower melting points than those from animals. Thus plant triacylglycerols are more likely to be oils, while those from animals are fats.

Fats and oils serve as energy reserves for the organism. Because they are in a lower oxidation state than carbohydrates, they provide more energy per gram when they are metabolized (see the Elaboration "Energy Content of Fuels" in Chapter 5).

Glycerophospholipids are an important class of compounds related to fats. They are also triesters of glycerol. However, in this case, two of the ester groups are formed from fatty acids, while the third is a phosphate ester that also has an ionic or very polar group. Glycerophospholipids are the major component of biological membranes. Their polar heads project into the aqueous solution, while the nonpolar tails form the bilayer membranes. An example of a glycerophospholipid is shown in the following structure. Again, the fatty acid components, as well as the polar part of the phosphate ester, can vary.

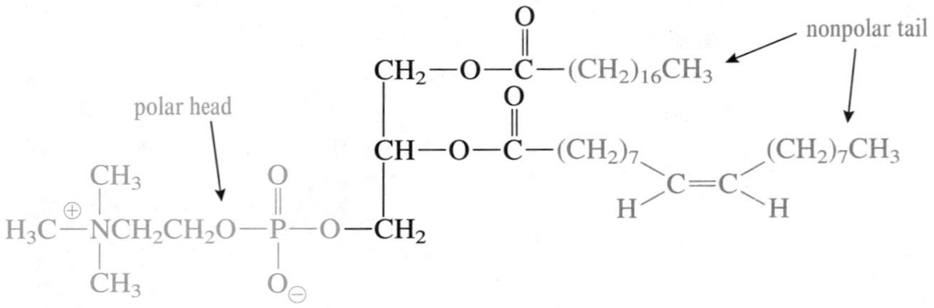

a glycerophospholipid

27.10 Prostaglandins

Prostaglandins are naturally occurring carboxylic acids that are related to the fatty acids. They contain the carbon skeleton of prostanoic acid, with various additional unsaturations and oxygen groups. One example is provided by PGE_2:

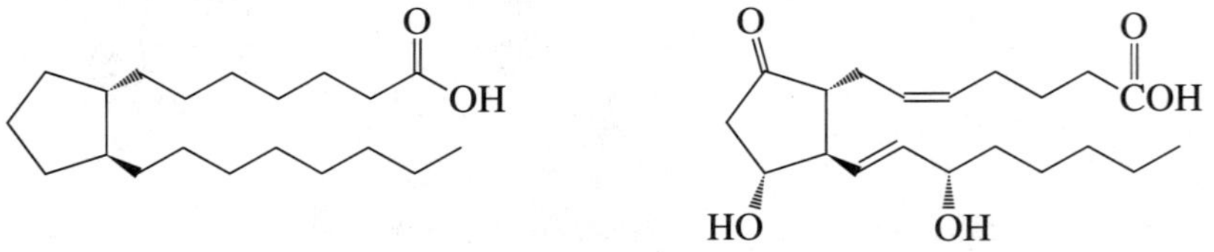

prostanoic acid PGE_2

Prostaglandins have been found to be involved in a number of important physiological functions, including the inflammatory response, the production of pain and fever, the regulation of blood pressure, the induction of blood clotting, and the induction of labor.

Prostaglandins are biosynthesized from arachidonic acid, an unsaturated fatty acid containing four double bonds. The enzyme prostaglandin endoperoxide synthase converts arachidonic acid to PGH_2, which serves as the precursor for prostaglandins and related compounds. Aspirin exerts its pharmacological effect by inhibiting this enzyme.

3.17

a) 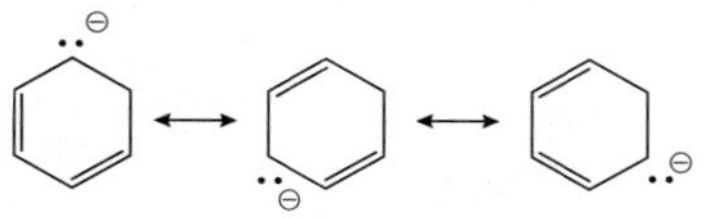 [resonance structures]

f) 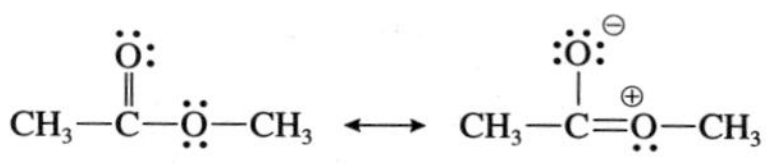 $CH_3-C-O-CH_3 \longleftrightarrow CH_3-C=O-CH_3$

e) 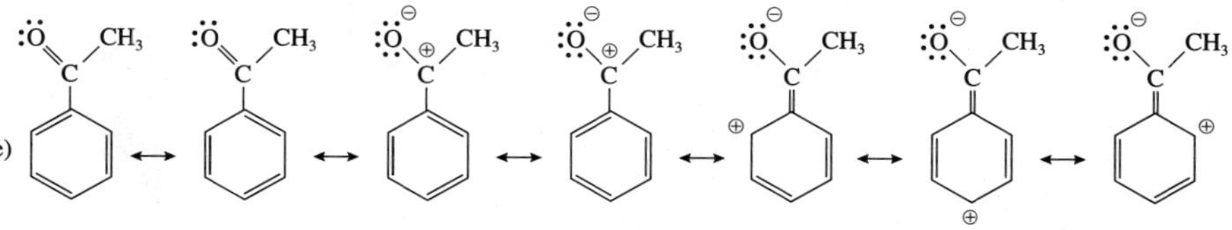 [resonance structures]

3.18

a)

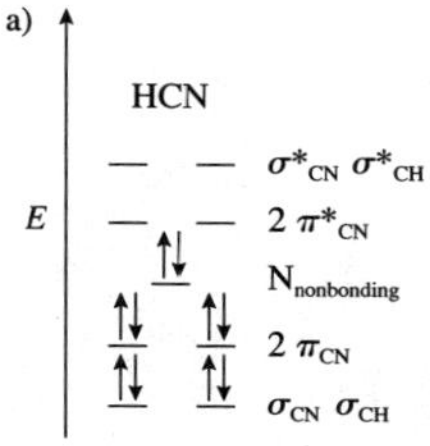

HCN

E

— — $\sigma^*_{CN}\ \sigma^*_{CH}$
— — $2\,\pi^*_{CN}$
↑↓ $N_{nonbonding}$
↑↓ ↑↓ $2\,\pi_{CN}$
↑↓ ↑↓ $\sigma_{CN}\ \sigma_{CH}$

b)

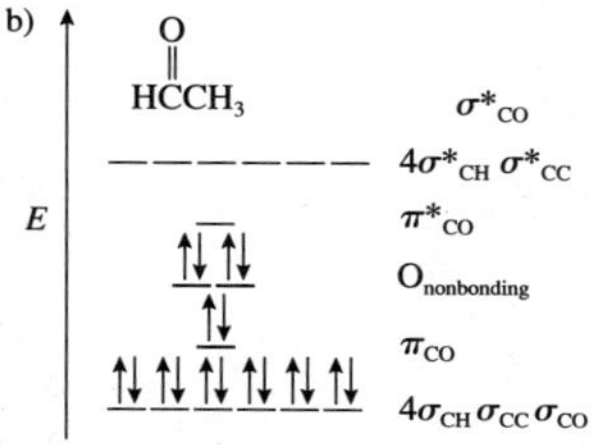

$\overset{O}{HCCH_3}$

E

— — — — — σ^*_{CO}
$4\sigma^*_{CH}\ \sigma^*_{CC}$
↑↓ ↑↓ π^*_{CO}
↑↓ $O_{nonbonding}$
↑↓ π_{CO}
↑↓↑↓↑↓↑↓↑↓↑↓ $4\sigma_{CH}\ \sigma_{CC}\ \sigma_{CO}$

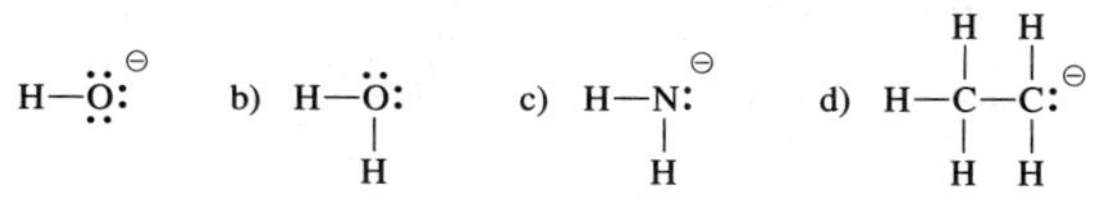

c)

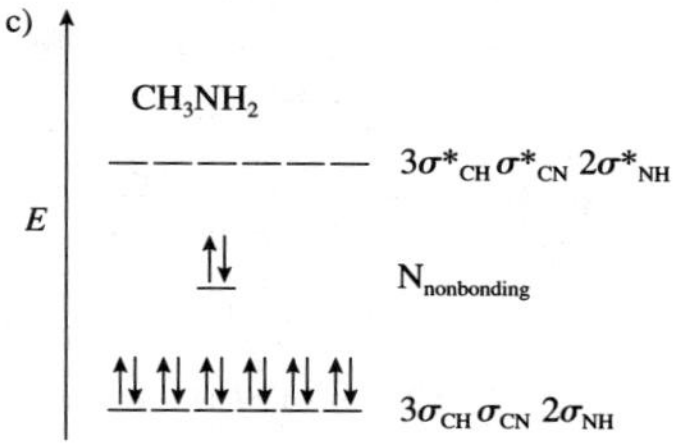

CH_3NH_2

E

— — — — — — $3\sigma^*_{CH}\ \sigma^*_{CN}\ 2\sigma^*_{NH}$
↑↓ $N_{nonbonding}$
↑↓↑↓↑↓↑↓↑↓↑↓ $3\sigma_{CH}\ \sigma_{CN}\ 2\sigma_{NH}$

Chapter 4

4.1 b) both c) acid d) base e) both f) both g) both

4.2 a) $CH_3-\overset{\oplus}{O}-H$ (with H below) b) $H-\overset{\oplus}{O}-H$ c) $CH_3-\overset{\oplus}{N}H_3$

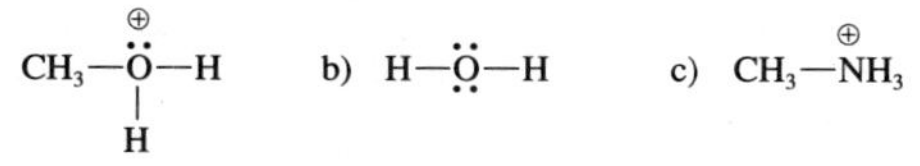

4.3 a) $H-\overset{\ominus}{O}:$ b) $H-\overset{..}{O}:$ (H below) c) $H-\overset{\ominus}{N}:$ (H below) d) $H-\overset{H\ H}{\underset{H\ H}{C-C}}:^{\ominus}$

4.4 a) $\overset{\ominus}{:NH_2}$ (base) $+$ $H-\overset{..}{O}-H$ (acid) $\rightleftharpoons$ $:NH_3$ (conjugate acid) $+$ $\overset{\ominus}{:O}-H$ (conjugate base)

b) $CH_3\overset{\ominus}{\overset{..}{O}:}$ $+$ $H-\overset{H\ \oplus}{\underset{}{O}}-H$ $\rightleftharpoons$ $CH_3\overset{..}{O}-H$ $+$ $H-\overset{..}{O}-H$

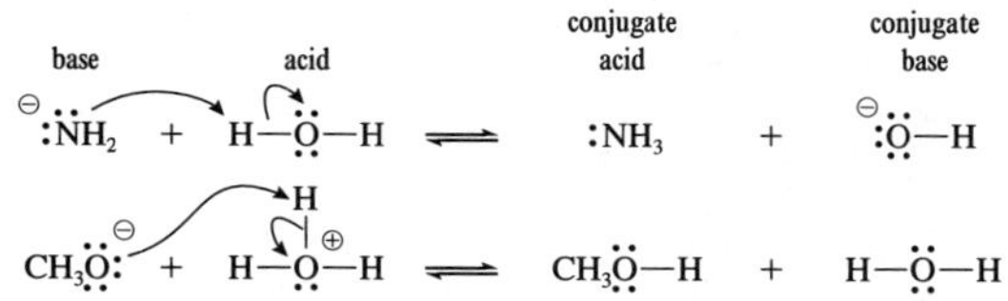

4.5 a) Lewis acid b) Lewis base c) Lewis acid d) Lewis base e) both

4.7 a) stronger b) weaker c) stronger d) weaker

4.8 a) stronger b) stronger c) weaker d) weaker

4.9 b) favors reactants c) favors products

4.10 a) favors products b) favors products

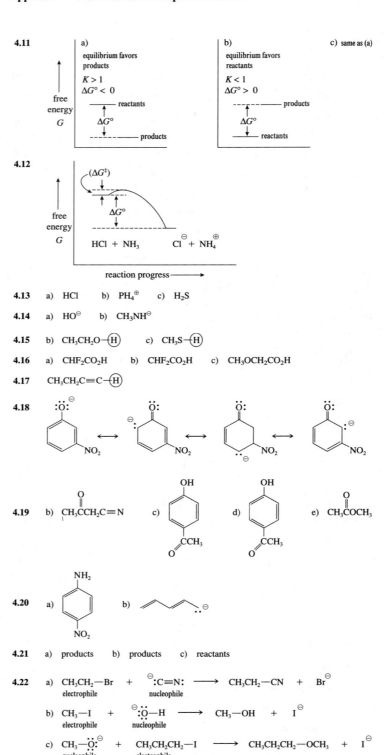

4.11   a) equilibrium favors products $K > 1$ $\Delta G° < 0$

free energy G

reactants

$\Delta G°$

products

b) equilibrium favors reactants $K < 1$ $\Delta G° > 0$

products

$\Delta G°$

reactants

c) same as (a)

4.12

free energy G

$(\Delta G^{\ddagger})$

$\Delta G°$

$HCl + NH_3$ $\overset{\ominus}{Cl} + \overset{\oplus}{NH_4}$

reaction progress ⟶

4.13   a) HCl   b) $PH_4^{\oplus}$ c) H_2S

4.14   a) $HO^{\ominus}$ b) $CH_3NH^{\ominus}$

4.15   b) $CH_3CH_2O\text{—}\boxed{H}$ c) $CH_3S\text{—}\boxed{H}$

4.16   a) CHF_2CO_2H b) CHF_2CO_2H c) $CH_3OCH_2CO_2H$

4.17 $CH_3CH_2C\equiv C\text{—}\boxed{H}$

4.18

4.19   b) $CH_3CCH_2C\equiv N$ c)   d)   e) CH_3COCH_3

4.20   a)   b)

4.21   a) products   b) products   c) reactants

4.22   a) $CH_3CH_2\text{—}Br$ (electrophile) + $^{\ominus}:C\equiv N:$ (nucleophile) ⟶ $CH_3CH_2\text{—}CN$ + $Br^{\ominus}$

b) $CH_3\text{—}I$ (electrophile) + $\overset{\ominus}{:}\!\overset{..}{O}\text{—}H$ (nucleophile) ⟶ $CH_3\text{—}OH$ + $I^{\ominus}$

c) $CH_3\text{—}\overset{..}{\underset{..}{O}}{}^{\ominus}$ (nucleophile) + $CH_3CH_2CH_2\text{—}I$ (electrophile) ⟶ $CH_3CH_2CH_2\text{—}OCH_3$ + $I^{\ominus}$

Chapter 5

5.1
a) 2-methylpentane
b) 2-methylpentane
c) 2,4-dimethylhexane
d) 5-ethyl-3-methyl-5-propylnonane
e) 3-ethyl-2,5-dimethylhexane
f) 3-ethyl-2,6-dimethylheptane

5.2   b) 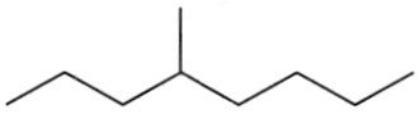   c)

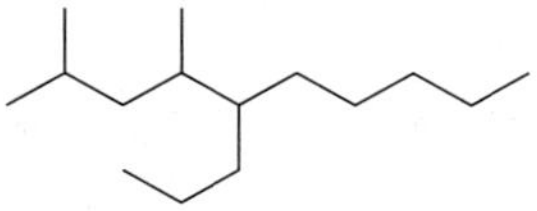

5.3   b) 4-ethyl-2,2-dimethylhexane   c) 2,2-dimethylpentane

5.4   b) (2-methylpentyl)   c) (1-methylpropyl)   d) (2,2-dimethylpropyl)

5.5   a) 4-(1-methylethyl)heptane   b) 5-(1,2-dimethylpropyl)decane

5.6   b)

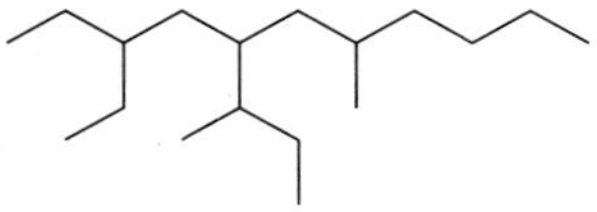

5.7   a) 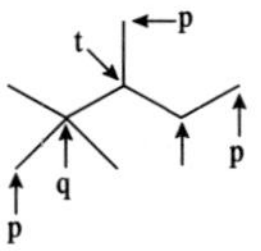   b) 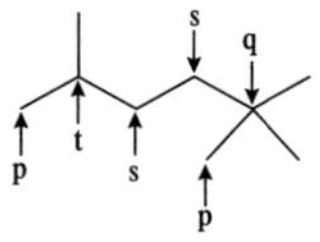   c)

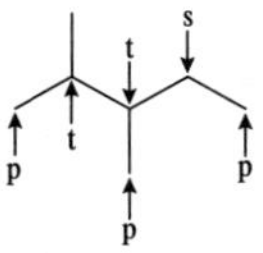

5.8   4-isopropylheptane

5.9

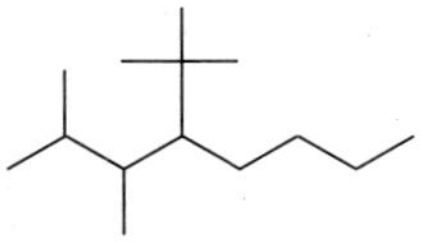

5.10
a) 1,2-dimethylcyclopentane
b) (1-methylpropyl)cyclohexane
c) 5-cyclopentyl-2-methylheptane
d) 1-ethyl-3,5-dimethylcyclooctane

5.11   a)    b)

5.12   b) 2,4-dimethyl-2-hexene   c) 2-methyl-1,3,5-cycloheptatriene
d) 3-ethyl-1,2-dimethylcyclopentene

5.13   a) 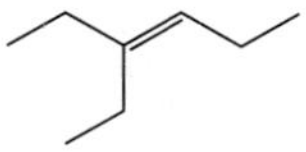   b)   c)

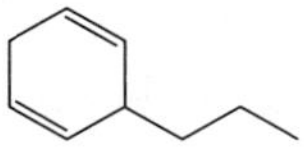

5.14   a) 3-isopropyl-1-hexyne   b) 2-methylpent-1-en-3-yne   c) 3-(2-methylpropyl)-1,4-hexadiyne

5.15 a) HC≡CCH₂CH₂CH₃   b)

5.16 a) 5-bromo-2,4,4-trimethylheptane   b) 1-chloro-3-ethyl-1-methylcyclopentane

5.17

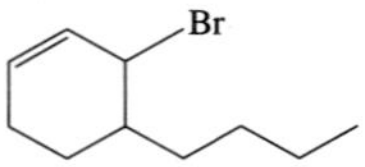

5.18 a) 2-butanol   b) 3-methyl-3-hexanol   c) 3-cyclopentyl-1-propanol
          d) 3-bromo-3-methylcyclohexanol

5.19 a) 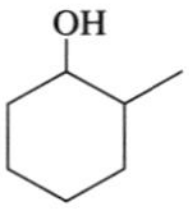   b)

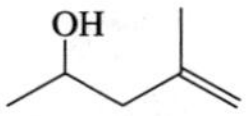

5.20 a) ethyl methyl ether   b) 1-chloro-3-methoxycyclopentane

5.21 a) propylamine   b) N-ethylcyclopentylamine

5.22 a) N,5-dimethyl-2-hexanamine   b) 5-amino-2-hexanol

5.23 a) 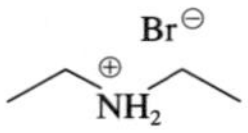   b)

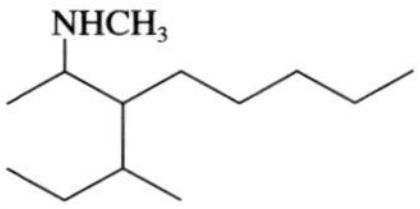

Chapter 6

6.1 a) 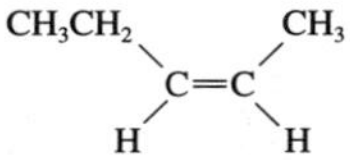
and 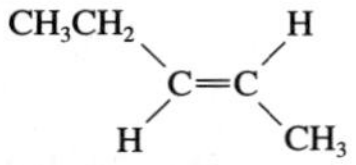   b) none

c) none   d) 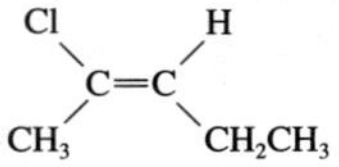
and

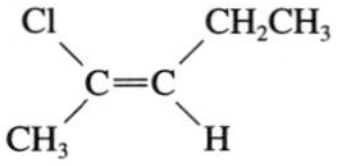

6.2 a) 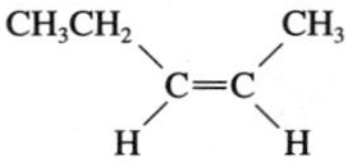
and

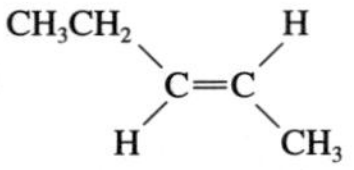

b) 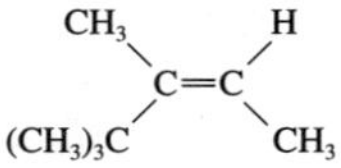
and

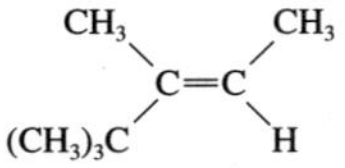

In both cases the right isomer is more stable because the larger groups are *trans*.

6.3 a) —CH₂CH₃   b) —C̈OH   c) —CH₂CH₂CH₂CH₃   d) —C≡N   e)

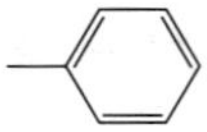

6.4 a) *E*   b) *E*   d) *Z*

6.5

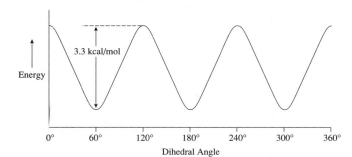

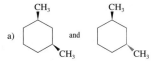

6.7

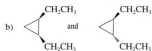

The conformation on the right, with the ethyl group equatorial, is more stable.

6.9   a) (CN)   b) (CH₂CH₃)   c) (Cl)

6.10   a) (CH₃ ... CH₃) and (CH₃ ... CH₃)   b) (CH₂CH₃ ... CH₂CH₃) and (CH₂CH₃ ... CH₂CH₃)

c) (Cl ... CH₃) and (Cl ... CH₃)

6.12   a) One methyl is axial and one is equatorial.

b) Both methyls are axial in one conformation and both are equatorial in the other.

c) The *trans*-isomer is more stable because it has a conformation with both methyls equatorial.

6.13   a) The conformation with both groups equatorial is the more stable conformation of the *cis*-stereoisomer. The conformation with the isopropyl group equatorial and the hydroxy group axial is the more stable conformation of the *trans*-stereoisomer.

b) The *cis*-stereoisomer is more stable than the *trans*-stereoisomer by 0.9 kcal/mol (3.8 kJ/mol).

6.14   b) The methyls are *trans;* the *t*-Bu is *cis* to the closer methyl; all are equatorial; the conformation shown is more stable; the stereoisomer shown is most stable.

c) The chlorines are *trans;* both are axial; the ring flipped conformation is more stable; the stereoisomer shown is more stable.

d) The groups are *trans;* the methyl is axial; the phenyl is equatorial; the conformation shown is more stable; the *cis*-stereoisomer is more stable.

6.15   a) achiral   b) chiral   c) chiral   d) achiral   e) chiral   f) chiral

6.16 a) not b) not c) 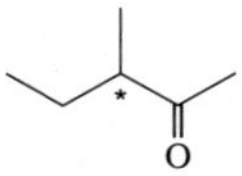 d) 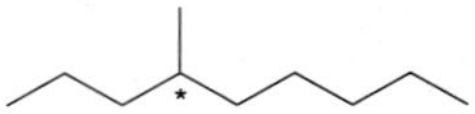

e) 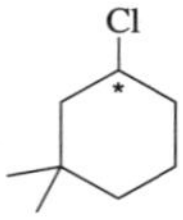 f) not

6.17 a) yes b) yes c) no d) yes e) no f) yes g) yes h) no

6.18 a) 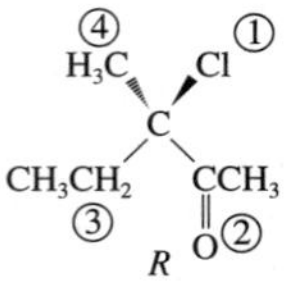 *R* b) 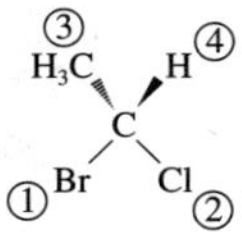 *R* c) 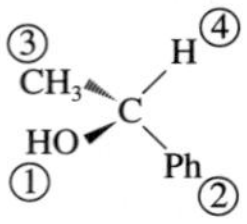 *R*

d) 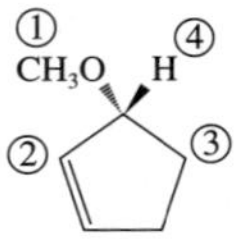 *R* e) 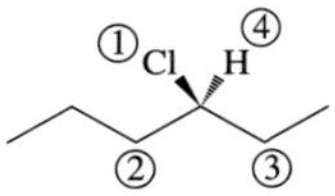 *S*

6.19 b) 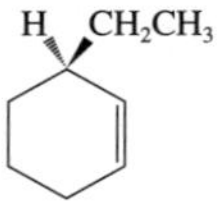 c) 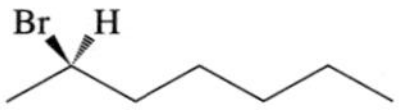

6.20 a) *R* b) *S*

6.21 (2*R*,3*R*)-2,3,4-Trihydroxybutanal is D-erythrose; L-erythrose is (2*S*,3*S*)-2,3,4-trihydroxybutanal.

6.22 a) false b) false c) true d) cannot be determined e) true

6.23

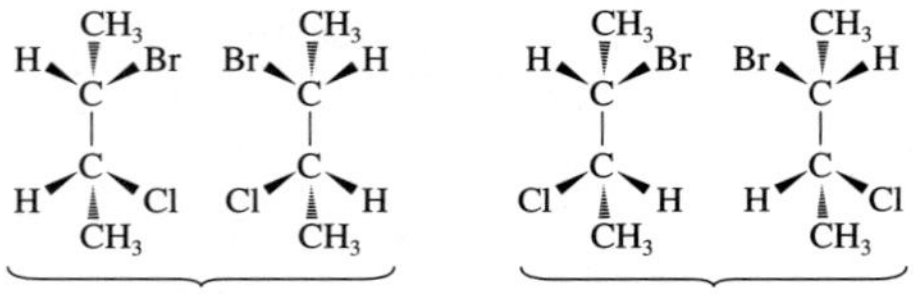

enantiomers enantiomers

diastereomers

6.24 a) 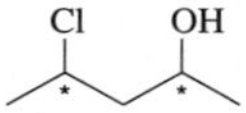 4 b) 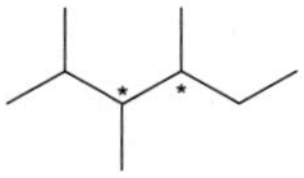 4 c) 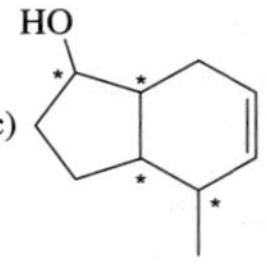 16 d) none

6.25

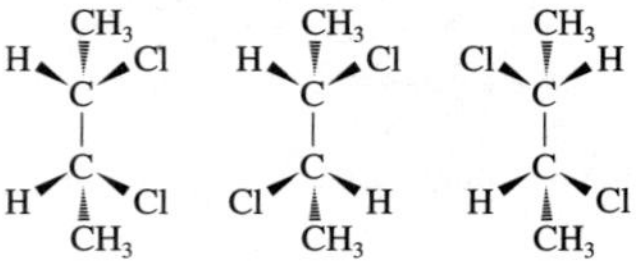

meso rotates rotates

6.26

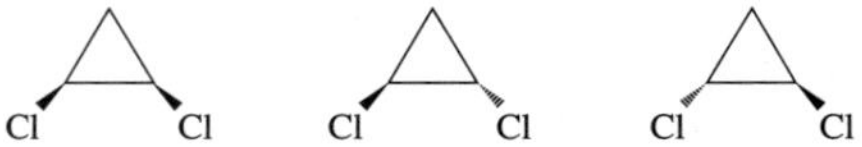

meso rotates rotates

6.27 b) yes c) no, meso d) yes e) no, meso

6.28   a) H—|—Cl   b) HO—|—H   c) H₃C—|—H

with CH₂OH, CH₃ (a); CO₂H, CH₃ (b); CHO, CH₂CH₃ (c) groups

6.29   b) *R*   c) *S*

Chapter 7

7.1   a) H₃C—C(OH)(Ph)(H)   b) cyclohexene with OCH₃ and H   c) cyclopentane with CH₃, CH₂CH₃, CH₂CH₂OH

7.2   a) The right compound reacts faster because it has less steric hindrance.

b) The right compound reacts faster because it has less steric hindrance.

c) The left compound reacts faster because of resonance stabilization of the transition state.

7.3

CH₃Cl > CH₃CH₂CH₂CH₂CH₂ > CH₃CH₂CHCH₂Cl > CH₃CH₂CH₂CHCH₃
fastest   (Cl)   (CH₃)   (Cl)   slowest

7.4

free energy (*G*) vs reaction progress

7.6   a) (H₃C)(CH₃CH₂)(OCH₃)C + HCl
(racemic, perhaps with some excess inversion)

c) CH₃CH₂—C(H₃C)(H)—CH₂—C(CH₃)₂—OCCH₃ + HCl

7.7   a) left; tertiary carbocation is more stable

b) left; resonance stabilized carbocation is more stable

c) left; resonance stabilized carbocation is more stable

d) right; methoxy group provides extra resonance stabilization of the carbocation

7.8

Ph group > Ph group > compound > compound
Cl   Cl   Cl   Cl
fastest   slowest

7.9   a) Primary substrates with a strong nucleophile (hydroxide ion) react by the S_N2 mechanism. The right reaction is faster because mesylate ion is a better leaving group than chloride ion.

b) Tertiary substrates react by the S_N1 mechanism. The right reaction is faster because iodide ion is a better leaving group than bromide ion.

7.10

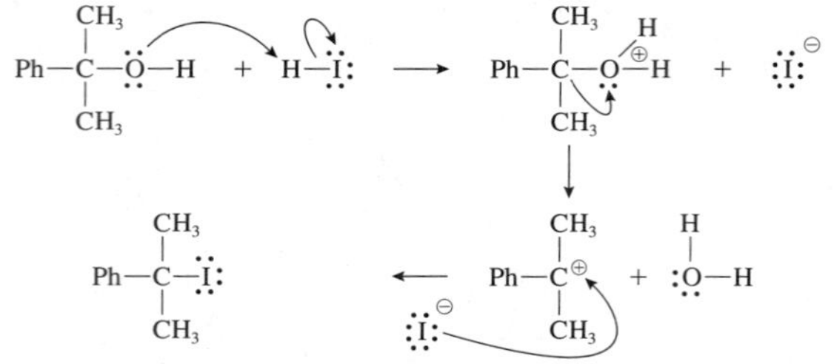

Ph—C—O—H + H—I: → Ph—C—O—H + :I:⊖

(CH₃ groups; arrows showing mechanism)

Ph—C—I: ← Ph—C⊕ + :O—H

7.11 b) Because the leaving group is on a tertiary carbon, the reaction proceeds by an S_N1 mechanism. The reactivity of the nucleophile does not affect the rate of an S_N1 reaction so both reactions proceed at the same rate.

c) Because the leaving group is on a primary carbon, the reaction proceeds by an S_N2 mechanism. The right reaction is faster because the nucleophile is stronger. (Nucleophile strength increases down a column of the periodic table.)

d) Because the leaving group is on a primary carbon, the reaction proceeds by an S_N2 mechanism. The left reaction is faster because the nucleophile is stronger. (It is a stronger base.)

7.12 a) S_N2 CH₃CH₂CHCH₃ + :SH⊖ → CH₃CH₂CHCH₃ + :Cl:⊖

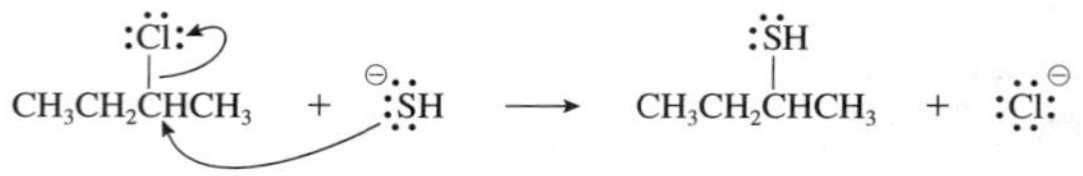

b) S_N1 CH₃CH₂CCH₃ → CH₃CH₂CCH₃ + :Br:⊖
(CH₃, :Br:, CH₃CH₂OH)

CH₃CH₂CCH₃
:O—CH₂CH₃ ← CH₃CH₂CCH₃
+ H—Br: :Br:⊖ , H—O⊕—CH₂CH₃

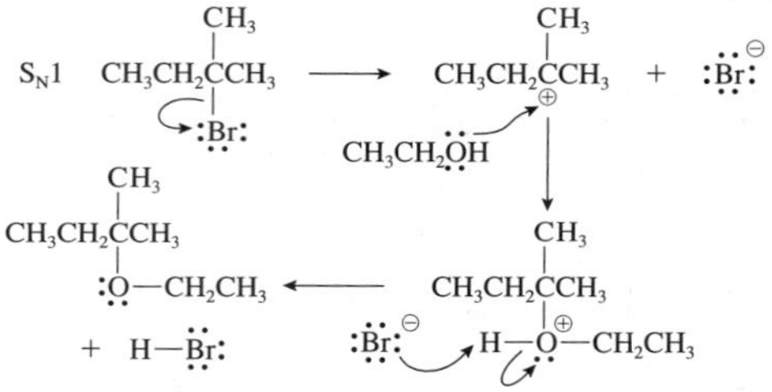

c) S_N2 CH₃CH₂CH₂CH₂CH₂ + :NH₂CH₃ → CH₃CH₂CH₂CH₂CH₂   TsO:⊖ ⊕NH₂CH₃

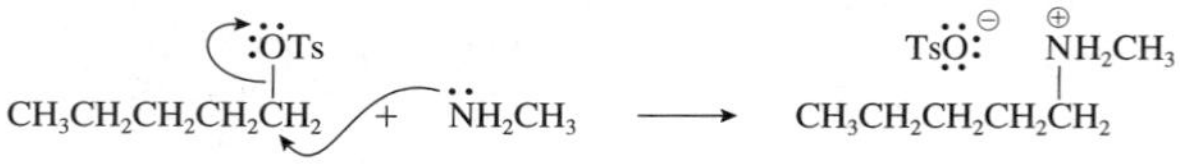

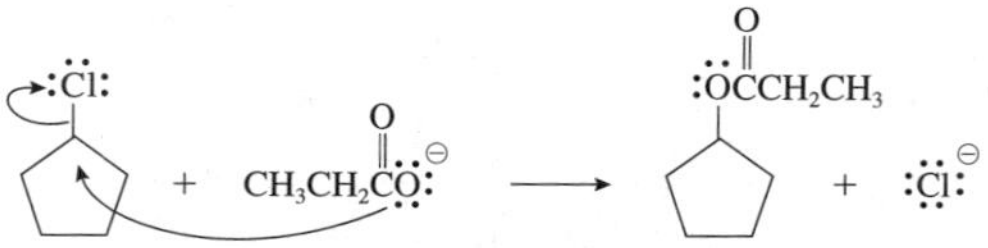

d) S_N2 + CH₃CH₂CO:⊖ →   + :Cl:⊖

7.13 CH₃
CH₃CCH₃ → CH₃CCH₃ + :Cl:⊖
(:Cl:, CH₃OH)

CH₃
CH₃CCH₃ ← CH₃CCH₃
:O—CH₃   :Cl:⊖ , H—O⊕—CH₃
+ H—Cl:

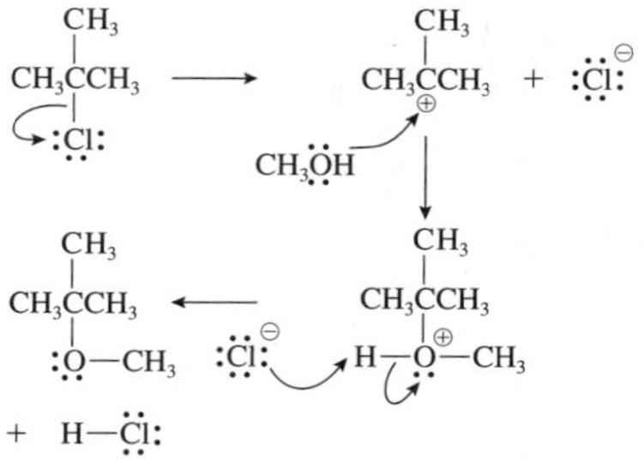

7.14 The conditions of Figure 7.7 have a more polar solvent (water v. acetic acid) and a weaker nucleophile (water v. acetate anion) than the conditions of Figure 7.8 so the carbocation has a longer lifetime under the conditions of Figure 7.7.

7.15 b) This S_N1 reaction is faster in the more polar solvent, methanol.

c) This S_N2 reaction, with a negative nucleophile, is faster in the less polar solvent, pure methanol.

d) This S_N2 reaction is faster in the aprotic solvent, DMSO.

7.16 b) tertiary, so S_N1

d) secondary, with a strong nucleophile and an aprotic solvent, so S_N2

e) unhindered substrate (less hindered than primary), so S_N1

f) allylic substrate, weak nucleophile and polar solvent, so S_N1

g) allylic substrate, strong nucleophile and aprotic solvent, so S_N2

h) secondary benzylic substrate, weak nucleophile and polar solvent, so S_N1

7.17 a)   b)

7.18 a)   +   b) $CH_3CH_2CCH_3$ + $CH_3CH=C$ + $CH_3CH_2CCH_3$

c)   +

7.19 a)   b) $CH_3CCH_2CH_2CH_3$ c)

7.20 a) $CH_3-C-CHCH_3$ + $CH_3-C-CHCH_3$ b)   +

c)   +

7.21 a)

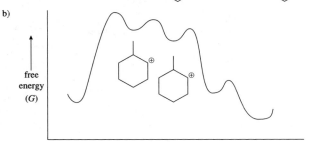

b)

reaction progress ⟶

free energy (G)

Chapter 8

8.1   a) $CH_3CH=CHCH_2CH_3$ b) 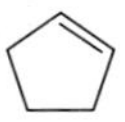   c) 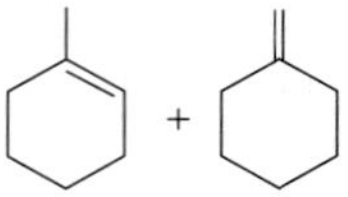   +

8.2

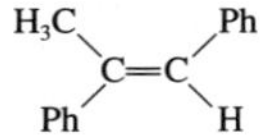

8.3   b) 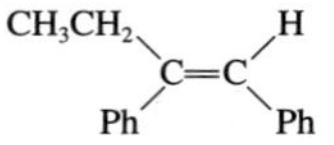   c)

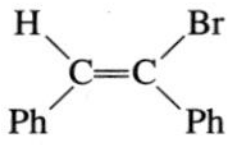

8.4   More of the *trans*-alkene is formed because the conformation that produces the *trans*-alkene is more stable because the bulky phenyl groups are anti.

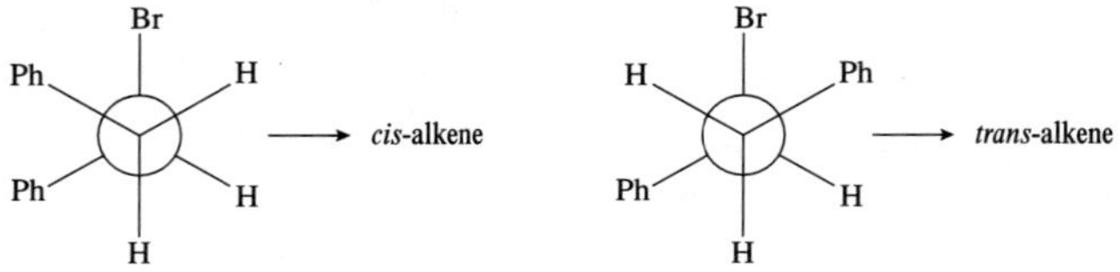

8.5   a) 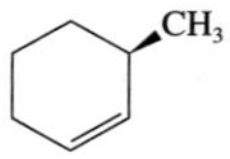   b) 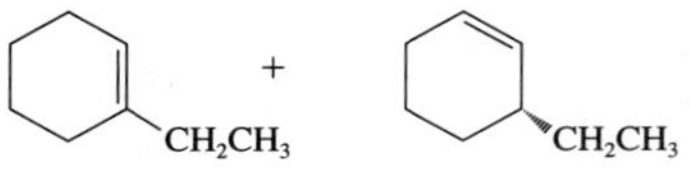   +

8.6   The bulky *t*-Bu group must be equatorial in both cases. The isomer on the left has the OTs group axial, so it can readily undergo E2 elimination. The isomer on the right has the OTs group equatorial, so it cannot readily react.

8.7   b) 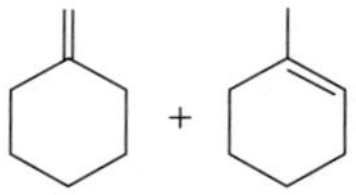 +   c)  +   d) +

                        major   *cis* + *trans*   major

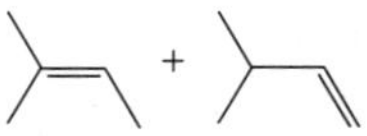

                                                            major

8.8   (*E*)-2-Butene is the major product because it is more stable and the conformation leading to it is more stable.

8.9   b) $CH_3CH_2CH_2CH=CHCH_3$

                       +

          $CH_3CH_2CH_2CH_2CH=CH_2$ major

      c) 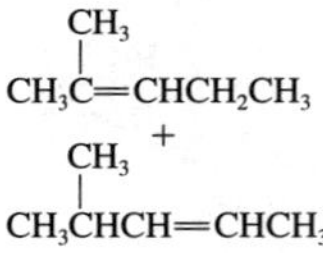 $CH_3\overset{\overset{\displaystyle CH_3}{|}}{C}=CHCH_2CH_3$

                     +

         $CH_3\overset{\overset{\displaystyle CH_3}{|}}{C}HCH=CHCH_3$ major

      d) 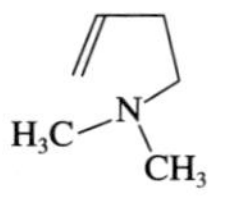   e) 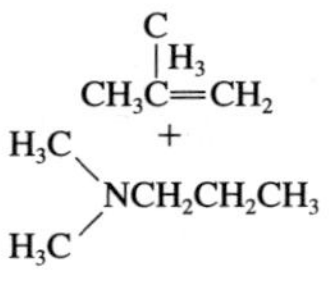   +

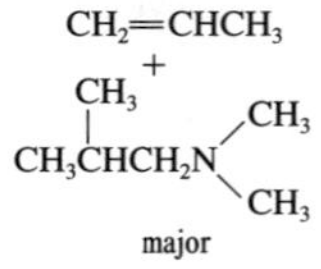

8.10   a) 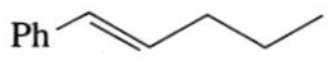   Ph   b)

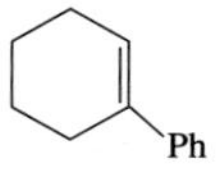

      c)  +   d) 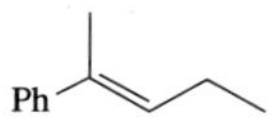   Ph

                          major

16.14 a) CH₃CH₂CCHCOEt   b) CH₃CH₂CHC—C—COEt

(structures for a, 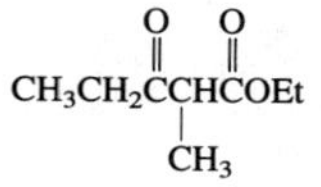 b, 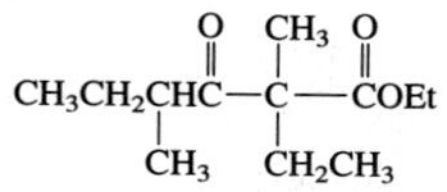 c, 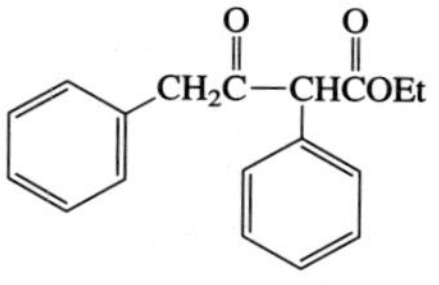 d)

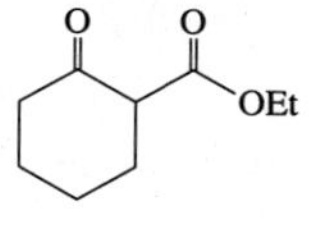

16.15 This product does not form because there is no acidic H between the two carbonyl groups so the equilibrium driving step cannot occur.

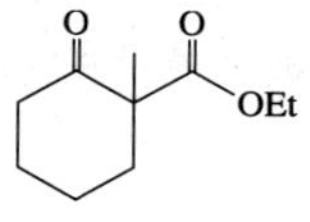

16.16 b) 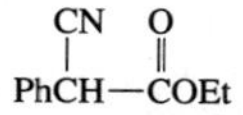 PhCH—COEt   c)

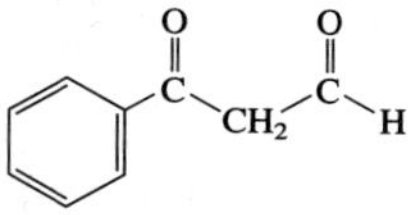

d) 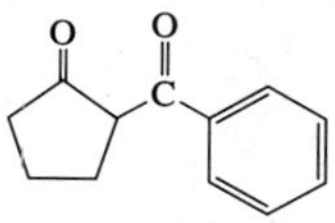   e)

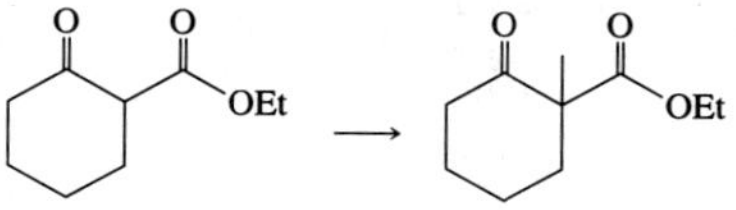

16.17

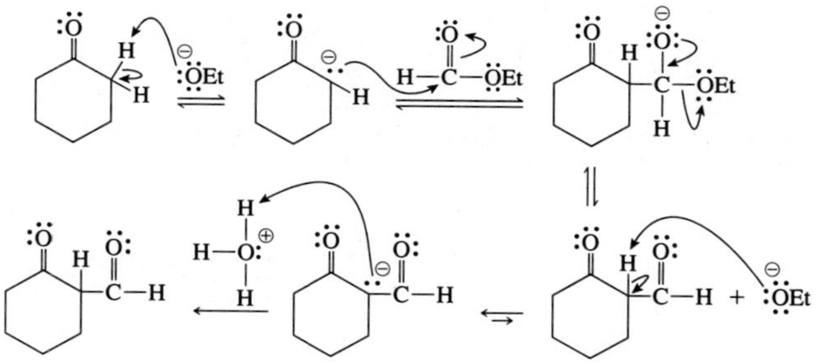

16.18 a)

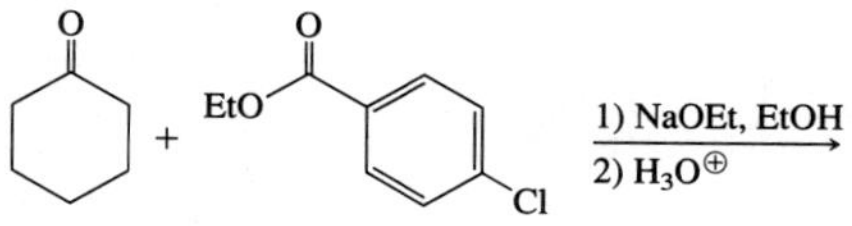

1) NaOEt, EtOH
2) H₃O⊕

b)

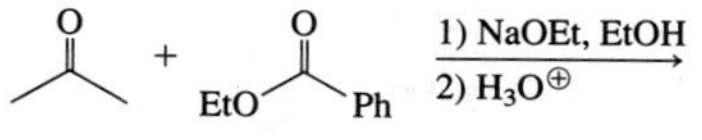

1) NaOEt, EtOH
2) H₃O⊕

c)

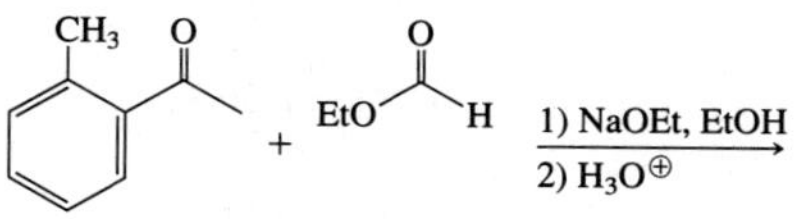

1) NaOEt, EtOH
2) H₃O⊕

d)

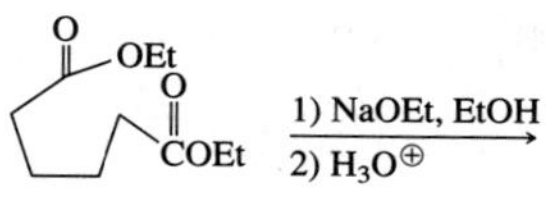

1) NaOEt, EtOH
2) H₃O⊕

e)

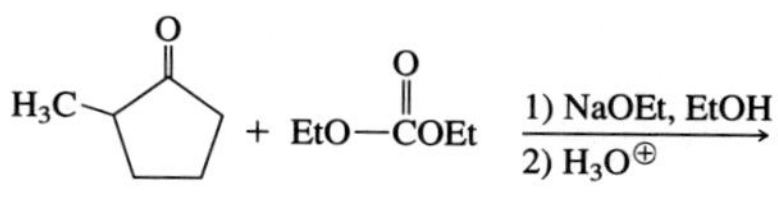

1) NaOEt, EtOH
2) H₃O⊕

16.19 a) 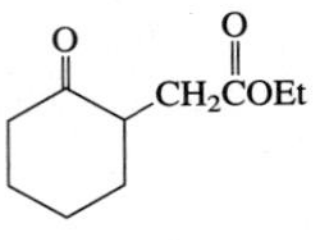   b) 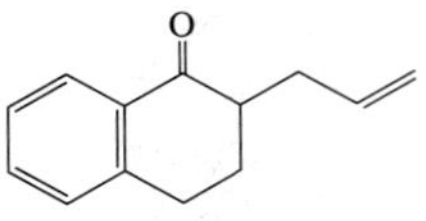   c)

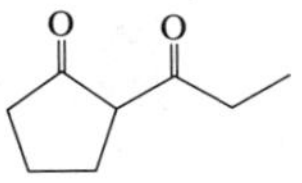

16.20 a) 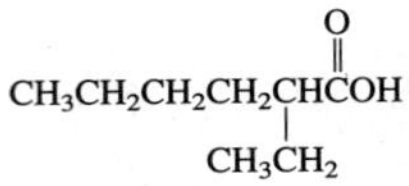 $CH_3CH_2CH_2CH_2CHCOH$ (with CH_3CH_2 substituent)   b) 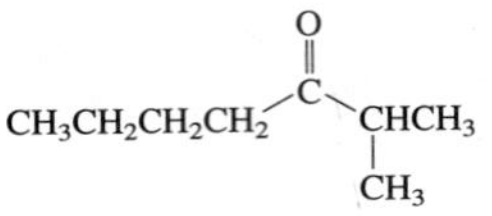 $CH_3CH_2CH_2CH_2$—C—$CHCH_3$ (with CH_3 substituent, O double bond)

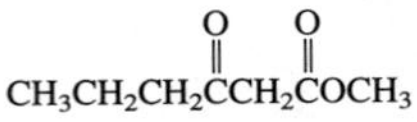

c) $CH_3CH_2CH_2CCH_2COCH_3$ d) 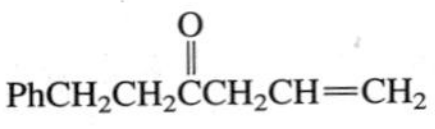 $PhCH_2CH_2CCH_2CH{=}CH_2$

16.21 a) 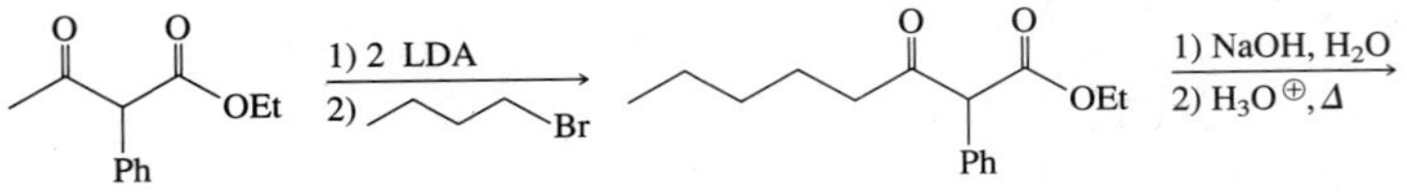 $\xrightarrow{\text{1) 2 LDA}}$ (2) butyl bromide) $\xrightarrow{\text{1) NaOH, H}_2\text{O}}$ (2) $H_3O^{\oplus}, \Delta$)

b) 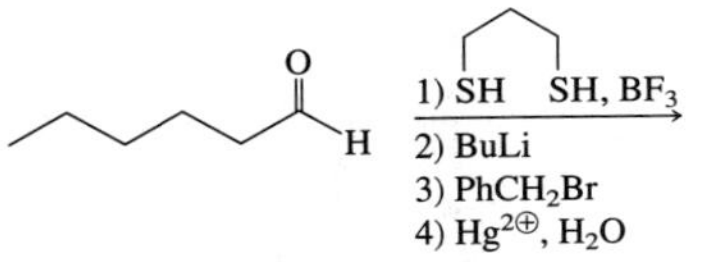
1) SH   SH, BF$_3$
2) BuLi
3) PhCH$_2$Br
4) Hg$^{2\oplus}$, H$_2$O

c) 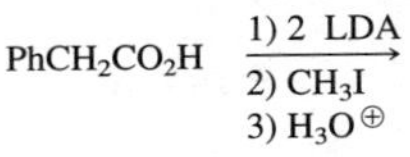 $PhCH_2CO_2H$
1) 2 LDA
2) CH$_3$I
3) H$_3$O$^{\oplus}$

16.22 b) 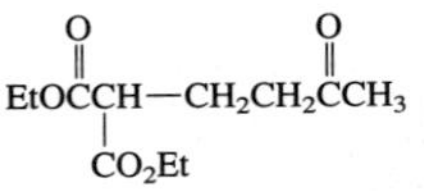 $EtOCCH{-}CH_2CH_2CCH_3$ (with CO_2Et)   c) 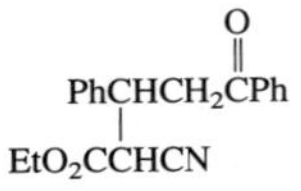 $PhCHCH_2CPh$ (with EtO_2CCHCN)   d) 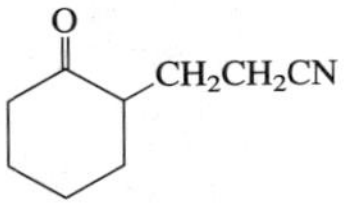 (CH_2CH_2CN)

16.23 The hydrogen on the α-carbon bonded to the phenyl group is more acidic because the resulting enolate anion is stabilized by resonance with the phenyl group.

16.24

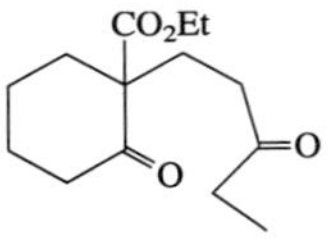

16.25 b)

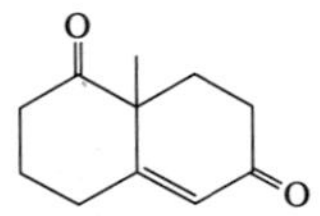

16.26 a) 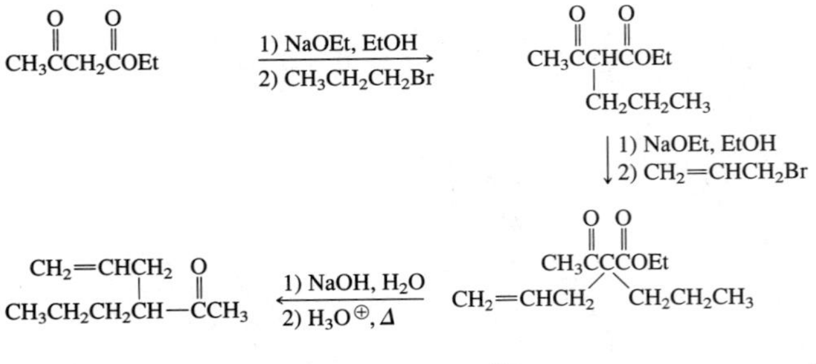 CH_3CCH_2COEt $\xrightarrow[\text{2) CH}_3\text{CH}_2\text{CH}_2\text{Br}]{\text{1) NaOEt, EtOH}}$ $CH_3CCHCOEt$ (with $CH_2CH_2CH_3$)

$\downarrow$ 1) NaOEt, EtOH 2) CH$_2$=CHCH$_2$Br

$CH_3CCCOEt$ (CH_2=CHCH$_2$ and $CH_2CH_2CH_3$)

$CH_3CH_2CH_2CH{-}CCH_3$ (with CH_2=CHCH$_2$) $\xleftarrow[\text{2) H}_3\text{O}^{\oplus}, \Delta]{\text{1) NaOH, H}_2\text{O}}$

b) 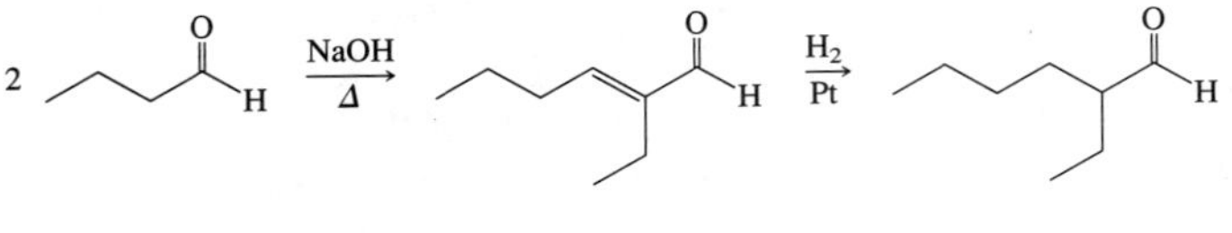 2 $\xrightarrow[\Delta]{\text{NaOH}}$ $\xrightarrow[\text{Pt}]{\text{H}_2}$

c)

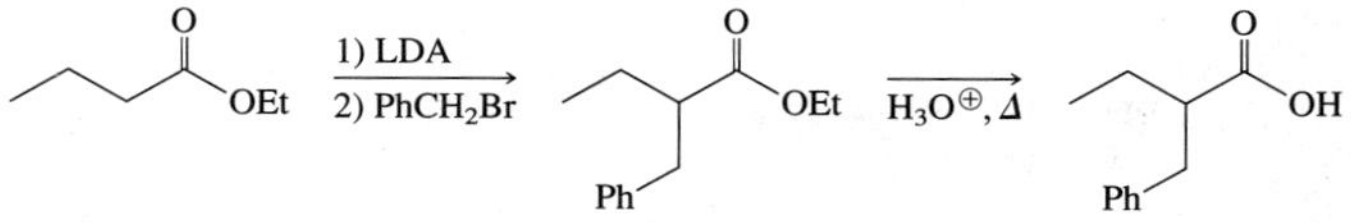

d)

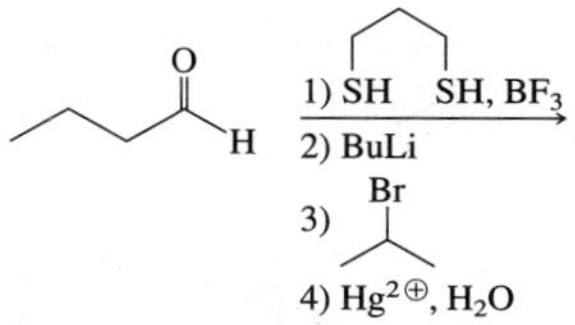

e)

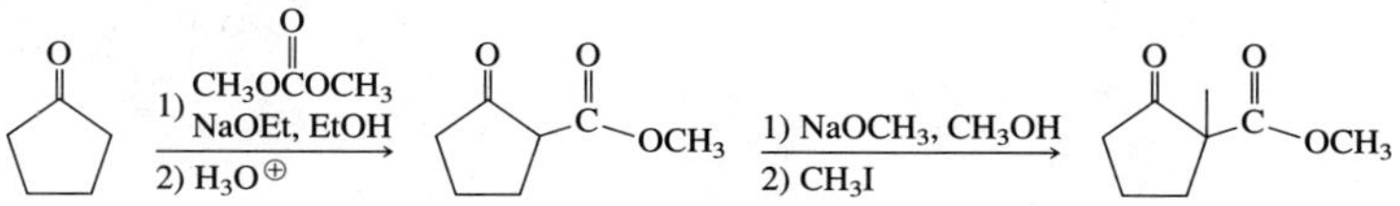

f)

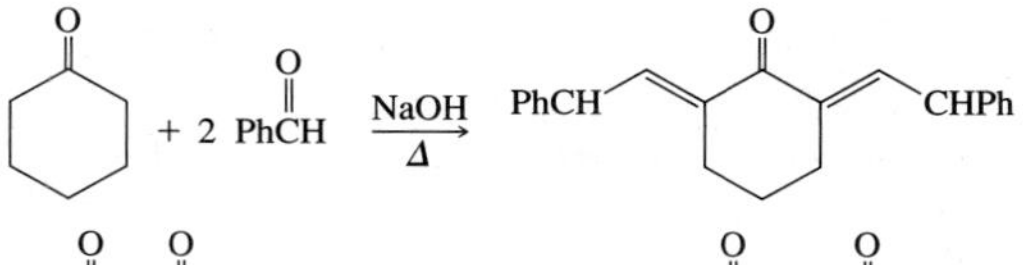

g)

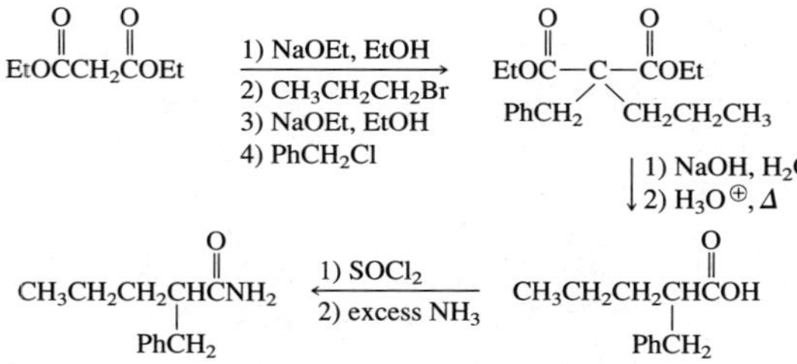

h)

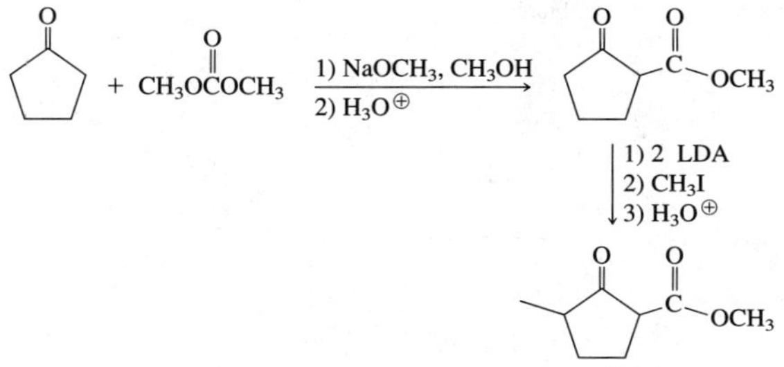

16.27   a) 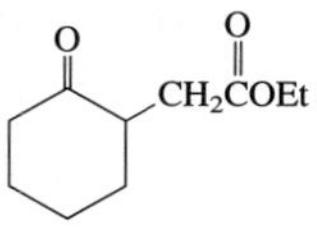   b) 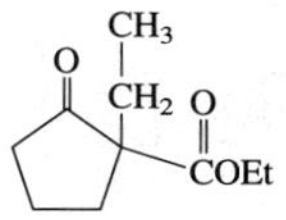   c)

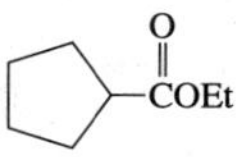

d) 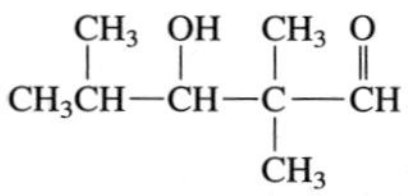 CH₃CH—CH—C—CH   e) 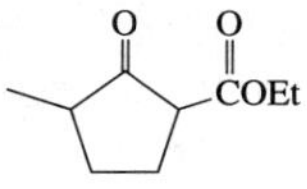   f)  CH₃CH₂CCH₂CH₂CH₂Ph

g) 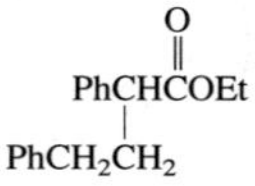 PhCHCOEt / PhCH₂CH₂   h) 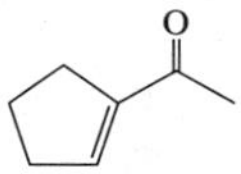   i) 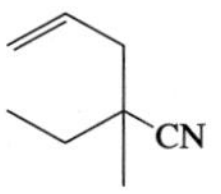   j)

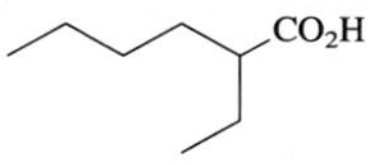

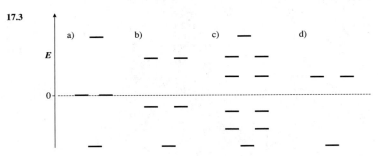

k) [structure] EtO₂C-CH(CH(CH₃)CH₂CH₃)-CO₂Et 1) NaOH, H₂O / 2) H₃O⁺, Δ → [product]

l) [structure] 1) NaOH, H₂O / 2) H₃O⁺, Δ → [acetylcyclopentane]

m) [structure]

n) [structure]

o) [structure]

p) [structure]

Chapter 17

17.1 a) [structure with Br, Br] b) [structure with Cl] c) [structure] d) [structure]

17.2 (5) (28.6 kcal/mol) − 80 kcal/mol = 63 kcal/mol; yes.
(5) (120 kJ/mol) − 335 kJ/mol = 265 kJ/mol; yes.

17.3

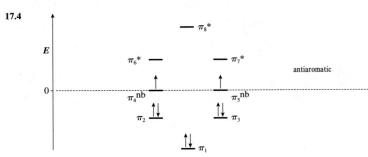

17.4

17.5

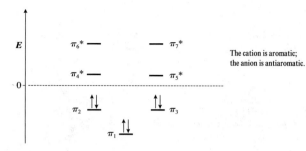

The cation is aromatic;
the anion is antiaromatic.

17.6

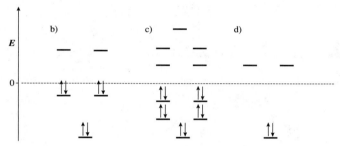

all are aromatic

17.7 a) 4 π electrons so antiaromatic   b) 6 π electrons so aromatic

c) 10 π electrons so aromatic   d) 2 π electrons so aromatic

17.9 a) 6 π electrons so aromatic   b) 6 π electrons so aromatic

c) 6 π electrons so aromatic   d) 6 π electrons so aromatic

e) 6 π electrons so aromatic   f) 8 π electrons so antiaromatic

17.10

The bond between C-9 and C-10 should be the shortest because it is a double bond in four of the five resonance structures.

17.11

17.12

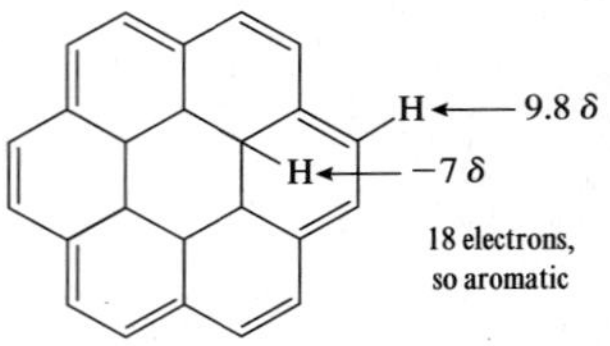

H ←— 9.8 δ

H ←— ~7 δ

18 electrons,
so aromatic

17.13   a)   left compound because its conjugate base is aromatic

b)   right compound because conjugate base of left compound is antiaromatic

c)   right compound because its conjugate base is aromatic

17.14   left compound because the carbocation that is formed from it is aromatic

Chapter 18

18.1

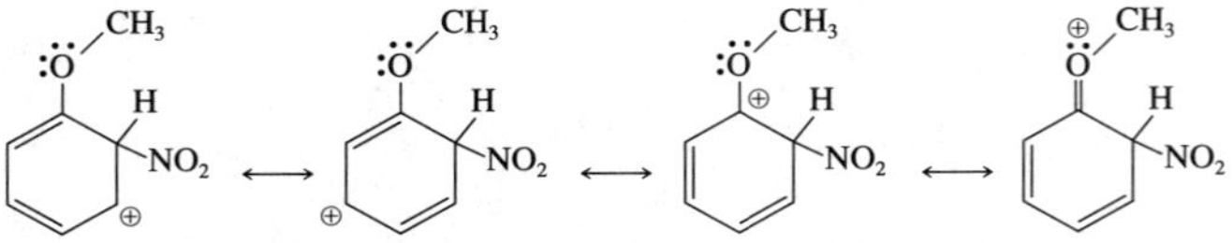

The last structure is especially stable because the octet rule is satisfied at all atoms.

18.2

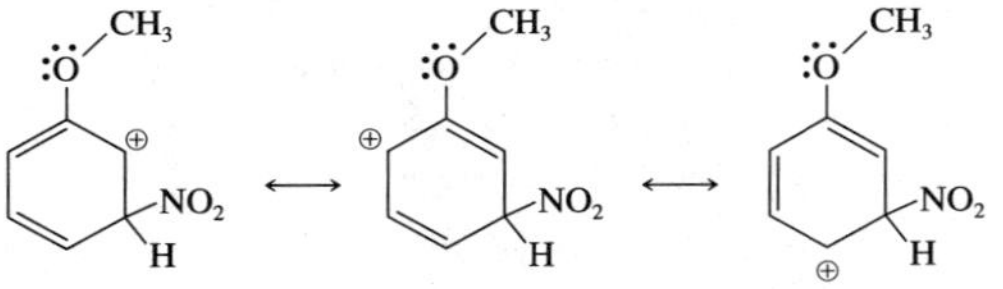

No especially stable resonance structure is formed.

18.3   a)   The electron pair on the N can
stabilize a positive charge on the
adjacent carbon by resonance.

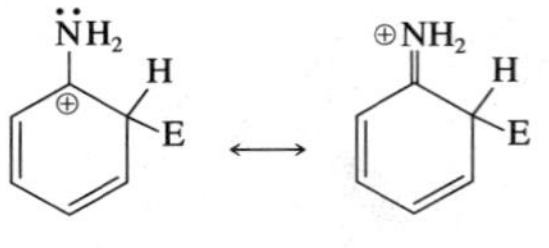

b)   The ethyl group stabilizes a
positive charge on the adjacent
carbon by hyperconjugation.

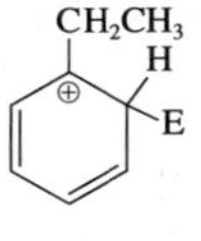

c)   The phenyl group can stabilize
a positive charge on the adjacent
carbon by resonance.

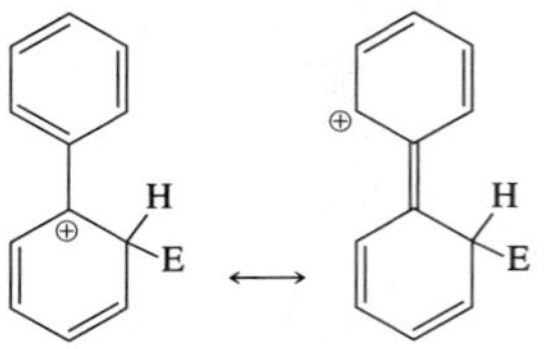

18.4   a)   The electronegative fluorines make the CF_3 group electron withdrawing.

b)   The positive nitrogen is an inductive electron withdrawing group.

18.5   a)   activating, o/p-director;   b)   deactivating, m-director;   c)   deactivating, m-director

18.6 b) 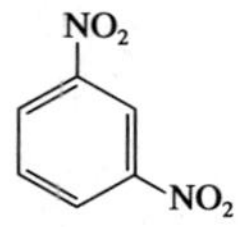 c) 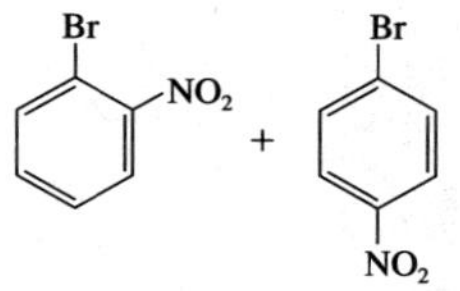 +

18.7 a) 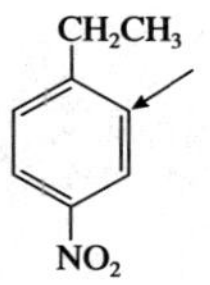 b) 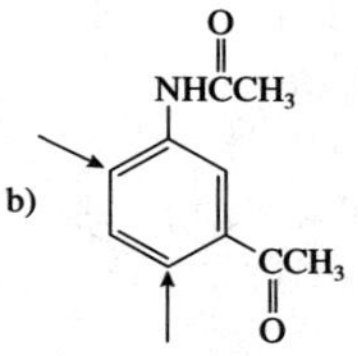 c)

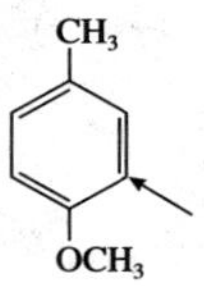

d) 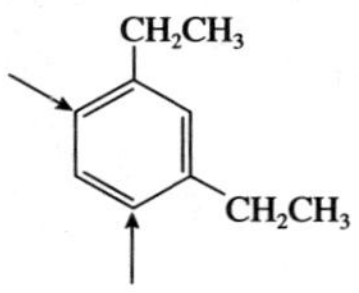 e) 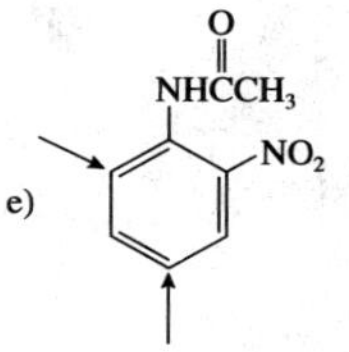 f)

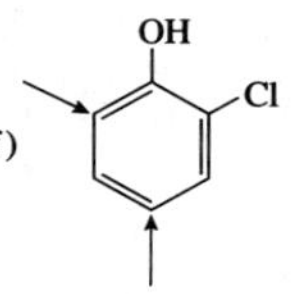

18.8 a) 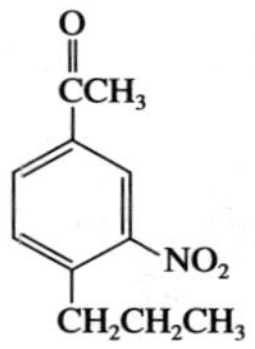 b) 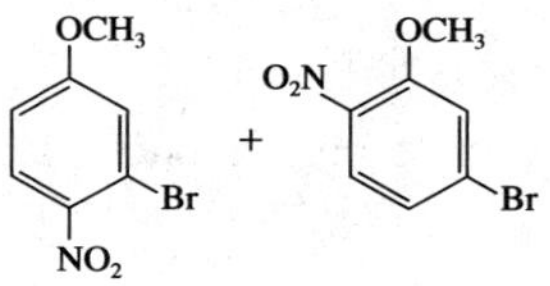 +

18.9 a) 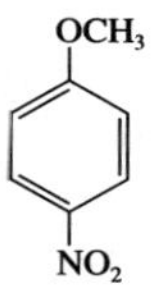 + ortho b) 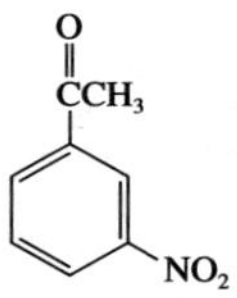 c) 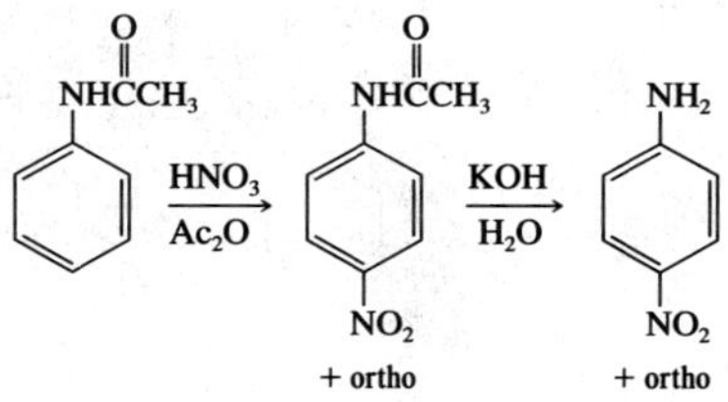 $\xrightarrow[\text{Ac}_2\text{O}]{\text{HNO}_3}$ + ortho $\xrightarrow[\text{H}_2\text{O}]{\text{KOH}}$ + ortho

18.10

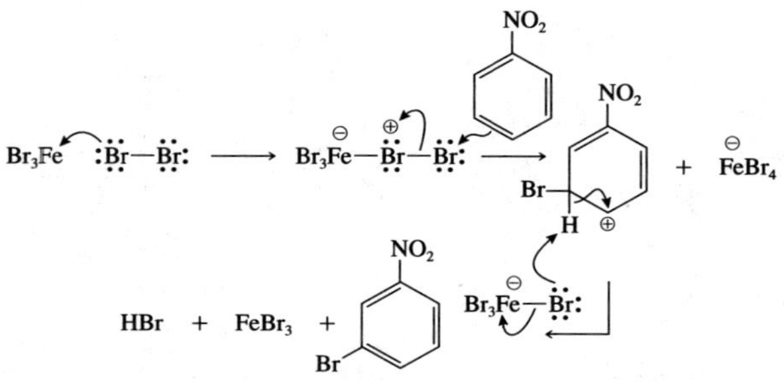

18.11 a) 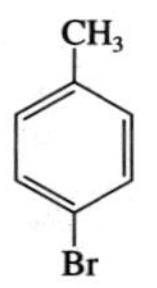 + ortho b) 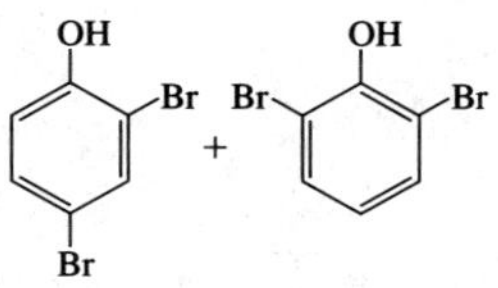 + c)

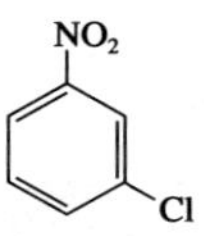

d) [structure: F on benzene, Cl para] + ortho   e) [structure: NO₂, Br, CH₃ benzene]   f) [structure: CH₃, Br, Br benzene]

18.12

18.13   a) [structure: benzene with SO₃H meta SO₃H]   b) [structure: CH₃ benzene SO₃H] + ortho   c) [structure: CH₃, SO₃H, C(CH₃)₃ benzene]

d) [structure: CH₃, OH, SO₃H benzene] + [structure: CH₃, OH, HO₃S benzene]

18.14   CH₃CH₂CH₂—Cl: + AlCl₃ ⇌ CH₃CH₂CH₂—Cl⁺—AlCl₃⁻ → CH₃CHCH₂⁺ + AlCl₄⁻
[mechanism scheme with benzene rings and intermediates]
CH₃CHCH₃ ... CH₃CHCH₃⁺ ... CH₂CH₂CH₃   CH₃CH₂CH₂

18.15   a) [structure: OCH₃, CH₃CHCH₃ benzene] + ortho   b) [structure: NHCCH₃ O, CH₃CCH₃ CH₃ benzene] + ortho   c) CH₃CHPh   d) [structure: CH₃CCH₃ CH₃ benzene]

e) [structure: CH₃CCH₂CH₃ CH₃ benzene]   f) [fused bicyclic structure with CH₃]   g) no reaction

18.16   a) [benzene] + CH₃C—Cl CH₃ CH₃ →AlCl₃   b) [benzene] + CH₃CHCH₃ Cl →AlCl₃

18.18 $CH_3(CH_2)_7CH(CH_2)_5CH_3$

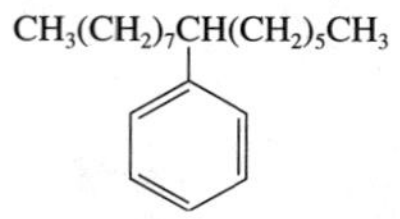

18.19   Both groups are weakly activating, but the position ortho to the methyl group is less sterically hindered than the position ortho to the isopropyl group.

18.20   a) 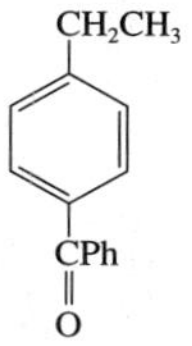   b) 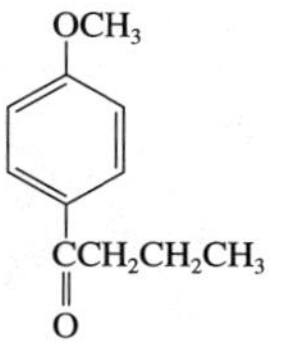   c) no reaction

d) 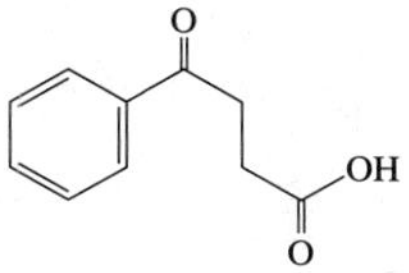   e) 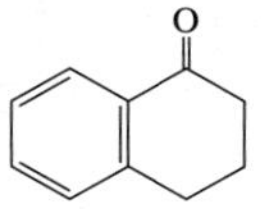   f)

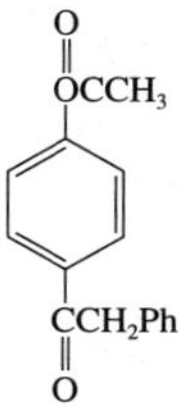

18.21   a) 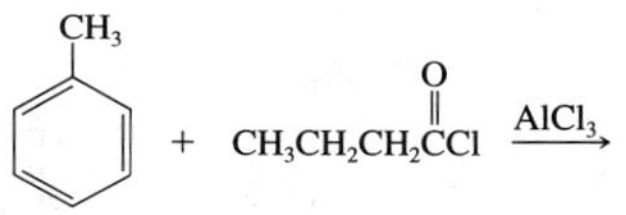 $+$ $CH_3CH_2CH_2\overset{O}{\overset{\|}{C}}Cl$ $\xrightarrow{AlCl_3}$ b) $+$ $CH_3\overset{O}{\overset{\|}{C}}Cl$ $\xrightarrow{AlCl_3}$

c) $+$ $\xrightarrow{AlCl_3}$ $\xrightarrow[\text{2) AlCl}_3]{\text{1) SOCl}_2}$

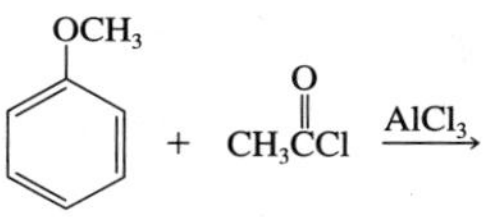

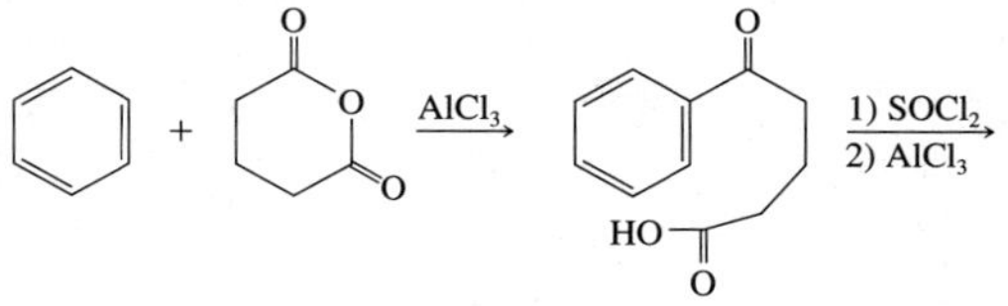

18.22

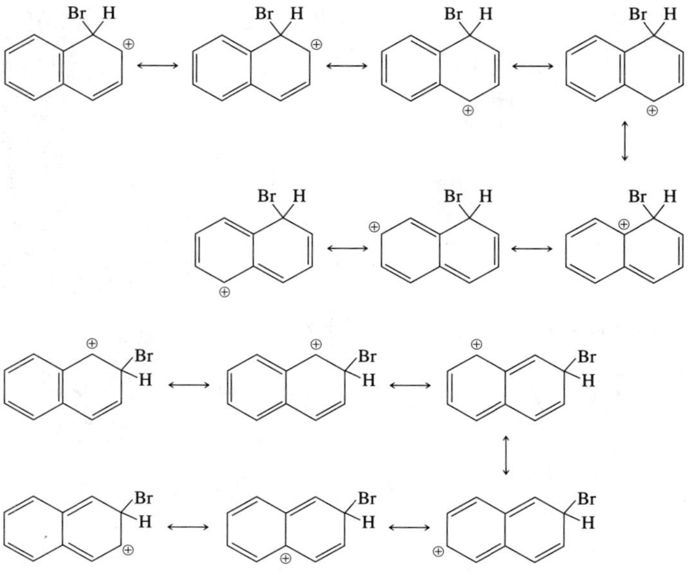

18.23 a) 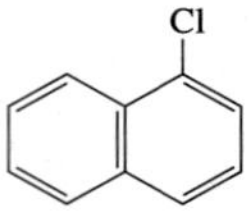 b) 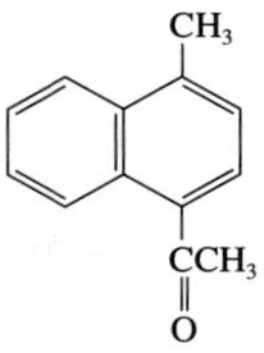 c)

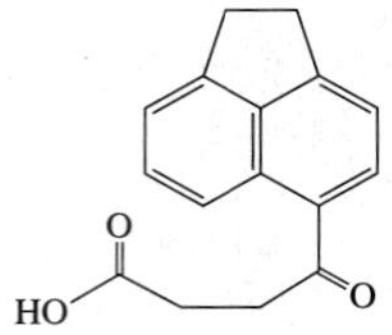

18.24 a) 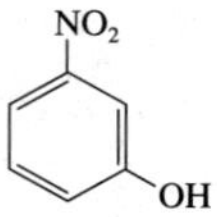 b) 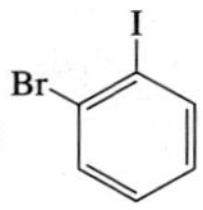 c) 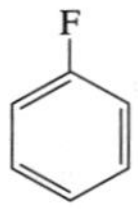 d)

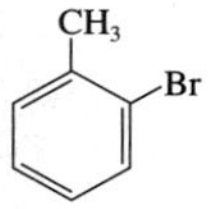

e) 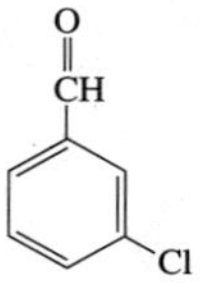 f) 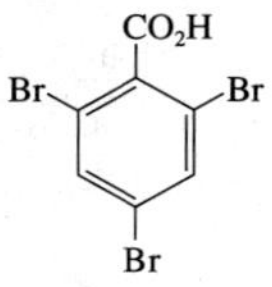 g) 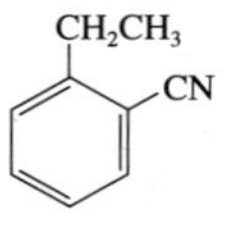 h)

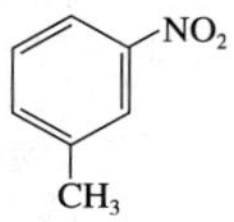

18.25

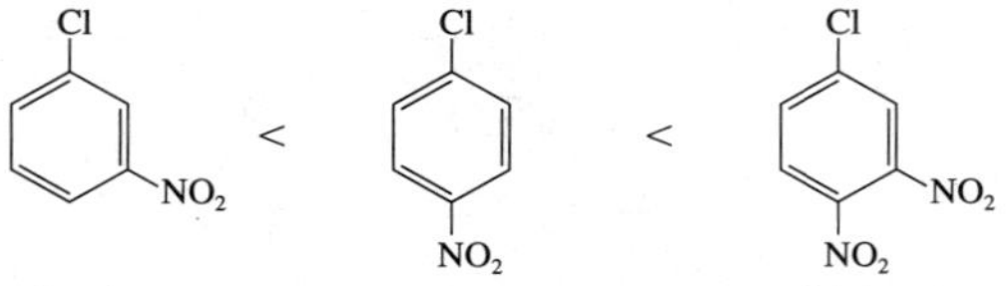

slowest < < fastest

18.26 a) 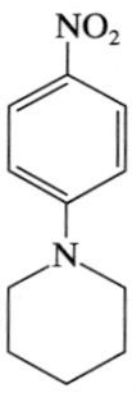 b)

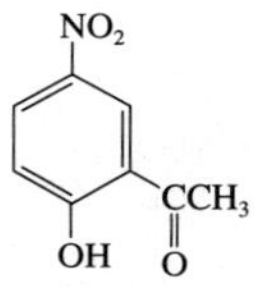

18.27 a) 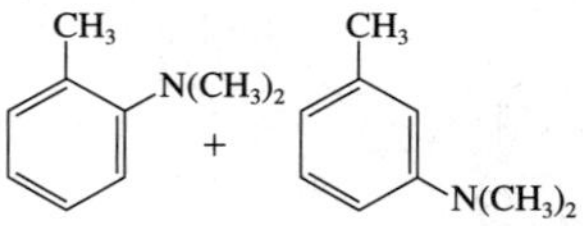 + b) 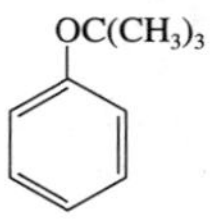 c) 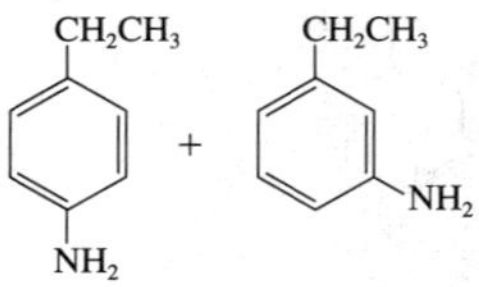 +

18.28

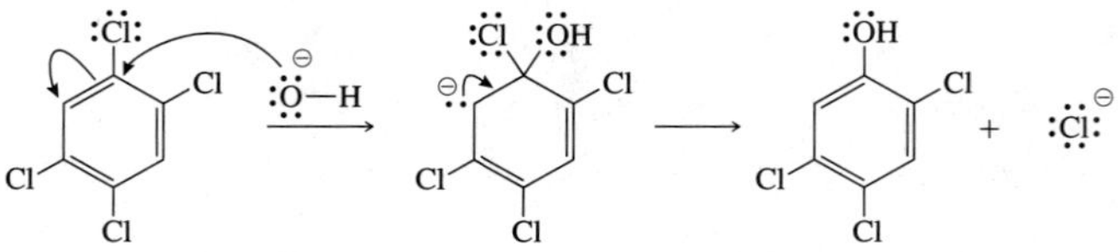

If a benzyne intermediate were involved, two products would be expected.

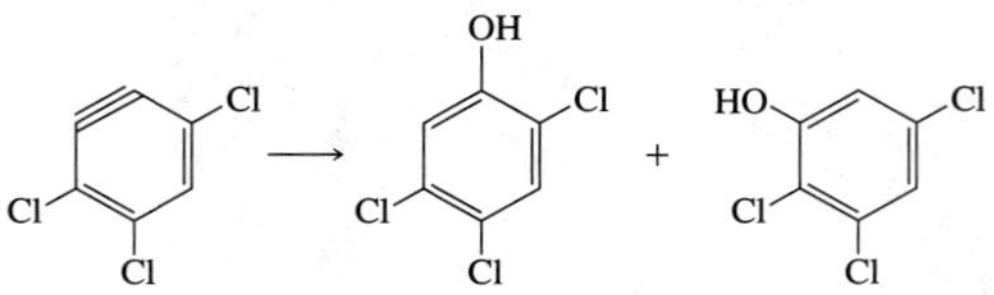

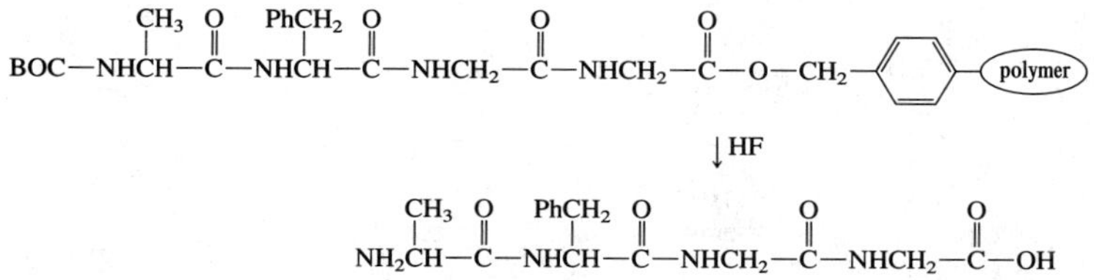

Chapter 26

26.1 a)

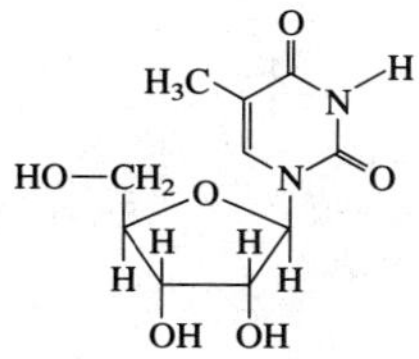

b)

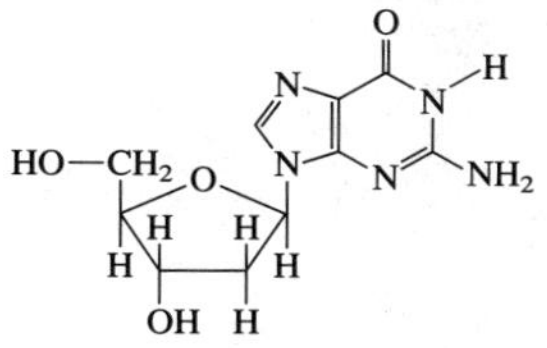

c)

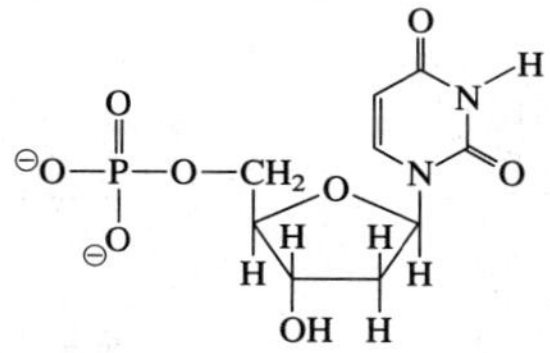

d)

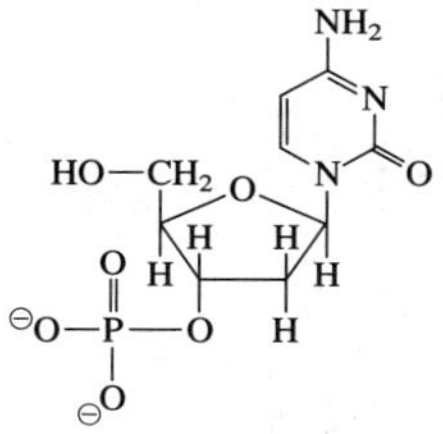

26.2 a) deoxyguanosine   b) uridine 5'-monophosphate

c) deoxycytidine 5'-monophosphate

26.3 Pyrimidine has six electrons in its cyclic *pi* system and therefore fits Huckel's rule. The unshared electrons on the nitrogens are not part of the *pi* system. The six-membered ring of purine is aromatic for the same reason as pyrimidine. The five-membered ring is also aromatic. Like imidazole (see Chapter 17), it has a total of six electrons in its cyclic *pi* system. The electrons on N-9 are part of the *pi* system, whereas the electrons on N-7 are not.

26.4 a)

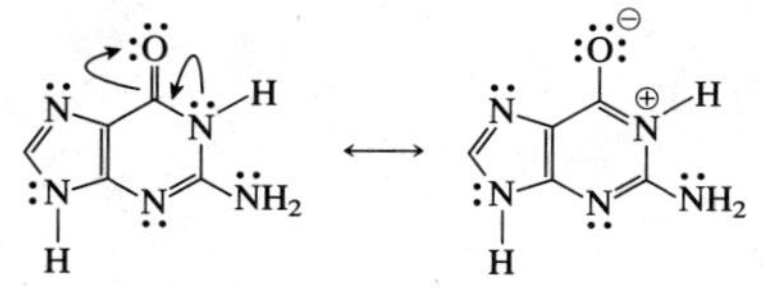

b)

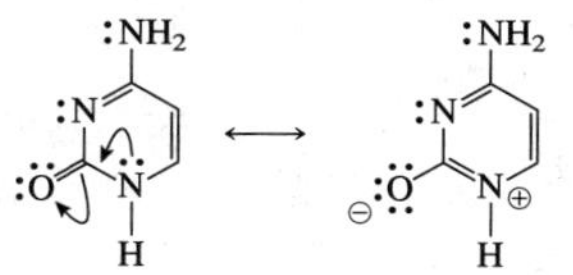

c)

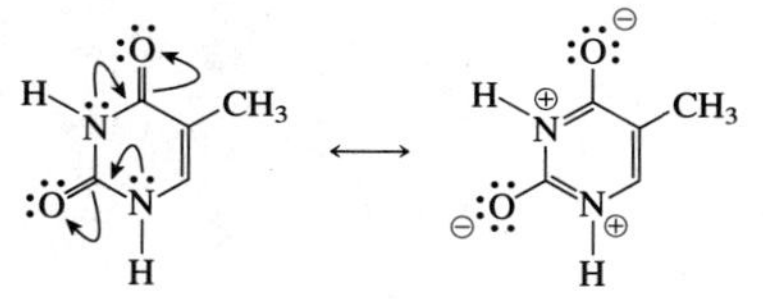

26.5 5'-end TAACGCTCG 3'-end

26.6

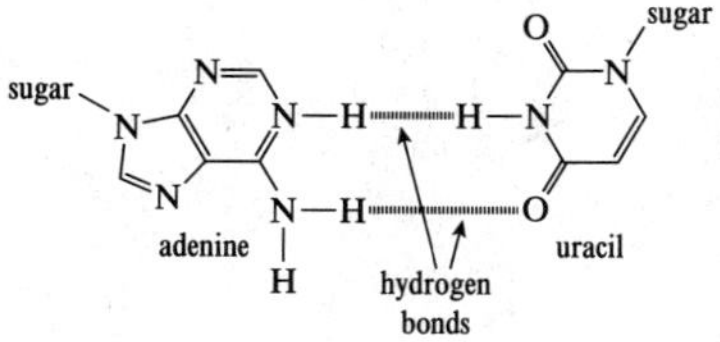

26.7

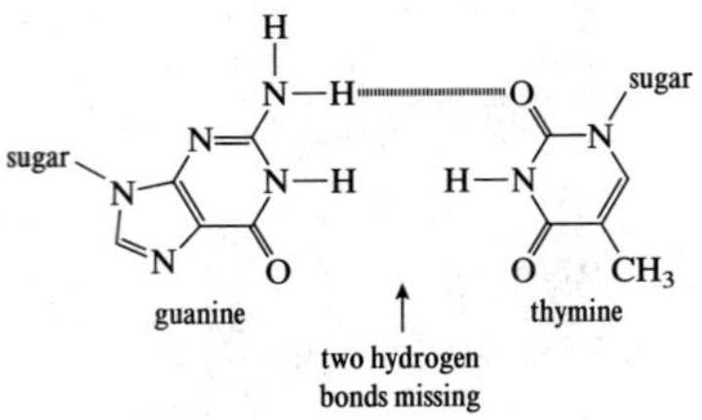

guanine   thymine

two hydrogen
bonds missing

26.8

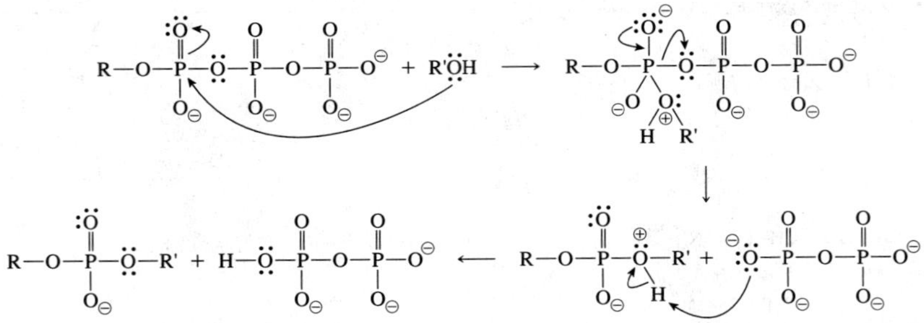

26.9

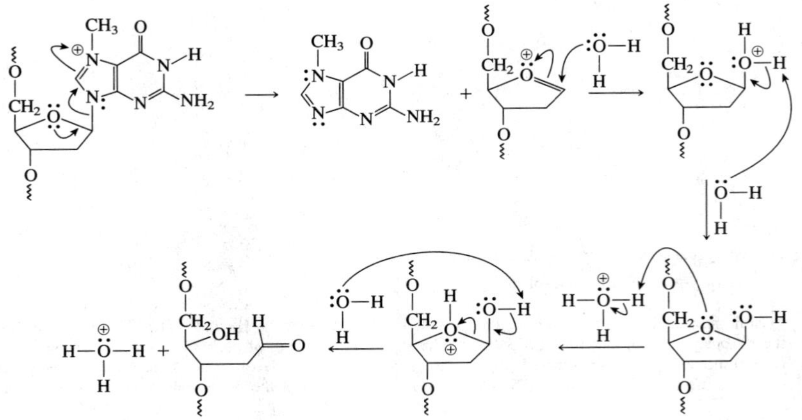

26.10

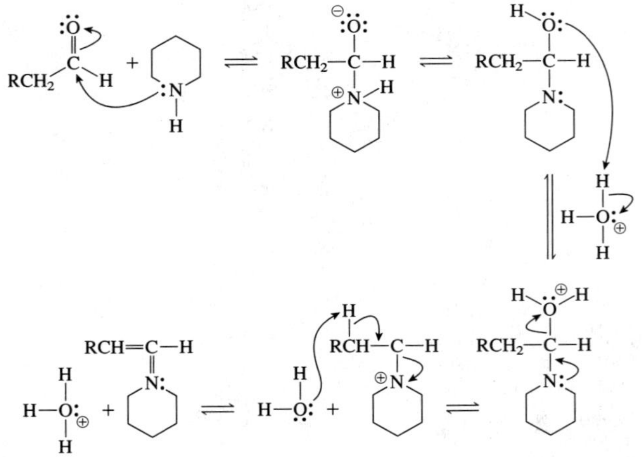

26.11

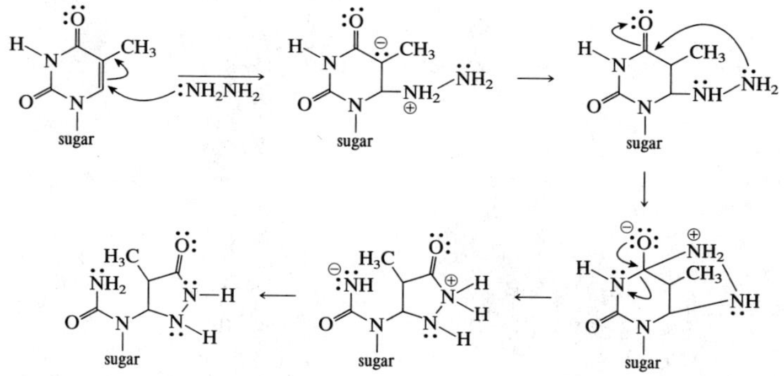

26.12

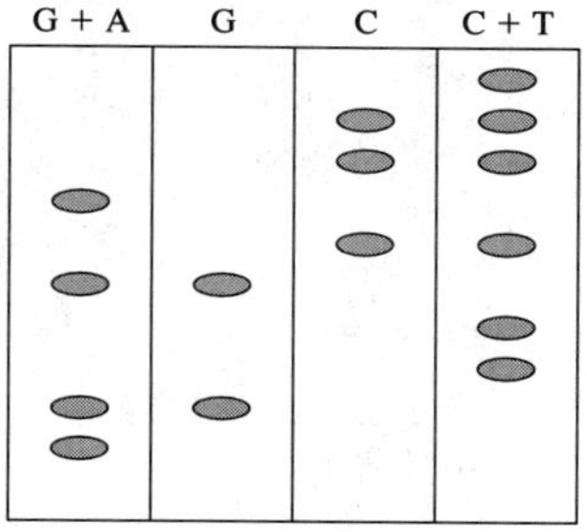

26.13

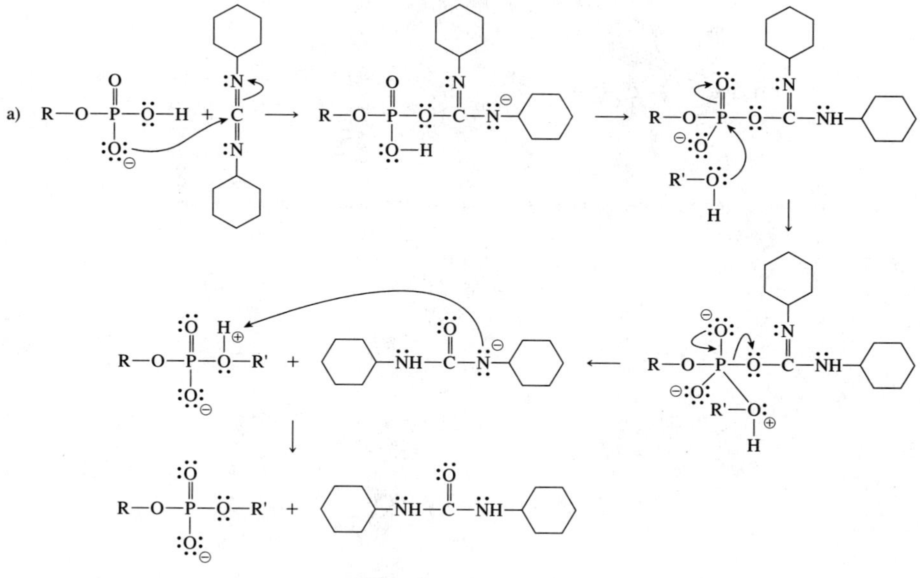

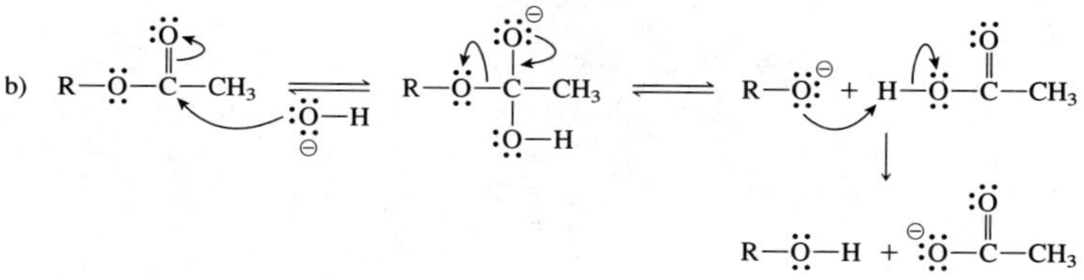

a)

b)

c) 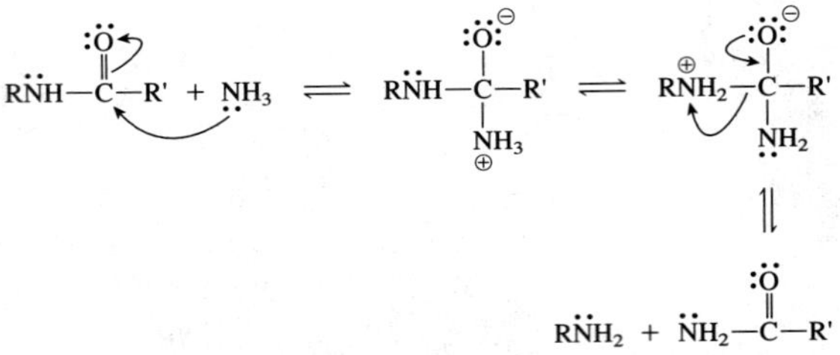 RNH—C—R' + NH₃ ⇌ RNH—C—R' ⇌ RNH₂—C—R'

RNH₂ + NH₂—C—R'

d) 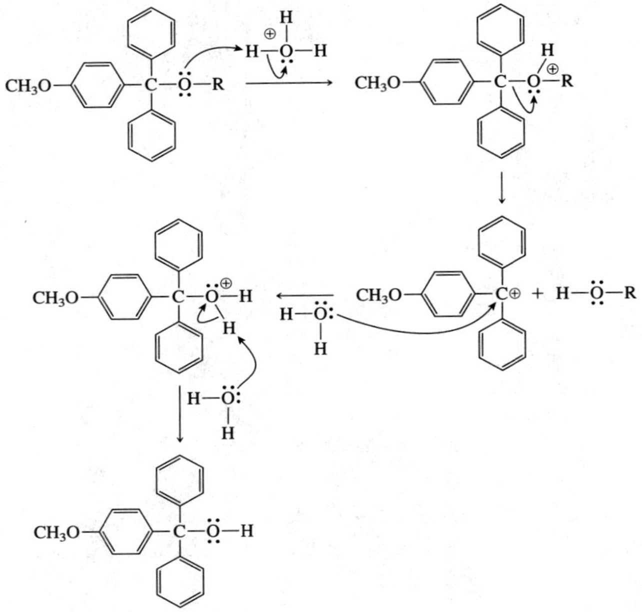 CH₃O—⟨C₆H₄⟩—C—O—R

26.14 The methoxy group provides additional resonance stabilization of the carbocation intermediate.

26.15

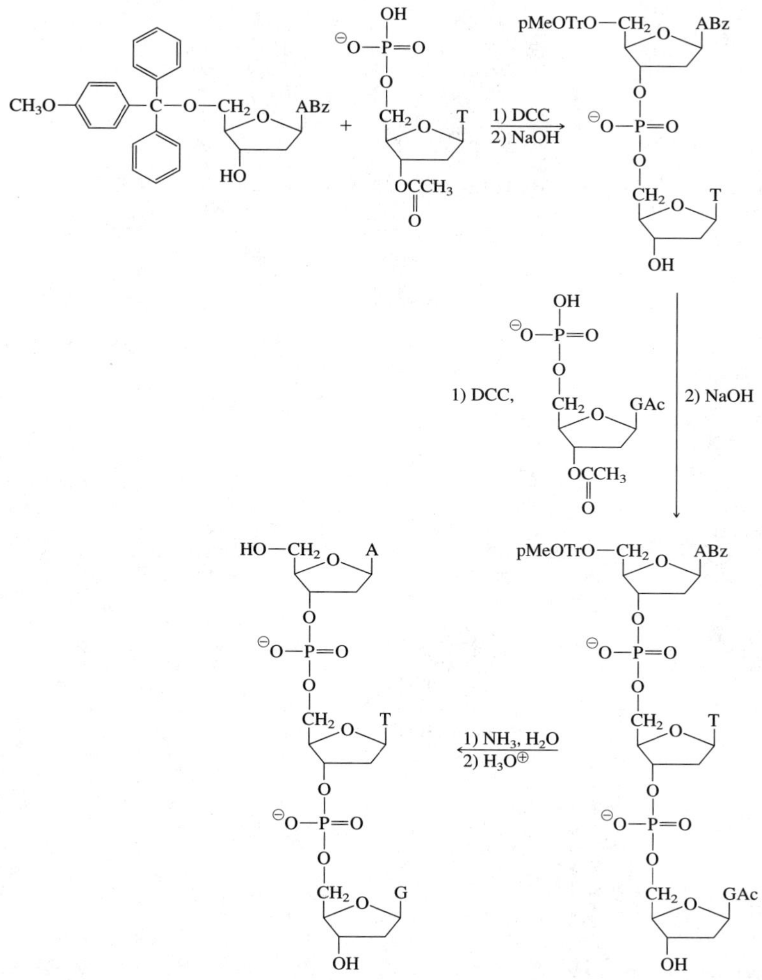

26.16

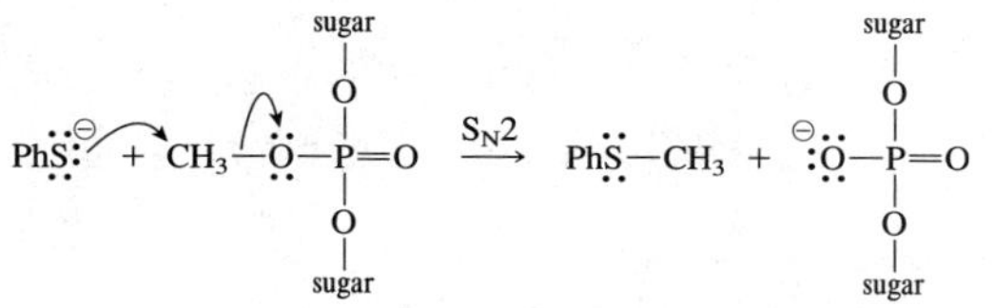

The S_N2 reaction is much faster at methyl than at the more hindered carbons of the other ester groups.

26.17 The DMTr protective group is cleaved more readily because it produces a more stable carbocation in the cleavage process.

26.18

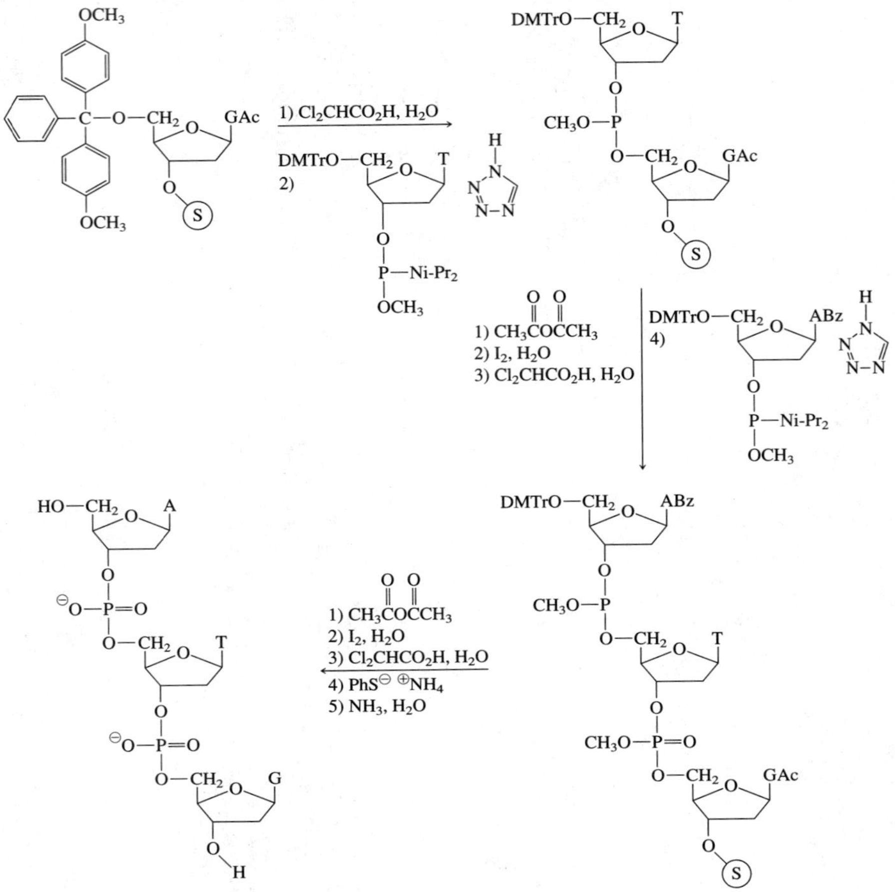

Chapter 27

27.1   a)

monoterpene

b)

sesquiterpene

c)

monoterpene

d)

triterpene

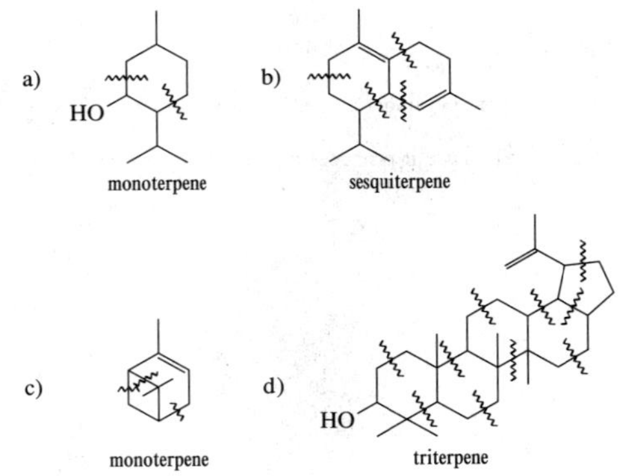

27.2

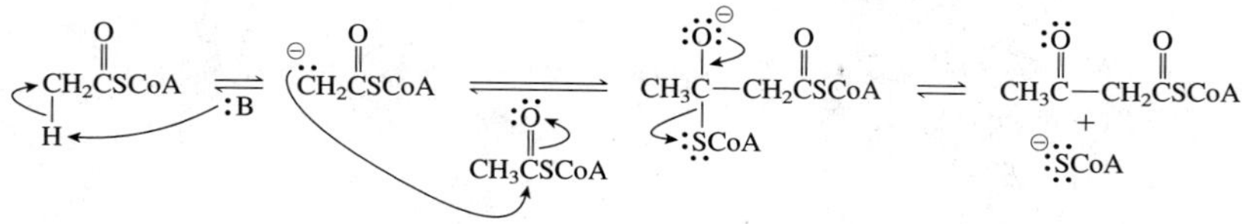

27.3

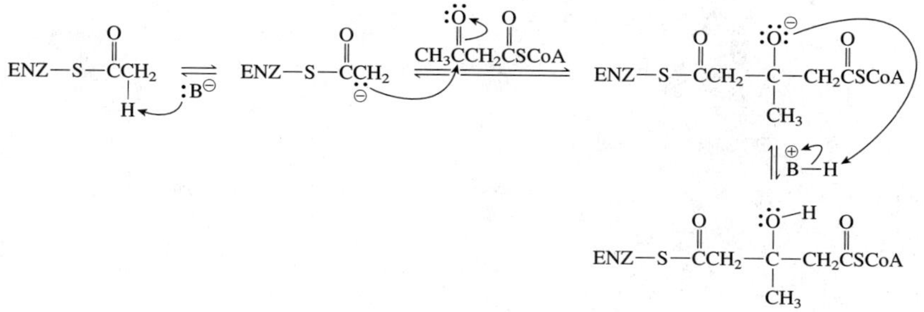

27.4

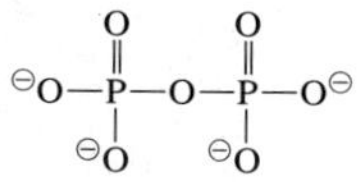

This is a good leaving group because it is a weak base.

27.5

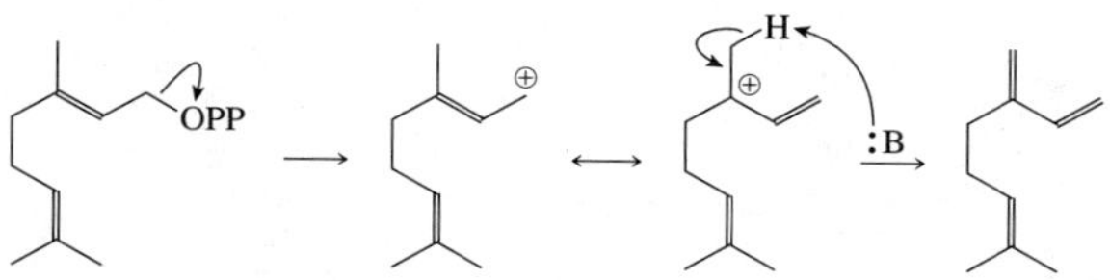

27.6

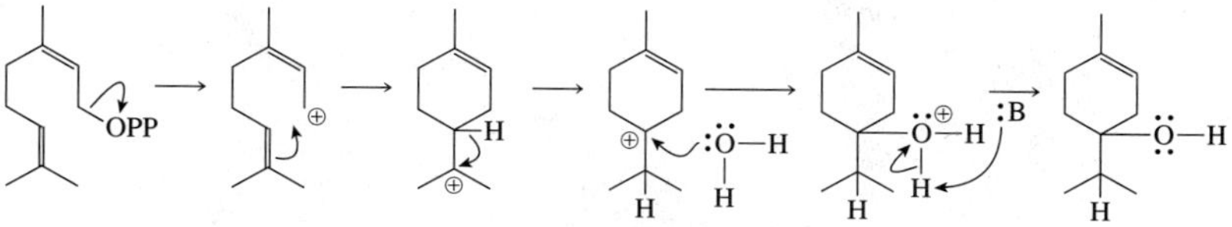

27.7 The cyclization occurs so as to form the more stable tertiary carbocation.

27.8 Path A is disfavored because it produces a secondary carbocation. It is favored because the 5-membered rings that are formed have lower ring strain. Path B is favored because it produces a tertiary carbocation. It is disfavored due to the strain in the 4-membered ring of the product.

27.9 The cyclization occurs so as to form the more stable tertiary carbocation.

27.10 The reaction proceeds as shown in Figure 27.4 to produce the following carbocation, which then cyclizes as shown.

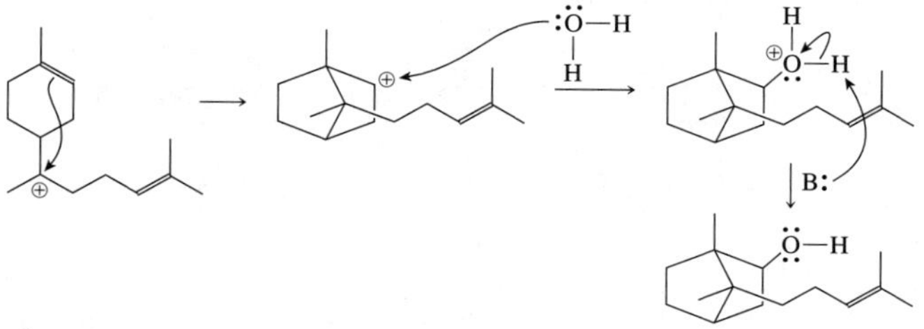

27.11 Both cyclizations occur so as to form the more stable tertiary carbocations.

27.12

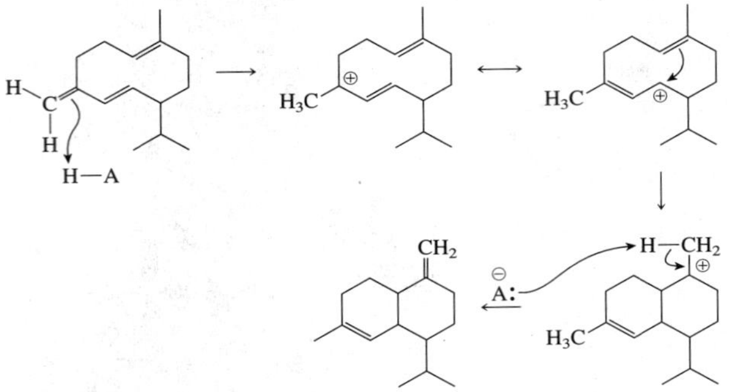

27.13

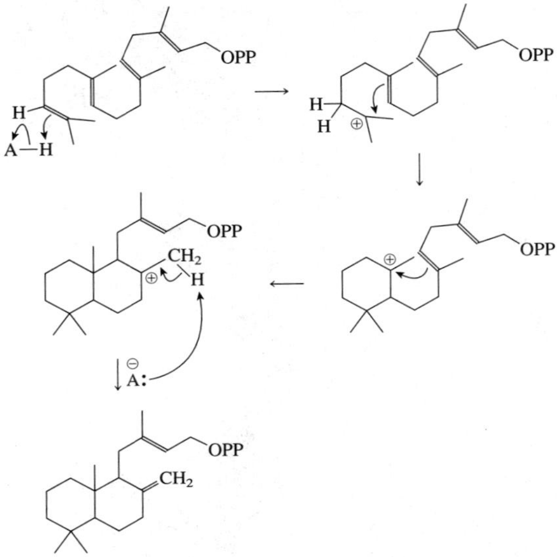

27.14

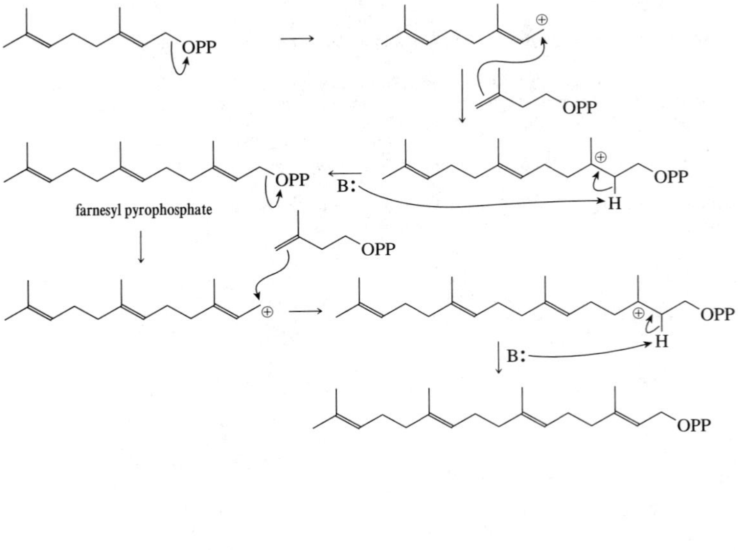

farnesyl pyrophosphate

London force, 45. *See also* Instantaneous dipole- induced dipole interactions
Low-density polyethylene (LDPE), manufacture of, 1081–1084
Lowest-energy state, 63
Lowest unoccupied molecular orbital (LUMO), 601–602, 604–606
    in cycloaddition reactions, 961–963
    in pericyclic reactions, 947–951
Low-resolution mass spectrometers, 610, 612–613
Lubricating oils, alkanes in, 150
Lucas test
    nucleophilic substitution reactions and, 291–292
    in preparation of alkyl halides, 357
Lupeol, structure of, 1220
Lycopene, structure and reactions of, 1231–1232
Lysergic acid, structure and effects of, 1242
Lysergic acid diethylamide (LSD), 1242
    structure of, 482
Lysine
    structure and acidity constant of, 1147
    structure of, 1242
D-Lyxose, Fischer projection of, 1107

Macromolecules, polymers as, 1070, 1073–1075
Magic acid, 139
Magnesium, organic compounds with, 647–649
Magnesium chloride, melting and boiling points of, 5
Magnesium oxide, formation of, 5
Magnetic quantum number, 60
Magnetic resonance imaging (MRI), medical uses of, 566–567
Magnetogyric ratio, 540
Magnets
    in magnetic resonance imaging, 566
    in mass spectrometers, 610, 611
    in nuclear magnetic resonance spectroscopy, 540–541
Ma-huang plant, (-)-ephedrine in, 232
Malonic ester synthesis, 761–765, 788, 944
    target compounds for, 1020
Maltose, structure and properties of, 1131
Malvalic acid, structure of, 210
D-Mannose, 1108
    in determining structure of glucose, 1125–1129

Fischer projection of, 1107
isomerization of, 1119–1120
structure of, 1125, 1126
Manufacturers, of organic chemicals, 1041–1042
Margarines, 162
    manufacture of, 1245
Marker, Russell, 1237
Markovnikov, Vladimir, 404
Markovnikov's rule, 404–405, 408, 418, 926
    hydroboration and, 424, 425
"Marsh gas," 3
Masses of isotopes, 612, 613
Mass spectrometers, 610–613
    resolution of, 610, 612–613
    schematic diagram of, 611
Mass spectroscopy, 608–610
    determining molecular formulas with, 610–616
    fragmentation of molecular ions and, 616–626
    gas chromatography and, 626–627
Mass to charge ratio, in mass spectroscopy, 610
McLafferty rearrangement, in mass spectroscopy, 624–625
Melting points. *See also* Physical properties
    of alkanes and cycloalkanes, 148
    of carboxylic acid derivatives, 480–482
    charge–charge interactions and, 44–45
    molecular structure and, 46–47
3-Menthene, in unimolecular elimination reactions, 329
Menthol
    formation of, 1223–1224
    properties of, 173
    structure of, 1220
Menthyl chloride
    in trans-diaxial elimination reactions, 316–318
    in unimolecular elimination reactions, 329–330
Mercaptans, naming of, 485–488
Mercuric acetate, in alkene hydration, 420–423
Mercurinium ion, 421, 422
Mercury, in alkene hydration, 420–423
Merrifield, R. B., 1177, 1179
Mesitylene, bromination of, 853
meso-Diastereomers, 235–236
    of cyclic molecules, 236–238
meso-Stereoisomers, 235–236

Messenger RNA (mRNA), in biological protein synthesis, 1199–1200
Mesylate ester, as a leaving group, 278–280
meta attack
    in nitrobenzene nitration, 844–845
    in toluene nitration, 840–842
meta directors
    in electrophilic aromatic substitution reactions, 845–846
    in Friedel-Crafts alkylation, 859
    in multiple-substituent benzene-ring reactions, 848–849
    in nitration of aromatic-ring compounds, 850–851
    in nitrobenzene nitration, 844–845
    in synthesis of aromatic-ring compounds, 884
Metal–carbene complexes, in metathesis reaction, 1052–1053
Metallic reagents, industrial uses of, 1048, 1052–1053
Metals. *See also* Organometallic nucleophiles
    in organometallic nucleophiles, 646–649
    in reduction reactions, 381
meta ring positions, 462, 465–467
    in benzyne reactions, 873
    in carbon-13 magnetic resonance spectroscopy, 572
meta substitution, in benzene-ring compounds, 840–848
Metathesis reaction, industrial uses of, 1052–1053
Methamphetamine, structure and formation of, 667
Methane, 3
    boiling point of, 48
    energy content of, 151
    formation of, 6
    greenhouse effect and, 538
    industrial uses of, 1042, 1063
    Lewis structure of, 67–68
    molecular shape of, 20–21
    orbital hybridization of, 69
    reaction between chlorine and, 913–915
Methanesulfonyl chloride, in nucleophilic substitution reactions, 280
Methanoic acid, structure of, 474
Methanol
    energy content of, 151
    industrial uses of, 1043, 1048
    orbital hybridization of, 69–70
    properties of, 173

of carbons and double bonds in
    alkenes, 163–165
of carbons in aldehydes and ketones,
    469–470
of carbons in alkanes, 155–158
of carbons in amines, 175–178
of carbons in aromatic-ring com-
    pounds, 462–463
of carbons in carboxylic acids, 473
of carbons in compounds with more
    than one functional group, 489
of carbons in cycloalkanes, 161–162
of carbons in nitriles, 478
of sigmatropic rearrangements,
    973–974
NutraSweet, as artificial sweetener,
    1118
Nylon 6, preparation of, 1055–1056,
    1090
Nylon 6,6, preparation of, 1055,
    1062–1063, 1089
Nylon 6,10, properties of, 1089–1090

OH group. *See* Hydroxy group
Octahedral molecules, 20
Octane
    infrared spectrum of, 507
    mass spectrum of, 618–619
    melting point of, 47
2-Octanol, in bimolecular nucleophilic
    substitution reactions, 281–282
Octet rule, 7, 11, 29
    compounds that do not satisfy, 30
    radicals and, 904
    in resonance structures, 83, 85
    stability and, 9, 10
Odd orbital systems, in pericyclic
    reactions, 949–951
Odors
    of aldehydes and ketones, 472
    of aromatic-ring compounds,
        461–462, 808–809
    of compounds with sulfur and phos-
        phorus, 486–488
    of organic compounds, 483–484
    of terpenes, 1223
Off-resonance decoupling, in carbon-13
    magnetic resonance
    spectroscopy, 570–571
Oil
    alkanes in, 148–149, 150
    as source of industrial organic chem-
        icals, 1043–1045
Oils, 1244, 1246
    industrial uses of, 1063
Olah, George, 139
"Olefiant gas," 3

Opsin, in chemistry of vision, 668–669
Optically active compounds, 228–230
    discovery of, 242
Oral contraceptives, development of,
    1239–1241
Orbital energies, of molecular orbitals,
    94–97
Orbitals, 4, 59–97. *See also* Atomic
    orbitals (AOs); Molecular or-
    bitals (MOs)
    degenerate, 61, 62–63
    quantization of, 499
    quantum numbers of, 60
Organic chemistry, 1, 2
    history of, 1–3
    industrial, 1041–1064
Organic compounds. *See also* Natural
    products
    acid constants of, 113
    acidity constants for selected,
        132–133
    bonds in, 29–53
    chlorinated, 169–170
    halogenated, 166–168
    naming of, 147–178, 461–491
    properties of, 1–3
    stereochemistry of, 183–248
Organic matter, deterioration of,
    921–922
Organic molecules
    functional groups in, 50–53
    structure of, 3–4
Organic synthesis, 1001–1035. *See also*
    Addition reactions; Elimination
    reactions; Substitution reactions
    of amino acids, 1153–1159
    of aromatic-ring compounds,
        876–880, 881–885, 889
    of carbon–carbon bonds,
        1019–1020, 1021–1022
    of carbon nucleophiles, 787–791
    chiral boranes in, 430–432
    designing good, 1010–1016
    of DNA, 1207–1212
    examples of, 1016–1019
    of peptides and proteins, 1170–1180
    preparation of functional groups in,
        1020–1035
    of progesterone, 1235–1236
    of prostaglandins, 1248
    protective groups for, 1001–1005,
        1006–1008, 1008–1010
    strategy of, 383–388
    substitution and elimination reac-
        tions in, 334–338, 345–388
Organolithium reagents, 647–648, 650,
    724–725, 751

Organomagnesium halides, 647–648
Organometallic compounds, 647
    in coordination polymerization,
        1078–1079
Organometallic nucleophiles
    addition reactions of, 649–653
    in conjugate addition reactions, 677
    preparation and properties of,
        646–649
    substitution reactions with,
        724–725
Ornithine, structure of, 1242
ortho attack
    in nitrobenzene nitration, 844–845
    in toluene nitration, 840–842
"*ortho*-like" products, of Diels-Alder
    reaction, 967–968
ortho/para directors
    in electrophilic aromatic substitution
        reactions, 845–846
    in Friedel-Crafts alkylation, 859
    methoxy groups as, 842–843
    in multiple-substituent benzene-ring
        reactions, 848–849
    in nitration of aromatic-ring com-
        pounds, 850–851
    in polymerization of phenol and
        formaldehyde, 1094
    in production of DDT, 1057
    ratio of *ortho* to *para* reaction prod-
        ucts and, 847–848
    in synthesis of aromatic-ring com-
        pounds, 882–883
ortho ring positions, 462, 465–467
    of benzene rings in biphenyls,
        247–248
    in carbon-13 magnetic resonance
        spectroscopy, 572
    in nucleophilic aromatic substitution
        reactions, 869–871
ortho substitution, in benzene-ring
    compounds, 840–848
Osmate ester, in hydroxylation,
    439–440
Osmium tetroxide, in hydroxylation,
    439–440
Overall dipole moments, of molecules,
    22–24
Oxanamide, synthesis of, 1018–1019
Oxaphosphetaine, in the Wittig reac-
    tion, 655–656
Oxetanes, structure and formation of,
    971
Oxidation reactions, 151, 380–383,
    384–385
    hydroboration as, 424–432
Oxidizing reagents, 382

GENERAL INTRODUCTION

The original volumes in the series *Who's Who in History* were well received by readers who responded favourably to the claim of the late C. R. N. Routh, general editor of the series, that there was a need for a work of reference which should present the latest findings of scholarship in the form of short biographical essays. Published by Basil Blackwell in five volumes, the series covered British history from the earliest times to 1837. It was designed to please several kinds of reader: the 'general reader', the browser who might find it hard to resist the temptation to go from one character to another, and, of course, the student of all ages. Each author sought in his own way to convey more than the bare facts of his subject's life, to place him in the context of his age and to evoke what was distinctive in his character and achievement. At the same time, by using a broadly chronological rather than alphabetical sequence, and by grouping together similar classes of people, each volume provided a portrait of the age. Presenting history in biographical form, it complemented the conventional textbook.

Since the publication of the first volumes of the series in the early sixties the continuing work of research has brought new facts to light and has led to some important revaluations. In particular the late mediaeval period, a hitherto somewhat neglected field, has been thoroughly studied. There has also been intense controversy about certain aspects of Tudor and Stuart history. There is plainly a need for fuller treatment of the mediaeval period than was allowed for in the original series, in which the late W. O. Hassall's volume covered the years 55 B.C. to 1485 A.D. The time seems also to be ripe for a reassessment of some Tudor and Stuart figures. Meanwhile the continued requests of teachers and students for the series to be reprinted encourages the authors of the new series to think that there will be a warm response to a fuller and more comprehensive *Who's Who* which will eventually include the nineteenth and early twentieth centuries. They are therefore grateful to Shepheard-Walwyn for the opportunity to present the new, enlarged *Who's Who*.

Following Volume I, devoted to the Roman and Anglo-Saxon period, two further books cover the Middle Ages. The Tudor volume, by the late C. R. Routh, has been extensively revised by Dr. Peter Holmes. Peter Hill and I have revised for re-publication our own volumes on the Stuart and Georgian periods. Between Edward I's conquest of Wales and the Act of Union which joined England and Scotland in 1707, the authors' prime concern has been England, with Scotsmen and Irishmen figuring only if they happened in any way to be prominent in English history. In the eighteenth century Scotsmen come into the picture, in the nineteenth Irishmen, in their own right, as inhabitants of Great Britain. It is hoped that full justice will be done to Scotsmen and Irishmen – and indeed to some early Welshmen – in subsequent volumes devoted to the history of those countries. When the series is complete, we believe that it will provide a comprehensive work of reference which will stand the test of time. At a time when so much historical writing is necessarily becoming more technical, more abstract, or simply more specialised, when textbooks seem so often to have little room to spare for the men and women who are the life and soul of the past, there is a place for a history of our country which is composed of the lives of those who helped make it what it was, and is. In contributing to this history the authors can be said to have taken heed of the stern warning of Trevor Roper's inaugural lecture at Oxford in 1957 against 'the removal of humane studies into a specialisation so remote that they cease to have that lay interest which is their sole ultimate justification'.

The hard pressed examinee often needs an essay which puts an important life into perspective. From necessarily brief accounts he may learn valuable lessons in proportion, concision and relevance. We hope that he will be tempted to find out more and so have added, wherever possible, the titles of books for further reading. Mindful of his needs, we have not however confined our attention to those who have left their mark on church and state. The man who invented the umbrella, the archbishop who shot a gamekeeper, a successful highwayman and an unsuccessful admiral find their place among the great and good. Nor have we eschewed anecdote or turned a blind eye to folly or foible: it is not the authors' view that history which is instructive cannot also be entertaining.

With the development of a secure and civilised society, the range of characters becomes richer, their achievements more diverse. Besides the soldiers, politicians and churchmen who dominate the mediaeval scene there are merchants, inventors, industrialists; more scholars, lawyers, artists; explorers and colonial pioneers. More is known about more people and the task of selection becomes ever harder. Throughout, whether looking at the mediaeval warrior, the Elizabethan seaman, the Stuart radical or the eighteenth century entrepreneur, the authors have been guided by the criterion of excellence. To record the achievements of those few who have had the chance to excel and who have left a name behind them is not to denigrate the unremarkable or unremarked for whom there was no opportunity to shine or chronicler at hand to describe what they made or did. It is not to deny that a Neville or a Pelham might have died obscure if he had not been born to high estate. It is to offer, for the instruction and inspiration of a generation which has been led too often to believe that individuals count for little in the face of the forces which shape economy and society, the conviction that a country is as remarkable as the individuals of which it is composed. In these pages there will be found examples of heroism, genius, and altruism; of self-seeking and squalor. There will be little that is ordinary. It is therefore the hope of the authors that there will be little that is dull.

GEOFFREY TREASURE

Harrow

PREFACE TO THE SECOND EDITION

The task I was invited to perform was to bring the late Mr Routh's text up to date. This has involved a major revision of the entries for the principal characters: kings, queens, and their ministers. I have also done my best to check all the other entries for errors, and to alter them in the light of the progress of historical research in the quarter century since the first edition was written. Needless to say, there will be biographers whose work I have omitted to consult, but I hope they are not many. I have brought the short bibliographies up to date as far as possible. Large parts of the book remain unaltered, however, and while these parts contain nothing with which I disagree strongly, they cannot, of course, reflect my own opinions.

P. J. HOLMES
September 1989

PREFACE TO THE FIRST EDITION

The second volume of *Who's Who in History* * covers the years 1485 to 1603 and deals with England and Wales. These years form one of the great creative periods in English history, comparable, for example, with the nineteenth century. Both experienced a tremendous revolution: in the Tudor age an ecclesiastical and doctrinal revolution, the Industrial Revolution in the Victorian age: both saw the British People travelling to the ends of their known world in search of wealth and trade: in both there was a great flowering of English literature: to each a woman on the throne bequeathed her name. The problem has been to know whom to include and whom to omit in this selection of Tudor biographies. Although the size of the book has been considerably increased beyond the original intention, space could not be found for all the men and women who might be thought to have strong claims to be included. Very reluctantly, in order to avoid scrappy biographies, I have left out all Scottish and Irish characters, the printers, the musicians, most of the writers and all the lesser statesmen or politicians. It may be possible to bring all these omissions together in another volume later on.

I wish specially to thank the Librarian and his staff of the Public Library at Stratford-on-Avon for their unfailing help, courtesy and co-operation. They have spared no trouble to meet my incessant and often baffling demand for books. No doubt there are many other towns as well catered for as Stratford-on-Avon: I have not met them.

Above all I must thank Professor Joel Hurstfield and Mr. Christopher Morris for their great kindness in reading this book in typescript. Their time and scholarship has been put at my disposal most generously and their conscientious reading and penetrating criticisms have saved me from innumerable errors. What errors remain are mine and not theirs. I am grateful also to Mr. William Baring Pemberton for reading the page-proofs.

Finally, it is a very great pleasure to me to put on record how much I have appreciated the unfailing patience, courtesy and friendship of Mr. J. A. Cutforth.

C. R. N. ROUTH
1964

* Basil Blackwell, 1964.

LIST OF ILLUSTRATIONS

The author and publisher wish to express their grateful thanks to the owners of the portraits listed below for permission to reproduce them in this book. Unless otherwise stated they are from the collection of the National Portrait Gallery in London.

Table I.—TUDOR AND STUART SUCCESSION

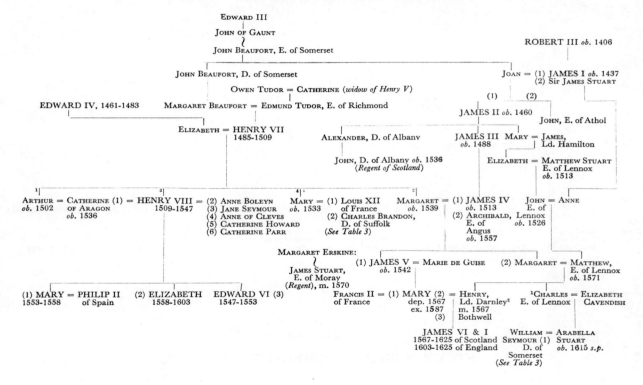